UNITED STATES ARMY IN WORLD WAR II

The European Theater of Operations

LOGISTICAL SUPPORT OF THE ARMIES

In Two Volumes
Volume II: September 1944–May 1945

by

Roland G. Ruppenthal

CENTER OF MILITARY HISTORY
UNITED STATES ARMY WASHINGTON,
D.C., 1995

Library of Congress Catalog Number: 53-60080

Reprinted by St. John's Press LLC
Alexandria, Va. 22301
ISBN 978-1-944961-14-5
2016
First Printed 1959—CMH Pub 7-3-1

For sale by the U.S. Government Printing Office
Superintendent of Documents, Mail Stop: SSOP, Washington, D.C. 20402

UNITED STATES ARMY IN WORLD WAR II

Kent Roberts Greenfield, General Editor

. . . to Those Who Served

Foreword

This volume completes the bridge between combat and services in the European theater for which the author laid the foundations in Volume I. It is as important a book for combat commanders as for those who have to plan and execute logistical operations. It will leave the nonmilitary reader in no doubt of the enormous weight and complexity of the administrative burden that the Army had to assume to assure the success of its ground and air forces, and the resourcefulness with which it managed that burden. On the other hand, those who have to think about the future can here study a test of the principle of a single service of supply supporting the national element of allied forces under a coalition headquarters and a supreme allied commander.

R. W. STEPHENS
Maj. Gen., U. S. A.
Chief of Military History

Washington, D. C.
15 June 1958

Preface

This volume completes the story of the logistic support of U.S. forces in the European theater, carrying the account forward from mid-September 1944 to the end of hostilities in May 1945. It follows the pattern, established in *Logistical Support of the Armies, Volume I,* of focusing on the influence which logistical support or lack of it had on the planning and the conduct of tactical operations. The inclination consequently has been to concentrate on the problem areas in logistic support, such as port discharge and transportation difficulties, and supply and manpower shortages. As explained in the Preface to Volume I, it was not intended to cover all aspects of logistics as the term is commonly defined. To avoid duplication, such subjects as hospitalization and evacuation, communications, and construction are purposely left to the technical service histories, where they can be given proper coverage. The one major exception is the account of the rebuilding of Cherbourg, which was so important to the development of the logistic structure in the summer and early fall of 1944 that it is presented as a case history in planning and execution. A substantial amount of space has been given to the discussion of theater command and organization because of the persistent influence which that problem had on logistic support and on the relations between the service and combat elements. In general, the topical treatment predominates, but within the boundaries of the two distinct periods of tactical developments. One major violation of chronology occurs in the treatment of local procurement (Chapter XVIII), which did not lend itself to division.

The author's work was again lightened by the use of preliminary studies prepared by members of the Historical Section, ETO. For Volume II these were: Robert W. Coakley's two-volume study of theater command and organization; John E. Henderson's study of the replacement problem; and George H. Elliott's study of the use of indigenous manpower. Once again Mr. Royce L. Thompson gave invaluable aid in running down records and in researching several thorny problems.

It is a pleasure to acknowledge the help of the several persons who contributed so cordially and generously in the final production of the volume: Mr. Joseph R. Friedman and Miss Ruth Stout for their expert editorial judgment; Mrs. Loretto Stevens for the laborious work of copy editing the manuscript; Mr. Wsevolod Aglaimoff for the excellent cartographic work;

and Miss Margaret E. Tackley for the fine selection of photographs. Again the author wishes to acknowledge the able assistance provided by Mr. Israel Wice and his staff of the General Reference Branch of the Office, Chief of Military History, and by the personnel of the Departmental Records Branch, Adjutant General's Office, particularly Mrs. Blanche Moore and Mrs. Lois Aldridge. Key staff officers and commanders associated with the logistic support of U.S. forces in the ETO provided first-hand knowledge of the events of 1944–45 as they did for Volume I. Generals John C. H. Lee, Raymond G. Moses, Ewart G. Plank, Morris W. Gilland, and Frank S. Ross read varying portions of the manuscript in draft form. Biographical sketches of the principal commanders and staff officers of the Communications Zone are included in Volume I.

ROLAND G. RUPPENTHAL

Washington, D. C.
15 June 1958

Contents

THE TYRANNY OF LOGISTICS, SEPTEMBER 1944–FEBRUARY 1945

Chapter | *Page*

I. LOGISTIC LIMITATIONS AS THE ARBITER OF TACTICAL PLANNING 3
(1) *The Pursuit's Effect on Logistic Plans* 3
(2) *Competing Tactical Plans* 8
(3) *The September Decisions* 11
(4) *Prospects at the End of September* 17

II. TACTICAL AND ORGANIZATIONAL DEVELOPMENTS . 22
(1) *Tactical Operations* 22
(2) *Organization and Command* 26

III. THE PORT DISCHARGE AND SHIPPING PROBLEMS . . 46
(1) *The Port Problem as Affected by the Pursuit* 46
(2) *The Beaches and Minor Ports of Normandy* 53
(3) *The Role of Cherbourg* 62
(4) *The Brittany Area* 89

IV. THE PORT DISCHARGE AND SHIPPING PROBLEMS (CONTINUED) 96
(1) *The Seine Ports—Le Havre and Rouen* 96
(2) *Antwerp and the Other Northern Ports* 104
(3) *Southern France* 116
(4) *The Shipping Tie-up* 124

V. TRANSPORTATION DEVELOPMENTS 134
(1) *Motor Transport: The Color Routes* 134
(2) *The Railways* 148
(3) *Air Transport* 161
(4) *Inland Waterways* 165

VI. FORWARD MOVEMENTS 169
(1) *The Tonnage Allocations System* 169
(2) *The Ardennes Counteroffensive and Its Effect on Movements* 181

Chapter		Page
VII.	SUPPLYING THE ARMIES: RATIONS, POL, AND COAL	188
	(1) *Rations*	188
	(2) *POL*	193
	(3) *Coal*	209
VIII.	SUPPLYING THE ARMIES: EQUIPMENT	215
	(1) *Class II and IV Shortages in General*	215
	(2) *The Case of the Winter Uniform*	218
	(3) *Weapons and Vehicles*	235
IX.	SUPPLYING THE ARMIES: AMMUNITION	247
	(1) *The October Crisis*	247
	(2) *Contention With the War Department*	257
	(3) *The November Offensive and the Bull Mission*	263
	(4) *Ammunition Supply in December and January*	269
X.	THE TROOP BUILD-UP, AUGUST 1944–MARCH 1945	276
	(1) *The Flow of Divisions*	276
	(2) *Service and Supporting Troops*	289
XI.	THE MANPOWER PROBLEM, AUGUST 1944–FEBRUARY 1945	304
	(1) *Rumblings of a Replacement Problem*	304
	(2) *The Storm Breaks, November–December 1944*	316
	(3) *The Theater Acts, January–February 1945*	326
	(4) *The Replacement System in Operation*	334
XII.	THE LOGISTIC STRUCTURE UNDER SCRUTINY	348
	(1) *The Communications Zone and the Field Commands*	348
	(2) *Expediting Supply Deliveries*	351
	(3) *Supply Planning Procedures*	352
	(4) *The Depot System and Records Keeping*	355
	(5) *Expedients*	359

THE LAST OFFENSIVE, FEBRUARY–MAY 1945

XIII.	TACTICAL, LOGISTICAL, AND ORGANIZATIONAL ASPECTS OF THE LAST OFFENSIVE	367
	(1) *Tactical Developments, 8 February–8 May 1945*	367
	(2) *Logistic Factors in Planning the Last Offensive*	372
	(3) *Command and Organization, February–August 1945*	378

Chapter		*Page*
XIV.	MOVEMENTS AND DISTRIBUTION: PORT DISCHARGE AND CLEARANCE	390
	(1) *Planning and Controlling Movements*	390
	(2) *Discharge and Clearance*	393
	(3) *Shipping*	398
	(4) *BOLERO in Reverse*	402
XV.	MOVEMENTS AND DISTRIBUTION: TRANSPORTATION AND FORWARD DELIVERIES	405
	(1) *The Railways*	405
	(2) *Motor Transport—XYZ*	415
	(3) *Inland Waterways*	424
	(4) *Air Transport*	424
	(5) *Forward Deliveries*	427
XVI.	SUPPLY IN THE LAST MONTHS	433
	(1) *Rations, POL, and Coal*	433
	(2) *Ammunition*	441
	(3) *Equipment*	452
XVII.	END OF THE REPLACEMENT PROBLEM	459
	(1) *The Turning Point*	459
	(2) *Withdrawals Are Stopped*	461
	(3) *Results of the Retraining Program*	467
XVIII.	LOCAL PROCUREMENT ON THE CONTINENT, JUNE 1944–AUGUST 1945	470
	(1) *Purpose and Policy*	470
	(2) *The Use of Nonmilitary Labor*	475
	(3) *Local Procurement of Supplies*	486
XIX.	RETROSPECT	503
GLOSSARY		513
CODE NAMES		517
BIBLIOGRAPHICAL NOTE		518
INDEX		523

Tables

No. *Page*

1. Beach Discharges, 1 July–17 November 1944 57
2. Discharge Performance of Normandy's Minor Ports 62
3. Discharge Performance of the Brittany Ports 95
4. Tonnages Discharged at Continental Ports, June 1944–April 1945 . . 124
5. Artillery Ammunition Expenditures, 15–21 October 1944 256
6. 12th Army Group Artillery Ammunition Expenditures, 6 June–22 October 1944, Compared With Day of Supply Rates 267
7. OVERLORD Divisional Build-up, D plus 90 to D plus 210 281
8. Divisional Build-up in the European Theater, 1942–1945 282
9. Theater Strength by Major Component, May 1944–April 1945 . . . 288
10. Battle and Nonbattle Casualties, June 1944–May 1945 317
11. Combat Zone Maintenance Factors, June–October 1944 430
12. Combat Zone Consumption Experience, 23 March–25 April 1945 . . 432
13. Ammunition Day of Supply by Type: Selected Planned Rates, September 1944–April 1945, and Actual Expenditures, June 1944–February 1945 446
14. Civilians Employed in the Communications Zone in Selected Weeks, 1944–45 477

Maps

1. Tactical Progress, 12 September 1944–9 February 1945 23
2. ETO Boundaries 30
3. COMZ Boundaries, November 1944–January 1945 39
4. The Port of Cherbourg 64
5. Highway Express Routes, September 1944–February 1945 136
6. Tactical Developments, 8 February–8 May 1945 368
7. COMZ Boundaries, April 1945 381
8. Movement Program, March 1945 391
9. Railways in Use by U.S. Forces East of Paris 411
10. XYZ Truck Routes, 25 March–8 May 1945 420
11. The POL Pipeline Systems 436

Illustrations

	Page
Maj. Gen. Arthur R. Wilson	40
Brig. Gen. John P. Ratay	41
Maj. Gen. Thomas B. Larkin	43
Troops Debarking Onto a Causeway	54
Bulldozer Stuck in the Thick Mud	56
The Port of Cherbourg	65
Destruction at Cherbourg	68
Marginal Wharf Construction	72
Laying Railway Tracks	73
Seatrain Unloading a Gondola	76
LST's Discharging Cargo on the Beach	91
Crane Lifting a Lock Gate	93
Dukws Transferring Cargo	101
Quay at Rouen Loaded With Incoming Supplies	103
Ships Discharging Cargo for Clearance by Rail	112
Remains of Decanting Site	116
Searching for Casualties in Wreckage	117
General Destruction at Marseille	120
Close-up of Damaged Dock Facilities	121
Trucks Loaded With Supplies	135
Red Ball Express Trucks	138
Truck-Tractor and Semitrailer Stuck in Thick Mud	142
Loaded 10-Ton Semitrailers	144
Truck Tractors, 5-Ton, Hauling 10-Ton Semitrailers	145
Four 750-Gallon Skid Tanks	147
American Locomotive Lowered by Crane	152
Maj. Gen. Frank S. Ross	157
Convoy of Trucks Carrying Essential Supplies	159
C–47's Airdropping Supplies by Parachute	164
Barge Convoy on Albert Canal	167
Supply Trucks Passing Through Bastogne	185
Portion of the Major Pipeline	196
Camouflaged Pumping Station	197
POL Storage Tanks	200
Thousands of Jerricans	202
Sign Appealing for the Return of Jerricans	206
U.S. First Army's POL Reserves	208
General Dwight D. Eisenhower	220
Serving a Hot Meal to Cold Infantrymen	224

	Page
Parsons Jacket 1941	227
Infantryman Wearing a Field Jacket M1943	231
Medium Tanks	240
Street Fighting in Aachen	255
"And me a Clerk-Typist!"	329
Officer Candidate Class, Fontainebleau	332
Orientation Lecture for Enlisted Men	339
Infantry Replacements Checking Equipment	341
Lt. Gen. Brehon B. Somervell	355
Ludendorff Railway Bridge	369
Stacks of War Matériel in Open Storage	397
Toot Sweet Express Ready To Leave Cherbourg	406
Bridge Across the Rhine at Wesel	408
Single-Track Railroad Bridge at Mainz	409
Truck-Tractor and 40-Ton Tank Transporters	417
Highway Bridge Over the Meuse River	423
C–47 Transport Planes Bringing in POL	439
Pershing Tanks M26	455
M24 Light Tanks	457
French Civilians Employed by U.S. Forces	480
Italian Service Unit Men Loading Cases of Rations	483
German Prisoners of War Filling 50-Gallon Oil Drums	487
Renault Plant	490
Belgian Workers in a Rubber Plant	491
Soldiers Equipping Medium Tank Tracks	493
First Division Troops Wearing Winter Camouflage Garments	495

The illustrations are from the files of the Department of Defense except for the cartoon by Sgt. Dick Wingert, page 329, courtesy of *The Stars and Stripes.*

THE TYRANNY OF LOGISTICS, SEPTEMBER 1944–FEBRUARY 1945

CHAPTER I

Logistic Limitations as the Arbiter of Tactical Planning

(1) The Pursuit's Effect on Logistic Plans

Three months after the landings in Normandy, the Allied armies, having pursued a disorganized enemy across northern France and up the Rhône valley, stood at the Dutch and German borders in the north, at the Moselle River in the center, and at the entrance to the Belfort gap in the south. That they were stopped there in mid-September was due in part to the increasing resistance which a reorganized enemy was able to offer from the prepared defenses of the West Wall and along the Moselle, and in part to supply shortages.

The supply shortages were the more exasperating because they occurred in the midst of spectacular advances and because they helped frustrate a short-lived hope that the war might quickly be brought to an end. Within a matter of days the deteriorating logistic situation led to one of the most reluctantly made, and most debated, decisions of the war. This was the decision which General Dwight D. Eisenhower made late in September to halt offensive operations on a large part of the front and to concentrate the bulk of the Allied resources on a relatively narrow front in the north.

The shortages which forced this decision were only a foretaste of a prolonged supply famine. For a period of almost three months logistic limitations largely dominated tactical planning, and U.S. forces learned to their dismay how supply, instead of holding her rightful position as the handmaiden of battle, could become war's mistress.

The inability to continue the pursuit of the shaken enemy forces was exasperating to combat commanders, and it was not to be expected that they would react calmly and objectively to the restrictions imposed by logistic difficulties. The necessity to halt the advance has been variously ascribed to the shortage of gasoline, to shortcomings within the Communications Zone (including the alleged preoccupation of service troops with luxurious living and black-market activities), to "high level politics," to undue favoritism toward the British, and, according to a British view, to the excessive strain on transport caused by Third Army's advanced position. It is not within the province of this volume to weigh all these charges. But the acrimonious debate which attended the supply difficulties early in September makes it appropriate to consider the larger question of the reasons why the Communications Zone was unable to meet the demands placed upon it, and to consider the purely logistic aspects of the various operational plans for which the field commands now claimed priority.

The reasons for the so-called "supply failure" become apparent enough when one recalls the invasion plan and compares the expected with the actual course of the operation. The predicted development of the OVERLORD operation was based on both tactical and logistical considerations. On its operational side the plan was predicated on an estimate that the enemy would make successive stands on the major water barriers across France and Belgium. In accord with this assumption it was expected that he would make a stand at the Seine and that that line would not be reached until D plus 90. Furthermore, plans had contemplated a fairly steady rate of advance rather than the pursuit of a disorganized enemy. While such a forecast of progress was admittedly conjectural, it necessarily formed the basis of logistic preparations. In the belief, for example, that the Seine ports would not quickly become available, great emphasis was placed on the development of the Brittany area, and a pause of at least a month at the Seine was expected to be necessary to develop an administrative base capable of supporting further offensives.

Even on these assumptions the margin of safety of the OVERLORD logistic plan was believed to be nonexistent. In an administrative appreciation prepared early in June, SHAEF planners had concluded that port capacity would suffice to support the planned build-up provided only that the ports were captured on the dates forecast, and then only by the narrowest margin. In fact, deficits in port discharge capacity were predicted beginning in the fifth month (October). They admitted that there were certain hidden assets, such as supply by air and other possible expedients. But there was also the possibility of additional liabilities. Any material variation in logistical planning factors, such as a higher rate of demolitions, or tonnage requirements larger than estimated, would impinge directly on the rate of build-up and capability of support. By the same token any unexpected acceleration in the advance would have a like effect in creating additional port discharge and transportation liabilities.[1]

Since the OVERLORD operation developed quite differently from what had been expected, the assumptions on which the schedules had been based were largely voided. For the first seven weeks the advance was much slower than anticipated, and the Allied forces were confined to a shallow Normandy bridgehead. From the viewpoint of logistic support, the lag in operations was not immediately serious, for it resulted in short lines of communication and gave the service forces added time to develop the port facilities at Cherbourg, whose capture had been delayed. But the long restriction to this area promised serious consequences for the future, for the port capacity of Cherbourg and the beaches severely limited the force which could be maintained during the fall and winter months.

Whatever temporary advantage ac-

[1] Adm Appreciation, SHAEF G-4 Post-NEPTUNE Opns, 17 Jun 44, SHAEF, 12 A Gp 370.2.

crued from the short lines of communication and low maintenance scales in Normandy quickly disappeared after the breakout at the end of July. By D plus 79 (24 August) Allied forces had closed to the Seine eleven days ahead of schedule despite a lag of approximately thirty days at the beginning of the breakout. Tactically, and to some extent logistically, the spectacular encircling drive of early August brought definite advantages to the Allied forces. It resulted in the almost complete destruction of the German *Seventh Army* and thus eliminated a large enemy force which later might have delayed the Allied advance to the Seine; it greatly accelerated the whole campaign and helped ensure a rapid advance to the enemy's border; and it facilitated the early capture of the Seine ports and Antwerp, making it possible to cancel plans for the capture of Quiberon Bay and southern Brittany ports.

From the point of view of logistic support the rapid advance to the Seine also had its less favorable aspects, and even at this early date (D plus 79) foreshadowed serious complications. The fact that the OVERLORD objective was reached on D plus 79 rather than D plus 90 was in itself not serious, for the supply structure was sufficiently flexible to accommodate itself to a variation of eleven days. The departure from the scheduled advance had actually been more serious. Because of the initial lag in operations, U.S. forces had been at the D-plus-20 line at D plus 49, and between D plus 49 and D plus 79, a period of thirty days, had actually advanced a distance which by plan was to have been covered in seventy days. The lines of communication obviously could not be developed beyond St. Lô in the period before the breakout, and in the subsequent period could not be developed at the speed with which tanks and other combat vehicles were able to race to the Seine. The result was that the armies had already used up their operational reserves by the time they reached the Seine. Since rail lines and pipelines could not be pushed forward quickly enough, motor transport facilities were strained to the breaking point attempting to meet even the barest maintenance needs of the armies. The Communications Zone consequently found it impossible to establish stocks in advance depots. Furthermore, none of the Brittany ports had as yet been captured, and only one major port—Cherbourg—was operational. It remained to be seen whether compensation could be gained by the earlier opening of either Le Havre or Antwerp.

But the arrival at the Seine marked only the beginning of supply difficulties. Despite the logistic complications which the rapid advance had already clearly foreshadowed, the decision was now made to cross the Seine, and a few days later to encircle Paris and to continue the pursuit without pause. On purely tactical grounds such decisions were logically indicated. The decisive victory in the Falaise–Argentan pocket and the disintegration of enemy resistance offered opportunities which it would have been folly to ignore. Furthermore, with forty-six divisions on the Continent, the Allies enjoyed a definite superiority in both armor and infantry, as well as in air power, and could move in almost any direction against a weakened enemy.

The situation in northern France, coupled with the Seventh Army's successful drive from the south, appeared so favorable, in fact, as to afford an opportunity to broaden the entire scope of the drive into Germany. Post-OVERLORD plans (beyond D plus 90) had contemplated only a secondary effort south of the Ardennes along the axis Reims–Verdun–Metz by a relatively small force. This force was to have the mission of diverting enemy resistance from the main thrust in the north and preventing the escape of enemy troops from southwest France by linking up with Seventh Army forces moving up the valley of the Rhône. This plan was now modified to provide for an attack along the subsidiary axis in greater strength than originally contemplated, although the main effort was still to be made in the north.

From the point of view of logistics these decisions to cross the Seine and continue the pursuit, and to augment the forces employed south of the Ardennes, constituted a radical departure from earlier plans. They carried with them a supply task out of all proportion to planned capabilities. They were much more far-reaching in their effects than the alteration in plans of early August by which the bulk of the Third Army's forces had been directed eastward rather than into Brittany. With the supply structure already severely strained by the speed with which the last 200 miles had been covered, these decisions entailed the risk of a complete breakdown.

The continued advances late in August and at the beginning of September consequently brought hectic days and sleepless nights to supply officers. All the difficulties which had already begun to appear during the approach to the Seine were now further aggravated. The main problem, as before, was the deficiency in transport, which only worsened as the lines of communication extended farther and farther eastward. Despite great efforts, rail reconstruction was unable to keep pace with the advance. Air supply repeatedly failed to match its predicted capacity. Motor transport therefore continued to bear the principal burden of forward movement and was unable to deliver daily maintenance needs, to say nothing of stocking intermediate or advance supply depots.

The unbearable supply task which the continued advance created can best be appreciated by comparing planned with actual developments. At D plus 90 it had been assumed that no more than twelve U.S. divisions would have to be supported at the Seine. Not until D plus 120 was it thought feasible to support twelve divisions in their first offensive action beyond that barrier, and not until D plus 150 was it contemplated that a "minor advance" might be supported from the Aisne River as a line of departure, seventy-five miles beyond the Seine.[2] At D plus 90 (4 September), however, the Communications Zone was already supporting sixteen divisions at a distance of 150 miles beyond the Seine. Within another ten days (mid-September) First Army forces were operating at the German border in the vicinity of Aachen, well over 200 miles beyond Paris. Since plans had not contemplated reaching that area until D plus 330 (May 1945), it was necessary to support U.S. forces at this distance approximately 230 days earlier than expected. Moreover, the city

[2] *Ibid.*, Annexure M.

of Paris had become an additional supply liability as the result of its liberation on D plus 80, 55 days earlier than expected. U.S. supply lines were now 450 miles long, leading exclusively from Cherbourg and the beaches, still the only points of intake.

In addition to overtaxing transport facilities, this extension of the lines of communication made unbearable demands on all types of service troop units. The service troop basis, like transportation facilities, was based on the more conservative rate of advance envisaged in the OVERLORD plan. When the tempo of operations accelerated in August, requiring the leapfrogging of depots and dumps and a high degree of mobility for supply stocks, available depot units were soon distributed thinly over most of northern France, and were unequal to the task.

At least some of the difficulties stemmed from the delay in capturing the Brittany ports, for port discharge and port clearance capacity were already proving inadequate. Scheduled to develop a discharge capacity of nearly 14,000 tons per day by D plus 90, the Brittany ports, with the exception of St. Malo, were still in enemy hands at this date. As a result, the entire capacity of the Brittany area, which had been counted on so heavily in logistic planning, was still unavailable to U.S. forces.

The delay in the capture of the Brittany ports was at least partially the fruit of the decision of early August by which the largest possible force was devoted to the exploitation of the breakthrough to the east. This decision almost inescapably involved postponement in the capture of the Brittany ports and was a decision for which the Communications Zone bore no responsibility. In fact, it was only because of the pressure of logistical planners that a full corps was devoted to the task. In the view of the chief of the G–4 Logistical Plans Branch at SHAEF, there was an element of poetic justice in the fact that the Third Army, whose mission it was to clear Brittany, later felt so acutely the shortage of supplies which resulted in part from the failure to acquire the Brittany ports.[3]

Contrary to plan, therefore, and as a direct consequence of the late August decisions, the Communications Zone within a matter of days suddenly had been faced with the task of supplying considerably greater forces at much greater distances than contemplated. This, despite a motor transport deficiency which had been predicted before D Day on the basis of even the conservative schedules of the OVERLORD plan; despite the failure to develop the port discharge capacity of the Brittany area, which had been regarded as essential to the administrative support of U.S. forces; and despite the premature assumption of responsibilities in connection with the civil relief of Paris.

In view of the Communications Zone's performances far in excess of what was believed feasible in the OVERLORD plan, it might be argued that estimates of logistic capabilities had been far too conservative. But these performances did not represent full-scale support, and were accomplished only by resorting to such expedients as immobilizing incoming divisions and other combat elements

[3] Col. William Whipple, "Logistical Bottleneck," *The Infantry Journal,* LXII (March, 1948), 14.

in order to provide additional truck companies, using army transportation for line-of-communications hauling, curtailing port clearance, and largely neglecting the armies' needs for replacement equipment and supplies. These were obviously makeshift arrangements which could not be continued indefinitely, and later exacted a big price. They were expedients, moreover, which were attended by such practices as hijacking supplies and "diverting" entire truck companies, and involved many other irregular practices which prevented an orderly and businesslike organization of the Communications Zone. They left deep scars and had a prolonged effect on its efficiency and on its ability to serve the armies.

According to a belief commonly held at the time, the armies might have rolled on had they only had sufficient gasoline. Such a view ignored the many other requirements of a modern army on wheels. By mid-September ordnance equipment—particularly combat vehicles and trucks—was already badly worn. Both armies had entered the Continent with new equipment, and in the first weeks maintenance had been a relatively simple matter. For six weeks following the breakout from Normandy, weather, terrain, and the disorganization and weakness of the enemy had presented ideal conditions for a gasoline-powered army. In that period the two armies made a grueling run across northern France without adequate maintenance. Forward reserves of major items and of spare parts were practically nonexistent.[4] It had been impossible to establish an adequate depot system, and the great bulk of all supplies on the Continent were still in the Normandy base area.

The whirlwind advances of August and early September thus left the Communications Zone in the condition of an immature athlete who has overexerted himself in his first test of endurance. And there was no time for true recovery. The task of delivering the increasing daily needs of the combat forces remained. At the same time the Communications Zone had to adjust itself to circumstances wholly unexpected a few weeks earlier, and try to build the muscle required to meet the strain of future extensions of the supply lines.

(2) *Competing Tactical Plans*

Although exasperated by the increasing difficulties over supply, field commanders did not immediately appreciate the full implications of the worsening logistic situation. A heady optimism still pervaded the Allied forces in the first days of September, and in at least two of the major field headquarters—Third Army and 21 Army Group—there were strong convictions that the war could be shortened if they were afforded priority in supply.[5]

The possibility of a quick drive across the Rhine by the Third Army was carefully investigated at Supreme Headquarters late in August. At the time there appeared little in the way of enemy forces to prevent such an advance, and it was believed by some that a bold thrust would induce an immediate surrender.

[4] FUSA Rpt of Opns, Bk. IV, p. 2.

[5] See Forrest C. Pogue, *The Supreme Command*, UNITED STATES ARMY IN WORLD WAR II (Washington, 1954), pages 250–60, for a fuller discussion.

Planners of 12th Army Group admitted that it could be carried out only by sacrificing the mobility of other forces, for transportation was already sorely strained. The Third Army by this proposal would be given priority on all available supplies. With a strength of not more than ten or twelve divisions, it was argued, this force could be maintained if other armies were held inactive; if bombers, in addition to troop carrier planes, were used for the transport of supplies; and if British forces were held at the Seine or shortly beyond that river. Even by these measures the advocates of the plan agreed that the force probably could be supported only a short distance beyond the Rhine, possibly as far as Frankfurt.

From both the strategic and logistic standpoints the plan had several weaknesses. A force of 10 or 12 divisions constituted but a small portion of the Allied forces then on the Continent (47 divisions at the end of August). It was also a relatively small force compared with the still-existing German Army in the west. A narrow thrust to the Rhine would not have impaired the strength of that force materially, and an advance in the center of the western front would have created exposed flanks of great length to both the north and south. In the north this flank would have extended approximately 300 miles through enemy territory, and would have rendered the Third Army lines of communication especially vulnerable to attack, particularly in view of the forced immobility of "quiescent" Allied divisions operating at reduced maintenance scales in the rear. Fighter cover would also be difficult to establish as far forward as the Rhine, for the establishment of advance fields required precious supplies and transportation.[6] Furthermore, Frankfurt was not an objective of prime importance, and the area which the advance would have occupied included neither the political nor economic heart of Germany.

Most important of all was the great gamble which such an undertaking would have entailed from the point of view of future logistic support. The concentration of all resources into a single thrust in the Third Army area would certainly have required indefinite postponement of any attempt to capture Antwerp. Without this port there was little hope of receiving, staging, and employing the new divisions arriving each month, and no possibility that the logistic potential would be great enough to allow the extension of the Third Army's operations beyond Frankfurt.

Finally, the entire proposal was predicated on the conviction that the enemy could be frightened into immediate capitulation. Herein lay the crux of the whole matter. Such a result was by no means assured at this time. While the enemy was badly disorganized at the moment, there was no certainty as to what was transpiring inside Germany despite the attempted assassination of Hitler in July. Should the enemy refuse to be shocked into immediate surrender, the operation, in the view of the logistic planners, would bring the Allied forces to the brink of administrative disaster.

[6] Competition was already keen for the use of advance fields for tactical use and the reception of air-transported cargo. See Roland G. Ruppenthal, *Logistical Support of the Armies, Volume I: May 1941–September 1944*, UNITED STATES ARMY IN WORLD WAR II (Washington, 1953), pp. 576–77. (Hereafter cited as *Logistical Support I*.)

The chance of success was a long one, therefore, and the possibility of failure too serious in its implications for future operations.[7] In general, this view represented informal staff reaction at Supreme Headquarters at the time, and General Eisenhower decided against the drive by the Third Army.

The proposed operation was not only hazardous; it ran counter to all the conclusions reached concerning the course to follow in the final drive into Germany. Allied planners had long ago decided that the major effort should be made in the north. Strategic, tactical, and logistical considerations had all favored such a plan of action. The northern route led most directly to the principal objectives in the enemy homeland—the industrial Ruhr and the governmental seat at Berlin. Tactically the terrain in the north was far more suitable than the southern approach for the employment of tanks. Logistically it was favored by close proximity to the channel ports and by excellent road and rail networks. The Combined Chiefs of Staff as well as the theater planners had long since favored this avenue for the main effort in the final advance into Germany.

Even in the north, however, Allied operations were being restricted by the means available. Early in September consideration was given to a 21 Army Group proposal for a rapid thrust to Berlin, a plan even more ambitious than the one just rejected. A study of the logistic implications of such an operation led the SHAEF G–4 staff to conclude that it could be carried out. But its conclusions were based on assumptions which soon proved completely unrealistic. The study assumed, for example, that the main forces of both army groups would have reached the Rhine by 15 September and that the thrust to Berlin could develop at that date. It established as a prerequisite, moreover, that by that date the Allies would already be discharging cargo at Antwerp to the extent of 1,500 tons per day. Other assumptions were made regarding the use of railheads at Brussels, Châlons-sur-Marne, and Paris, and the reduction of maintenance scales and port clearance to save transportation. But the overriding need was for transportation, and to obtain sufficient lift, the study concluded, would require the widest possible use of air transport and the most thoroughgoing marshaling of motor transport yet attempted.

It was estimated that the thrust could be made by three British and two U.S. corps. The support of such a force required the equivalent of 489 truck companies. At the moment there were only 347 available, leaving a shortage of 142. This deficit, it was proposed, might be made up in part by air transport, which was believed capable of achieving a lift of 2,000 tons per day, the equivalent of 60 truck companies. But the largest part of the deficit was to be met by the wholesale grounding of divisions. By diverting their organic truck companies and by forming provisional companies from both U.S. and British units it was estimated that 181 companies could be made available. Combined with the airlift, an equivalent of 241 companies was thus believed attainable. By these measures it was thought possible to support five corps in the operation, three driving to

[7] Whipple, *op. cit.*, p. 12.

Berlin, one British corps to the Bremen–Hamburg area, and one U.S. corps to the area of Frankfurt–Magdeburg. Ten U.S. divisions (1 in Paris, 9 in the Cotentin) would have to be grounded, and an additional 12 U.S. divisions relegated to a "quiescent" state (6 in Brittany, 3 in the Frankfurt–Metz area, and 3 in the area of Ruhr–Koblenz). Only one U.S. corps of three divisions could actually be supported as far forward as Berlin, and even these divisions on reduced maintenance.

Only on the basis of the optimum conditions outlined above was the operation considered at all possible. Since the assumptions on which the plan was based—reaching the Rhine and using Antwerp by 15 September—proved invalid, it appears that such an operation was quite infeasible.[8]

Logistic limitations at the beginning of September thus made it inadvisable to attempt either of the two schemes outlined above. While the thrust in the south was logistically feasible, its reward was uncertain and even of dubious value, and the operation was most hazardous from the point of view of future needs. The thrust in the north, while in accord with long-range plans, for the moment was clearly beyond available means.

(3) The September Decisions

While reluctantly concluding that any effort involving a major extension of the lines of communication was out of the question, SHAEF nevertheless had continued to examine alternative possibilities of maintaining the offensive. Specifically, General Eisenhower had hoped that a force might be built up east of Paris for an additional drive on the subsidiary axis south of the Ardennes. As early as 24 August the Supreme Commander had tentatively concluded that such a project appeared impossible and decided on the necessity to push northeastward in accordance with long-standing strategic plans, meanwhile completing the conquest of Brittany so as to provide the ports required for the accelerated flow of divisions.[9]

The uninterrupted success of the Allied armies in the following weeks continued to nourish the hope that a two-pronged offensive might yet be carried out. In the first week of September General Eisenhower decided that such simultaneous drives to both the Ruhr and the Saar were still within the Allies' capabilities, and he accordingly authorized an advance across the Siegfried Line by both U.S. armies. He admitted that the supply organization by this time was stretched to the breaking point, and that such an operation therefore involved a gamble. But he believed it was a gamble worth taking in order to profit fully by the disorganized state of the German armies in the west.[10] On 10 September Lt. Gen. Omar N. Bradley gave the First and Third Armies equal priority for supply for the operation, subject only to the higher priority accorded the capture of Brest.

Supply of the armies was touch and go at this time, and it was necessary to keep a constant finger on the logistic pulse to

---

[8] Plng Paper, Logistical Implications of a Rapid Thrust to Berlin, Sep 44, SHAEF G–4 Logistical Forecasts, Folder 13.

[9] Cbl, Eisenhower to Marshall, 24 Aug 44, OPD Exec Office File 9.

[10] Cbl FWD–14376, Eisenhower to CCS, 9 Sep. 44, SHAEF SGS 381 Post-OVERLORD Planning.

determine whether operations could continue on the scale desired. On 12 September General Bradley met with his army commanders and their G–4's, and with the Commanding General, Advance Section, and the G–4 of the Communications Zone to discuss the relation between the supply situation and the tactical moves then in progress. Both armies at this time reported sufficient ammunition and gasoline to carry them to the Rhine. In view of the current tactical commitments and the relatively good state of supply at the moment, General Bradley decided to permit the simultaneous attacks by the two armies to continue. Supply capabilities were clearly unequal to the support of sustained operations by both armies against determined opposition, however, for deliveries were being made at the rate of only about 3,300 tons per day to the First Army and 2,500 tons to the Third. The dual offensive was supportable, therefore, only if it could continue at its previous pace and achieve quick success. Lt. Gen. George S. Patton, Jr., Commanding General, Third Army, accordingly was told that unless he was able to force a crossing of the Moselle with the mass of his forces within the next few days he was to discontinue the attacks and assume the defensive along his southern flank.

This condition could not be met. Within the next ten days increasing resistance in both the First and Third Army sectors brought about a shift of effort and of available resources to the north, as was suspected might be necessary.

By mid-September, then, the Allied forces had definitely become the victims of their own success; logistic limitations had clearly come to dominate operational plans. Indeed, a survey of supply capabilities at this time indicated that logistic restrictions might determine the scale of the Allies' efforts for some time to come. U.S. cargo discharge was averaging less than 35,000 tons daily. This was insufficient to clear the arriving shipping, with the result that over a hundred Liberty ships were already awaiting discharge early in September.

Even this tonnage was more than could be cleared from the ports by the available personnel and transport. The number of truck companies available for port clearance had dropped to sixteen and sometimes less as a result of the demands for line-of-communications hauling, and supplies were accumulating in the ports and in the beach areas. Inadequate transportation plus the deficit in port discharge capacity thus threatened to create a bad congestion both offshore and in the Normandy base area.[11]

These basic deficiencies were bound to restrict the number of divisions supportable in active operations and hence limit the scale of combat. At D plus 90 (4 September) the build-up of U.S. divisions on the Continent totaled twenty-one as planned (plus the southern forces), despite the acceleration of July. After the middle of August, however, it had been impossible to maintain all divisions adequately. By early September three had been immobilized and their motor transport used to form provisional truck companies. Two new divisions—the 94th and 95th—were to begin arriving on 10 September. But Brig. Gen. Ray-

[11] Ltr, Whipple to CAO, sub: U.S. Troop Flow To Support a Maximum Effort, n.d. [early Sep], SHAEF G–4 Logistical Forecasts, Folder 13.

mond G. Moses, the army group G–4, doubted whether they could be supported east of the Seine, and surmised that they might also have to remain in the lodgment area and be made available to the Communications Zone in order to increase the latter's hauling capacity.[12]

Logistic planners estimated that there would be twenty-nine divisions on the Continent by 1 October (in addition to the DRAGOON forces in southern France) but thought it unlikely that more than twenty of these could be maintained in combat as far forward as the Rhine at that date on the basis of the current logistic outlook. Any extension of the lines of communication beyond the Rhine promised to reduce further the number of divisions supportable in combat.[13]

This concern over the size of the forces which could be supported beyond the Rhine reflected the optimism which still pervaded the higher headquarters in mid-September and which proved quite unrealistic. But while crossing the Rhine and seizing the vital Ruhr objective in the north were still considered feasible, the gloomy logistic forecasts served to underscore one conclusion which had already been accepted by Supreme Headquarters if not the lower echelons. This was that, even should it prove possible to capture both the Saar and Ruhr objectives, these areas were at the absolute maximum distance at which Allied forces could be supported for the time being and it would be absolutely imperative to develop additional logistic capacity before attempting a power thrust deep into Germany. General Eisenhower felt that the supply organization was now stretched to its absolute limit both as to port intake and inland distribution. He believed that the line at which administrative difficulties were expected to impose a period of relative inaction, originally expected to be the Seine, had certainly now been reached.[14]

The situation in mid-September clearly indicated an urgent need to shorten the lines of communication. The problem was actually a dual one, for there was a parallel need for additional port capacity. The maximum force which could be supported through Cherbourg and the beaches was rapidly being reached, and new capacity was required to compensate for that lost in Brittany.

The obvious solution to this dual requirement lay in the development of the Seine ports and Antwerp. Even at the time the Allied forces were crossing the Seine, when the capture of the Brittany ports was still considered of prime importance, General Eisenhower had emphasized the imperative necessity in the drive to the northeast to secure a "permanent and adequate base at Antwerp." [15]

The further extension of the lines of communication in the next few weeks served to enhance the importance of Antwerp even more. Conferences with his top commanders between 9 and 11 September sustained the Supreme Commander in his previous conviction that the early seizure of deepwater ports and the improvement of base facilities were essential prerequisites to a final drive

[12] Memo, Moses for CofS 12 A Gp, sub: Use of Divs on Line of Communications, 5 Sep 44, 12 A Gp G–4 Memos 1944, Folder 56, Drawer 11.

[13] Ltr, Whipple to CAO (early Sep).

[14] Ltr, Eisenhower to Marshall, 14 Sep 44, OPD Exec Office File 9; Cbl FWD–13792, OPD Cbl Files.

[15] Cbl, Eisenhower to Marshall, 24 Aug 44.

into Germany. His analysis of the situation is clearly revealed in cables which he addressed to both the army group commanders and the Combined Chiefs of Staff on 13 September. In the Supreme Commander's view the port position at this time was such that a stretch of a week or ten days of bad Channel weather—a condition which became more and more probable with the aproaching fall months—would paralyze Allied activities and make the maintenance of the armies even in a defensive role exceedingly difficult. Distribution of supplies on the Continent was approaching solution, he felt, through the improvement in the railway system. But the most immediate objective, and one which had been foreseen as essential from the very inception of the OVERLORD plan, was winning deepwater ports and improving communications.[16]

The Allied forces at this time had before them two possible courses of action. (1) They could concentrate all resources behind a single blow on a narrow front directed toward the center of Germany (a proposal favored by the 21 Army Group commander). (2) They could advance along the entire front with the aim of seizing suitable positions on the German frontier where they could regroup, establish maintenance facilities, and mount a broad drive into Germany. The first course, conceived of as a "knife-like thrust" to Berlin, had already been rejected by the Supreme Commander on both tactical and administrative grounds.[17]

Logistic resources were likewise lacking for the full implementation of the second course, for they were not sufficient to permit simultaneous attacks along the entire front. The decision, as announced by the Supreme Commander, provided that the Allies were to push forward to the Rhine, securing bridgeheads over that river, seize the Ruhr, and concentrate in preparation for a final nonstop drive into Germany. Because of the limited logistic capabilities available, the timing of the Allies' efforts toward the attainment of immediate objectives along the entire front now became of the utmost importance. Implementation of this plan consequently required a succession of attacks, first by the 21 Army Group, then by First Army, and then by the Third Army, with supply priority shifting as necessary. Tactical operations, to paraphrase an old maxim, had now definitely become the art of the logistically possible.[18]

The paramount influence which logistic considerations were to have in any operations undertaken in the near future went beyond the determination of the scale and timing of attacks. Future logistic needs also figured large in the determination of immediate objectives, for General Eisenhower specified that additional bases must be secured simultaneously with the attacks eastward. Accordingly, the Supreme Commander

[16] Cbl FWD–14764, Eisenhower to CCS and Major Comds, 13 Sep 44, 12 A Gp 371.3 Mil Objectives, and SHAEF AG 381–3 SHAEF to AGWAR Rpts on OVERLORD.

[17] Cbl FWD–13889, Eisenhower to Montgomery, 5 Sep 44, Eyes Only Cbls, Smith Papers, Dept of Army Library.

[18] For a fuller discussion of the debate on strategy in September see Pogue, *The Supreme Command*, pp. 288–98.

gave Field Marshal Sir Bernard L. Montgomery, the 21 Army Group commander, the mission of securing the approaches to Antwerp or Rotterdam and the capture of additional Channel ports; and he ordered General Bradley to reduce Brest as quickly as possible in order to accommodate the staging of additional divisions, and to make physical junction with the forces from the south so that the supply lines leading from Marseille might assist in the support of the 12th Army Group as soon as any surplus capacity could be developed.[19]

It was in accord with the above decisions that the 21 Army Group was given preference in the allocation of the available administrative means for the combined U.S.-British airborne operation known as MARKET-GARDEN, which was launched on 17 September with the intention of winning a bridgehead over the Rhine and turning the flank of the enemy's fortified defense line in the north. MARKET-GARDEN was a limited objective operation, however, as General Eisenhower later found it necessary to re-emphasize. At a meeting with his principal staff officers and top commanders held on 22 September he took pains to make clear his desire that all concerned "differentiate clearly between the logistic requirements for attaining objectives covered by present directives, including the seizure of the Ruhr and breaching the Siegfried Line, and the requirements for the final drive on Berlin." In this connection he demanded general acceptance of the fact that the possession of an additional major deepwater port on the north flank was an indispensable prerequisite for the final drive into Germany.[20] Even the present operation in the north, he noted in a separate communication to Field Marshal Montgomery, was a bold bid for a big prize in view of the current maintenance situation. The Supreme Commander considered the operation amply worth the risk. But he took this additional opportunity to stress once again the conviction that a large-scale drive into the "enemy's heart" was unthinkable without building up additional logistic potential. He indicated that this desideratum was now taken for granted in his own mind by closing with the remark, "Of course, we need Antwerp." [21]

The dilemma in which the Allies found themselves at this time was a direct outcome of the August and early September decisions by which logistic considerations had been repeatedly subordinated to the enticing prospects which beckoned eastward. General Eisenhower himself admitted that he had repeatedly been willing to defer the capture of ports (referring obviously to Brittany) in favor of the bolder actions which had taken the Allied armies to the German border. But such deferments could no longer be made in view of the approaching bad weather and the resistance the enemy was beginning to offer in fortress defense.[22]

The necessity for building up the logistic potential before attempting major offensives was more and more widely accepted toward the end of September. Since the overriding necessity was for a shortening of the lines of communication

---

[19] Cbl, Eisenhower to CCS and Major Comds, 13 Sep 44.

[20] Min, Mtg SHAEF War Room, 22 Sep 44, SHAEF SGS 381 Post-OVERLORD Planning.

[21] Ltr, Eisenhower to Montgomery, 24 Sep 44, SHAEF SGS 381 Post-OVERLORD Planning.

[22] Ltr, Eisenhower to Marshall, 14 Sep 44.

and additional port capacity, the development of Antwerp offered the best possible solution of the problem. The effect which the pursuit and the decisions of early September had on the logistic structure was therefore momentous, for it rendered earlier plans largely obsolete by necessitating the shift of the main administrative base northeastward months earlier than had been anticipated.

Logistic planners realized early that these developments would require a complete recasting of administrative plans.[23] Tactical operations had been supported all the way across northern France and Belgium without the Brittany ports, and with the front lines now 400 to 500 miles distant, the value of these ports greatly diminished. By 9 September the Supreme Commander had decided that Quiberon Bay and the ports of Lorient, St. Nazaire, and Nantes were no longer essential for the support of U.S. forces, and he informed the 12th Army Group commander that it would not be necessary to reduce these ports by force of arms, and that the enemy garrisons might simply be contained.[24] At this time the capture of Brest still held the highest priority, but that port was also destined to be abandoned before long. Logistic planners, foreseeing this possibility, suggested in the first days of September that U.S. port development resources be used to supplement British units in developing the ports north of the Seine.[25]

The speed with which the northern ports could be brought into operation was to have an important bearing on the employment of Allied, and particularly U.S., forces. Estimates made late in September indicated that Antwerp might not begin operating before 1 November. There was every prospect, therefore, that U.S. forces would have to depend on lines of communication reaching all the way back to Normandy, aided somewhat by the capacity of the Seine ports. As General Bradley noted to the Third Army commander in explaining the reasons for the decisions of mid-September, the total tonnages which the Communications Zone could guarantee to deliver were sufficient to support the attacks of only one of the American armies if all the other U.S. forces reverted to the defensive. Even such commitments required the postponement of many essential administrative measures, such as moving air units and replacements forward, building advance airfields, winterizing troops and equipment, and replacing worn-out equipment.[26] The priority now held by operations aimed at the Ruhr inevitably placed the burden of the sacrifice on the 12th Army Group forces operating south of the Ardennes—that is, the Third Army.

Fortunately, it was possible to relieve the strain on the attenuated lines of the 12th Army Group somewhat by shifting a portion of the burden to the south. The 6th Army Group, commanded by Lt. Gen. Jacob L. Devers and now embracing all the forces which had built up in the southern lodgment, possessed an in-

[23] Memo, Whipple for CAO, sub: Latest Log Devs, 3 Sep 44, SHAEF G–4 381 War Plans General I.

[24] Cbl FWD–14376, Eisenhower to CCS, 9 Sep 44.

[25] Memo, Whipple for CAO, 3 Sep 44.

[26] Ltr, Bradley to Patton, 23 Sep 44, 12 A Gp 371.3 Mil Objectives I; Memo, Moses for Col William L. Barriger, sub: Confirmation of Telephone Conversation This Date, 9 Sep 44, 12 A Gp G–4 Memos 1944, Folder 56.

dependent line of communications. In the decisions which the Supreme Commander had just announced it was authorized to continue its operations northward into Alsace without restriction since its operations did not divert resources from the north. In fact, the southern line of communications possessed surplus capacity, and measures had already been taken to divert through Marseille three divisions coming from the United States in October.[27]

General Eisenhower was most anxious to take advantage of the logistic potentialities of the southern line of communications, and when General Devers reported at the 22 September meeting at SHAEF that the 6th Army Group could immediately maintain three additional divisions, the Supreme Commander promptly directed the 12th Army Group to release the XV Corps, then operating under the Third Army and consisting of the 2d French Armored and 79th Infantry Divisions, to the southern army group. Seventh Army accordingly took over the XV Corps, along with the sector it then occupied, before the end of the month.[28] A few days earlier another division—the 7th Armored—was taken from the Third Army to strengthen the forces in the north. By these various actions, as General Eisenhower admitted, "things were being stretched thin in the middle." [29] Additional measures taken to relieve the strain on the 12th Army Group supply lines included diverting shipping lying off the Normandy beaches to Marseille and allotting to U.S. forces certain port capacity in excess of British needs at Le Havre. Preparations were also made to discharge LST's on the Pas-de-Calais beaches as an emergency measure.[30]

The decisions of mid-September thus reflect a full realization of the extent to which logistic limitations had come to straitjacket tactical plans. The Supreme Commander's directives on the 13th reveal a determination to maintain the offensive in accordance with earlier strategic plans, but such plans now had to be tailored to severely restricted logistic capabilities. Any thought of carrying out a power thrust aimed at objectives deep inside Germany was definitely abandoned with the realization that any sustained drives would require a major orientation of the entire logistic structure—that is, a shift to shorter lines of communication based on the northern ports.

(4) *Prospects at the End of September*

Detailed studies of logistical capabilities made at both SHAEF and Headquarters, Communications Zone, toward the end of September confirmed the earlier doubts regarding the scale on which combat operations could be conducted in the near future. Their gloomy forecasts stemmed directly from the poor prospects with regard to port discharge and inland transportation.

SHAEF planners had little hope that any material advantage would be gained from the use of ports in the low countries by early October, and predicted that both the 21 and 12th Army Groups would

[27] See below, Ch X, Sec. 1.

[28] Min, Mtg SHAEF War Room, 22 Sep 44, and Ltr, Eisenhower to Montgomery, 24 Sep 44.

[29] Ltr, Eisenhower to Montgomery, 24 Sep 44.

[30] 12 A Gp AAR, Sep 44.

be operating on very extended lines of communication for some time to come. For administrative reasons, therefore, they concluded that it would be desirable to withhold an advance into Germany beyond the Ruhr until late October in order to permit the development of the Antwerp–Rotterdam port capacity and to ensure the establishment of secure advance bases near Antwerp for 21 Army Group and in the region of Metz–Nancy for 12th Army Group. In considering several alternative courses the planners estimated that it would be possible to support "limited forces" in an advance in October, but only on the basis of conditions which were unlikely to be met: if infantry divisions were not motorized; if air supply were made available to the extent of 2,000 tons per day or some thirty truck companies were withdrawn from quiescent divisions to assist in the forward movement of supplies; if opposition were slight; if forward reserves were accumulated at about half normal rates; if not more than twenty-five divisions were employed forward on the Rhine; and if objectives were reached within about a fortnight.[31]

At the end of the month the Communications Zone, in response to an inquiry from the army group G–4, General Moses, presented figures on its delivery capabilities which revealed even more clearly the impossibility of supporting large-scale operations east of the Rhine. General Moses had indicated to Brig. Gen. James H. Stratton, the theater G–4, that the field forces should be able to count on daily maintenance at the rate of 650 tons per division slice. Added to the needs of the Ninth Air Force and the Advance Section, this brought the total maintenance requirements in the forward areas to 18,800 tons per day in the first half of October, assuming the employment of 22 divisions, 20,750 tons per day in the second half of the month, with 25 divisions, and 22,700 tons by 1 November, when the strength of the 12th Army Group would reach 28 divisions. General Moses estimated, however, that the field forces would require the delivery of approximately 100,000 tons of supplies over and above these daily maintenance requirements in order to meet deficiencies in equipment and establish minimum reserves of about three days in all classes of supply. He used 1 November as a target date for the beginning of discharge operations at Antwerp. On the assumption that the daily maintenance requirements could be met, he requested the Communications Zone to estimate the time necessary to deliver the additional 100,000 tons and also to establish depots in the Advance Section.[32]

The COMZ reply was discouraging indeed. General Stratton was even less optimistic about the prospects of quick dividends from the port of Antwerp. He did not believe that discharge there could be counted on before the middle of November, although he estimated that

[31] SHAEF G–3 Appreciation, Factors Affecting Advance into Germany After Occupation of the Ruhr, 24 Sep 44, and Memo by Plng Stf, summarizing above, same date, SHAEF SGS 381 Post-OVERLORD Planning.

[32] Memo, Moses for Stratton, sub: Supply Estimate, 25 Sep 44, SHAEF G–4 400 Supplies General 44 III.

some tonnage would begin to move forward from Le Havre and Rouen by mid-October and might average 3,000 tons per day in the last half of the month. On this basis he estimated that it would be approximately sixty days before any substantial tonnages could be built up in the forward areas. In fact, he pointed out that for the entire month of October COMZ deliveries would not even meet daily maintenance requirements.

The Communications Zone planned to build small reserves—totaling about 11,000 tons—in the First Army area in the last two weeks of October, but this was to be accomplished only at the expense of falling short of the daily maintenance needs of the other armies. Delivery capacity was expected to exceed daily needs by a few hundred tons early in November, but the build-up of reserves was contemplated in only one area—the First Army's—and even then at the sacrifice of some of the Ninth Army's daily requirements. Less than 30,000 tons of reserves were expected to be built up—all in the First Army area—by mid-November. Not until then did the Communications Zone expect its port and transportation situation to improve sufficiently to begin building reserves over and above daily needs in the Third and Ninth Army areas and in the depots of the Ninth Air Force and the Advance Section. In the last two weeks of November it hoped to deliver approximately 100,000 tons of reserve supplies to the forward areas. These stocks it planned to set down in the Huy–Aachen area for the First Army, in the Longwy–Metz–Nancy area for Third and Ninth Armies, and initially at Reims for the Ninth Air Force. The Advance Section's dumps were to be distributed between the two army areas.[33]

The outlook for the next six to eight weeks was thus a depressing one. There appeared no escaping the prospect that the forces which the 12th Army Group could maintain actively operational would either have to be reduced in size or continue on the starvation scales which had characterized their support for the past several weeks. At the beginning of October the 12th Army Group comprised 20 divisions (10 in First Army, 8 in Third, and 2 in Ninth). In addition, there were in the Cotentin and Brittany a total of 7 divisions, of which 1 was engaged in a tactical mission and the remaining 6 were either grounded or in the process of marrying up with their equipment. By mid-October, it was estimated, 25 divisions could be made operationally available to the 12th Army Group. The army group commander naturally wanted to use the additional units becoming available, and General Bradley in fact indicated his intention of moving 6 divisions forward from the Cotentin between 10 and 20 October. Allowing for the transfer of 1 division (the 44th) to the 6th Army Group, this would result in a net increase of 5 divisions and raise the strength of the group to 25 divisions.[34]

G–4 planners at SHAEF had estimated that the provision of daily maintenance for this number of divisions and the building of a reasonable minimum of re-

---

[33] Memo, Stratton for Moses, sub: Supply Estimate, 1 Oct 44, ADSEC AG 400 Supplies General, or SHAEF G–4 400 Supplies General 44 III.

[34] Memo, Col C. Ravenhill, Deputy Chief Log Plans, for G–4, 10 Oct 44, SHAEF G–4 Maintenance of British and U.S. Forces 153/2/GDP–1, Box 1, Folder 42.

serves would require deliveries of more than 22,000 tons of supplies per day.[35] The forwarding of such tonnages was obviously out of the question in view of the Communications Zone's announced capabilities. Deliveries in September had averaged only 8,000–10,000 tons per day, and the maximum tonnage which the Communications Zone estimated it might deliver in the last half of October was 20,000. With this tonnage it was estimated that all twenty-five divisions could be maintained only if scales of maintenance were kept below those considered sound. The support of the 12th Army Group would therefore be on a hand-to-mouth basis and without the accumulation of any reserves. This was a risk that was unacceptable to supply officers in view of the imminence of winter weather (which would require the forward movement of substantial tonnages of heavy clothing and winterization equipment), the need to build up bridging equipment for the expected river crossing operations, the heavier requirements for ammunition, and the accomplishment of higher echelon repair of vehicles which had already been long delayed. On the basis of the COMZ estimates the SHAEF G–4 planners concluded that logistic limitations would not permit the employment of more than twenty divisions in the 12th Army Group. Even this allowed the "mobile" engagement of only two thirds of the total, and the accumulation of reserves for only the one army having first priority for its effort—that is, First Army, with ten divisions.[36]

On 11 October SHAEF informed General Bradley of this limitation on the size of the maintainable force, noting that his proposal to move additional divisions forward from Normandy consequently would be feasible only if a corresponding number of divisions was withdrawn from the line and rested.[37] Within the next ten days, nevertheless, additional units were moved forward and committed, bringing the line strength of the 12th Army Group to twenty-three divisions.

Meanwhile, supply deliveries had actually been sufficient to support only thirteen divisions adequately, to say nothing of maintaining the twenty on which recent plans had been based.[38] The prospect of providing adequate support to the growing number of divisions thus appeared even more dismal toward the end of October, providing abundant proof of the necessity for a strict rationing of the meager resources available. In the course of the month Supreme Headquarters therefore prepared to in-

[35] On the basis of 560 tons of maintenance per division slice at regular scales, 280 tons of reserves build-up per day (one half day per day), totaling 840 tons per division per day or 21,000 tons for the three armies. In addition, the army group would require civil affairs supplies, raising the total to more than 22,000 tons. SHAEF G–4 Allocation of Tonnages 1, for period 15–28 Oct, 8 Oct 44, SHAEF G–4 400 Supplies General 44 III.

[36] SHAEF G–4 Allocation of Tonnages 1 for Period 15–28 Oct, 8 Oct 44, and covering Ltr, Whipple to Crawford, sub: Tonnage Allocations, 8 Oct 44, SHAEF G–4 400 Supplies General 44 III; Memo, Whipple for G–3 Plng Stf, sub: Advance Across the Rhine, 14 Oct 44, SHAEF G–4 381 War Plans General II.

[37] Cbl S–61798, SHAEF to Bradley, 11 Oct 44, SHAEF AG 381–3 SHAEF to AGWAR Reports on OVERLORD.

[38] Memo, Whipple for G–3 Plans, sub: Maintenance of 12 A Gp Divs, 21 Oct 44, SHAEF G–4 Maintenance General, Box 1, Folder 47.

stitute a detailed allocation of supplies, specifying the scale of maintenance and reserves build-up for each army and distributing available tonnages to each army in line with the expected scale of activity of its divisions, depending on the missions assigned.

The full effect of the strain which the overextension of the lines of communication had imposed on the logistic structure now became apparent. It was also clear that the maintenance of large-scale operations would remain unsatisfactory until the port of Antwerp and adequate rail lines of communications were made available. The opening and development of that port consequently remained a matter of transcendent importance.[39] For the next few months the military operations of the 21 and 12th Army Groups were to be dominated by the necessity to develop a new administrative base in closer proximity to the theater of action.

[39] Cbl FWD–16181, Eisenhower to CCS, 29 Sep 44, SHAEF SGS 381 Post-Overlord Planning; Cbl S–64375, Eisenhower to Major Comds, 23 Oct 44, SHAEF AG 381–3 SHAEF to AGWAR Reports on Overlord; Memo, Whipple for G–3 Plng Stf, 14 Oct 44.

CHAPTER II

Tactical and Organizational Developments

(1) Tactical Operations [1]

Between September and February the theater's command and organizational structure, both tactical and administrative, took its final form. Allied tactical operations, meanwhile, were to contrast markedly with those of the pursuit period. Hampered by the lack of adequate logistic support, and faced with a reorganized and reinforced enemy, the Allied armies for the next several months either marked time or measured their gains in yards, at heavy cost in casualties and matériel. In one instance, the enemy's December counteroffensive, they sustained a severe setback.

At no time was the impact of logistics on tactical operations more evident than in late September and October. Though SHAEF was able to marshal limited American logistical assistance for the big airborne operation in the Netherlands in support of the 21 Army Group (Operation MARKET-GARDEN), the diversion added to the supply famine in the 12th Army Group. The First Army, for example, pierced the Siegfried Line at Aachen but found a combination of inhospitable terrain, renewed German tenacity, and supply limitations too restrictive to permit exploitation of the breach. (*Map 1*) Artillery ammunition and replacement tanks were in particularly short supply. At the same time the Third Army, though establishing bridgeheads over the Moselle River near Metz and Nancy, felt a similar pinch. The commanders of these forces would have been more inclined to accept the supply limitations with equanimity had the 21 Army Group, in the meantime, been able to exploit Operation MARKET-GARDEN to the extent projected. But the airborne operation and the concurrent ground advance in the north, designed to put the British across the major water obstacles of the eastern part of the Netherlands and into position for a thrust against the Ruhr, fell short of its objectives.

By late September the full effect of the transportation and supply shortages was all too apparent. General Eisenhower

[1] For detailed accounts of tactical operations during this period, see H. M. Cole, *The Lorraine Campaign*, UNITED STATES ARMY IN WORLD WAR II (Washington, 1950), and the following volumes in preparation for the series: H. M. Cole, The Ardennes Campaign; Charles B. MacDonald, The Siegfried Line Campaign; and James D. T. Hamilton and Robert Ross Smith, Southern France and Alsace.

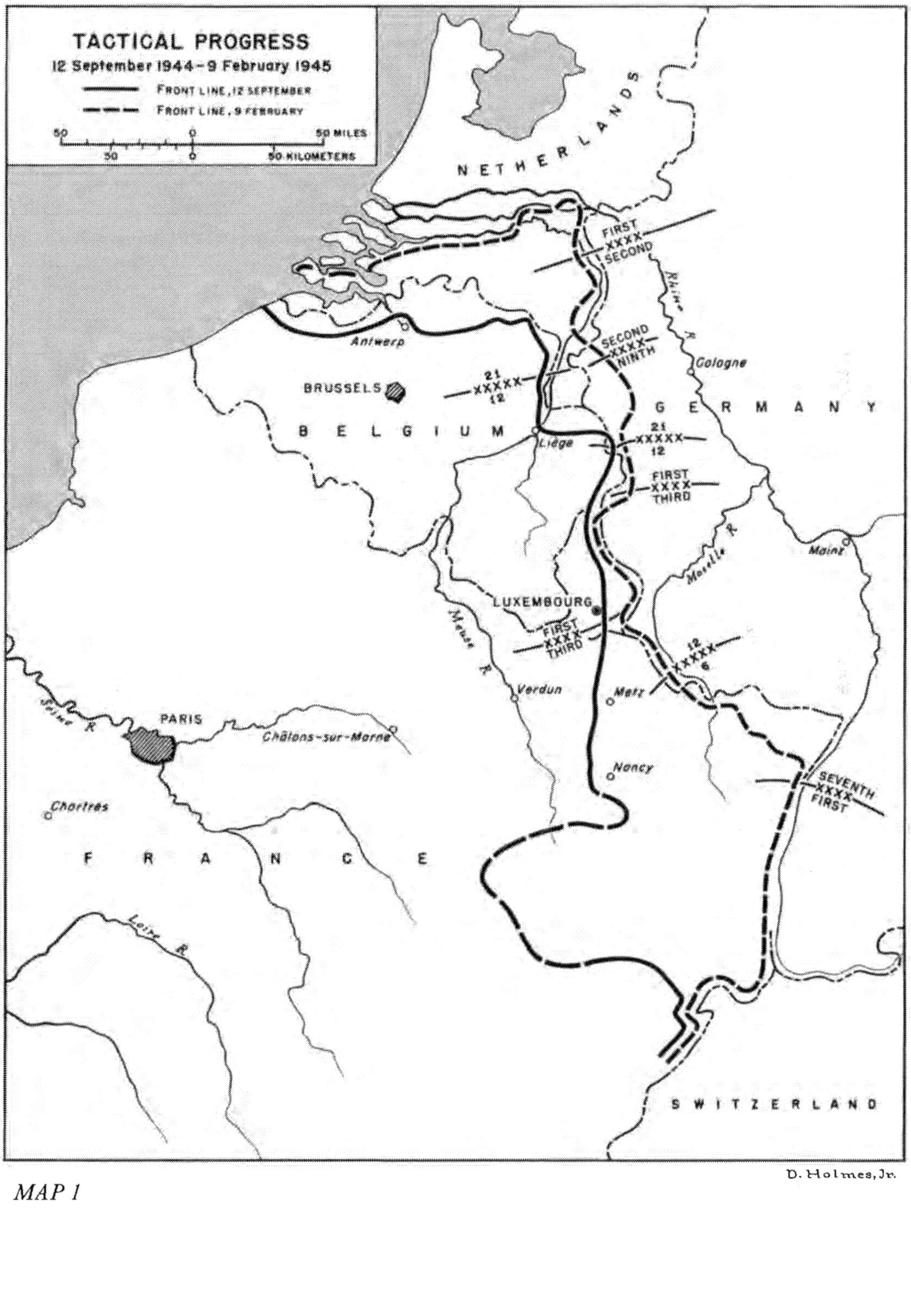

D. Holmes, Jr.

MAP 1

had little choice but to concentrate almost all available resources behind only one portion of the front lest his entire advance bog down. In keeping with original planning, he chose the north. The 21 Army Group, while retaining its general objective of enveloping the Ruhr from the north, was at the same time to open the port of Antwerp as a matter of great urgency. The neighboring First Army could do little except conduct limited operations in general support of the 21 Army Group and in preparation for the day when the supply situation might improve and the push toward the Rhine be renewed. Operating on an axis of advance which had been denied first priority, the Third Army felt even more stringent supply restrictions than did the First. The only major achievement by either army during the logistically drab days of October was the First Army's reduction of Aachen.

In the extreme north the commander of the 21 Army Group, Field Marshal Montgomery, tried at first to continue his push toward the Ruhr while his left wing, in the form of the First Canadian Army under Lt. Gen. H. D. G. Crerar, opened Antwerp. But even when strengthened by a temporary shift of some territory to the 12th Army Group, Montgomery found the dual assignment too demanding. The enemy's position along both sides of the Schelde estuary, which denied Allied access to Antwerp from the sea, proved particularly strong. Even after the 21 Army Group in mid-October turned undivided attention to opening Antwerp, almost a month of severe fighting under the most adverse conditions of weather and terrain remained before the approaches to the port were clear. Not until 28 November was the first Allied ship to drop anchor in the harbor.

Meantime, on the extreme right flank of the western front, operations had assumed much the same character. After the link-up of Seventh Army with the 12th Army Group in mid-September, the pursuit of the enemy from the south had halted for the same reasons it had in the north, despite the existence of a separate line of supply leading from Marseille. Almost coincidentally with the end of the pursuit, the southern forces, consisting of the Seventh U.S. Army and the First French Army, came under a new headquarters, the 6th Army Group. Until this time controlled by AFHQ in the Mediterranean, these forces now came under the direct command of General Eisenhower.[2] When directed by the Supreme Commander to continue the drive toward the Rhine, the 6th Army Group tried to push northeastward through the Saverne and Belfort gaps, only to be forced to the unavoidable conclusion that, under the existing logistical situation, this plan, like the plans in the north, was far too ambitious.

The six weeks' fighting between mid-September and the end of October had brought little noticeable advantage to the Allies as far as the enemy, terrain, or weather was concerned. Meanwhile, the logistic situation, even without use of Antwerp, had shown some improvement. In the Supreme Commander's opinion, sufficient supplies had been accumulated, when superimposed upon the promise of Antwerp, to warrant a resumption of the offensive along the entire front. This is not to say that all was well from a logistical standpoint—the U.S. armies, for example, still had to ration artillery ammunition drastically—but in keeping with the general principle of giving the enemy no rest, General Eisenhower deemed a renewal of the attack advisable.

Viewed in retrospect, the November plans, while aimed only at closing to the Rhine, were far too optimistic. The 12th Army Group, with twenty-seven divisions now available, planned major efforts both north and south of the Ardennes; the 6th Army Group with fourteen divisions, taking advantage of the maintenance resources available from the Mediterranean, was to advance to the Rhine while protecting the 12th Army Group's right flank. Only the 21 Army Group, still engaged in freeing the approaches to Antwerp, was to remain relatively quiescent.

The Third Army struck the first blow in the new offensive on 8 November, while on 13 and 14 November, respectively, the Seventh U.S. and First French Armies joined the assault. After awaiting favorable weather for a major air bombardment, the First Army and the newly formed Ninth U.S. Army under Lt. Gen. William S. Simpson took up the attack in the north.

Adverse weather, which set new rainfall records, and terrain favoring the defender, plus a remarkable German resurgence, militated more against the renewed Allied drive than did logistics. Whatever the primary cause, the drive fell short. The Third Army and the 6th Army Group made the most impressive advances, including capture of Metz, arrival at the Siegfried Line in front of the Saar industrial region, and, with the exception of a big enemy pocket hinged on the city of Colmar, reaching the Rhine along a broad front in the extreme south. In the north, the First and Ninth Armies by mid-December had gotten no farther than the little Roer River, less than fifteen miles beyond Aachen. The 21 Army Group, in the meantime, had begun to assist by attacks along the American left flank, but the necessity to clear a stubborn enemy force which held out in constricted terrain west of the Maas River still limited British participation.

Mid-December thus found the Allied armies still attempting to close to the Rhine in execution of the first phase of the November offensive, when, on 16 December, the enemy seized the initiative with a powerful counteroffensive in the Ardennes area of eastern Belgium and northern Luxembourg. Aimed at crossing the Meuse and recapturing Antwerp, thereby splitting the Allied forces and possibly annihilating those north of the Ardennes, the offensive was launched with twenty-four divisions which the enemy had secretly marshaled.

Attacking on a fifty-mile front in the center of the First U.S. Army's sector, the enemy struck in an area which had been lightly held throughout the fall in order that U.S. forces might concentrate for offensive operations elsewhere. Though the Germans quickly broke through, tenacious resistance by isolated American units and quick reaction to the attack by Allied commanders halted the thrust short of its first objective, the Meuse River, The end result of the enemy's effort was a big bulge in the American line. Because Antwerp had begun operation as a port not quite three weeks before the enemy struck, the impact of the bulge on the U.S. logistical structure never became critical. Existence of the bulge nevertheless imposed certain communications difficulties, which prompted the Supreme Commander to place the Ninth Army and those elements of the First Army north of the enemy salient under the operational control of Field Marshal Montgomery.

On 23 December an attack by the Third U.S. Army established tenuous contact with a beleagured American garrison in the communications center of Bastogne, which had held out against repeated German assaults. Three weeks later, on 16 January, drives by the First and Third U.S. Armies linked up at the road center of Houffalize. After severe fighting in the face of winter weather, rugged terrain, and skillful enemy defense, the two armies by the end of the first week of February had completely erased the enemy's earlier gains and in some sectors had followed the withdrawing Germans through the West Wall. The net effect of the German effort so far as the Allies were concerned was to delay Allied offensive operations toward the Ruhr and the Saar about six weeks.

The first week of February also saw the enemy's holdout position on the extreme right flank of the Allied line around the city of Colmar eliminated, but not before the Germans had launched a smaller-scale counteroffensive in this sector. Though the attacks achieved some initial success and for a time threatened recapture of Strasbourg, adroit withdrawals followed by counterattacks with strong reserves soon contained them all. On 20 January First French Army troops led an attack to drive the last Germans from the west bank of the Rhine in the 6th Army Group's sector. Joined later by a corps of the Seventh Army, on 9 February they achieved their goal. Thus by early February the entire western front was stabilized and the enemy's offensive capabilities west of the Rhine were eliminated.

(2) *Organization and Command*[3]

While the command and administrative structure of ETOUSA never achieved complete finality because of the repeated necessity to adapt itself to the changing tactical situation, in August and September 1944 it took the final form envisaged in the OVERLORD plan.

The first week in August had been an important one in the evolution of command and organization in both the tactical and administrative fields. On the tactical side, a second army (the Third) was introduced on 1 August and an army group organization under the command of General Bradley was established.[4] On the administrative side, meanwhile, the implementation of the organizational plan had suddenly accelerated. On 2 August, after several delays, the Communications Zone finally achieved legal status on the Continent by the drawing of a rear boundary for the armies. Contrary to plan, the movement of Headquarters, Communications Zone, was ad-

[3] As pointed out in the Preface to *Logistical Support I,* the command and organizational developments of the pursuit period and the command decisions of early fall 1944 are discussed here in order to adhere to the topical-chronological organization adopted for both volumes.

[4] Marking the inauguration of Phase III of the command plan, on D plus 56 instead of D plus 41 as planned. See *Logistical Support I,* 208ff.

vanced a full month. On 7 August General Lee's headquarters itself took over operation of the continental Communications Zone. Two entire phases in the OVERLORD command plan were thus virtually eliminated so far as the administrative structure was concerned, for the Advance Section held a position independent of the First Army as the operative communications zone on the Continent for only five days during the first week in August, and Forward Echelon, Communications Zone, did not become operational at all.

The month of August nevertheless represented a transitional stage. Since the advance element of Supreme Headquarters had not yet moved to the Continent, 12th Army Group was to remain under the over-all command of 21 Army Group for the next month, and for a period of three weeks General Bradley, as the senior U.S. field force commander, continued to exercise supervision over the development of the Communications Zone, as he had as commanding general of First Army. In this capacity he retained the authority to prescribe levels of supply in the Communications Zone and army depots, assign priorities for supply, and regulate the apportionment of service troops between the armies and the Communications Zone. The 12th Army Group commander had always desired such an arrangement, for it was his belief that the senior field force commander should also control his line of communications, as in British practice.

This situation prevailed until the end of August, when the final stage of the command setup was ushered in by the arrival of Supreme Headquarters. SHAEF began moving to the Continent in mid-August. On 1 September it assumed operational control of all forces, bringing the 12th Army Group under its direct control, and placing the Communications Zone directly under the command of General Eisenhower as theater commander. The Communications Zone thereby attained a position at least co-ordinate with the 12th Army Group. Unfortunately the effect of this development was to perpetuate the friction between the Communications Zone and the field forces which had developed over General Lee's position in the United Kingdom, for Lee's headquarters continued to exercise some of the independence and authority of a theater headquarters by virtue of the presence there of the theater's general and special staff divisions.[5]

While command relationships were relatively final by 1 September, the gradual accession of additional forces necessarily changed the organizational complexion of the theater somewhat. Early in September the 12th Army Group was strengthened by the addition of a third army—the Ninth, under the command of General Simpson. Ninth Army headquarters had been formed from the Fourth Army headquarters at San Antonio, Texas, and had started its movement to the United Kingdom in April 1944.[6] Its first duties in the theater were to receive and train various units arriv-

[5] [Robert W. Coakley] Organization and Command in the ETO, Pt. II of the Administrative and Logistical History of the ETO, Hist Div USFET, 1946, MS (hereafter cited as Organization and Command) II, 153–59.

[6] The new headquarters came to the United Kingdom as the Eighth Army, but was redesignated the Ninth Army in May to avoid confusion with the British Eighth Army.

ing in England, and it was not expected to move to the Continent until the port of Brest was in operation. Late in August, however, Ninth Army headquarters moved to France and was assigned to the 12th Army Group, and on 5 September was given command of the VIII Corps in the Operations against Brest. Headquarters were established in close proximity to those of Brittany Base Section, which had been set up at Rennes, in order to provide the necessary communications facilities and to facilitate co-ordination with the Communications Zone in matters of supply and maintenance. In addition to its operational mission the Ninth Army was given the task of administering the reception and accommodation of the various corps and divisions then arriving in Normandy. After the capture of Brest it redeployed northeastward and eventually participated in the November offensive, as already related.[7]

In the meantime another type of army had been organized to co-ordinate and control Allied airborne operations. Early in August SHAEF created a British-U.S. combined airborne headquarters for this purpose, and in mid-August designated this organization the First Allied Airborne Army and placed it under the command of Lt. Gen. Lewis H. Brereton, who had been Commanding General, Ninth Air Force. For command and administrative purposes the U.S. components of this new army were organized into the XVIII U.S. Airborne Corps under the command of Maj. Gen. Matthew B. Ridgway, former Commanding General, 82d Airborne Division. With certain exceptions the operational control of the First Allied Airborne Army remained with Supreme Headquarters.[8]

The most significant augmentation of the tactical command structure occurred in mid-September with the incorporation of the Allied forces in southern France, shortly after SHAEF assumed direct operational control on the Continent. Operation DRAGOON had long been intended as a corollary to OVERLORD and had been planned to be launched simultaneously with the landings in Normandy. But the necessity to transfer some of the Mediterranean troop lift to the United Kingdom for OVERLORD led to the postponement of the operation. The final decision to launch the southern operation was not made until 10 August 1944, and the actual assault was made five days later.

DRAGOON was mounted from the North African theater, and consequently came under the direction of the Supreme Commander in that area, General Sir Henry Maitland Wilson. Make-up of the assault force was to be predominantly American, and the operational control in the early stages was given to the Seventh U.S. Army, which was reactivated for this purpose and placed under the command of Lt. Gen. Alexander M. Patch. A French army command was to come into being as soon as the build-up of French forces was large enough to justify it. General Jean de Lattre de Tassigny was to be in command of the first French troops to go ashore, the intention being that he should eventually command the French army.

[7] Organization and Command, II, 162–67.

[8] *Ibid.*, II, 168–70.

In anticipation of the time when there would be two armies in the south, provision was also made for the organization of an army group command. On 1 August the 6th Army Group was activated at Bastia, Corsica, under the command of General Devers, who at the time was commanding general of the North African theater and Deputy Supreme Commander under General Wilson. Since the army group command was not to become operational until such time as two armies were in existence, an Advance Detachment, AFHQ, was organized to exercise over-all control in the first stages. This forward echelon of General Wilson's headquarters, also commanded by General Devers, was actually composed of basically the same personnel which made up the 6th Army Group headquarters, thus providing the desired continuity in command.

Operation DRAGOON was a striking success, and within two weeks of its launching General Wilson contemplated the transfer of control of the southern France forces to SHAEF, the intention being that 6th Army Group would become operational at the same time. The size of the force under General Devers' command at the time (nine divisions) hardly justified the latter step, but General Wilson felt it was required by both the nature of operations with French forces and the length of the lines of communication, and negotiations were therefore undertaken to effect the transition. Tactical developments and the absence of essential communications facilities actually delayed the transfer of command somewhat. But the time finally appeared ripe after the junction of French forces from the south with elements of the Third U.S. Army near Sombernon on 11 September, and it was agreed that SHAEF should take over operational control of the southern forces on 15 September. On that date the control of AFHQ over the DRAGOON forces came to an end.

At the same time French Army B, which had become operational on 21 August, was redesignated as the First French Army, and the 6th Army Group, now consisting of the French First and Seventh U.S. Armies, became operational, absorbing Advance Detachment, AFHQ.[9] The XII Tactical Air Command, which supported the DRAGOON forces, also passed from the control of the North African theater to the Ninth Air Force in the north.

Planning and launching DRAGOON had entailed two territorial revisions in the North African and European theaters. Southern France had been included in the European theater in 1943. Because DRAGOON was to be launched from Mediterranean bases, the War Department redrew the theater boundaries in February 1944, removing southern France, plus Switzerland, Austria, and Hungary, from the European theater, and placing them within the North African theater. On 18 September, after SHAEF had assumed control over the southern forces, the boundaries of the European theater were once more extended to include all of France and Switzerland. (*Map 2*) [10]

One more major addition to the U.S. tactical command structure occurred

[9] General Devers was subsequently relieved of his other assignments in the Mediterranean, Lt. Gen. Joseph T. McNarney becoming Commanding General, NATOUSA, and Lt. Gen. Mark W. Clark the Deputy Supreme Commander.

[10] Organization and Command, II, 223–32.

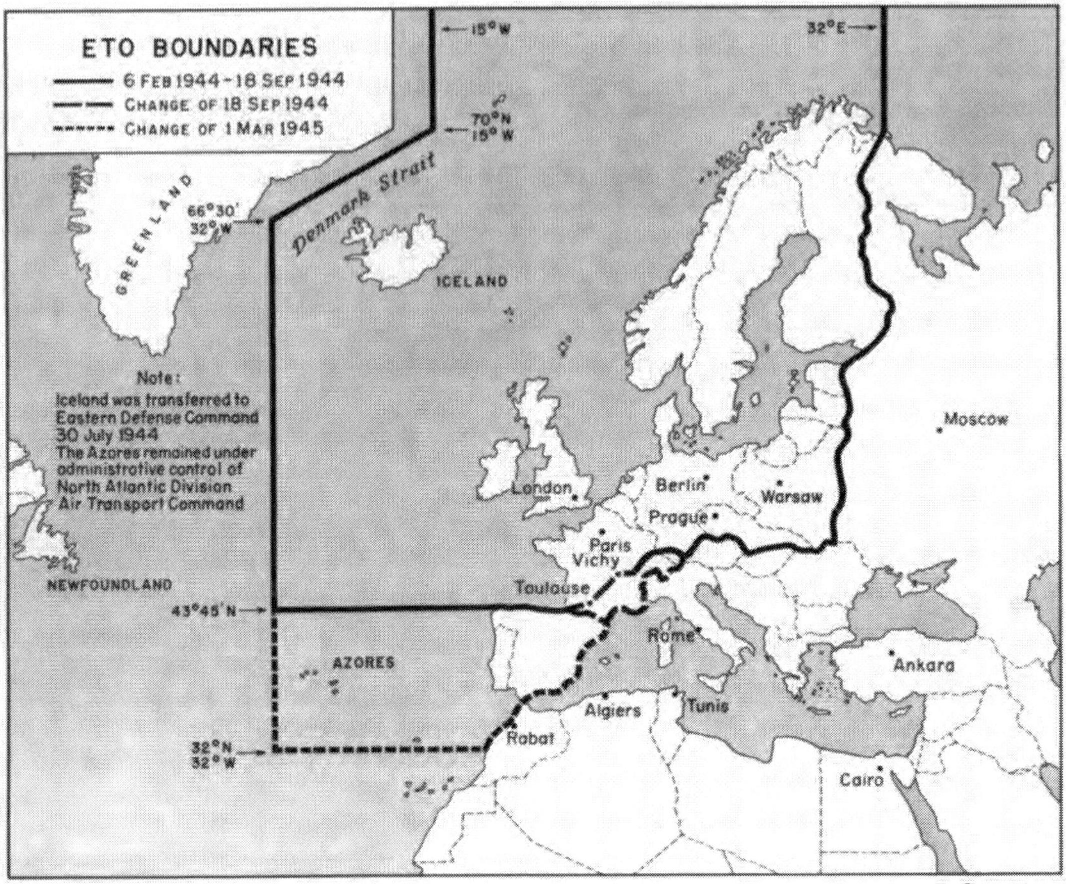

MAP 2

late in 1944. On 27 December the Fifteenth Army was activated and assigned to the 12th Army Group. Fifteenth Army at first consisted only of a headquarters, and its functions initially were purely administrative. Its first responsibility was to serve as a headquarters for the control of U.S. units in the SHAEF reserve, a task it began to perform early in January 1945. Shortly thereafter it was given responsibility for staging, equipping, and training new units entering the Continent, for the reorganization and re-equipping of units returning from the combat zone, and for planning the occupation, particularly in the 12th Army Group sector of the Rhineland. The command of the army went temporarily to Maj. Gen. Ray E. Porter, but on 14 January passed to Maj. Gen. Leonard T. Gerow, former commander of the V Corps.[11]

In the meantime the logistic structure of U.S. forces on the Continent had

[11] *Ibid.*, II, 261–64.

undergone its logical development, its evolution being governed largely by the rate at which ground was recaptured. The movement of Lt. Gen. John C. H. Lee's headquarters alone was a process which took several weeks and was not completed until mid-September. A portion of the COMZ headquarters remained in England for some time. The base section organization developed over a period of some months, according to need.

The core of the COMZ headquarters merged with the Forward Echelon near Valognes, where engineers constructed new tented quarters for approximately 11,000 persons and about 560,000 square feet of hutted office accommodations.[12] But its location there was destined to be of short duration. Within a matter of days the capture of Paris appeared imminent, and General Lee contemplated an early move to that city.

The order for the move to Paris was actually issued by Brig. Gen. Royal B. Lord, and without Lee's knowledge.[13] On 1 September the Chief of Staff dispatched a forward echelon to Paris, and within the next two weeks the entire headquarters moved to the French capital, a portion of it going there directly from London, and the remainder from Valognes. Using precious motor and air transportation at the very time of the critical supply shortages at the front, the movement naturally produced strong criticism from combat commanders. In the view of Lt. Gen. Walter B. Smith, the SHAEF Chief of Staff, the premature move was also largely responsible for the delays and difficulties in communications with the War Department which caused much embarrassment in September. SHAEF depended on the Communications Zone for such communications, and the movement made it impossible to extend reliable long-range communications from the rear.[14]

General Eisenhower himself disapproved of the move to Paris, and on 13 September sharply reproved the COMZ commander, notifying Lee that his headquarters was not to be located there.[15] The Supreme Commander modified this order a few days later, realizing that the heavy shipments of personnel and supplies to the French capital made it impossible to shift the COMZ headquarters for the present without interfering with higher priority tasks. Paris, after all, was the logical location for the headquarters of the theater's administrative organization, for it contained a concentration of supply depots, hospitals, airstrips and airfields, railway stations and marshaling yards, and inland waterway offloading points which were vitally important to the supply structure of the theater.

General Eisenhower nevertheless considered the movement of General Lee's headquarters extremely precipitate and believed the influx of Americans into Paris to be unwise. He was obviously disturbed over the reports he had received about the blackmarket activities of U.S. personnel in Paris, and in his second communication to General Lee

[12] *Final Report of the Chief Engineer, European Theater of Operations, 1942–45* (Paris, Hervé et Fils [c. 1946], prepared in Office of Chief Engr ETO, 1946, p. 326.

[13] Interv with Lord, 9 Aug 51, and Col Talley, 7 Mar 51.

[14] Cbl FWD–16198, Smith to Lt Gen Thomas T. Handy, 29 Sep 44, OPD Cbl Files.

[15] Cbl FWD–14637, Eisenhower to Lee, 13 Sep 44, SHAEF SGS 014.1 Civil Affairs in Paris I.

made pointed reference to reports that the dress, discipline, and conduct of Americans in the French capital were "little short of disgraceful." While the Supreme Commander now authorized the COMZ headquarters to remain in Paris, he directed General Lee to stop immediately the entry into that city of all personnel not needed there for essential duty, and to institute a survey at once of the units already located there with a view toward removing all whose presence was not absolutely required. One of the main purposes of the reduction was the desire to use Paris primarily as a leave center.[16]

Efforts to break the "Paris fever" which had seized Americans of all ranks did not meet with spectacular success. On 20 October General Smith met with SHAEF and COMZ staff officers to re-emphasize the Supreme Commander's policy on movements to the French capital and to discuss ways of transferring additional units without detriment to the war effort. Some units had already moved outside the city, but the number of new requests for accommodation within Paris more than made up for the removals. The fact that about 90 percent of the hotels in Paris had been taken over by the Americans produced an unfortunate public reaction and gave rise to comments that Allied demands in several respects exceeded those of the Germans. The COMZ headquarters alone occupied 167 hotels, the Seine Base Section headquarters another 129, and other organizations such as SHAEF, the Office of Strategic Services, the U.S. Navy, and the Air Transport Command were using approximately 25 more. The Seine Section and Communications Zone at this time estimated that between them they could release approximately 150 hotels, but the extent to which this was carried out is not known.[17]

These efforts to keep to a minimum the number of U.S. troops in Paris did nothing to reverse the decision by which that city became the permanent headquarters of the Communications Zone. In September General Lee's headquarters quickly became the nerve center of the theater's administrative activities.

Meanwhile the Communications Zone's territorial organization also made progress. As outlined elsewhere the OVERLORD planners had contemplated the establishment of a sectional organization on the Continent and before D Day had actually activated Base Sections No. 1 and No. 2, intended for operations in Brittany and Normandy, respectively.[18] In addition, they had tentatively formed Base Section No. 3 for use as an intermediate section between the Advance Section and the base sections. Like other plans, the scheme for the sectional organization was based on a forecast of tactical operations. Since these did not proceed as expected, the planned organization underwent considerable change.

Early in July plans still called for Base Sections No. 1 and No. 2 to develop the Brittany and Cotentin Peninsulas.

[16] Cbl FWD–15033, Eisenhower to Lee, SHAEF SGS 400.3/1 Supply Problems of Allied Advance; Memo, Col Allen Sibley to Maj Gen Robert W. Crawford, 12 Sep 44, SHAEF SGS 014.1 Civil Affairs in Paris I.

[17] Conf Min, Hq ETO, 20 Oct 44, EUCOM 337 General Conferences 44 II.

[18] See *Logistical Support I*, 216–18.

Base Section No. 3 was now tentatively assigned the area of the Seine ports. In addition, an intermediate section and a Paris section were contemplated, the former to take over the area from Le Mans to Reims, the latter to take over the administration of the headquarters command in Paris. Personnel for the additional sections was to come from the U.K. sections, which were to be disbanded generally in accordance with the Reverse BOLERO process, and the United Kingdom organized into a single U.K. Base.

Developments in July began to bring about important alterations in the missions of some of the sections. Early in the month the Advance Section decided to subdivide its territory into area commands. On 11 July it organized the port of Cherbourg as Area No. I under the command of Col. Cleland C. Sibley, commander of the 4th Major Port. A few days later the Advance Section set up the rest of its territory as Area No. II. Confined for an unexpectedly long period to the Normandy beachhead area, Brig. Gen. Ewart G. Plank's organization was now handling a heavy load on the Continent, acting in the capacity of both an advance and a base section. The creation of area commands was obviously an attempt to decentralize some of its activities, particularly the specialized functions of a major port which was to begin operating within the next few days.

As it turned out, the organization of area commands was the first step in the creation of a new base section. On 21 July the Advance Section took another step in this direction, redesignating Area No. I the Cherbourg Command (Provisional) and placing it under Col. Theodore Wyman, formerly a district commander in Western Base Section. The Cherbourg Command remained under the control of the Advance Section, but Colonel Wyman was given most of the powers of a base section commander, and within the next few days his organization was reinforced with officers drawn from the provisional Base Section No. 3 in the United Kingdom.

These developments soon raised the question as to whether the Cherbourg Command might not take over the mission originally assigned to Base Section No. 2—i.e., operations in Normandy. The position of Base Section No. 2 consequently became more and more uncertain, and in the last two weeks of July its movement to the Continent was twice scheduled and twice postponed. The decision not to use Base Section No. 2 for the purpose originally intended finally came as the result of a tour of inspection which General Lord, the COMZ Chief of Staff, made of that area toward the end of July. Satisfied with the progress of operations at Cherbourg, General Lord concluded that it would be inadvisable to impose another headquarters to supervise a job already well under way. He proposed rather that Colonel Wyman's organization be built up and that it be allowed progressively to take over additional activities as the Advance Section could relinquish control over them, the idea being that the Cherbourg Command should eventually expand in size to include the beaches as well as the port of Cherbourg, and that Colonel Wyman should become the base section commander in that area. General Plank concurred in this view.[19]

By 1 August it was definitely decided, therefore, that the Cherbourg Command would be the forerunner of a full-fledged base section. For this purpose it was now to be reinforced by the bulk of the Base Section No. 3 headquarters. The transition was very rapid. On 7 August, the same day on which Headquarters, Communications Zone, opened on the Continent, the area around Cherbourg was detached from Advance Section and became operational as the Cherbourg Base Section. In the succeeding weeks its responsibilities were increased rapidly by the turning over of additional territory in the vicinity of the beaches and by the release of supply dumps by Advance Section in the beach area. On 16 August Colonel Wyman's command was finally redesignated Normandy Base Section. By this time it was rapidly assuming the enormous task of handling the reception, staging, and dispatch of troops and supplies coming in over the beaches and through the minor ports and Cherbourg.[20]

The creation of a base section organization in the Cotentin on 7 August brought to realization the situation visualized in the OVERLORD plans whereby the Advance Section and a second COMZ section should become contiguous.[21] Until that time progress in the development of the continental organization had been retarded. With the advent of the COMZ headquarters proper on the Continent early in August, and with the sudden increase in size of the lodgment which followed the breakout from Normandy, this process accelerated noticeably.

The next area to be organized was Brittany. Base Section No. 1 had already arrived off UTAH Beach on 3 August and in accordance with its planned mission immediately proceeded to the vicinity of Rennes to assume the task of supporting the Third Army in Brittany and developing that area. An amendment in plans now took place, however, to permit the employment of Base Section No. 2 in the Brittany Peninsula also. The expansion of the Cherbourg Command into a base section had left Base Section No. 2 without a mission, and it was now decided to divide the responsibilities in Brittany between Base Sections No. 1 and No. 2, leaving to the former the task of developing the Quiberon Bay area and supporting the Third Army in Brittany, and holding the latter in reserve for use in developing the port of Brest when it was captured.

Base Section No. 1 became operational on 16 August, redesignated as Brittany Base Section, thus giving the Communications Zone two full-fledged base sections in an operational status on the Continent, one operating in the rear of the Advance Section in Normandy and the other in the rear of the VIII Corps in Brittany. The plan to use Base Section No. 2 at Brest proved abortive in view of the disappointing turn of events at that port, and Brittany Base Section eventually took control of all Brittany, as originally intended. The mission of Brig. Gen. Leroy P. Collins' Base Section No. 2 consequently had to be changed once more. On 5 September it was finally activated as the Loire Base

[19] Ltr, Lord to Plank, 25 Jul 44, and Plank's reply, 28 Jul 44, EUCOM 381/2 War Plans General.
[20] Organization and Command, II, 178–87.
[21] See *Logistical Support I*, 208.

Section, with headquarters at Le Mans. Loire Base Section embraced an area approximately 130 miles long extending from Laval to Orleans, and assumed the role of an intermediate section, largely in the support of Third Army.[22]

Several days earlier the capture of Paris had hastened the activation of a section to organize the administration of Paris and its environs. The Communications Zone had an organization tailor-made for this task in Base Section No. 5, the provisional headquarters organized under the command of Brig. Gen. Pleas B. Rogers, the perennial headquarters commandant of the London area. The new section was officially activated as Seine Base Section on 24 August. Within the next few days General Rogers was in Paris urgently requesting the dispatch of the remainder of his headquarters and, above all, an MP battalion.

Seine Section's mission initially included only the administration of metropolitan Paris and did not include responsibility for the operation of line-of-communications depots. Its principal tasks at first were the administration of civil relief, the rehabilitation of the city insofar as was necessary to aid in military operations, and preparation for the reception of the COMZ and SHAEF headquarters. Within the next few weeks Seine Section's area of responsibility was enlarged somewhat by the transfer of additional territory from Advance Section. At the same time Seine Section's activities were greatly expanded, for in taking over the COMZ installations, such as depots, in this area, it began to perform the more normal functions of a COMZ section, in addition to administering a large headquarters and leave center.[23]

The rapid activation of sections on the Continent in August was accompanied by the liquidation of the U.K. establishment. By mid-August plans were complete for the consolidation of the base sections in the United Kingdom into a single U.K. base, the old base sections becoming districts of the new command. Late in June Brig. Gen. Harry B. Vaughan, Jr., had been relieved of his job as Forward Deputy Commander of the Communications Zone in anticipation of his assignment as the commander of the new base. In August the entire liquidation process was speeded up, and, as planned, Western Base Section became Western District in the new U.K. Base Section, Southern Base Section became Southern District, and Central Base Section became the London District.[24]

General Vaughn took command of the newly activated U.K. Base on 1 September. By that time the old headquarters had been largely stripped of their personnel to form the new continental sections, and their commanders were given new assignments. In this way Col. Roy W. Grower of Eastern Base Section had become the commander of Base Section No. 1 (Brittany), General Rogers of Central Base Section had become commander of the Seine Section (Paris), and General Collins of Northern Ireland Base Section had organized Base Section No. 2, which in a few days was to be activated as the Loire Section. Col.

---

[22] Organization and Command, II, 184–89.

[23] *Ibid.*, II, 191–94.

[24] Eastern Base Section had been deactivated at the end of April and absorbed by Western Base Section as District VIII.

Fenton S. Jacobs of Western Base Section had initiated planning for Base Section No. 3 (intended as an intermediate section) and Brig. Gen. Charles O. Thrasher of Southern Base Section was tentatively scheduled to be the deputy commander of the new U.K. Base.[25]

In September plans called for the creation of two additional sections to complete the continental COMZ organization. One of these was to be the Oise Section, which was scheduled to take over the area east of Paris as territory was relinquished by the Advance Section; the other was to be Channel Base Section, scheduled to develop the Le Havre–Rouen area. Both sections were activated and began operations in September, although the assignment of their respective missions was attended by some confusion.

It will be recalled that originally Colonel Jacobs had tentatively been given command of Base Section No. 3 and a planning mission for the Continent. Early in August the transfer of the bulk of his headquarters to Cherbourg to build up Colonel Wyman's organization had left Colonel Jacobs without a command. Later in the month another headquarters, known as Base Section No. 4, was constituted with Colonel Jacobs in command, and tagged as an intermediate section. On 3 September this headquarters was officially redesignated Oise Section and opened at Fontainebleau.

A week later Channel Base Section was activated under the command of General Thrasher, who was given this assignment rather than that of deputy commander of the U.K. Base. Before either Oise or Channel could actually become operational, however, it was decided to switch their missions because Colonel Jacobs' organization was considered better qualified by reason of its personnel and experience for port development and operation. Consequently on 15 September the original Oise Section was redesignated Channel Base Section, under the command of Colonel Jacobs, and Channel Base Section was renamed Oise and placed under the command of General Thrasher.[26]

Neither of the two newly activated sections was immediately assigned area responsibility. Channel Base Section presented a special problem because it was to operate ports lying in British territory, north of the boundary between the 21 and 12th Army Groups. Its operations thus required crossing British lines of communication and violated the principle established by the OVERLORD planners that U.S. and British lines of communication should be kept completely separate. The resulting situation paralleled closely the relationship between British and American forces at ports in the United Kingdom, and naturally called for close co-operation. Early in October Channel Base Section was given control of Le Havre and the immediate vicinity except for civil affairs functions, thus becoming a small enclave within British-controlled territory. U.S. activities in the British territory—transportation, for example—were han-

[25] Organization and Command, II, 86–92, 190–91.

[26] At the same time an exchange of engineer sections was made between Colonel Jacobs' headquarters and Brittany Base Section. The latter's engineer staff had prepared specifically for the development of Brest, and the decision not to develop that port left this staff without a mission to match its ability.

dled by Normandy Base Section west of the Seine and by Channel Base Section east of that river.

Oise Section's development, meanwhile, was somewhat arrested, and not strictly in keeping with original intentions. The drawing of boundaries and the assignment of a definite territory to General Thrasher's headquarters proved infeasible at first because of the lack of troops to carry out area responsibilities. For the first weeks, therefore, Oise Section functioned in territory actually assigned to the Advance Section. Communications Zone redrew the latter's boundaries early in October, and Oise at that time assumed area control of the territory between the Advance Section and Seine Section. Its domain subsequently grew several times as the changes in Advance Section's boundaries conformed with the forward movement of the armies. The bad logistic situation in the fall meanwhile made it impossible for Oise Section to develop into a true intermediate section as intended. Almost all supplies forwarded in September and October were immediately consumed, with the result that few intermediate depots could be established. Consequently Oise Section's functions were limited mainly to rail and road maintenance and the supply of units stationed within its own borders.[27]

By mid-October, with three base sections, three intermediate sections, and one advance section in operation, the Communications Zone's administrative organization in support of the 12th Army Group was relatively complete, although there were to be consolidations and boundary changes to meet changing needs and the shifting tactical situation. The territorial organization had developed only roughly as intended, but plans had proved sufficiently flexible to permit the necessary adaptations. Activation and phasing in of the various continental section headquarters had lagged at first, and had then accelerated in conformance with the sudden change in pace of tactical operations.

The principal changes in plans had stemmed from the fact that the main line of communications had developed not from Brittany as expected, but from Cherbourg and the beaches, which assumed an inordinately greater importance than anticipated. This had resulted first in the growth of a base section organization in the Cotentin in a manner not contemplated, and second, in the development of only relatively minor activity in Brittany and along the lines of communications extending eastward therefrom.

The small scale of activity in the Brittany area eventually led logically to the disbandment of both the Loire and Brittany sections. The first step was taken on 1 December, when Brittany Base Section absorbed Loire Section as a district. Even this consolidation left the Brittany Base Section with only limited responsibilities, which hardly warranted the maintenance of a separate base section headquarters. On 1 February 1945 Brittany Base Section was in turn absorbed by Normandy Base Section as a district. At this time Normandy also expanded eastward, taking over territory from Channel Base Section north of the Seine to the new boundary between the 21 and 12th Army Groups. This gave Normandy Base Section control over an area ex-

[27] Organization and Command, II, 194–202.

tending all the way from Brest to Dieppe, including responsibility for the operation of the ports of Le Havre and Rouen.[28]

Channel Base Section's main preoccupation, meanwhile, had become handling U.S. activities at Antwerp, which finally opened at the end of November, and operating U.S. supply lines to the Advance Section. At Antwerp, as at Le Havre and Rouen. Channel Base Section operated within British territory, but to the south of the U.S.-British boundary Channel Base Section was given responsibility in southern Belgium extending forward to the ADSEC rear boundary. *(Map 3)*

In the course of these consolidations and repeated boundary changes, command in the base sections remained relatively stable. The only important changes occurred in Normandy Base Section, where dissatisfaction with the development of Cherbourg's capacity led to the relief of Colonel Wyman at the end of October 1944. He was succeeded by Maj. Gen. Lucius D. Clay, who came to the theater on temporary loan from the ASF. In mid-December General Clay was in turn succeeded by Maj. Gen. Henry S. Aurand, who also came from an assignment in the ASF. Except for these changes, command in the base sections was stable, all the commanders having had similar assignments in the United Kingdom.[29]

In some respects the development of the communications zone in southern France was simpler than in the north. There was no question there of the development of alternate port areas or bases, since there was but one line of communications possible, leading directly northward from Marseille up the valley of the Rhône. The major organizational problem in the south was the integration of its logistic structure with that in the north. This integration did not occur simultaneously with SHAEF's assumption of operational control of the 6th Army Group. It was gradual, and it was effected only after a long transitional period, during which the southern forces continued to maintain close ties with the North African theater in matters concerning their support.

Logistic planning for both French and U.S. forces in the DRAGOON operation had been done by the SOS, North African Theater of Operations, commanded by Maj. Gen. Thomas B. Larkin, whose headquarters was with AFHQ at Caserta, Italy. After the landing in southern France the communications zone there had developed very much as it had in the north. In the early stages an organization known as Coastal Base Section (COSBASE), similar to the Advance Section in the north, was attached to the Seventh U.S. Army and charged with the initial development of the communications zone. Coastal Base Section had been activated early in July 1944 at Naples under the command of Maj. Gen. Arthur R. Wilson. It accompanied the Seventh Army on the landings, and for the first three weeks concerned itself mainly with the operation of the beaches. On 10 September an army rear boundary was drawn, and

---

[28] Within Normandy Base Section the British retained a small enclave in the area of their D-Day landings.

[29] Organization and Command, II, 265–70.

U.K. BASE
LONDON
Antwerp
CBS
Aachen
Lille
CBS
Liege
Namur
Frankfurt
NBS
ADSEC
Mainz
CHANNEL
OS
Luxembourg
Cherbourg
BASE
SECTION
OISE
COMBAT
NORMANDY
COMZ
Reims
Metz
Karlsruhe
BASE
Verdun
SECTION
SECTION
ZONE
Brest
PARIS
SS
Nancy
SEINE
Chartres
BRITTANY
BASE
SECTION
SECTION
COMBAT ZONE
SOLOC
CONTINENTAL
Dijon
ADVANCE SECTION
CAS
DELTA
BASE
SECTION
COMBAT
ZONE
DBS
Marseille
COMZ BOUNDARIES
November 1944 - January 1945
ARMY REAR BOUNDARY
SECTION BOUNDARIES
INTER-ALLIED ADMINISTRATIVE BOUNDARY
BRITISH AREA IN WHICH NBS WAS RESPONSIBLE FOR US ACTIVITIES
BRITISH AREA IN WHICH CBS WAS RESPONSIBLE FOR US ACTIVITIES
ADVANCE SECTION
CBS CHANNEL BASE SECTION
50 0 50 100 MILES
50 0 50 100 KILOMETERS
F. Temple

MAP 3

MAJ. GEN. ARTHUR R. WILSON, *Commanding General, Continental Advance Section.*

COSBASE, redesignated Continental Base Section (CONBASE), assumed area responsibility for the territory to the rear thereof.

As in the north, the period during which CONBASE was the operative communications zone in southern France was short. The early capture of Marseille and the increasing length of the lines of communications soon necessitated that CONBASE be freed of its base section duties to move forward in support of the 6th Army Group. On 1 October, therefore, a new organization, known as Delta Base Section, was formed under the command of Brig. Gen. John P. Ratay to take over the operation of Marseille and other installations in the base area. CONBASE was now given its final designation of Continental Advance Section (CONAD), in keeping with its role as an advance section moving forward in direct support of the 6th Army Group, the same role which the Advance Section in the north performed for the 12th Army Group.

In order to exercise effective control over the two sections which were now operating in the southern lodgment AFHQ decided shortly thereafter to dispatch an advance echelon of the COMZ headquarters in Caserta to southern France. Such an advance echelon, under the command of Brig. Gen. Morris W. Gilland, moved to southern France; and by the end of October was operating the communications zone there from headquarters at Dijon under the general direction of General Larkin's headquarters in Italy.[30]

The problem of incorporating the southern France administrative structure into ETOUSA had been under discussion for some weeks. ETOUSA appeared anxious to extend its administrative control over the southern forces as soon as possible, for it expected that problems would soon arise on such matters as railways, civil affairs, labor, and procurement, in which there should be a uniform policy. Such problems would become more and more difficult to coordinate through AFHQ once the 6th Army Group forces joined with those of the 12th Army Group. Furthermore, it was thought that important economies might result from pooling logistic resources once the road and rail communications of the two forces were joined.

On the other hand, it was also desirable that the Communications Zone have sufficient time to make all arrangements to take over the requisitioning and shipping schedules from the SOS, North African theater. SHAEF planners believed the supply responsibility probably could not be transferred until most supplies were coming directly from the United States.[31] A major consideration in postponing the transfer of responsibility was the fact that a substantial reserve of supplies existed in the North African theater in excess of those required for the support of other American forces there, and it was felt that the southern France forces should be supplied through that theater at least as long as this situation obtained.[32]

Representatives of the North African theater, AFHQ, 6th Army Group, and COMZ-ETOUSA discussed the entire problem at a conference on 29 September. COMZ-ETO again urged the earliest possible absorption of the southern system—by 1 November at the latest—arguing that this was necessary in order to achieve a balance in supplies in the whole theater. The SHAEF staff favored leaving the southern line of communications separate for the time being, however, thus utilizing to the full the capabilities of the supply organization in the Mediterranean during the period when the full operating capabilities of COMZ-ETO were required to support the 12th Army Group. SHAEF officials believed this could best be accomplished under the existing setup, for General Larkin's organization could more easily draw personnel and supplies from the Mediterranean as long as it was part of the North African theater. Serving the southern France organization put it in the position of drawing on another theater.

BRIG. GEN. JOHN P. RATAY, *Commanding General, Delta Base Section.*

The conference at the end of September produced no immediate decision, but the problem continued under review. Within the next few weeks a compromise solution was worked out by which it was decided to place the southern communications zone under the ETO for "administration," but to

[31] Ltr, Whipple to G–4, Transfer of Adm Responsibility, DRAGOON Forces, 1 Sep 44, SHAEF G–4 381 ANVIL 44 I.

[32] Cbl FX–91666, Wilson to SHAEF, 4 Sep 44, SHAEF AG 381–3 SHAEF to AGWAR Reports on OVERLORD; Cbl FWD–14276, SHAEF to CCS, 6 Sep 44, SHAEF SGS 381 Post-OVERLORD Planning.

leave the responsibility for logistic support with the North African theater.[33] This arrangement took effect on 1 November, on which date all NATOUSA units in the southern lodgment were reassigned to ETOUSA, the latter at this time assuming responsibility in all purely administrative matters such as personnel (including replacements), finance, and other matters handled through the adjutant general, provost marshal general, judge advocate general, and inspector general. SHAEF at this time assumed responsibility in civil affairs matters.

By this arrangement it was contemplated that logistic support would remain the responsibility of the North African theater for an indefinite period. Late in October, however, before the above agreement actually went into effect, the scheme was further modified to permit the transfer of control in logistic matters to the ETO within the next few weeks. To facilitate this transfer a special vehicle was created, known as Southern Line of Communications, or SOLOC. SOLOC was to be a subcommand of the Communications Zone, ETO, interposed between General Lee's headquarters and the two southern sections. But it was to retain the right to communicate directly with the North African theater on matters concerning personnel and shipping from the Mediterranean. The granting of such authority thus met some of the SHAEF objections to the immediate absorption of the southern logistic structure by COMZ-ETO. General Larkin became the commander of the new organization, and was also named Deputy Commander of the Communications Zone, ETO. Early in November he moved his main headquarters from Caserta, Italy, to Dijon, France, where the advance echelon of Communications Zone, Mediterranean Theater of Operations (MTOUSA), was already operating the communications zone in southern France.[34]

In effect, the advance echelon of COMZ-MTOUSA became SOLOC, and since General Larkin was able to bring most of his headquarters staff with him to southern France, the new headquarters was simply a continuation of the parent headquarters at Caserta. When SOLOC became operational on 20 November, therefore, COMZ-ETO took over a relatively intact and operating supply organization from COMZ-MTOUSA.[35]

The actual integration of the supply system of the southern armies and supporting air forces into the Communications Zone, ETO, was not accomplished without difficulties. These arose not only from the necessity of SOLOC to adjust itself to the standing operating procedures of the ETO, which in some respects differed from those in the Mediterranean, but also from the rather unusual position which SOLOC held as a quasi-independent and intermediate command between Headquarters, Communications

[33] Conf, Hq NATOUSA, sub: Transfer of Adm and Supply Coincidental to the Change of Theater Boundary, 27 Sep 44, and Aide Memoire by Lee, 30 Sep 44, ETO 381/200 Transfer . . .; SHAEF Memo, sub: Policy on Transfer of Adm Responsibility for DRAGOON Forces, 8 Oct 44, SHAEF G–4 Diversion of Service Troops, DRAGOON, Box 1, Folder 51.

[34] North African Theater (NATOUSA) was redesignated Mediterranean Theater of Operations (MTOUSA) in October.

[35] Organization and Command, II, 235–40.

Zone, on the one hand and CONAD and Delta Base Section on the other.

In his letter of instructions to the SOLOC commander, General Lee had specified that General Larkin, as Deputy Commander of the Communications Zone and as his representative on the ground, was to represent the COMZ commander in all dealings with the southern forces. Direct communication with SOS, MTOUSA, was authorized on all matters affecting Mediterranean shipping, and General Larkin was also permitted to call on the COMZ staff sections for any aid he might require in discharging his duties. Conversely, members of Headquarters, Communications Zone, were forbidden to communicate with the southern sections except through the SOLOC headquarters.

In order to avoid upsetting existing supply arrangements, General Lee announced that current programs for the phasing in of supplies, personnel, and equipment from the Mediterranean and the United States would remain in effect, subject only to adjustment directed by Headquarters, Communications Zone, after consultation with SOLOC, and that any extension of the program would also be made only after such consultation.

While the implementation of the directive generally proceeded smoothly, nevertheless it was attended by certain difficulties and required adjustments and adaptations. Disagreements arose, for example, as to the limits of SOLOC's jurisdiction and as to the role it was to have in the operations of the next higher and lower echelons of command. The Communications Zone, it appears, tended to treat SOLOC as a less independent command than SHAEF intended it to be. The principal complaints arose over the Communications Zone's close scrutiny of SOLOC's requisitions. The COMZ staff, it was charged, made demands for requisitions in much greater detail than was actually needed, delayed the forwarding of requisitions to the New York Port, and in some cases exceeded its authority in determining what SOLOC could request. General Larkin considered these practices unwarranted in view of the responsibility he himself avowedly possessed for editing the requisitions which he forwarded.

MAJ. GEN. THOMAS B. LARKIN, *Commanding General, Southern Line of Communications, and Deputy Commander of the Communications Zone, ETO.*

Difficulties arose within SOLOC itself, resulting in part from misunderstanding over its rather special status as a separate command, and from the supervision which it exercised over the operations of CONAD. Some of SOLOC's own staff officers tended to overlook their organization's separate command status and had to be cautioned against encouraging higher echelons to correspond directly with subordinate commands, and vice versa. General Wilson, commander of CONAD, had opposed the organizational changes by which his organization had become purely an advance section and been deprived of control over the base area. He had proposed, without success, that the activities of the two sections in support of the 6th Army Group be co-ordinated by his own command, which, he argued, was in a better position than any other headquarters to anticipate and meet the requirements of combat troops. CONAD found it difficult to adjust to the new situation and complained about SOLOC's close supervision.

It is likely that such a complaint against supervision from above would have been made even if there had been no intermediate command such as SOLOC. Quite understandably, the Advance Section in the north had reacted in the same way initially, resenting the loss of control over certain activities in the territory it relinquished to the base sections. Nevertheless, the activities in the advance and base sections were divided in both the north and south, and in the latter area the supervision over communications zone operations was exercised by SOLOC. This took the form mainly of editing CONAD's requisitions and arranging with Delta Base Section for shipment of supplies northward, but also of closely scrutinizing the operations of both sections. CONAD occasionally protested the orders of its superior headquarters, claiming "interference," as for example when SOLOC, concerned over the wasteful use of transport facilities, stepped in to direct CONAD to obtain prior approval before sending its motor transport outside its own territory. But General Larkin regarded it fully within his authority and responsibility to exercise such supervision in the interest of the most efficient use of resources along the entire southern line of communications.

SOLOC's continued ties with the Mediterranean theater were also a source of some confusion, arising mainly from the situation whereby shipping was handled through the Mediterranean and requisitions through ETOUSA. A heavy flow of both troops and supplies from the United States in November created considerable difficulties along the southern line. They led General Larkin to report that the New York Port and the War Department were apparently confused as to SOLOC's position, for they were continuing to ship all supplies in bulk to the Mediterranean, including those not only for Italy but also for southern France, which was now under another theater's jurisdiction. General Larkin desired that shipping for the southern forces be set up for ETOUSA and then allocated to SOLOC as necessary.

Such problems were eventually ironed out, but they pointed up the difficulties attending the transition of control from one theater to another, as well as the difficulty of employing an intermediate headquarters to effect such a transfer.

By early 1945 the integration of the southern supply system had progressed sufficiently to render SOLOC superfluous. SOLOC had served a useful purpose in facilitating the transfer of supply functions from MTOUSA to ETOUSA, but in the view of the 6th Army Group its continued existence contributed to the slowness of procedure involved in going through so many headquarters.[36] Pressure to abolish the organization consequently increased, and on 6 February 1945 SOLOC was finally dissolved, CONAD and Delta Base Section then coming directly under the command of General Lee's COMZ headquarters in Paris.

By February, then, the continental Communications Zone consisted of three base, two intermediate, and two advance sections. In the north the Advance Section operated in direct support of the First and Third Armies in the 12th Army Group and of the Ninth U.S. Army in 21 Army Group. In the south CONAD performed the same function for the 6th Army Group. The Advance Section for the most part drew its support directly from the Normandy and Channel Base Sections, which controlled the entire coastal area and operated the ports. Similarly in the south Delta Base Section organized the base area and operated the port of Marseille. Besides these base and advance organizations two sections operated in the north, Oise Section controlling transportation facilities in the forwarding of supplies from the base to the advance section, and Seine Section concerning itself mainly with the administration of the headquarters in and around Paris. The U.K. Base, meanwhile, was also a part of the Communications Zone, feeding supplies and personnel from the United Kingdom and handling certain residual administrative functions there for SHAEF and ETOUSA-COMZ.[37]

[36] Cbl BX-23473, 6 A Gp to SHAEF, 26 Jan 45, SHAEF SGS 400.3/1 Supply Problems of Allied Advance.

[37] Organization and Command II, 203, 241-48.

CHAPTER III

The Port Discharge and Shipping Problems

(1) The Port Problem as Affected by the Pursuit

In the course of the endless calculations involved in the logistic planning for OVERLORD, an exasperated staff officer summed up his frustrations over the port problem in a parody of the invasion plan known as "Operation OVERBOARD." "The general principle," he wrote, "is that the number of divisions required to capture the number of ports required to maintain those divisions is always greater than the number of divisions those ports can maintain."

Logistic planners had ample reason to suspect that this statement contained an element of truth. From the start, port discharge capacity had been the major single cause for concern in the planning of OVERLORD. Logistic planners had predicted a deficit beginning at D plus 120, even assuming that operations proceeded as scheduled. In this matter, certainly, they were not excessively conservative, for their fears were to be largely realized. In the search for adequate port discharge capacity in the summer and fall of 1944 they eventually gave consideration to approximately thirty-five French and Belgian ports and beaches.

In July the port discharge problem appeared to constitute the very root of future supply difficulties.[1] Plans had provided that by the end of that month U.S. forces should be supported in part through Cherbourg and the Normandy beaches and minor ports, and in part through St. Malo and Quiberon Bay in Brittany. At the time of the breakout late in July, however, Cherbourg had been in operation only a few days, nearly 90 percent of all U.S. supply support was still coming in via the beaches, and the total U.S. discharge on the Continent averaged only 20,000 to 25,000 tons per day as against previously estimated requirements of about 30,000.

The deficiency had not been immediately serious because requirements had not been as large as predicted. But weather was expected to close out the beaches in late September, and the delay in capturing the Brittany ports was already a fact. An even more serious deficit therefore loomed ahead than was originally predicted. Before the end of July, the planners had already taken steps to compensate for this expected loss

---

[1] See *Logistical Support I*, 463ff.

by projecting an increase in the capacity of Cherbourg and the minor Normandy ports and by seeking a larger allotment of coasters so that small ports could be used to fuller advantage. They also considered opening the smaller Brittany ports, which had not seriously entered into the original plans.

Nothing had happened thus far to alter the previous conviction that both Brest and Quiberon Bay were absolutely essential for the support of American forces. While doubts had already arisen as to the feasibility of developing Quiberon Bay, mainly because of the towing problem in the fall, it was still expected at the end of July that the Brittany area would be developed as originally intended.

The effect which the pursuit was to have on the port discharge problem was not immediately apparent. In the first half of August neither SHAEF nor COMZ planners contemplated major changes in port development plans. The Communications Zone reaffirmed its intention of pushing the Quiberon Bay project vigorously. It also advocated doubling the planned capacity of Brest and, in addition, opening the Loire ports (Nantes and St. Nazaire) as soon as resources permitted.[2]

In the last two weeks of August confidence that Brest and Quiberon Bay could be relied on to meet U.S. port needs finally began to wane. The continued delay in the capture of the main Brittany ports, coupled with the knowledge that bad weather would soon render the beaches unusable, prompted logistic planners to resume the search for alternate port discharge capacity to meet at least the interim requirements until the major Brittany ports could be opened. The capture of Le Havre and Rouen did not yet appear imminent. Consequently attention once again turned to the smaller Brittany ports, which had received only sporadic consideration in the past. On 25 August the Communications Zone decided to go ahead with the development of Morlaix, St. Brieuc, St. Malo, and Cancale, and at the same time issued instructions to the base section commander to develop the port of Granville on the west Normandy coast. A target of 20,000 tons total capacity was set for the five ports, three of which were scheduled to meet their goals within the next ten days.[3]

The decision to open these minor ports hardly dispelled the growing anxiety over the port situation. Meeting port discharge requirements was not a simple matter of adding up the total capacity of every little inlet along the coast and balancing this against the total tonnages it was desired to import. Port capacity not only had to be adequate in quantity but of the kind suitable for handling various types of shipping and cargo. On paper the Allies had sufficient port capacity to handle all the imports scheduled for the next few weeks. But they were actually very short of capacity of the type suitable to handle the unloading of such commodities as coal, boxed vehicles, and heavy lifts. In the first two or three months of operations this was not a

[2] Ltr, Deputy G–4 Mov and Tn Br G–4 SHAEF to CAO, 11 Aug 44, App. 5 of Mov and Tn Study, sub: MULBERRY, Aug 44, SHAEF G–4 War Diary/Jnl; Ltr, Hq COMZ to SAC, sub: Port Capacities, 10 Aug 44, SHAEF G–4 825.1 Piers, Wharves, Docks, and Berths 1944, III.

[3] G–4 History, I, 43; 12 A Gp Tn Sec Jnl, 2 Sep 44.

major problem. All vehicles, for example, were brought across the Channel assembled in MT ships,[4] LST's, or LCT's. This was possible because the stowage factor on so short a voyage was relatively unimportant, the vital thing being to achieve the fastest possible turnaround. Beginning in September, however, a larger and larger portion of U.S. supplies was scheduled to come directly from the United States, loaded so as to reduce broken stowage to a minimum and thus use space more economically. Such a schedule meant that practically every Liberty ship coming from the United States would contain either boxed vehicles or other heavy lifts.

Naturally it was desirable that such awkward loads be discharged at ports where suitable shore cranes were available. In August the only port in Allied hands which possessed facilities even partially adapted to handling such cargo was Cherbourg, and it was obvious that that port lacked sufficient capacity to handle all the shipping coming directly from the United States. In fact, proposals to handle heavy troop movements through Cherbourg (again because of Brest's unavailability) already threatened to cut into the limited cargo-handling capacity of that port. It was inevitable that some boxed vehicles and other awkward loads would have to be received at the beaches and minor ports, where the handling of such cargo would be extremely difficult.[5] But eventually the major portion of U.S. port discharge requirements could be met only through the development of the larger deepwater ports. The capture of such facilities had already been delayed, and anxiety increased at the end of August as to the condition in which ports such as Brest and Lorient would be found if and when they were finally captured.[6]

Tactical developments within the next week radically altered the entire outlook on the port situation, and eventually led to a recasting of the entire port development program. On 4 September British forces captured the great port of Antwerp with most of its facilities intact. On 12 September the stubbornly defended and badly damaged port of Le Havre, 225 miles to the rear, also fell to British troops. Rouen had been occupied on 30 August.

Preinvasion planning had accepted it as essential from the logistic point of view that both Le Havre and Rouen should be captured and used. In fact, it contemplated that a crossing of the lower Seine and seizure of these ports would be the first operation attempted after the capture of the lodgment area. The purpose in seizing the Seine ports, however, was not to serve U.S. needs, but to relieve the British forces from dependence on the beaches. U.S. lines of communication were to be based on Cherbourg and Brittany. It was not until much later that the transfer of Le Havre to the Americans was contemplated.[7]

The delay in capturing Brest and the

[4] Liberties outfitted to handle motor transport.

[5] Memo, Capt N. H. Vissering, Mov and Tn Br G–4 SHAEF, for Whipple, sub: Comments on Stf Study 13, 12 Aug 44, SHAEF G–3 381 War Plans General; Shipping Note for CAO, 26 Aug 44, SHAEF G–4 Mov and Tn Br War Diary/Jnl.

[6] Shipping Note for CAO, 26 Aug 44.

[7] SHAEF Plng Study, sub: Post-NEPTUNE Course of Action After Capture of the Lodgment Area, Sec. II, 30 May 44, SHAEF SGS 381 Post-OVERLORD Planning.

unexpected early seizure of both the Seine ports and Antwerp changed all this. The advantages which these ports offered over those of Brittany were obvious, and logistic planners had turned their attention to them as a possible solution of the discharge problem as soon as their capture appeared probable. On 3 September Col. William Whipple noted that it would be unprofitable for U.S. forces to devote their resources to the development of the geographically remote Brittany ports if the Seine ports, which were 200 miles farther forward on the line of communications, could be developed instead. Every 5,000 tons discharged at Le Havre rather than the South Brittany ports, he observed, would save an equivalent of seventy truck companies.[8]

Antwerp had still greater advantages, even when compared with the Cherbourg line of communications. It was only 65 miles from Liège, while Cherbourg was more than 400 miles from that advance depot area. Even Nancy, the forward depot area on the Third Army line of communications, was only 250 miles by rail from Antwerp, but more than 400 miles from Cherbourg. The matter of rail lines was particularly important because the rail capacity from Cherbourg and the beaches was only about 10,000 tons per day as against a discharge rate of 20,000, with the result that motor transportation bore a heavy transportation burden at great cost to equipment. In terms of the forces supportable it was estimated that only 21 divisions could be provided with daily maintenance on the Cherbourg route (6 of them by motor transport), while 54 could be similarly supported via the northern line of communications (all of them by rail). In effect, therefore, the effort required to support a division via Antwerp would be only one third that required to transport a division via Cherbourg. Contrasted with both the Brittany and Normandy areas, moreover, Antwerp was virtually undamaged and possessed unmatched cargo-handling facilities.[9]

For a moment, at least, the capture of Antwerp dissipated the darkest cloud on the logistical horizon, thus contributing to the otherwise unbounded optimism of these early September days. Logistic planners at COMZ headquarters were so encouraged that they were ready to abandon not only the Brittany ports, but the Seine ports as well, and advocated concentrating all efforts on the Belgian and Dutch ports. The condition of Brest (which had not yet been captured), they argued, did not warrant development for cargo discharge, and the long rail haul would place a serious strain on the transportation system. Except in emergency, the value of the smaller ports did not justify the expenditure of manpower and equipment to develop their relatively small capacity. Le Havre, they observed, was badly damaged, and its development to major capacity could not be accomplished for the period in which its use was required. Weighing these disadvantages against the realization that the Belgian and Dutch ports had more than ample capacity for both British and American forces, COMZ officials decided to place a bid with SHAEF for alloca-

[8] Memo, Whipple for CAO, sub: Port Dev, 3 Sep 44, SHAEF G–4 825.1 Piers, . . . 1944, III.

[9] G–4 History, I, 47–48.

tion of a portion of the Antwerp facilities.[10]

The initial enthusiasm for a complete shift to Antwerp as the answer to port discharge needs was soon tempered by more sober realization concerning both the speed with which the new port capacity might become available and the current plight of the Allied forces. The Allied Naval Commander-in-Chief, Expeditionary Force, immediately gave warning that both Antwerp and Rotterdam were highly vulnerable to blocking and mining, and that if the enemy was successful in these operations no estimate could be made of the time it would take to open these ports.[11] The Allies needed additional capacity immediately; the pursuit was at its height, and maintenance of the armies was stretched to the limit. All transportation was fully committed, and port clearance was already largely sacrificed for the sake of line-of-communications hauling. Meanwhile port discharge had shown no improvement, averaging only 25,000 tons per day. "Altogether," Lt. Gen. Sir Humfrey M. Gale, the chief administrative officer of SHAEF, concluded, "the administrative situation remained grim." [12]

Logistic officers at SHAEF, consequently, did not share the view that Antwerp would immediately meet all the Allies' needs for port discharge, although they agreed that at least some of the Brittany projects could be canceled. As early as 3 September the Logistical Plans Branch had recommended that the South Brittany ports be abandoned.[13] On the 7th SHAEF announced that neither Nantes, St. Nazaire, nor Lorient would be developed, and also decided finally that the much-debated Quiberon Bay project would be abandoned.[14]

While attention thus definitely shifted from Brittany, no final decisions had yet been made regarding either Brest or the ports which had just been uncovered. In view of the great changes which tactical developments had brought about, necessitating a recasting of port plans, General Lee on 14 September summarized the entire port situation for the Supreme Commander and offered his recommendations for meeting future requirements. General Lee's analysis led him to conclude that the development of Brest as well as the other western ports to the tonnages originally planned was no longer sound. Le Havre, he noted, was reported to be seriously damaged, making it unlikely that large tonnage capacity could be developed there in the near future. In his opinion, moreover, its location did not materially shorten the lines of communication. He believed it advisable, therefore, to carry out only a limited development of Le Havre as an interim port with a capacity of between 8,000 and 10,000 tons per day, and to do this as rapidly as possible with a minimum expenditure of reconstruction effort.

General Lee voiced the now generally accepted opinion that the bulk of Allied

[10] Daily Jnl, G-4 Plans Br COMZ, 10 Sep 44, ETO Adm G-4 145C.

[11] Cbl, ANCXF to SHAEF, 3 Sep 44, SHAEF AG 323.3-2 Captured Ports.

[12] Mil Shipments Priority Mtgs, 2 and 9 Sep 44, SHAEF AG 337-18 Mil Shipments Priority Mtgs; Shipping Note for CAO, 9 Sep 44, SHAEF G-4 500 Transportation General 1944, II.

[13] SHAEF G-4 War Diary/Jnl, 3 Sep 44.

[14] Cbl FWD-14066, SCAEF to 12 A Gp, 7 Sep 44, SHAEF AG 323.3-3 Port Capacities.

port requirements would eventually be provided by the north coast ports of Antwerp, Rotterdam, and Amsterdam. Their capacity, it was agreed, was more than sufficient to meet both British and U.S. needs, and, because of their location, would also alleviate the desperate transportation problem. Lee therefore recommended that port development be limited to Cherbourg, Le Havre, and the north coast ports. Even Cherbourg, he advised, should not be developed beyond the 20,000-ton capacity it was expected to achieve by early October because of the heavy demands on both rail and motor transport which the use of that port entailed.[15]

Supreme headquarters concurred in these recommendations with one exception. It was not yet ready to abandon Brest in view of the need for reception facilities to handle the accelerated flow of divisions. Furthermore, the condition of the port was unknown even at this time, for it was still in enemy hands. SHAEF for the moment therefore directed that Brest should be developed to the extent needed to receive troops and their organizational equipment, and left to General Lee's discretion which of the smaller ports should be kept open.[16]

On 27 September the COMZ commander outlined these decisions to his staff and section commanders: In the Normandy area Cherbourg was to continue as the major point of intake and was planned to receive the maximum tonnage in both cargo and POL which it could handle pending the availability of Antwerp. Grandcamp-les-Bains was to be closed, but the other minor ports of Normandy—Barfleur, St. Vaast-la-Hougue, and Isigny—were to continue working on second priority for coasters until Antwerp developed a satisfactory discharge, and Port-en-Bessin was to continue the intake of POL. The beaches were to continue to operate at maximum capacity, although discharge was expected to drop to 10,000 tons per day at OMAHA and 4,000 tons at UTAH in October. The extent to which Brest was to be utilized was still undetermined and was to be decided after its capture on the basis of a survey by representatives of Brittany Base Section, the Navy, the chief engineer, and chief of transportation. St. Brieuc and Granville were to continue their development for the reception of coal. Morlaix was to develop its maximum unloading capacity from Liberty anchorage, but in a reversal of the plans made late in August both St. Malo and Cancale were now eliminated and actually never operated on U.S. account. The Seine ports were still thought of as providing only interim relief, with Le Havre scheduled to develop a capacity of about 8,500 tons and Rouen 3,000. Plans for the north coast area were of necessity somewhat less specific, but they were made at this time to share the use of Ostend with the British for the import of POL, to survey both Calais and Boulogne with a view to assigning one or both to U.S. forces, and to reconnoiter the coast northeast of Le Havre for suitable beaches at which LST's might be offloaded.

---

[15] Ltr, Lee to SAC, 14 Sep 44, SHAEF SGS 800 Harbors, Opening, Use, Construction; see also G–4 COMZ Plans and Communications Diary/Jnl, 18 Sep 44, ETO Adm 145C.

[16] Ltr, SHAEF to Lee, sub: Alloc of Ports, 19 Sep 44, SHAEF SGS 323–3 Ports—Allocation and Development.

The most important project of all was the development of Antwerp as the major joint U.S. British port on the Continent.[17] Antwerp was now almost universally looked upon as the early solution to the most fundamental logistic problem facing the Allies. It was contemplated that there should be a gradual closing down of activity in the ports farther to the rear as Antwerp's capacity developed. The extent to which the ports in the rear were to be developed and used from now on was therefore predicated on the progress in bringing Antwerp into operation.

Pending the development of new capacity, meanwhile, supply officials were faced with the hard fact that in the immediate future the port situation would continue to deteriorate. Port discharge had improved slightly in the third week of September, averaging nearly 37,000 tons per day. But in the following week it dropped to less than 28,000 tons, and for another full month was to average barely 25,000 as against the originally estimated requirement of 38,500.[18]

The drop in performance at the end of September was caused in part by inclement weather, which hampered operations at the beaches.[19] But discharge had also fallen off at Cherbourg because of the handling of troop convoys with organizational equipment, which put an additional strain on the port's facilities and out into normal unloading.[20] It had always been planned that Brest should handle the reception of personnel and organizational equipment. But the report of the survey group sent to examine Brest upon its capture on 25 September finally confirmed the fears which had been held regarding that port's condition. The report disclosed that extreme demolitions, mining, and damage to quay facilities had rendered the port useless and estimated that even limited unloading of cargo and troops would not be possible for seventy-five days.[21] In view of the more urgent commitments to clear and rehabilitate Le Havre and Rouen, Brest was now given the lowest priority,[22] and was eventually abandoned altogether.

At the end of September, therefore, the port situation remained grave. All hopes now centered on Antwerp, the opening of which, as the Supreme Commander had recently noted, would have "the effect of a blood transfusion" on the entire maintenance situation.[23] His concern, which was widely shared, was expressed in a memo to the Chief of Staff at this time. "As you know," he said, "I am terribly anxious about Antwerp, not only the capture of its approaches, but the getting of the port to working instantaneously thereafter." [24] Antwerp

[17] Ltr, Lord to Chiefs of Gen and Special Stf Secs and Sec Comdrs, sub: Dev of Continental Ports, 27 Sep 44, EUCOM 400 Supplies, Services, and Equipment 1944, V, or SHAEF AG 323.3-2 Captured Ports.

[18] OMAHA District Summary of Opns, 25 Nov-2 Dec, Pt. IV, ETO Adm 231.

[19] Mil Shipments Priority Mtg, 29 Sep 44, SHAEF AG 337-18.

[20] Memo, Deputy Chief of Mov and Tn Br G-4 SHAEF for G-4, sub: Port Info, 29 Sep 44, SHAEF G-4 825.1 Piers, . . . 1944 III.

[21] Cbl 281255, CTF 125 to ANCXF, 29 Sep 44, SHAEF SGS 800 Harbors, Opening, Use, Construction.

[22] Mil Shipments Priority Mtg, 29 Sep 44, SHAEF AG 337-18.

[23] Ltr, Eisenhower to Marshall, 21 Sep 44, OPD Exec Office File 9.

[24] Memo, Eisenhower for CofS, 30 Sep 44, SHAEF SGS 800 Antwerp.

had already been in Allied hands for four weeks. It was thought at the time of its capture that it would surely be in operation some time in October. At the end of September General Moses, the 12th Army Group G–4 made an estimate which proved far more realistic when he suggested that it would be better not to plan on the port's opening until 1 December.[25]

(2) *The Beaches and Minor Ports of Normandy*

One of the outstanding features of logistic support in the first six months of operations was the unexpected extent to which U.S. supplies and personnel were funneled through the Normandy ports. OVERLORD plans had envisaged the Normandy area primarily as an interim base pending the development of Brittany. Since the beaches were expected to have a short-lived usefulness and Cherbourg a relatively small capacity, a maximum discharge of less than 26,000 tons per day was counted on for the Normandy beaches and ports.[26] This maximum was to be attained at about D plus 90, at which time Normandy's facilities were to account for about 55 percent of the total U.S. port capacity on the Continent. Discharge through the Normandy ports was scheduled to decline to about 13,000 tons per day and account for only 30 percent of the total intake by early November.

At D plus 90 (4 September) the daily discharge was actually averaging upwards of 28,000 long tons and frequently exceeding 30,000 tons despite the fact that MULBERRY A, the artificial port at OMAHA Beach had been abandoned. Except for relatively minor unloadings across the beaches at St. Michel-en-Grève in Brittany, Normandy's facilities comprised the entire discharge capacity available to U.S. forces on the Continent. Early in November (D plus 150), despite the virtual cessation of operations at the beaches, the intake through the Normandy ports still averaged about 17,000 long tons per day and accounted for 63 percent of the total U.S. discharge.

The performance of the Normandy ports in terms of percentage of total discharge is explained by the failure to develop the Brittany area. Their performance in terms of tons discharged is explained first by the unexpected capacity of the open beaches, and second, by the development of Cherbourg to a capacity far beyond that contemplated in plans.

The capacity of the beaches proved a godsend in view of the delayed opening of Cherbourg. In the first seven weeks they constituted practically the only intake capacity on the Continent. After the opening of Cherbourg and the development of other port capacity the relative importance of the beaches naturally declined. Nevertheless they continued to account for a significant percentage of the personnel, vehicles, and cargo brought to the Continent until the end of October. After the big storm in June operations at the beaches settled down to a normal routine. Both beaches were soon operating as well-organized ports and, except for occasional bad weather which halted or slowed down operations for brief periods, enjoyed relative stabil-

[25] Memo, Moses for Bradley, 30 Sep 44, 12 A Gp G–4 Memos of Gen Moses 1944.

[26] Granville is not included as a Normandy port in these comparisons for reasons explained later.

TROOPS DEBARKING ONTO A CAUSEWAY *at Omaha Beach, 4 August 1944.*

ity. The establishment of beach transfer points, improvements in the road networks, the landing of additional truck companies, and the general improvement in unloading and clearance methods brought increased discharge and movement of supplies, with the result that targets were consistently exceeded. In the week after the storm both beaches surpassed all previous performance, OMAHA averaging 13,000 tons as against a target of 10,000, and UTAH averaging 7,200 tons compared with its goal of 5,700.[27]

Unloading continued at a good rate in July, although fog and high winds again interfered in the period from the 20th to the 23d. On the last day of the month an all-time high of 25,853 tons was unloaded at the two beaches, OMAHA handling 15,834 tons, 158 percent of its target, and UTAH discharging 10,019 tons, 175 percent of its rated capacity. This record was almost duplicated on the following day, when 25,303 tons were offloaded, and again on 8 August, when 25,563 tons were brought ashore.

The landing of personnel kept pace, although the record was more erratic. The largest number of personnel debarkations at OMAHA Beach, except in the initial assault, had taken place on 23 June, when 24,425 men came ashore. That day also established a record for combined debarkations at the two beaches, totaling 30,916. OMAHA nearly equaled this performance again on 23 July, when 24,068 men went ashore.

[27] [Clifford L. Jones] NEPTUNE: Training for and Mounting the Operation, and the Artificial Ports, Pt. VI of the Administrative and Logistical History of the ETO, II, 137–39, MS, OCMH.

UTAH's record was achieved on 19 July, when it handled 22,780 men. These performances were exceptional, of course, but the 10,000-mark was reached frequently in the first three months.

July was also a good month for vehicle discharge, the two beaches handling a total of slightly more than 100,000 vehicles of all types for an average of 3,283 per day. OMAHA's best single day came on 5 July, when it received 3,837; the highest discharge rate was achieved at UTAH on 25 July, when a total of 4,256 was passed ashore.[28]

Another storm struck the beaches on the night of 1 August and cut deeply into the unloading rate on the 2d and 3d. But the two beaches quickly resumed normal operations on 4 August. In fact, the demonstrated capacity of the beaches in July led supply officials to raise the tonnage target of UTAH from 5,700 to 10,000 tons per day, and that of OMAHA from 10,000 to 15,000 tons. Raising the tonnage targets provided new goals for the beach organizations, but did not affect the rated capacity of the beaches. UTAH met its new target the very next day—6 August—with a discharge of 10,500 tons, and two days later achieved the best performance of the entire period of its operations by unloading 11,577 tons. Thereafter UTAH met its new target only once—on 29 September, when it handled 10,612 tons. OMAHA exceeded its new goal on 9 and 10 August, and again on four successive days from 17 to 20 August. On 25 August it set a record for discharge over either beach of 16,078 tons.

Many factors affected the discharge record of the beaches. Shortages of transportation often restricted discharge at first, and the improved performance in July could be attributed in large part to the provision of additional trucks. Congestion in the dumps also proved a limiting factor. Late in July an inspection of a Class V dump behind UTAH Beach revealed 100 loaded trucks standing idle in the sorting and receiving bays for lack of adequate personnel to handle and store the ammunition. According to one estimate this resulted in a 30 to 40 percent loss in efficiency in dukw operations and a reduction of 25 percent in tonnage discharged. Early in August the assignment of additional troops to the dumps and improvements in supply handling methods at least temporarily eliminated this bottleneck. But the attainment of perfect balance between the various functions and facilities involved in the unloading of a ship, movement over a beach, and clearance to a dump under the unpredictables prevailing in Normandy was next to impossible and was achieved for only short periods, if at all.[29]

September was still a good month at the beaches, although the discharge record did not quite equal that of July and August. Heavy seas restricted operations on the 2d and 3d, and on 7 September rough weather again interfered with unloading for a while. A more serious interruption occurred a few days later, when a storm cut rather deeply into the discharge rate for three days beginning on the 11th. Finally, high seas and fog again hampered operations, particularly at OMAHA, toward the end of the month.

[28] *Ibid.*, II, 140–43.

[29] *Ibid.*, II, 143–45.

BULLDOZER STUCK IN THE THICK MUD *on a road near Marigny, France, July 1944.*

On the whole, however, unloadings in September exceeded expectations, and the restrictions on discharge were caused as much by the unavailability of loaded craft and by shortages in trucks, dukws, ferry craft, and men, as by bad weather. The last two days of September saw substantial tonnages discharged at both beaches, totaling 20,933 tons on the 29th and 18,575 on the 30th.

But 1 October marked the beginning of a definite decline from which the beaches never recovered. As before, factors other than the weather contributed to the falling off in discharge, notably the problem of vehicles. It became increasingly difficult to keep adequate numbers of dukws in operation. Most of them had been running continuously since the early days of the landings and were being deadlined for repairs an increasing percentage of the time. The availability of trucks was always unpredictable, and early in October a substantial withdrawal of vehicles to bolster the Red Ball Express led to a downward revision in the tonnage goals for both beaches, the OMAHA target reverting to the original 10,000 tons and UTAH's to 4,000.

The most serious limiting factor by this time was the weather. The fog and storms of August and September had interfered with operations for only short periods, and, with one exception, had never closed down all activity at the beaches, despite the tendency of the beach organizations to report "all operations stopped" in these periods. The unloading of beached coasters had usually continued, and full advantage was taken of temporary breaks in the weather, so that substantial unloadings were registered even during the storm periods. In October, however, conditions were consistently bad, bringing operations to a complete halt four times at UTAH Beach and twice at OMAHA. On the last day of the month both beaches for the first time reported no discharge whatever, and for the entire month unloading at the two beaches averaged only 6,243 tons per day.

Bad weather also made clearance more and more difficult, for rain fell practically every day and turned the roads behind the beaches into quagmires. Most roads were poorly drained and required constant maintenance. The problem of mud had actually begun to give trouble much earlier. By the first of October it had reached serious proportions. Clay also accumulated on metaled roads, causing drivers to spin their wheels and puncture tires as they broke through to the hard, rough foundations.

By the end of October it was obvious that the beaches were nearing the end of their usefulness. At that time the Navy withdrew its ferry craft from UTAH, leaving the unloading entirely to dukws. Within another two weeks it did the same at OMAHA. Conditions deteriorated steadily in these weeks, and unloading finally came to an end on 13 November at UTAH, and on 19 November at OMAHA, after 167 days of operation.[30] In the final three weeks the scale of activity dropped rapidly, discharge totaling only a few thousand tons.

Fortunately it was possible to operate

TABLE 1—BEACH DISCHARGES: 1 JULY–17 NOVEMBER 1944 [a]

[Long Tons Weekly] [b]

	Week	Total	OMAHA	UTAH
	1–7	133,506	79,652	53,854
	8–14	112,953	74,674	38,279
July	15–21	116,548	84,095	32,453
	22–28	108,625	67,773	40,852
	29–4	115,879	73,435	42,444
	5–11	160,444	95,440	65,004
August	12–18	117,061	75,166	41,945
	19–25	97,307	68,174	29,133
	26–1	134,983	90,783	44,200
	2–8	87,819	64,623	23,196
	9–15	77,434	48,446	28,988
September	16–22	111,005	70,726	40,279
	23–29	80,986	37,527	43,459
	30–6	53,511	30,125	23,386
	7–13	50,584	35,059	15,525
October	14–20	57,755	37,406	20,349
	21–27	24,610	11,390	13,220
	28–3	15,118	6,105	9,013
November	4–10	11,576	3,998	7,578
	11–17	1,579	1,579	0

[a] For June 1944 figures see *Logistical Support I*, 416.
[b] Vehicles and bulk POL not included.
Source: NEPTUNE: Training for and Mounting the Operation, Part VI, Vol. II, 175–78.

the beaches considerably longer than originally expected, and their over-all record was a spectacular one. In the twenty-four weeks of their operation they received approximately 2,000,000 long tons of cargo, which constituted about 55 percent of the total tonnage brought onto the Continent up to that time. In addition, they had discharged 287,500 vehicles and debarked 1,602,000 men.[31] Table 1 summarizes the tonnages discharged by week.

The official demise of the beach or-

[30] *Ibid.*, II, 151–52, 154–58.

[31] Monthly Progress Rpt, OCofT COMZ ETO, Jun 45, ETO Adm.

ganizations did not come until 4 December, when the OMAHA Beach Command was finally dissolved. Several administrative changes had taken place which had altered somewhat the shape and status of the beach organizations since their arrival in the first days of the invasion. In the main, they reflected the evolution of the Communications Zone's organization on the Continent. On 7 August the OMAHA Beach Command, the 1st, 5th, and 6th Engineer Special Brigades, and key assigned and attached units were released from assignment to First Army and were temporarily assigned to the Advance Section, to which they had been attached in mid-June. Within another week they were assigned to the Normandy Base Section, which assumed control of the Normandy area. The engineer brigades and attached units in the OMAHA area had already been organized into the OMAHA Beach Command, and the UTAH area was now similarly organized into the UTAH Beach Command, although the brigades retained their original identity.

In September the areas controlled by the beach commands were enlarged, and a final expansion took place in October when Normandy Base Section was subdivided into districts, the OMAHA and UTAH Commands respectively becoming OMAHA and UTAH districts under the reorganization. As the headquarters of COMZ subdivisions the original brigade organizations had thus assumed roles far removed from the specialized functions they had been trained for, and the change was not universally welcomed. First Brigade troops in particular were somewhat resentful, for as experts in amphibious operations they had half-expected to move on to the Pacific, and they did not relish their rear-echelon service role.[32]

Another administrative change, effected in August, had little importance in the over-all organization of the beaches, but had its significance for the troop units involved. On 14 August the 531st Engineer Shore Regiment (of the 1st Engineer Special Brigade) was reorganized and redesignated the 1186th Engineer Combat Group. In effect, this change ended the Army's recognition of the 531st Engineer Shore Regiment as a specialized organization trained for amphibious work, and also ended the recognition of this unit as one with an unusual record of achievement. The 531st had participated in four invasions and consisted of veterans with a professional pride in their organization and their specialty. The change consequently brought an inevitable letdown in morale. Officers and men alike felt they had lost their distinctive identity.[33]

In summarizing the operations at OMAHA and UTAH it is appropriate to include a note on the British MULBERRY at Arromanches-les-Bains since that installation operated on U.S. account for about a month. MULBERRY B was operated by British forces throughout the summer and was of unquestioned importance as a source of discharge capacity because of the delay in capturing suitable ports. Early in August the British Chiefs of Staff, encouraged by the favorable developments following the breakout, questioned the future value and importance of the MULBERRY in view of the prospects for the early cap-

[32] NEPTUNE: Training for and Mounting the Operation, II, 146–48, 152–54.

[33] *Ibid.*, II, 145–46.

ture of Brest, which, they suggested, might permit the transfer of Cherbourg to the British in the near future.[34]

This proposal found little support at Supreme Headquarters, where it was pointed out that the requirement for Liberty ship discharge facilities made it imperative that the Arromanches MULBERRY be employed to maximum capacity as long as the weather permitted.[35] In fact SHAEF officials were already concerned over the slow progress being made in winterizing the MULBERRY, which involved strengthening the existing units and emplacing additional caissons (known as PHOENIXES) then under construction in the United Kingdom.[36] Early in September General Gale, the SHAEF chief administrative officer, once more emphasized the urgency of this program, pointing out that the Allies had not gained a single Liberty ship berth since the capture of Cherbourg more than two months earlier. He expressed the belief that the MULBERRY might still "save our lives." [37]

Disagreement nevertheless persisted over the advisability of attempting to extend the life of MULBERRY B. In September the capture of Antwerp and Le Havre raised hopes that the discharge shortage might soon be eliminated. Furthermore, winterization was costly. The British Chiefs of Staff, when they first questioned the future value of the MULBERRY, had noted that the construction of additional PHOENIXES required the use of dockyards badly needed for the repair of damaged shipping. Movements and transportation officers at 21 Army Group pointed out that the continued use of the port entailed considerable work on the roads, which were beginning to break up, and noted that winterization would not guarantee a port throughout the winter.[38] So desperate was the need for discharge capacity, however, that the expenditure of effort was considered justified and was therefore permitted to continue.[39]

By mid-October MULBERRY B had assumed more importance to the Americans than to the British, for it had been decided to discharge U.S. Liberties there. The first unloading of U.S. ships had actually begun on the 13th.[40] The port was definitely a wasting asset now, and 21 Army Group was anxious to withdraw personnel and floating equipment for use in the Belgian ports. In view of the bad condition of the roads back of the port and the limited capacity of the railway which came to within about twelve miles of the beach, the sole value which the MULBERRY now had even for U.S. forces was that it relieved the shipping backlog which had developed off the coast. The Communications Zone

[34] Ltr, COS Com to CofS SHAEF, 9 Aug 44, SHAEF SGS 800.1 MULBERRY O/CS II.

[35] Ltr, Smith to Secy COS Com, 13 Aug 44, SHAEF SGS 800.1 MULBERRY O/CS II; Min, CAO Mtg, 12 Aug 44, SHAEF AG 337–14 CAO Mtgs.

[36] Cbl, ANCXF to SHAEF, 15 Aug 44, ETO 381/430 Tonnage, OVERLORD, and documents in SHAEF G–4 825.1 MULBERRY; Cbl FWD–12919, Gale to War Office for VQMG, 16 Aug 44, SHAEF SGS 800.1 MULBERRY Case A.

[37] Ltr, Gale to Maj Gen A. R. Godwin-Austen, VQMG, 9 Sep 44, SHAEF SGS 800.1 MULBERRY Case A.

[38] Memo, DQMG Mov and Tn, 21 A Gp for MGA, sub: Policy on the Winterization of MULBERRY, 17 Sep 44, SHAEF G–4 825.1 MULBERRY II.

[39] Min, Mtg at Arromanches, with Maj Gen H. B. W. Hughes, Chief SHAEF Engr, presiding, 10 Oct 44, SHAEF SGS 800.1 MULBERRY Case A.

[40] Conf Notes, 29 Sep 44, G–4 COMZ Plant and Communications Diary/Jnl, ETO Adm 145C.

was inclined to agree on the low value of the MULBERRY, and favored a proposal to use the newly constructed PHOENIXES at Le Havre, where they could be used to greater advantage, in part because clearance would be directly by rail.

On 16 October the whole matter of the MULBERRY's future was reconsidered at a meeting at SHAEF over which General Gale presided. The conference recommended that winterization be abandoned. Five PHOENIXES then en route to the port were to be installed; but no more blockships were to be sunk off Arromanches, and ten of the new PHOENIXES were to be sent to Le Havre, the remainder being held in the United Kingdom as a reserve. The unloading of cargo at MULBERRY B was to continue until 31 October, and MT ships were to be accepted as long as conditions permitted. Except for the salvaging of removable equipment, the port would be left to disintegrate after the end of October. The SHAEF Chief of Staff approved these recommendations on 18 October.[41]

The rapid deterioration at OMAHA and UTAH in the last half of October, coupled with the fact that Le Havre and Rouen were developing much more slowly than had been expected, compelled the Communications Zone to make the most of the British MULBERRY. Unloading of Liberties necessitated the use of pierheads and entailed a risk of losing part of the WHALE bridging making up the pier at Arromanches in case bad weather suddenly broke up the port. In view of the requirement for the maximum possible discharge of cargo, however, Maj. Gen. Robert W. Crawford, the SHAEF G-4 believed the risk justified, and he recommended on 2 November that certain bridging equipment be left at Arromanches.[42] One pier and pierhead were left for U.S. use, therefore, and the unloading of U.S. cargo continued for a few more weeks.[43] The entire matter of winterization was reconsidered again early in November, Admiral Sir Bertram H. Ramsay recommending that winterization be continued because experiments with the PHOENIXES at Le Havre had not gone well.[44] But the earlier decision stood firm, and MULBERRY B ceased operating on 19 November, the same day on which discharge at OMAHA Beach came to an end.

The cargo which U.S. forces received through the British MULBERRY was actually negligible in quantity, for the total intake in the five-week period was a bare 20,000 tons, consisting chiefly of ammunition from ships lightened at OMAHA.[45]

Meanwhile, the smaller Normandy ports made their contribution to the total tonnage discharged on the Continent, although it was rather short-lived and small in terms of the total cargo unloaded. Grandcamp, Isigny, Barfleur, St. Vaast, and Carentan were all tidal

[41] Min Mtg at SHAEF, 16 Oct 44, and Cbl S-62823, SHAEF to ANCXF, 21 A Gp, and COMZ, 18 Oct 44, SHAEF SGS 800.1 MULBERRY O/CS II; CAO Mtg, 20 Oct 44, SHAEF AG 337-14 CAO Mtgs; COMZ G-4 Plant and Communications Diary/Jnl, 16 Oct 44, ETO Adm 145C.

[42] Ltr, Crawford to CAO, sub: Winterization of MULBERRY, 2 Nov 44, SHAEF G-4 825.1 MULBERRY II.

[43] Memo, Brig W. E. Blakey, Deputy Tn SHAEF, sub: MULBERRY, 6 Nov 44, and Ltr, Gale to CofS SHAEF, sub: Winterization of MULBERRY, 6 Nov 44, SHAEF G-4 825.1 MULBERRY II.

[44] Ltr, Gale to CofS SHAEF, 6 Nov 44.

[45] G-4 History, I, 53.

ports which were practically useless at low water; in addition, Carentan was a locked harbor at the end of an eight- to ten-mile long channel connecting it with the sea, and Isigny was also several miles inland. None could accommodate deep-draft vessels and their use was therefore restricted to receiving coasters. OVERLORD plans had contemplated the development of these ports, with the exception of Carentan, to a combined capacity of less than 3,000 tons.

All five ports were found in good condition, except for mines, roadblocks, sunken craft, barbed wire, and other obstructions. The removal of sunken vessels and the cleanup of the debris was largely completed by the first of July, and rehabilitation was continued to develop the planned capacities of the ports. Grandcamp and Isigny were the first to be captured and began to receive cargo on 23 and 24 June, respectively. St. Vaast discharged its first supplies on 9 July, and Carentan and Barfleur opened on the 25th and 26th.

Up to the time of the breakout the cargo discharged by these small ports totaled 48,343 long tons, the equivalent of about three good days at the beaches. All these ports were capable of greater development, however, and the targets for both Grandcamp and Isigny had already been raised at the end of June. Late in July port plans were completely re-examined in view of the threatened deficit in discharge capacity, and the decision was made to develop the minor ports to their maximum capacity, then estimated at 17,000 tons per day.[46]

At Barfleur and Carentan the additional work which the new targets entailed was completed in the first week of August, raising the capacity of the former from 1,000 to 2,500 tons, and giving the latter a capacity of 4,000 tons. ADSEC engineers also carried out additional rehabilitation at Grandcamp, Isigny, and St. Vaast, except for the dredging which was necessary to the development of their full capacity. Since these harbors dried out at low tide, a dredge capable of resting on the bottom was required. Such a vessel—the French bucket dredge *Divette*—was found in the British sector, was towed to Isigny, and after some repairs began dredging the silt from that port.[47] A few days of operation, however, revealed that craft could not enter the harbor while the dredge was working. This, plus the prospect that Cherbourg would soon be handling large tonnages, led to the cancellation of the entire program for the three ports, leaving them somewhat short of their maximum development.[48]

Failure to develop these ports to their full capacity actually entailed no loss, for at no time in the course of their operations did their combined discharge equal their rated capacity of 12,000 tons. The utilization of these ports depended entirely on the availability of barges and small coasters, which were forced to handle a much greater portion of the tonnage than planned because of the long delay in developing deepwater berths for the direct discharge of Liberty ships. A variety of difficulties plagued

[46] Normandy Base Section, Engineer Section History, 8, ETO Adm 596.

[47] This area was under the control of Normandy Base Section by this time.

[48] Normandy Base Section, Engineer Section History, pp. 26–27; Port Construction and Repair, Hist Rpt 11, OCE ETO, pp. 69–70, ETO Adm.

the operations of the minor ports, the chief of which was the fact that they were all tidal. In some cases, particularly during the low neap tides, it was necessary to lighten even coasters before they could enter.

Plans for the use of Carentan proved the most unrealistic. The long narrow channel which separated that port from open water was particularly troublesome. In July three vessels either sank or ran aground in this channel through various causes.[49] A succession of such difficulties finally led to the conclusion that the operation of that port was more trouble than it was worth. Carentan consequently had a short life as a cargo port, discharging for a period of only seven days, from 25 to 31 July. Its total intake for that single week amounted to a mere 2,114 tons for an average of about 300 per day, far short of its rated capacity of between 2,000 and 4,000 tons.

While Carentan made the poorest showing, none of Normandy's minor ports met their targets even for short periods. Only one, St. Vaast, achieved an average of more than 1,000 tons. (*Table 2*) All five ports had ceased operating by mid-October, and were turned back to the French on 9 November.[50] Taken together, the five ports discharged a grand total of 330,600 long tons, comprising 10 percent of all the cargo offloaded on the Continent up to mid-October. Their performance was of course overshadowed from the beginning by that of the beaches, and was later put completely in the shade by the development of Cherbourg's capacity. Their tonnage record, however, like that of the airlift during the pursuit, is hardly a fair measure of their value, which was considerable when viewed in the light of the desperate shortage of discharge capacity at the time.

TABLE 2—DISCHARGE PERFORMANCE OF NORMANDY'S MINOR PORTS

[Long Tons]

Port	Date opened	Date closed	Days operated	Total tons discharged	Average daily discharge
Grandcamp	23 June	19 September	88	58,382	675
Isigny	24 June	15 October	114	85,431	740
St. Vaast	9 July	16 October	100	117,185	1,172
Carentan	25 July	31 July	7	2,114	300
Barfleur	26 July	16 October	84	67,491	803

(3) The Role of Cherbourg

Considering that the port problem dominated logistic planning for OVERLORD and involved such meticulous preparations it is ironic that the actual development of port capacity should have proceeded so differently from that planned. This was first evident at the beaches, which fortunately revealed potentialities far beyond expectations and remained in operation longer than anticipated. Performance at the minor ports roughly matched the pre-D-Day estimates, although it fell short of the goals set for them after their capture. Meanwhile, the port of Cherbourg also played a far different role in the logistic support of U.S. forces from what had been expected.

As the only large port in the area of the landings Cherbourg was the first

[49] History of TC ETO, IV 11th Port, pp. 12–13.

[50] [E. Cutts] American Port Plans, August to November 1944, prep by Hist Sec ETO, pp. 20–23, MS in OCMH.

major objective of the U.S. forces in the Normandy landings and was scheduled for early capture and rehabilitation. But Cherbourg, which in peacetime was primarily a passenger port and naval base, was expected to develop a capacity of less than 9,000 tons and to bear a sizable portion of the discharge burden for only a short time. Contrary to these plans, and despite the delay in capturing the port and bringing it into operation, Cherbourg achieved a discharge rate more than double the goal originally set, and until Antwerp was finally brought into operation was the mainstay of the port system supporting U.S. forces.

On the day of its capture late in June there was little indication of the great role that Cherbourg was to have, for the picture which the harbor presented when the last major resistance in the arsenal area collapsed was discouraging indeed. As a port Cherbourg had been destroyed with Teutonic thoroughness. Reconnaissance of the harbor, which began before the last resistance ended, showed that 95 percent of the existing quayage capable of handling deep-draft shipping was destroyed; many of the harbor buildings, particularly in the arsenal area, were demolished; and dozens of sunken ships and smaller craft, ranging from a 550-foot whaler to tiny fishing boats, blocked the entrance channels leading to the various basins and docks. In addition, native Frenchmen and captured prisoners told of wholesale mining of the harbor. Adolph Hitler himself appeared well satisfied with the way in which his naval commandant at Cherbourg had carried out his mission, commending Konteradmiral Walther Hennecke for his "exemplary destruction" of the harbor.[51]

The scale of demolitions at Cherbourg was actually no greater than anticipated, but the opening of the port within a period of three days of capture, as originally scheduled, was clearly out of the question. Some conception of the problems which confronted both Army and Navy units at Cherbourg may be gained by a brief survey of the damage and obstructions in the main port areas, beginning at the eastern end of the Petite Rade, or inner roadstead, and proceeding clockwise around the harbor. *(Map 4)* The eastern end of the port was the least-developed area. There the terraces known as the Reclamation and the Terre Plein simply formed anchorage for shallow-draft vessels and had suffered little damage. The masonry sea walls there were intact, although they were heavily fortified and backed by tank traps and roadblocks. Just west of the Terre Plein, however, some of the worst demolitions on the entire port were found at the quays forming the great deepwater Darse Transatlantique. This dock, built by the Germans between 1923 and 1935 as a World War I reparation, was 800 feet wide and 2,000 feet long. Forming the eastern side of the dock was the Quai de Normandie, which had not been completed, about 1,000 feet of it consisting of unfilled caissons. Before the demolitions one crawler crane and five large gantry cranes weighing about thirty tons

---

[51] Cherbourg Port Reconstruction, prep by OCE ETO, 1944, Annex B to History of the Normandy Base Section, ETO Adm 596; Cherbourg—Gateway to France: Rehabilitation and Operation of the First Major Port, prep by Hist Sec ETO [1945], Ch. I, p. 12, Ch. III, pp. 4–5, Ch. IV, p. 3, MS in OCMH.

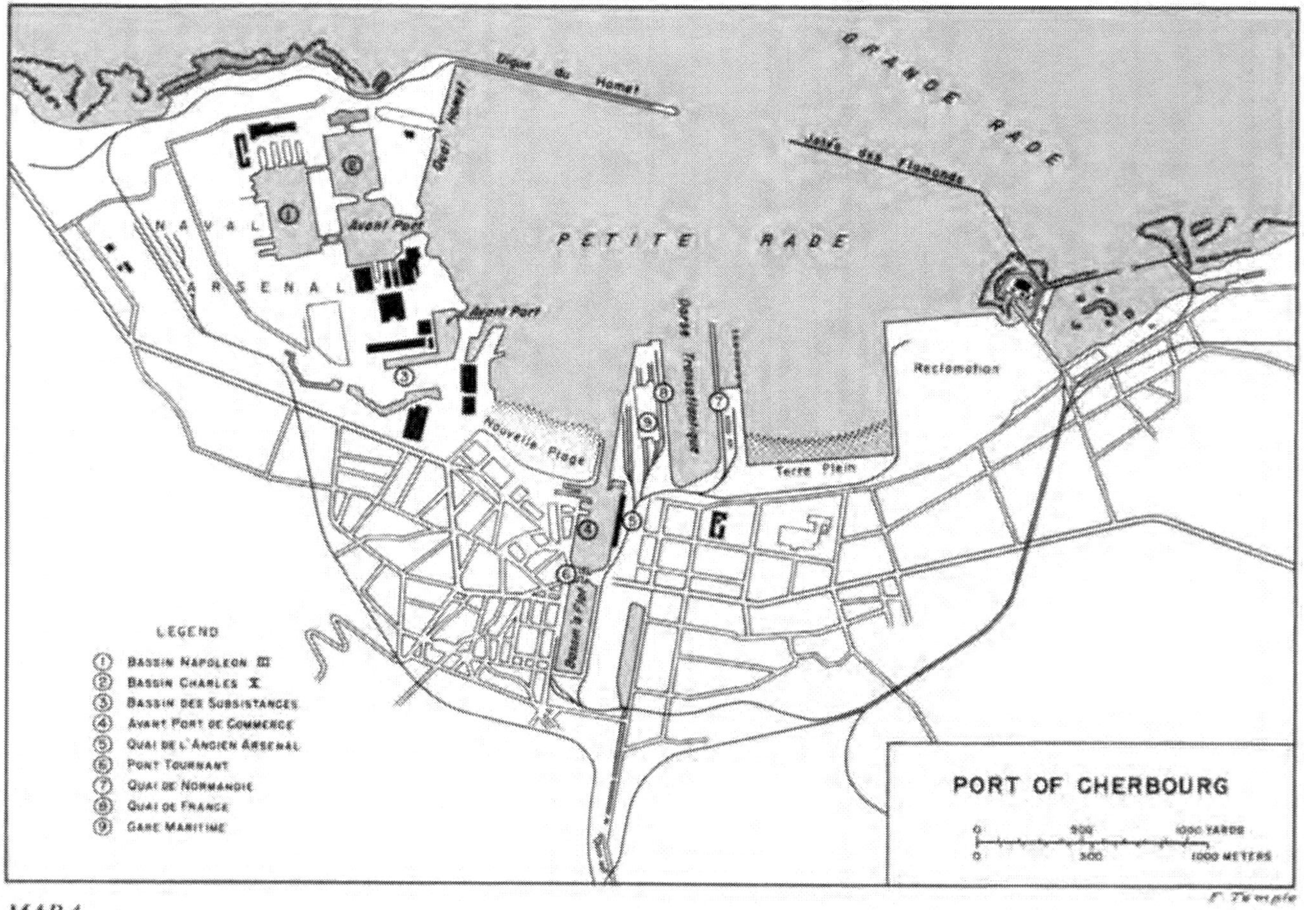
PORT OF CHERBOURG
GRANDE RADE
PETITE RADE
Digue du Homet
Jetée des Flamands
Quai Homet
Avant Port
Avant Port
NAVAL ARSENAL
Nouvelle Plage
Darse Transatlantique
Reclamation
Terre Plein
Bassin à Flot
LEGEND
1 Bassin Napoleon III
2 Bassin Charles X
3 Bassin des Subsistances
4 Avant Port de Commerce
5 Quai de l'Ancien Arsenal
6 Pont Tournant
7 Quai de Normandie
8 Quai de France
9 Gare Maritime
0 500 1000 YARDS
0 500 1000 METERS
F. Temple

MAP 4

THE PORT OF CHERBOURG

had been tipped into the water and the quay walls then blown in on top of them. Forming the western side of the Darse was the new Quai de France, a modern wharf capable of berthing the largest ocean liners and including the pride of the city, the fine Gare Maritime, where passengers could alight from transatlantique liners and under the same roof board streamlined trains for Paris. Demolitions here followed the same pattern as at the adjacent Quai de Normandie, although there was much more to destroy. More than 15,000 cubic yards of caisson masonry had been blown from 2,000 feet of the quay wall into the Darse. The railway station, a reinforced concrete building 787 feet long, while not totally destroyed, was badly damaged, and its utilities, such as the heating plant and electric control apparatus, were completely demolished.

Immediately to the west of the great pier forming the Quai de France and the Gare Maritime a channel led directly south into a tidal basin and wet dock—the Avant Port de Commerce and the Bassin à Flot—lying in the very heart of the city. No damage had been inflicted on the innermost Bassin à Flot, but in the Avant Port the eastside Quai de l'Ancien Arsenal was entirely in ruins, and a swing bridge known as the Pont Tournant, which spanned the channel between the two basins and was on the principal artery connecting the two halves of the city, had been wrecked, half of it lying in the channel and the other half mined. Together with the installations surrounding the Darse Transatlantique this area formed the commercial part of the port.

To the west lay the Nouvelle Plage, a 400-yard beach, and beyond it a seaplane base. The Nouvelle Plage, since it contained no man-made installations, suffered no damage, although it had been fortified with barbed wire and fences. At the seaplane base, however, all hangars and other buildings were wrecked, and all cranes had been blown into the harbor.

Off the western side of the Petite Rade lay the most elaborate installations of the entire port—the arsenal and naval dockyard, containing both tidal and nontidal basins and drydocks, workshops, barracks, and storage facilities. Destruction to buildings was particularly widespread in this area, some of it caused by Allied air attacks in the preceding years. All drydock gates were wrecked, every bridge except one was demolished, including a retractable railway bridge, and all portal cranes had been blown up. Where facilities remained standing they had been prepared for demolition by 500-pound bombs. Only quay walls were intact in this entire area.

Forming the northwest edge of the Petite Rade was the Quai Homet, a berth which apparently had been used by coal coasters. This quay had been cratered in nine places and the explosions had thrown much of the wall out of line. Finally, the Digue du Homet, the 1,100-yard mole which formed the western breakwater of the Petite Rade, had also been systematically demolished. The Digue du Homet carried both railway tracks and oil lines and was quayed on the port side. All these facilities were unusable, for the breakwater was blown out on the quay side at eleven places, and in two additional places had craters more than a hundred feet long extending the full seventy-foot width of the mole. Opposite the Digue du Homet the eastern breakwater, known as the Jetée des Flamands, had no value so far as cargo discharge was concerned and was untouched.[52]

The cratered and crumbling quays, the toppled cranes, the blown bridges, and the demolished buildings represented only the most evident damage to the port. Intelligence had already revealed that the black waters of the Petite Rade and the various basins had been rendered treacherous by hundreds of mines, and the first reconnaissance disclosed that every passage or channel in the harbor was blocked by sunken ships. These mines and ships, as it turned out, proved to be the chief obstacles delaying the opening of the port. A complete catalog of the ships, cranes, and other miscellaneous wreckage with which the harbor floor was strewn cannot be detailed here, but a few examples will illustrate the extent to which the various channels were blocked.

The principal sinkings had taken place in the entrance channels leading to the Darse Transatlantique, the Port de Commerce, and the arsenal area. The Darse Transatlantique, for example, was completely blocked off to any deep-draft vessels by two large ships which had been sunk across its entrance. One was a 550-foot whaler of 7,000 to 8,000 tons, the *Solglint,* which lay on its side with its bow against the north end of the Quai de Normandie. The other was a rotted coaster, the 325-foot *Granlieu,* which completed the closure of the Darse's entrance, extending across to the tip of the Quai de France.

In the Avant Port de Commerce and the Bassin à Flot about a dozen small craft had been sunk. These presented no serious trouble and were removed by floating cranes. But in the channel leading to these basins lay sixteen vessels which denied entrance to the Port de Commerce to everything but small craft, such as barges. The major obstacle was a large coaster, the *Normand,* the other fifteen consisting of old tugs of various sizes.

The arsenal area had by far the largest concentration of sunken vessels. The entrance to the first basin—the Avant Port —was effectively blocked by two barges and an old German-built submarine lifting vessel of about 1,000 tons. Floating the latter proved impossible, so that eventually it was necessary to cut up the vessel and remove it piecemeal, a task which required almost four weeks. In the Avant Port itself lay eight other vessels, the larger ones at the entrances to the inner basins—the Bassin Napoléon III to the west, and the Bassin Charles X to the north. One of the sunken barges in the Avant Port carried a deadly cargo of sixty-five to seventy contact, magnetic, and acoustic mines, each of which had to be gingerly removed by divers. Additional vessels had been sunk in the passages which led to the inner basins—a large trawler in the passage leading to the Bassin Napoléon III, and five vessels in that leading to the Bassin Charles X. The latter contained an additional eight craft, including a 110-ton floating crane, and the Bassin Napoléon III was littered with another fifteen barges, tugs, and trawlers, which denied access to the

[52] Cherbourg Port Reconstruction, pp. 13–25.

DESTRUCTION AT CHERBOURG. *Gare Maritime, left. Note damage left of pier and the* Normand, *right, blocking entrance to the Avant Port de Commerce. Right, the* Solglint *sunk off the Quai de Normandie.*

quays. The passage connecting the two inner basins was completely blocked by four craft and a demolished swing bridge. Since this passage was too narrow to be of much service no attempt was made to clear it.

These sinkings constituted the bulk of those obviously calculated to render the port unusable, but there were many other craft scattered about the harbor. An armed trawler had been sunk in the big battleship drydock in the northwest corner of the Petite Rade, and several trucks had also been dumped into the dock; two barges and three large tugs had been sunk alongside the Digue de Homet; a coaster and ten-ton floating crane lay to one side of the Nouvelle Plage; and the small basins at the southern end of the arsenal area, particularly the Bassin des Subsistances, were cluttered with motor launches, barges, trawlers, tugs, and floating cranes.[53]

The clearance of some of this wreckage was obviously one of the first steps required to bring the port into operation. This task could get under way as soon as agreement was reached on the order in which various sections of the port should be rehabilitated. Navy salvage officers and Army engineer and transportation officers had entered the port on the day of its capture and after a quick reconnaissance established priorities on 28 June for the reconstruction of certain areas. In every case both salvage by the Navy and shore reconstruction work by the Army engineers were required to permit the start of discharge operations. Almost all salvage operations,

[53] *Ibid.*, pp. 26–29.

DESTRUCTION AT CHERBOURG. *Wrecked hangars at seaplane base, left. Right, partially sunken craft and demolished swing bridge in the narrow passage to Bassin Charles X, naval arsenal area.*

however, required the use of lifting craft, pontons, or large floating cranes, and none of these could be brought in until the approaches to the various quays and channels were clear of mines. Mine clearance consequently took precedence over even the removal of obstacles.

The enemy had done as masterful a job in mining the harbor as he had in demolishing its onshore facilities and sinking obstacles in the approaches. According to Commodore William A. Sullivan, an experienced salvage officer who surveyed the port upon its capture, the mine-sweeping problem which it created was the most complicated yet encountered in any harbor clearance work. Various types of magnetic, acoustic, contact, and "Katy" mines were uncovered, either by sweeping operations or through accidents to shipping. The first three types were already well known and were found in both the outer and inner roadsteads and entrances thereto, in the arsenal basins, and in the Darse Transatlantique. The Katy mine was new, however, and proved the most troublesome. Normal mine-sweeping would not set it off, for it was usually planted in the growth of the harbor floor. Most of the Katies were sown in the shallow anchorage of Querqueville Bay at the western end of the Grande Rade or outer roadstead, and just off the Reclamation area at the eastern end of the Petite Rade.

Virtually no part of the harbor was clear, although the mining of the harbor apparently had been far from completed, as evidenced by the sunken barge in one of the arsenal basins that contained the mines. Twenty-four carloads were eventually removed from the debris in the Gare Maritime. A narrow lane had been left free inside the central outer breakwater—the Grand Digue—from the western entrance to a point opposite the entrance to the Petite Rade. Starting with this slender opening, sweeping operations got under way on 30 June, the day after the enemy finally surrendered the forts on the outer breakwater. Both U.S. and British vessels participated in the clearance operations, all under the direction of Commander John B. G. Temple of the Royal Navy. A large, roughly rectangular anchorage in the Grande Rade was cleared first, and the sweepers then moved into the Petite Rade and Darse Transatlantique. Lack of maneuver space hampered operations throughout, and before the dangerous work was completed ten vessels, including three mine sweepers, were sunk by mines and three others were damaged. Most of the casualties were attributed to the unfamiliar Katy mine, which vessels "sat on" and detonated when anchoring or moving about at low tide. Contrary to all expectations mine clearance proved the major factor in delaying the opening of the port, and eventually required more than three and one half months to complete.[54]

The start of discharge operations did not of course await the completion of the demining task. Naval salvage work and Army engineer construction started immediately to ready for cargo reception four areas of the port which had been named in the priority program on 28 June: (1) the Nouvelle Plage, suitable for dukws; (2) the Bassin à Flot or wet basin of the commercial port, for barge discharge; (3) the Reclamation area, for railway rolling stock and LST's; and (4) the Digue du Homet, for Liberty ships and seatrains.

Salvage work actually got under way simultaneously with demining operations where it could be undertaken without the benefit of lifting craft, pontons, and cranes. In many cases it was found possible to raise undamaged vessels simply by pumping the water out of them and then floating them away. Others were patched at low water and then floated at high tide. In this way four of the sixteen vessels blocking the entrance to the Port de Commerce were removed before the arrival of the heavier salvage equipment, and would have permitted access to the inner basins within a week of the port's capture had it not been for the delay in mine clearance. No attempt was made to remove the biggest obstacle in the channel, the large coaster *Normand,* which lay on its port side at right angles to the west side of the Gare Maritime. Its starboard side was approximately level with the pier of the Gare, and therefore provided a perfect foundation for a pier at which small coasters could unload.

Similar use was made of the two vessels which had been sunk across the entrance to the Darse Transatlantique. The whaler *Solglint* lay on its starboard side at right angles to the Quai de Normandie, its port side level with the latter's deck and thus forming an excellent foundation for an additional pier. Only its superstructure was removed so that Liberty ships could moor on both sides. The coaster *Granlieu* was utilized in the same way, but had to be moved since it rested between the stern of the sunken

[54] *Ibid.,* p. 26; Cherbourg—Gateway to France, Ch. III, pp. 1–12.

Solglint and the northeast corner of the Quai de France, completely blocking the entrance to the Darse. Approximately seventy-five tons of concrete from the demolished buildings of the Gare Maritime lay atop the *Granlieu* and had to be removed first. Then, by the use of compressed air and tows, the *Granlieu* was swung out in a 90-degree arc so that it formed a continuation of the quay. An army tug promptly sank in the exact spot from which the coaster had been moved, but it was raised within a few days. On 18 September, eleven weeks after the port's capture and seventy-six days later than planned, the entrance to the Darse was finally clear.

Clearance of the arsenal area also started, beginning with the removal of the submarine lifting craft and barges blocking the entrance to the Avant Port. Obstacles were literally piled one on top of another in this area, and floating cranes and lifting craft eventually removed forty tugs, barges, and cranes from the three basins. The arsenal area, like the Darse Transatlantique, was low on the salvage priority list, and it was not until 7 September, sixty-six days after the capture of the port, that the first Liberty ship could be brought into the Bassin Napoléon III, and 21 September before the first Liberty could dock in the Bassin Charles X. The salvage task at Cherbourg was finally completed on 29 September.[55]

The development of shore facilities had proceeded simultaneously with salvage operations under priorities designed to open as quickly as possible those areas of the port which could receive cargo with the least delay. In accordance with the priorities established on 28 June, work immediately started on the Nouvelle Plage, the Bassin à Flot, the Digue du Homet, and the Reclamation area for the reception of various types of craft, ranging from dukws to Liberty ships. The establishment of priorities was designed to get work under way immediately and did not constitute a detailed plan for the port's reconstruction. The ADSEC engineer worked out such a plan in the succeeding days and presented it to the theater engineer on 4 July. Two days later it was approved by the commanding general of the Communications Zone.

The plan's most striking feature was the doubling of the original discharge target for Cherbourg, raising it from 8,800 tons to 17,000 tons per day. Most of the additional capacity was to be provided by the construction of marginal wharves along the undamaged Terre Plein and Reclamation area and sea walls, and by the construction of "fingers" for additional Liberty berths along the Digue du Homet and along the Quai de France and Quai de Normandie. Within a few weeks the growing concern over the port situation led to the realization that Cherbourg in all probability would have to bear an even larger portion of the port discharge burden. On 24 July, in the first major amendment to the development plan, the port's reception capacity was raised another 4,000 tons, to 21,800, by the decision to develop more fully the quays in the Darse Transatlantique and by a greater development of the arsenal area than was originally contemplated.

Reconstruction had started under the

[55] Cherbourg Port Reconstruction, pp. 34–39; Cherbourg—Gateway to France, Ch. III, pp. 12–19.

MARGINAL WHARF CONSTRUCTION *along the Terre Plein, Cherbourg, 24 July 1944.*

direction of the 1056th Engineer Port Construction and Repair (PC&R) Group. The main working force consisted of several engineer general and special service regiments, although a variety of specialist units, such as engineer dump truck companies, engineer fire fighting platoons, bomb disposal squads, and port repair ships were also attached. An advance party of the 1056th PC&R Group, including its commander, Col. James B. Cress, arrived at Cherbourg on 27 June, and on the following day began work on the Nouvelle Plage and the commercial port.

Development of the Nouvelle Plage was not a complicated or elaborate reconstruction job, since this beach was simply to be converted into a landing point for dukws and LCT's. Except for barbed wire and other scattered debris there was no heavy wreckage to clear. Engineers blasted three exits in the sea wall, graded the beach, and built three concrete roads. This work was completed in eight days, and would have permitted the reception of cargo had it been safe to bring ships into the harbor. Not until 14 July, however, were the western ends of the outer and inner roadsteads declared free of mines. Finally on 16 July four Liberty ships loaded with construction supplies and vehicles needed in the rehabilitation of the port entered the harbor and anchored in the Grande Rade. Late that afternoon a dukw driven by Pvt. Charles I. Willis of the 821st Amphibious Truck Company brought the first load of supplies, consisting of Signal Corps wire,

LAYING RAILWAY TRACKS *for reception of supplies, Bassin à Flot, Cherbourg, 13 July 1944.*

to the Nouvelle Plage, where a crane transferred the cargo to a waiting truck driven by Pvt. William G. O'Hair of the 3884th Quartermaster Truck Company (Transportation Corps). A few minutes later the first supplies discharged at Cherbourg were on their way to a depot five miles south of the city. Port operations were finally under way.[56]

Additional construction carried out at the Nouvelle Plage during the next month, consisting mainly of a concrete loading platform and a ramp similar to the hards built in the United Kingdom, brought the beach's dukw discharge capacity to a rated 2,000 tons per day. Most of this work was carried out by the 342d Engineer General Service Regiment, using captured enemy supplies.[57]

Less than twenty-four hours after the first cargo arrived at the Nouvelle Plage unloading also began in the Bassin à Flot, or wet basin of the commercial port. The heaviest damage in this area had been to the eastern quay of the Avant Port de Commerce (the Quai de l'Ancien Arsenal) and to the Pont Tournant, which bridged the channel between the two basins. The demolished swing bridge could not be salvaged, and was quickly cut up with torches. In its place engineers improvised a retractable bridge, using Bailey bridging and an electric hoist.

[56] Cherbourg—Gateway to France, Ch. II, p. 2, Ch. IV, pp. 6–7, Ch. VI, pp. 5–6.

[57] Cherbourg Port Reconstruction, p. 44.

Before attempting to restore the badly demolished eastern quay, engineers constructed a paved LST ramp in the northeast corner of the Avant Port so that railway rolling stock could be brought ashore there. Two railway tracks were laid to accommodate two LST's at the same time. But the basin was found to be too small and too crowded with barge traffic to permit maneuvering LST's, and the facilities were never used for the purpose intended.

Late in August attention shifted to the eastern quay, the Quai de l'Ancien Arsenal, where the concrete pile wharf had been completely demolished. After the removal of much of the debris this quay was reconstructed, part of it of concrete and steel, and the remainder of timber. The Avant Port eventually could accommodate 8 barges and the Bassin à Flot 13 barges and 6 coasters. Together the two basins had a capacity of about 2,000 tons per day.

The seaplane base, where a concrete ramp, or hard, already existed, provided facilities similar to those at the Nouvelle Plage. The ramp there was simply widened considerably and its craters were filled. In addition, rail lines were built down to the water's edge so that rolling stock could be discharged directly from LST's. When completed the seaplane base was wide enough to accommodate six LST's discharging simultaneously, and could also be used for dukws.

Similar accommodations were constructed in the Reclamation area, at the eastern end of the harbor. A concrete sea wall was first removed there, and a concrete apron 80 feet by 270 feet was then laid, wide enough to accommodate six LST's. Three berths were intended for the discharge of vehicles, and three for railway rolling stock, for which purpose rail lines were laid to the water's edge. The first delivery of rolling stock at this point was made by converted LST's on 31 July.

One of the most profitable and ambitious construction projects was carried out in the basin bounded by the Reclamation area, the Terre Plein, and the Quai de Normandie. This area had never been developed for unloading operations, and its sea walls were undamaged, although a few craters along the terrace had to be filled and pillboxes removed. Once this was accomplished a timber platform forty-two feet wide was built on wood piles directly over and straddling the sea wall to form a wharf at which barges could discharge. When completed this platform extended along all three sides of the basin and was 4,200 feet long. It could accommodate forty-one barges at a time, and was served by fifty-two stiff-leg derricks, which could transfer cargo directly to railway cars. At low tide barges dried out in this area.

Rehabilitation of the northwest corner of the harbor, namely the Digue du Homet and Quai Homet, had been given high priority, and work began there within a week of the port's capture. The Digue du Homet was particularly valuable, for it promised to provide badly needed deepwater berths for Liberty ships, which could discharge directly to the rail lines running the entire length of the mole, and also berths for train ferries bringing in locomotives and rolling stock. The Digue itself was badly cratered and in two places completely breached, permitting the water to flow through. The first priority was to fill these craters and clear away the debris. This was completed in four days, after which the repair of the railway was undertaken. Since the quay side of the Digue had an underwater shelf, engineers constructed five pile and timber platforms to serve as unloading wharves for Liberty ships. Later they filled in the gaps between them to create a continuous quay along 2,700 feet of the 3,300-foot mole. The first Liberty ship docked at one of the platforms on 9 August.

At the shore end of the Digue a portion of the Quai Homet was selected as a pierhead to accommodate the Twickenham Ferry, a British train ferry specially built to carry locomotives and rolling stock. Two berths were provided, one of them consisting of a "seat" which could accommodate a ramp lowered into position by the ferry, permitting rolling stock to roll from the ferry to the quay, and the other providing a site where locomotives could be lifted from the ferry to the quay by means of an overhead crane which was part of the vessel itself. The Twickenham made its first delivery—several 65-ton diesel electric locomotives and other rolling stock—on 29 July, all of the stock being unloaded by means of the overhead crane.

Early rehabilitation plans had not provided for any substantial development of the arsenal area. Upon the port's capture, however, ADSEC engineers immediately included the arsenal in their reconstruction plan and counted on this area to provide at least one fourth of the port's total capacity. Contrasted with other sections of the harbor, the quay walls in the arsenal area were undamaged, which meant that this area could receive ships as soon as the basins were demined and cleared of the many vessels and cranes which littered the waters. Additional berths were provided by constructing timber trestles across the various openings, such as submarine and shipbuilding pens and drydocks, and by bridging boat slips with standard timber piling wharf to provide continuous quays. In this way the arsenal area eventually provided berths for eleven Liberty ships and five coasters. It was 7 September, however, before the first Liberty could enter the arsenal area, and 21 September before the Bassin Charles X could discharge a deep-draft ship.

Last of the port areas to come into operation was the Darse Transatlantique. This was the most modern and best-developed area of the port, but was also one of the most thoroughly demolished. The Gare Maritime, for example, was almost completely useless, and tremendous quantities of debris had to be removed before the adjoining quay could be reconstructed. Utilization of the Quai de France and of the Quai de Normandie on the other side of the Darse entailed tremendous engineer construction projects. At both quays reconstruction proceeded in two phases. T-head ramps at Liberty hatch spacing were first constructed, and standard pile timber wharf was then filled in to form continuous quays. The full length of the Quai de France—2,000 lineal feet—was reconstructed in this way, providing berths for four Liberties when completed. At the Quai de Normandie not only the original quay was rebuilt, but the caissons of the uncompleted extension were leveled off and the gaps bridged to form an additional 1,200 feet of quay. This work went on round the

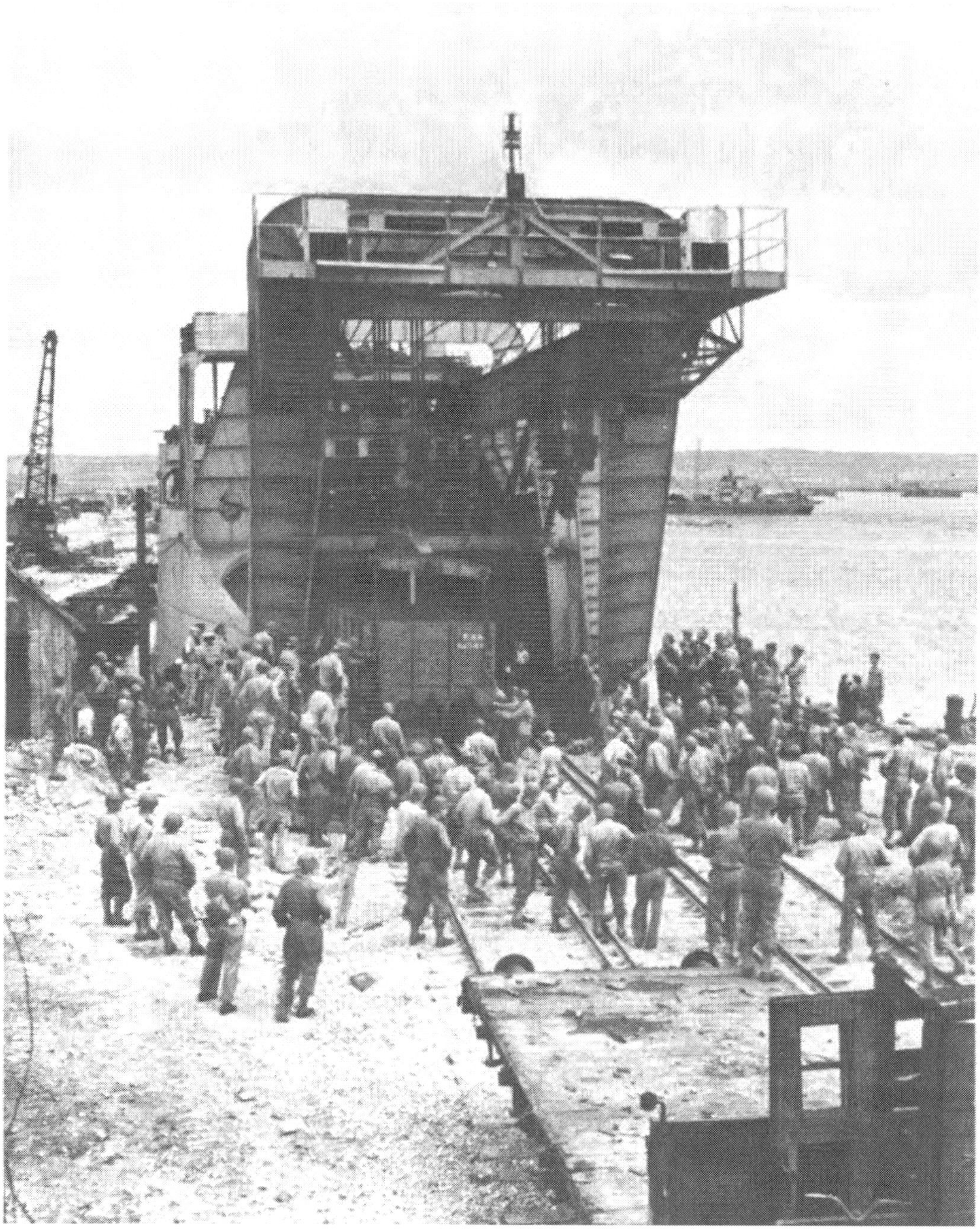

SEATRAIN UNLOADING A GONDOLA *onto rail lines laid to the water's edge, Cherbourg, August 1944.*

clock for eleven weeks, and when completed provided six Liberty berths.[58]

Port reconstruction, mine clearance, salvage, and cargo discharge had gone on simultaneously throughout the summer. In the course of this work the reconstruction plan and discharge target for the port had been amended still further. In mid-August the commander of the PC&R Group submitted plans which would have increased the port's capacity another 10,000 tons—from 21,800 to 31,900—by the construction of a PHOENIX pier, a fuller development of the arsenal, and further extensions to the Quai de France and Quai de Normandie. But these projects were found to be infeasible for one reason or another. Early in September a less ambitious proposal was made, calling for the development of Cherbourg's capacity to 26,650 tons with a minimum of additional construction. This proposal underwent further modifications in the course of its examination. The principal feature of the plan finally adopted and carried out was the addition of marginal wharves for two Liberty ships on either side of the battleship drydock off the Quai Homet. This was to bring the total capacity of the port to 28,300 tons through the provision of berths for 28 Liberty ships, 14 LST's, 75 barges, 13 coasters, 2 train ferries, and 1 tanker (the latter at the Digue de Querqueville).[59]

Target dates had been established in July for bringing the various areas of the port into operation, and a goal of 20,000 tons capacity had been set for development by the middle of September. Rehabilitation consistently fell short of these targets, in many cases because of delays in mine clearance and removal of sunken obstacles. Operations at the Nouvelle Plage, for example, which were scheduled to begin on 9 July, did not get under way until a week later. Barges and coasters were to start discharging in the Bassin à Flot on 26 July, but it was 11 August before coasters could enter. Use of the tanker berth at the Digue de Querqueville was postponed more than two weeks with the result that it was 25 July, the date of the breakout at St. Lô, before the POL Major System could come into operation. Similarly, use of the Twickenham Ferry berths at the Quai Homet began two weeks later than planned.

Deep-draft cargo ship berths were the last to come into operation. Not until 9 August, nineteen days later than planned, was the first Liberty ship berth —at the Digue du Homet—ready to begin discharge. In the arsenal area it was another full month before Liberties could berth in the Bassin Napoléon III, and 21 September before the Bassin Charles X was ready. The delay in opening the Darse Transatlantique to deep-draft vessels illustrated most pointedly of all how effective the enemy's mining and demolitions had been in denying the Allies the use of the port. Plans initially called for the completion of two Liberty berths there by 26 July. But the Darse was not even clear of mines until 21 August, and an access channel was not opened until 18 September. The

[58] Cherbourg—Gateway to France, Ch. IV, pp. 6–16; Cherbourg Port Reconstruction, pp. 44–130.

[59] Normandy Base Section, Engineer Section History, p. 28; Cherbourg Port Reconstruction, p. 40.

first Liberty ship finally berthed in the Darse on 8 October.[60]

The port fell considerably short of its goal of 20,000 tons by mid-September. At that time the rehabilitation was reported 75 percent complete. But the uncompleted 25 percent consisted of all-important Liberty berthing, or "alongside berths," where cargo could be transferred directly to waiting trucks or freight cars. Only five Liberty berths were then available. The port did not reach its projected development of 28,300 tons for another three months. As late as mid-November seven of the planned Liberty berths had still not opened. Planned facilities were essentially complete by 15 December, but even at that date two berths in the Darse Transatlantique were not in full use because of difficulties in dredging to the required depth.

The port had easily surpassed the discharge targets established before D Day. By the first week in August, three weeks after the start of operations, it was discharging approximately 6,000 tons per day. By mid-September it was handling double this volume. Port operations reached a plateau at that time, and for a full month thereafter Cherbourg's performance was uneven, at times exceeding 15,000 tons and occasionally dropping to 6,000 and 8,000.

Despite elaborate plans and preparations, the inauguration of discharge operations at Cherbourg, as at the beaches, was attended by many difficulties. Operations went through a considerable shakedown period before they became routine. The operation of Cherbourg was the responsibility of the 4th Major Port (Transportation Corps), which had operated the Mersey River ports around Liverpool, under the command of Colonel Sibley. It was originally intended that there should be a single command in the Cherbourg area, including the operation of the port and the administration of the surrounding area. For about two weeks, however, Colonel Sibley was left free to reconnoiter the port and organize it for discharge operations, and the administration of the area was assumed by the deputy commander of the Advance Section, Col. Claude H. Chorpening. On 11 July the original command plan went into effect: Colonel Chorpening left Cherbourg and Colonel Sibley was designated the commander of the newly created Area No. I of the Advance Section. This arrangement was short-lived. Because of the tremendous engineering task which lay ahead, the decision was made only a few days later to separate the functions of port operations and area command. On 21 July, as related earlier, Colonel Wyman took over the Cherbourg Provisional Command, Colonel Sibley reverted to his more limited role as commander of the 4th Port.[61]

On 25 July Colonel Sibley's organization was augmented by the attachment of the 12th Port (Col. August H. Schroeder), which moved up from St. Vaast. Colonel Sibley used the personnel of this headquarters mainly to form a Provisional Port Troop Command Headquarters to handle training, administration, and discipline of the six port battalions and other attached units which

[60] Cherbourg—Gateway to France, Ch. IV, pp. 17–18; Cherbourg Port Reconstruction, pp. 39, 42.

[61] Cherbourg—Gateway to France, Ch. II, pp. 4–8.

then comprised the 4th Port organization.[62]

The operation of a port requires the same high degree of synchronization and balancing of capacities as do other logistic activities. All steps in funneling supplies through a port are closely linked, and each is a potential bottleneck. The lack of sufficient cranes, for example, can render dozens of ships idle. Cherbourg had its share of such hitches before it became a smoothly run port. Many of them were almost exact duplicates of the early troubles at the beaches.

The proper co-ordination of Army and Navy functions, for example, was worked out only after considerable trial and error. Bringing a vessel into the port was a Navy responsibility, and as often as possible the Army's wishes were followed in determining a vessel's berth. Port officials wanted craft carrying suitable dukw cargo anchored in the Petite Rade, for example, to avoid long hauls from the outer roadstead. In many cases shore facilities left no choice as to the berthing of a vessel. But communications and liaison were faulty at first, and port officers often were unaware of the arrival of a vessel until it was about to be berthed. The logical remedy was to station port representatives in the office of the Naval Harbor Master, whence information on arriving vessels was telephoned to the appropriate port officials.

A more basic difficulty, one which had also plagued operations at the beaches, was the lack of information as to when ships were to arrive and what cargo they carried. Such information was needed so that sufficient numbers of port battalion personnel could be provided for hatch operations, so that freight cars or trucks could be spotted for the prompt clearance of quays, so that the proper cargo-handling equipment, such as cranes, could be provided, and so that technical service representatives could alert depots and dumps for the reception of supplies. Part of the difficulty arose over the failure of manifests to arrive in time. At first the port normally received information on scheduled arrivals from the Advance Section. But many vessels appeared in the crowded harbor before any news of their arrival was received, and consequently no preparation for handling them was made. Shortly after discharge operations started the port refused entry to one of these "ghost" ships only to find that it carried top priority cargo. Lacking manifests and stowage plans, port personnel frequently had to board vessels to determine what they carried and how it was stowed. Naval officials, who were in direct communication with London, were in the best position to have advance information on future arrivals. Again, but only after some trial and error, the problem was partially resolved by maintaining constant liaison with the Naval Harbor Master.[63]

Discharge operations were made difficult by the initial lack of deepwater berths. Until such facilities were available all ships had to anchor in the roadstead and discharge to lighters (dukws and barges) for movement to the beaches or the basins of the commercial port, where supplies were transferred to trucks or freight cars. Because of the double handling involved, such operations were

[62] History of TC ETO, IV, 4th Port, 7.

[63] Cherbourg—Gateway to France, Ch. VI, pp. 15–19, Ch. VII, pp. 2–3.

costly in labor and equipment, and in time. Moreover, the volume of cargo which could be handled in this way was always limited. Dukws had a rated capacity of only two and one half tons, although as in the case of the 2½-ton truck orders were given in mid-August to overload by 100 percent.[64] But the dukw was still suitable for only relatively small items and packaged goods. Barges had greater capacity, but could not handle the most awkward equipment.

Lighterage operations called for the closest planning and supervision to make the best possible use of the available tugboats, barges, and port battalion personnel. Dukws could go ashore at the Nouvelle Plage at all times, but barges had to be tied to stake boats in the harbor and await favorable tide conditions, for they could be towed into the basins of the commercial port and later to the Terre Plein area only during a few hours at high tide. The control of tugs and of hatch operations under this type of unloading was difficult, since adequate communications were initially found wanting. Both the Navy's blinker system and the use of a shuttle boat were too slow and inefficient for this purpose. The answer was finally found in the use of a small Signal Corps radio which had been designed for combat but which was found to be excellent for the control of all offshore activities. Two systems were installed, one for the control of tugs and one for hatch operations.

Lighterage operations also suffered more from the hazards of weather than did ship-to-shore discharge. Bad weather frequently prevented dukws and barges from venturing out into the harbor, and during one storm on 21 August a 30-ton floating crane and two car ferries, each with a barge, were beached by high winds.

All these handicaps indicated an urgent need for deepwater berths, for only by direct ship-to-shore operations could relatively uninterrupted discharge be assured and the largest volume of tonnage handled.[65]

A variety of other difficulties plagued the early operations of the port. There were shortages of all kinds initially—of tugs, barges, cranes, and of all types of gear. The 4th Port's gear had been loaded on twelve ships in the United Kingdom and was scheduled to be brought in at Cherbourg. Instead the vessels were sent to UTAH Beach, where much of the heavier gear was unloaded and later had to be searched. Its arrival in Cherbourg was delayed considerably. Hatch crews repeatedly lacked ropes, slings, nets, or other unloading gear. In an effort to keep them supplied, three dukws cruised about the harbor at first, taking gear from vessel to vessel. Much of the equipment was lost when ships left the port without returning it.

Most serious of the equipment problems was the shortage of cranes. Crane operation had been handled largely by civilian workers in the United Kingdom, and inadequate numbers of military personnel had been trained for the continental ports. An intensive training program had to be initiated on the spot, conducted by two sergeants who had had experience in the United Kingdom. Operation of the cranes by inexperienced workers took its toll in damaged equip-

[64] *Ibid.*, Ch. VI, p. 26.

[65] Cherbourg—Gateway to France, Ch. VI, pp. 1–5, 11, 18.

ment, just as it did in the case of motor transport operations, and the shortage of spare parts and skilled mechanics contributed to the prolonged deadlining of equipment. In this way, for example, a $15,000 Koehring crane vitally needed in port operations was deadlined for three weeks because of a broken water pump valued at $2.00. At times as many as half of all the assigned cranes were inoperative from such causes.[66]

Achieving a satisfactory discharge rate was only half the problem at Cherbourg. At least equally important in the long run was the problem of port clearance, for it was this aspect of Cherbourg's operations which proved to be the factor preventing the maximum utilization of the port's intake capacity. In one sense a port represents the narrow neck of an hourglass. Thousands of tons of supply converge upon it from the sea and must be passed through this defile and then distributed to dumps and depots. Within the port itself two principal operations are involved—the unloading of ships and the loading of freight cars and trucks. Adequate discharge facilities must be available if the maximum tonnage is to be received and if ships are to be emptied promptly and not allowed to stand idle in the harbor. Equally vital, however, are the complementary facilities required to dispatch cargo from the port in order to keep quays clear. Essentially this means that ample transportation must be available, and an adequate depot structure provided to receive the cargo.

It was desirable that clearance be accomplished by rail, for rail transport was more economical than motor transport and capable of handling much larger tonnages. Plans had of course been made to rehabilitate the existing railway facilities at Cherbourg. Damage to them had been fairly extensive, but no greater than expected. Just south of the city a 400-foot tunnel on the main Cherbourg–Paris trunk line had been blown shut, and between this tunnel and the Gare de l'Etat in the city the switches and frogs had been systematically destroyed. A roundhouse just south of Cherbourg was also largely demolished, mainly as the result of friendly artillery fire.

Within the port itself there were only about fifteen miles of trackage, branching out from the Gare de l'Etat to serve the Digue du Homet, the arsenal area, the Gare Maritime, the Amiot Aircraft Works, and the Bassin à Flot. The worst damage was evident at the Gare Maritime, in the arsenal, and in the area of the aircraft works.

All these facilities were required for the clearance of the port, and rehabilitation began immediately. But the decision to double Cherbourg's intake capacity made it apparent that the existing rail complex would be far from adequate. Additional lines were needed at the new quays which were being constructed, so that discharge could be directly from ship to rail car, and the capacity of storage and marshaling yards had to be greatly expanded to handle the tremendous volume of rail traffic. Cherbourg was not equipped for large-scale freight handling operations. Just as new discharge facilities had to be developed, therefore, a sizable railway construction program had to be undertaken.

[66] History of TC ETO, IV, 4th Port, 5, V, 4th Port, 4; Cherbourg—Gateway to France, Ch. VI, p. 23, Ch. VII, pp. 10–11.

New construction and the restoration of the existing system were carried on simultaneously. Existing facilities were augmented considerably in several places, as for example in the yards between the tunnel and the Gare de l'Etat, where 11,500 feet of new track were laid, and along the Digue du Homet, where three lines were extended the entire length of the mole. Additional trackage was also built along the barge wharves in the Terre Plein area, and along two sides of the Bassin à Flot. In the arsenal area a considerable augmentation of existing facilities was carried out by the construction of lines to the inner quays, and by the construction of additional yards.

The most urgent need created by the new tonnage target was for storage and marshaling yard capacity. At the time of its capture Cherbourg possessed storage capacity for only 350 cars, and a marshaling yard capacity of only 400. Brig. Gen. Clarence L. Burpee, Director General of the 2d Military Railway Service, estimated that the projected daily discharge of 20,000 tons would require the loading of 2,000 freight cars per day. Since a two-day supply of empty cars was required on hand at all times, 4,000 cars would be required. Storage facilities for 4,000 cars and marshaling yard facilities for 2,000 cars were therefore required to clear 20,000 tons of freight from the port each day.

Transportation Corps officials proposed to meet part of this need by expanding existing yards. But the larger part of the requirement had to be met by new construction. Plans now called for the building of three new yards, one at the Terre Plein with a capacity of about 700 cars, one at Couville, six miles south of the port, with a capacity of 1,400 cars, and another at Sottevast, five miles farther south on the main rail line, capable of holding 2,600 cars. All together, these projected expansions were to provide a gross storage capacity of about 4,600 cars and marshaling facilities for nearly 2,700.

Construction of the Couville and Sottevast yards was undertaken largely at the insistence of Maj. Gen. Frank S. Ross and was among the most ambitious construction projects undertaken by engineers in the Normandy area. The Couville installation was to be a true marshaling and classification yard, where trains were made up and given track clearance for the eastward run. The facilities there were entirely of new construction and involved a tremendous earth-moving job—287,000 cubic yards of cut, and 177,000 yards for a seventy-foot fill. The heavy equipment needed for the task was lacking when construction began on 2 August. But the project had high priority, and the urgency of the job was further emphasized after 20 August when a heavy rain revealed how easily the entire area could be transformed into a quagmire. Every effort was therefore made to augment the organic equipment of the engineer general service regiment assigned the task, and work was stepped up, including the operation of a night shift. Mud produced by autumn rains also forced a change in plans for ballasting the tracks and made it necessary to open a rock quarry, repair its machinery, and haul hundreds of carloads of crushed rock to stabilize rail beds. The first yard at Couville was completed on 18 September, and received the first train five days later. On 3 November, three months after work had begun, the engineers turned over the completed project to the Transportation Corps. More than 780,000 man-hours of labor eventually went into the project, which comprised 16 miles of track and had a capacity of 1,740 cars, considerably more than originally planned.

The Sottevast yard, a few miles farther south, involved an even greater expenditure of effort. This yard was intended primarily as a storage and classification yard for empty cars, which could be dispatched to Cherbourg on call. Construction started on 15 August and continued until 12 December, although a portion of the facilities was ready for use in mid-October. Work on the Sottevast project was carried on under much the same conditions as at Couville. The major handicap was the heavy rains, which at one time inundated portions of the area to a depth of eighteen inches and necessitated extensive rock ballasting.[67] More than 2,000 men, including several companies of prisoners, were employed on this project in mid-November, and upwards of 1,300,000 man-hours went into it. The requirement for marshaling facilities had already declined somewhat by early December, and the original plans for the yard were therefore modified somewhat. When construction stopped in mid-December the yard contained eighteen miles of track and had a capacity of 2,280 cars.[68]

Until these facilities were ready, clearance capacity was a serious limiting factor in developing the maximum flow of cargo through the port of Cherbourg. As long as U.S. forces were confined to a small Normandy bridgehead, and dumps and depots were within easy reach of the port, motor transport was of course the more efficient means of clearance. But the railways eventually were expected to handle the bulk of the tonnage. It was hoped at first that 60 percent of the cargo could be shipped by rail by early August, and by mid-September approximately 88 percent—or 17,500 of the 20,000 tons discharged each day.

These targets could not be met. In the first week of August less than 10 percent of the cargo was being dispatched by rail.[69] In mid-August the Transportation Corps was called on to make emergency shipments totaling 25,600 tons via rail to the Le Mans area for the Third Army. At the close of the month even larger shipments were ordered to the Chartres area. These demands put a tremendous strain on the available supply of both locomotives and freight cars and left little rolling stock for port clear-

[67] Similar difficulties arose at the small marshaling yard built to serve OMAHA Beach. The site of this yard, along the main rail line between Lison Junction and Bayeux, was extremely low and difficult to drain. September rains reduced this area to a quagmire before proper ballasting could be accomplished, with the result that the rails gradually sank into the mud, and access roads almost disappeared. Derailments consequently became frequent, a record twenty-one occurring in one eight-hour period. Normandy Base Section, Engineer Section History, pp. 24–25; Normandy Base Section Transportation Corps History, pp. 20–21.

[68] Normandy Base Section, Engineer Section History, p. 25; History of the Couville and Sottevast Marshaling Yards, App. XIV of Normandy Base Section History; Normandy Base Section, Transportation Corps History, p. 21; History of TC ETO, IV, 4th Port, 9–10, and V, 4th Port, 12.

[69] History of TC ETO, IV, 4th Port, 18; Cherbourg—Gateway to France, Ch. VII, pp. 15–16.

ance.[70] For the entire month of August clearance by rail consequently accounted for only 38 percent of the total tonnage moved inland from the port.[71]

Port clearance continued to bear the main sacrifice imposed by the greater urgency of long-distance hauling. Far from getting motor transport to compensate for the inadequate rail capacity, port clearance actually lost trucks to line-of-communications hauling. By the end of August only three companies remained against a requirement for five or six times this number. By the end of August the quays at Cherbourg were piled high with cargo which could not be moved, averaging upwards of 70,000 tons. To relieve the congestion several of the services were authorized to establish temporary subdepots or dumps at the Terre Plein.[72]

Difficulties at the depots contributed to the port clearance problem. The depots were rarely ideally located or equipped to receive cargo. Many were established in open fields, which became muddy in rainy weather; many lacked trucks for internal movements, and also cranes.

Lacking transportation, the depots frequently insisted that trucks from Cherbourg distribute cargo at several unloading points, thus delaying their return to the port. On 1 August the port began round-the-clock operations. Some depots, lacking either lighting facilities or adequate personnel, were slow to adopt the twenty-four-hour schedule and thus contributed further to the delay in releasing transportation. Depot locations were frequently changed, and the services did not always give the port prompt notification, with the result that drivers at times returned to the port with fully loaded trucks after searching unsuccessfully for their proper destinations.

A common complaint, heard frequently during the pursuit, arose from the practice of ordering drivers after they had arrived at their designated destinations to deliver their loads to another location farther forward. From the point of view of port clearance this naturally aggravated the transportation situation.[73]

Port clearance continued to be the limiting factor at Cherbourg for some time, mitigating to some extent the failure to meet discharge goals. The port failed by about 8,000 tons to meet the discharge target of 20,000 tons per day by mid-September. Colonel Sibley, the 4th Port commander, had expressed doubt in July that Cherbourg could meet the higher tonnage targets established at that time. He cited in particular the difficulties over mine clearance and obstacle removal, and the probable inadequacies of the rail system, especially the shortage of rolling stock, which he thought would probably delay the maximum development of the port's capacity.[74] Unfortunately these fears were largely substantiated.

The seriousness of Cherbourg's deficiencies was increasingly highlighted by the delay in bringing Antwerp into use, and by the declining performance at the

[70] Cherbourg—Gateway to France, Ch. VI, pp. 16–17.

[71] History of TC ETO, IV, 4th Port, 22.

[72] History of TC ETO, IV, 4th Port, 22; Cherbourg—Gateway to France, Ch. VII, pp. 17, 21–22.

[73] History of TC ETO, V, 4th Port, 11; Cherbourg—Gateway to France, Ch. VII, pp. 20–23.

[74] Cherbourg—Gateway to France, Ch. I, p. 18.

beaches, which became a wasting asset in October. Cherbourg, it must be remembered, was the only major deep-water port which the Allies possessed in operating condition at the time. By the end of September its disappointing performance became of sufficient concern to attract the attention of higher echelons in the theater, and led to some reorganization of the port. On 29 September Colonel Sibley was relieved of command of the 4th Port and was succeeded by Col. James A. Crothers.[75]

The new commander promptly submitted a request to Normandy Base Section for additional personnel and for more rolling stock, which he regarded as the two principal deficiencies. The main troop requirements listed were twelve port companies; an engineer general service company trained to operate cranes, derricks, and other equipment; additional prisoners of war to handle tonnage at quayside; and a battalion of infantry to guard prisoners. He asked for 1,310 freight cars per day, sufficient to clear 17,000 tons per day. Another 4,000 tons, he estimated, could be moved to local dumps and depots via motor transport, and 3,000 tons, consisting of vehicles, would move out on their own power. Additional cargo-handling equipment, such as nets and slings, was also requested. With this additional complement of personnel and equipment Colonel Crothers believed that Cherbourg could work forty-four ships simultaneously,[76] each averaging 500 tons per day for a total of 22,000 tons. This tonnage, along with another 2,000 tons brought in via LST's and car ferries, would enable the port to unload and clear 24,000 tons, which the Communications Zone had recently established as the new target.[77]

Meanwhile officers from both SHAEF and the Communications Zone arrived to investigate. All of them found much that needed correction or improvement. The SHAEF G–4 representatives—Capt. L. A. Thackrey of the U.S. Navy and Col. N. H. Vissering—concluded that neither the port's facilities nor its labor force were being used to best advantage. Unloading personnel, for example, were working twelve-hour shifts, and morale was suffering accordingly. Port headquarters and port companies were composed almost wholly of inexperienced and inadequately trained men. Almost without fail, it seemed, once units had learned their job they would be transferred elsewhere, leaving the port with untrained labor. Port companies generally suffered from poor supervision, and the lack of supervisory personnel in turn was the main deterrent against changing to eight-hour shifts. Furthermore, none of the port companies then employed at Cherbourg had their full allowances of per-

[75] History of TC ETO, V, 4th Port 1. Colonel Sibley was relieved by Col Benjamin B. Talley, who commanded Normandy Base Section during Colonel Wyman's temporary absence. There was some disagreement over the justification for the relief. Colonel Talley himself regarded Sibley as an able officer although unsuited for the kind of organization required at Cherbourg, and agreed that he be relieved without prejudice. General Ross expressed his confidence in Sibley by giving him an important assignment in the U.K. Base Section. Interv with Talley, 6 Mar 51, and Ltr, Ross to Harold Larson, 22 Jan 51, OCMH.

[76] Twenty-three at quays, 4 working dukws, 12 working barges, and 5 coasters discharging coal.

[77] Ltr, Crothers to CG Normandy Base Sec, sub: Additional Requirements, Port of Cherbourg, 3 Oct 44, EUCOM 800 Rivers, Harbors, and Waterways, I.

sonnel, with the result that there were not enough hatch gangs in operation.[78]

In addition to noting personnel deficiencies, both SHAEF and COMZ observers felt that the facilities of the port were not being efficiently utilized. Col. W. E. Potter, logistic planner from the Communications Zone, observed that there was considerable misuse of quayage. Engineer Class IV ships, for example, were being discharged at the convenient Quai de Normandie, while high priority cargo like ammunition was being unloaded at less suitable basin quays, which were difficult to work by rail. A few days of good weather invariably brought congestion on the quays. Planning had also been deficient, particularly in minor details. Even standing operating procedures were lacking on certain functions which might have been reduced to routine, such as the provision of empty cars at loading sites.[79]

One of the principal bottlenecks which all the inspections recognized was the lack of adequate rolling stock needed to clear the port. But this deficiency, in the view of observers, was needlessly aggravated by poor co-ordination between port authorities and the railways in providing the proper types and numbers of cars for loading, and by inefficient operations at the depots, where the excessive time required to unload trucks resulted in exorbitantly long turnarounds.[80]

Magnifying the entire clearance problem, meanwhile, was the fact that depots in the Cherbourg area were carrying on retail supply operations. Requisitions from the armies and even subordinate headquarters were being filled in detail from dumps and depots in the Cherbourg area, entailing much sorting and switching of freight cars and segregation of small loads, thus adding to the already overtaxed facilities of the port area and contributing to the congestion there. Base areas were not intended to carry on retail issue, except on a small scale to units located in the immediate vicinity. This was another of the unorthodox practices forced on the Communications Zone in the pursuit period, when lack of forward depots left no choice but to fill requisitions directly from base dumps and depots. The port area was ill-equipped for such operations, and it was imperative that its operations be limited to wholesale supply, so that complete train loads of various classes of supply could be dispatched to depots farther forward in the Communications Zone.[81]

The SHAEF representatives who visited the port early in October concurred in general with the new port commander's request for additional troops and equipment. General Ross, the chief of transportation, immediately took steps

[78] Memo, Thackrey for G–4 SHAEF, sub: Port Discharge Conditions at Cherbourg Which Can Be Improved: Recommendations Regarding Their Improvement, 9 Oct 44, EUCOM 800 Rivers, Harbors, and Waterways, I.

[79] *Ibid.;* Col W. E. Potter, Report of Inspection of Cherbourg, 23 Oct 44, EUCOM 800, Rivers, Harbors, and Waterways, I; COMZ G–4 Br Chief's Mtg, 24 Oct 44, ETO Adm 145C.

[80] Potter, Rpt of Inspection, 23 Oct 44, and Memo, Thackrey for SHAEF G–4, 9 Oct 44.

[81] Memo, sub: Cherbourg, 20 Oct 44, attached to Ltr, Rear Adm Alan G. Kirk to Smith, 21 Oct 44, SHAEF SGS 800 Harbors, Opening, Use, Construction.

to send eighteen additional port companies to Cherbourg.[82] But the SHAEF representatives were frankly skeptical over the prospects that the port would soon perform at the rate which Colonel Crothers had predicted. Captain Thackrey pointed out that the port had never worked more than eighteen ships at a time despite the fact that there were berths for thirty-five, and that only half of them had averaged 400 tons on a single day. Furthermore, he confirmed what had been suspected earlier, that the discharge figures for September, which averaged 10,000 tons per day, had included about 4,000 tons of coal and railway rolling stock. The actual deadweight cargo discharged had averaged only about 6,000 tons.[83] Such figures, as SHAEF logistic planners had pointed out earlier, were worse than useless unless carefully interpreted.[84] Regarding clearance capabilities, Captain Thackrey estimated that between 8,000 and 9,000 freight cars would have to be put into service between the forward depots and Cherbourg in order to provide the 1,300 cars needed at the port each day. He doubted that this number could be made available before the beginning of December. He was not very hopeful, therefore, that the port could achieve either discharge or clearance of 24,000 tons. With the additional hatch gangs then being organized, and with additional supervisory personnel and trucks, he believed that discharge and clearance could be raised to about 15,000 tons per day by the end of October.[85]

This estimate actually proved the more realistic, for Cherbourg's discharge rate at the end of October was averaging only 13,000 to 14,000 tons per day. Its performance had been highly erratic throughout the month, averaging only 11,750 tons. This was partly attributable to bad weather, for high winds and rough seas repeatedly hampered or completely suspended barge, dukw, LST, and even crane operations. The shortage of freight cars also held up unloading at times.[86]

Cherbourg was hardly less important to the Allies at the end of October than it had been a month earlier, even though Le Havre and Rouen had been opened in the meantime. Continued efforts were therefore made to eliminate the deficiencies which prevented it from realizing its maximum potential performance. At the very end of the month an organizational problem which had been recognized for some time but gone unremedied, was finally solved. General Stratton, the COMZ G–4, had called attention early in October to a defect in command organization which he considered to be at the very root of the difficulties at Cherbourg. In his view too many people were interfering with the port commander, who, in a sense, had come to have the position of a mere executive to the base section commander, located in the same city.[87] Essentially, the difficulty lay in the Normandy Base

[82] [E. Cutts] American Port Plans, August to November, 1944, p. 16.

[83] Memo, Thackrey for SHAEF G–4, 9 Oct 44; COMZ G–4 Plant and Communications Diary/Jnl, 4 Oct 44, ETO Adm 145C.

[84] Memo, Whipple for Current Opns Br, sub: Plng Factors, 1 Sep 44, SHAEF G–4 Supplies General.

[85] Memo, Thackrey for G–4 SHAEF, 9 Oct 44.

[86] SHAEF G–4 Weekly Logistical Summaries, Oct, Nov 44, SHAEF G–4 War Diary/Jnl, Oct, Nov, Apps.

[87] COMZ G–4 Plant and Communications Diary/Jnl, 4 Oct 44, ETO Adm 145C.

Section's retention of control over rail operations, with the result that the port commander was unable to exercise the centralized control necessary to co-ordinate all the functions involved in passing cargo through the port.

This situation was finally remedied in the first week of November, after a change in command. On 30 October Colonel Wyman was relieved as commander of Normandy Base Section and was succeeded by General Clay,[88] who came to the theater on loan from the Army Service Forces on the Supreme Commander's request. General Clay quickly recognized the defect described above and granted the port commander the authority he needed.[89] The first important change occurred with the transfer of control of all rail movements involved in the clearance of the port. This entailed the control of railway operations only as far south as the Couville and Sottevast yards, but it now gave the port Commander the authority he had previously lacked to co-ordinate all the functions connected with discharge and clearance.[90]

These changes undoubtedly accounted at least in part for the prompt improvement in Cherbourg's performance. Unloadings averaged 14,600 tons per day in the first week of November, and in the third week the port achieved its best performance with an average discharge of 15,600 tons. November proved to be Cherbourg's best month, averaging about 14,300 tons per day. Meanwhile clearance operations also showed remarkable improvement, averaging 12,930 tons in November.

Cherbourg had hardly achieved this increased efficiency when its importance began to decline. The opening of other ports, notably Antwerp at the end of November, relieved Cherbourg of the heavy responsibility it had had for many weeks. It was logical of course that railway rolling stock, of which there was a never-ending shortage, should not be tied up in hauls of more than 400 miles when it could be used to so much better advantage on the shorter hauls from Le Havre and Antwerp. Beginning in December, therefore, Cherbourg's discharge targets were gradually lowered, first to 12,000 tons, and in the middle of the month to 7,000 tons per day.[91] For the next two months unloading actually averaged about 8,200 tons per day.

Efforts to improve the efficiency of the port continued to be made. On 2 December, for example, port operations were organized into two ten-hour shifts in place of the twelve-hour shifts previously in effect. Port clearance, the most persistent limiting factor, continued to be given close attention. But all projects for the physical improvement of the port, such as laying additional track and salvaging berths, and construction at Sottevast, were canceled.

Before the end of the year Cherbourg lost both personnel and equipment to other ports.[92] By the end of December,

[88] History of Normandy Base Section, p. 12.

[89] Lucius D. Clay, *Decision in Germany* (Garden City: Doubleday & Company, Inc., 1950), p. 2. General Clay commanded the base section barely four weeks, for on 26 November he was given a new assignment by General Eisenhower. Col. Eugene M. Caffey commanded the base section pending the arrival of General Aurand on 17 December.

[90] History of TC ETO, V, 4th Port, 12.

[91] Cherbourg—Gateway to France, Ch. X, p. 3.

[92] History of TC ETO, V, 4th Port, 25–26.

for example, thirty-nine crawler cranes, eight port companies, a dukw company, and a harborcraft company were being transferred. Truck and trailer companies were also being released, the need for vehicles having diminished as the railways accounted for a larger and larger percentage of total clearance. The labor force employed dropped from the November average of 5,300 to 2,900 in the case of port battalion personnel (mostly hatch gangs), from 1,000 to less than 800 in the case of civilian workers, and from 3,800 to 3,600 in the case of prisoners of war.[93]

Cherbourg's usefulness by no means came to an end in December but it was never again to operate at full capacity. The port never met its goal of 24,000 or even 20,000 tons, although it approximated the earlier target on one day—4 November—when 19,955 tons were discharged. But the emphasis given above to Cherbourg's endless difficulties and shortcomings need not obscure the significance of its accomplishment. Originally scheduled to develop a capacity of less than 9,000 tons and to provide but a fraction of the total port needs, Cherbourg in November alone averaged 14,300 tons per day and received a total of more than 430,000 tons of cargo.[94] For many weeks it handled fully 50 percent of all the U.S. tonnage brought to the Continent. As the only major deepwater port available in the period of the greatest logistic stress, Cherbourg therefore served as the mainstay of the entire continental port system through which the American forces were nourished.

(4) *The Brittany Area*

Ironically, the ports which U.S. forces had counted on so heavily in the Brittany area—that is, Brest, Quiberon Bay, Lorient, and St. Malo—were never put to use, and the only ports that proved of value were those which had either never been considered or had been eliminated from the OVERLORD plan after brief consideration.

As indicated earlier, enthusiasm for development of the Brittany area fluctuated with the prospects of opening the Seine ports and Antwerp. After the review of the entire port situation in mid-September General Lee had decided to go ahead with the development of three of the five smaller Brittany ports—St. Brieuc, Granville, and Morlaix. Cancale offered little more than an anchorage, and was dropped from plans because of bad tidal conditions. St. Malo at first appeared to offer fair prospects, and rehabilitation of the port had actually begun late in August. But damage was extensive and, late in September, on the basis of a discouraging report on the condition of the St. Malo–Rennes canal, it was decided that the return would not warrant the effort required to restore the port. Work was discontinued, therefore, and on 21 November St. Malo was turned back to the French.[95]

The first of this group of ports to be uncovered after the breakout at the end of July was Granville. Geographically, Granville is not part of Brittany, but its consideration here is logical both because of the date of its capture and because it was customarily grouped with

[93] *Ibid.*, V, 4th Port, 25.

[94] See Table 4, p. 124, below.

[95] History of TC ETO, Ch. IV, 5th Port, pp. 1–5; [Cutts] American Port Plans, pp. 7–9.

the other small Brittany ports in plans after mid-July. Granville was mainly a fishing port in peacetime; like Cherbourg, it had little importance as a freight handler, its prewar intake averaging less than 60,000 tons per year. It was a completely artificial port, its harbor consisting of two basins—an outer tidal basin known as the Avant Port, and an inner wet basin, the Bassin à Flot —formed by two jetties and a mole. A locked channel 223 feet long connected the two basins. The Bassin à Flot could maintain a water depth of nineteen feet by means of the locks, had quays with cranage facilities all around, and offered quayside accommodations for vessels up to 4,000 tons. Pre-D-Day plans called for developing a capacity of approximately 2,500 tons per day.

Granville was captured on 30 July, five weeks later than scheduled. A reconnaissance party representing the ADSEC Engineer, the 1055th PC&R Group, and the 11th Port, which was to operate Granville, immediately surveyed the port. As expected, they found it badly damaged. The gates of the lock channel had been totally destroyed and the channel itself blocked by about 7,000 cubic yards of masonry blown into it from the walls. Seven traveling cranes around the Bassin à Flot had been destroyed and some of the wreckage toppled into the basin. Craters had been blown in both jetties forming the basins, and all berths in the two basins were obstructed by damaged tugs, barges, and other craft. Finally, all cargo-handling facilities and rail spurs had been rendered useless. Nevertheless it was concluded that the port could provide sixteen coaster berths and sufficient unloading facilities to handle 5,000 tons, double the earlier estimate. Plans were made to make Granville the number one coal port on the Continent.[96]

The 1055th PC&R Group initiated work at Granville on 3 August by clearing the debris from the streets and starting the repair of the craters in the westernmost jetty forming the Avant Port. On 12 August Normandy Base Section relieved the Advance Section of responsibility at Granville, and two weeks later the 1058th PC&R Group replaced the 1055th. Clearing the lock channel and stabilizing its torn walls proved one of the largest projects. No attempt was made to repair the lock gates, but the channel itself had to be cleared to permit the passage of vessels into the Bassin à Flot, which was to be utilized as a drying basin like the Avant Port. The enemy had blown five tremendous craters in the massive granite block retaining walls which in some places reached to the very channel floor. The channel was choked with masonry and clay backfill from these demolitions, and the removal of this debris was a time-consuming task, for tidal conditions made it impossible to work more than about six hours each day in the basins. The task required approximately seven weeks, after which there still remained the job of stabilizing the channel's ruptured retaining walls. Parts of the wall were rebuilt with rubble masonry; the remainder was repaired by building sandbag revetments. Because of the strong current which the

[96] Port of Granville History, Annex A to Normandy Base Section, Engineer Section History, pp. 1–4, ETO Adm 596.

LST's Discharging Cargo on the Beach *at St. Michel-en-Grève, near Morlaix, 5 September 1944.*

ebb and flow of the tide caused in the channel, it was also necessary to line the channel with timber fenders to provide stout bumpers as protection against vessels moving along the channel. Tidal conditions restricted this work to a few hours a day.

At the same time engineers had begun to clear berths and construct cargo-handling facilities in both the outer and inner basins. Sunken vessels did not present a great problem; they were simply patched at low tide, pumped out, and then floated on the high tide and towed out. The biggest problems in the Avant Port were the removal of debris from the craters blown in the jetties which enclosed the harbor, and the repair of the craters themselves. The whole basin was littered with the debris from these demolitions, which had to be removed because of the danger of puncturing the hulls of vessels as they settled down on the harbor floor with the receding tide. Repair of the craters presented problems similar to those in the lock channel. Work initially was restricted to periods of low tide, and on stormy days the high tide frequently washed out the preceding day's progress.

Meanwhile engineers also repaired existing rail facilities and laid additional track to serve berths in the western part of the harbor, and finally installed lighting facilities along the quays and at the Granville railhead to make night operations possible. Rehabilitation of the port was finally completed on 6 November, by which time sixteen coaster berths had been provided, as planned, offering a discharge capacity of 5,000 tons per day. More than 363,000 man-hours of work went into developing the port to this capacity.[97]

[97] *Ibid.*, pp. 4–13; Normandy Base Section, Engineer Section History, p. 27.

As was the case at all the ports restored, the reception of cargo at Granville did not await the final completion of all construction projects. Four berths were ready by 15 September and on that date the first coaster—a Swedish vessel—entered the port.[98] Coal deliveries were rather insignificant at first, partly because the better berths in the inner basin were not ready, but chiefly because vessels had not been dispatched to the port owing to draft limitations.[99] Discharge began to average about 1,000 tons per day toward the end of October after the opening of additional berths in the Bassin à Flot.[100] But stormy weather that month kept coaster sailings down and prevented the port from coming anywhere near realizing its 3,000-ton potential.[101]

The inability to utilize Granville's facilities for coal reception more fully led to the suggestion that other types of cargo be sent to the port. It was estimated that beaching, storage, and rail facilities were such that about 6,000 tons of general cargo could be handled there without interfering with the discharge of coal. But other ports already existed farther forward on the line of communications—Rouen, for example—which could receive all the available coaster shipping. Consequently there was little point in using Granville for any other purpose than coal.[102]

Maintenance of equipment was a never-ending problem at Granville, in part because of a lack of skilled labor, in part because of the lack of tools. This problem, plus the difficulties that bad weather created in bringing vessels into the shallow harbor, kept the port's performance disappointingly low.[103] Granville occasionally topped the 3,000-ton mark, but its average discharge rate from the time of its opening to the end of January 1945 was under 1,300 tons per day.

The first cargo discharged in Brittany was brought in via open beaches at St. Michel-en-Grève, near Morlaix. The discharge of cargo there was purely an improvisation to meet an emergency requirement for supplies for the VIII Corps in its operations against Brest. Representatives of the 16th Port, which had been designated to operate the Brittany ports, had followed closely on the heels of the advancing forces in the peninsula early in August to reconnoiter all the port facilities along the northern coast. They found St. Michel suitable as a landing beach, and, lacking usable port facilities, immediately made plans to bring LST's in at this point.

The first three LST's arrived on 11 August, and unloading began on the afternoon of the 12th under the proper tidal conditions. A shortage of trucks

[98] Port of Granville History, p. 9, and Figure 38.

[99] Min, Mtg on Port Situation with CofS COMZ, G-4, TC, *et al.*, 13 Oct 44, G-4 Plant and Communications Diary/Jnl, ETO Adm 145C.

[100] At that time the 11th Port headquarters moved to Rouen and the operation of Granville was turned over to the 4th Port. History of TC ETO, V, 4th Port, 9, and 11th Port, 1.

[101] G-4 Plant and Communications Diary/Jnl, 26 Oct 44, ETO Adm 145C.

[102] Ltr, Col John H. Judd, QM Normandy Base Sec, to CO Normandy Base Sec, sub: Full Utilization of Granville as a Port, 24 Oct 44, with Inds Normandy Base Sec and Hq ETO, 8 Nov 44, EUCOM 800 Rivers, Harbors, and Waterways, I.

[103] CAO Mtgs, 14 and 17 Nov 44, SHAEF AG 337-14; History of TC ETO, V, 4th Port, 9-10.

CRANE LIFTING A LOCK GATE *from the water at Morlaix, 23 August 1944. Engineers watching the operation are from the 1057th PC&R Group.*

halted operations temporarily, but unloading resumed with the arrival of additional transport from the VIII Corps, and the discharge of 1,500 tons of rations, POL, and ammunition was completed on the following day. After a lapse of a few days LST's began to arrive fairly regularly.[104] On 19 September the 16th Port turned over St. Michel to the 5th Port, which continued to operate the beaches until the end of September.[105]

In the course of their operation the St. Michel beaches handled about 60,000 tons of supplies, much of it consisting of the emergency ammunition shipments made at the behest of Lt. Gen. Troy H. Middleton's VIII Corps for the siege of Brest. St. Michel's record was not spectacular in terms of tonnage, but the beaches served their purpose in a time of pressing need.

The port of St. Brieuc, forty miles farther east, at first was believed to possess good potentialities, for it had a well-protected landlocked harbor, and was well situated with respect to the rail and road network. It was estimated at first that the port should handle 3,500 tons per day, at least half of this in coal. But St. Brieuc, like all the small ports, could not receive Liberty ships for direct ship-to-shore discharge. Shortly after its opening in mid-September its operations were

[104] TUSA AAR, II, G-4, 11; History of TC ETO, IV, 16th Port, 1, 5-6.

[105] History of TC ETO, V, 5th Port, 1.

limited entirely to the intake of coal, which was to be used for the local generating plants and railways. In mid-October the shortages of coasters led to the decision to close the port.[106] Its discharge record had been poor, the entire cargo handled during its month of operation totaling less than 10,000 tons. On 9 November St. Brieuc was turned back to the French.[107]

Morlaix and Roscoff, the westernmost of the Brittany ports utilized by U.S. forces, like St. Michel and St. Brieuc did not figure in plans as of D Day, although they had been considered earlier. After a reconnaissance in mid-August, however, both ports showed sufficient promise to warrant their restoration and use. Roscoff and Morlaix were two separate ports, the former situated at the tip of the Penlam peninsula and the latter about twelve miles up the Dossen estuary, but they were consistently linked in all plans and were restored and operated by one headquarters. Roscoff was strictly tidal, while Morlaix, like Granville, had both a drying-out and a locked wet basin. Both had the disadvantage of all minor ports in that they could not accommodate deep-draft shipping except to provide anchorage.

The rehabilitation of the two ports consisted in the main of dredging the estuary, removing sunken craft, repairing lock gates, constructing POL reception and storage facilities, and erecting floodlights. This work was carried out by the 1057th PC&R Group. When completed, Morlaix-Roscoff provided anchorage for six Liberty ships, which were discharged into schuits, LCT's, and barges. The area initially came under the control of the 16th Port, which operated the other small Brittany ports, but discharge operations did not actually get under way until after the 5th Port assumed control on 5 September.[108] Morlaix and Roscoff remained in operation until the middle of December and were the last of the Brittany ports to be closed. In the three months of their operations the two ports averaged approximately 2,100 tons per day, although they often exceeded their target of 3,000 tons. Their average receipts therefore exceeded those of all the other minor ports, and their entire intake totaled over 200,000 tons.

Had there been any choice in the matter it is unlikely that the smaller Brittany ports would have been opened at all. In many respects they were uneconomical to operate. By the time they were sufficiently repaired to begin receiving supplies (mid-September) the front line had advanced several hundred miles eastward. None of them could discharge deep-draft ships except by lighters, of which there was never an adequate number to take full advantage of the smaller ports' capacity. Brittany's ports, which by plan were to have developed a capacity of 30,050 tons by early November, were discharging a mere 3,000 tons per day by that time. This represented but

[106] Min, Mtg on Port Situation with CofS COMZ, G-4, TC, *et al.*, 13 Oct 44, G-4 Plans and Communications Diary/Jnl, ETO Adm 145C.

[107] [Cutts] American Port Plans, pp. 5-6.

[108] Port Reconstruction and Repair, Hist Rpt 11, OCE ETO, pp. 70-72; History of TC ETO, V, Brittany Base Section, 9, IV, 5th Port, 1-2; [Cutts] American Port Plans, 5-6.

TABLE 3—DISCHARGE PERFORMANCE OF THE BRITTANY PORTS

Port	Date Opened	Date Closed	Day Operated	Long Tons Discharged	Average Daily Discharge
Granville	15 September 1944	21 April 1945	219	272,562	1,244
St. Michel	12 August1944	30 September 1944	81	60,343	745
St. Brieuc	16 September 1944	15 October 1944	30	9,521	317
Morlaix-Roscoff	5 September 1944	14 December 1944	101	212,636	2,105

10 percent of the 30,000 tons then being unloaded daily as compared with the 70 percent of total capacity which they were expected to provide. They were useful in support of the forces operating in the peninsula, of course, but these forces were relatively insignificant in number after September. The Brittany ports consequently met an almost purely interim need pending the development of deep-draft capacity farther east. Nevertheless, they contributed in relieving the deficit in port capacity at a time when approaching bad weather threatened to close the Normandy beaches, and during the period when Cherbourg's reconstruction was in progress. Their discharge performance is summarized in Table 3.

CHAPTER IV

The Port Discharge and Shipping Problems (Continued)

(1) The Seine Ports—Le Havre and Rouen

The logistic support of U.S. forces reached its lowest ebb in the month of October. At no other time during the eleven months of continental operations did the supply situation appear so unfavorable in all its aspects. This can be attributed in large measure to the unsatisfactory port situation.

In October, as expected, bad weather at last had its adverse effect on operations at the beaches and, by preventing the dispatch of shallow-draft shipping, on the full utilization of the smaller ports particularly Granville. Cherbourg's reconstruction was far from complete, and its worst bottleneck, clearance, remained unsolved. The small Brittany ports were making only a minor contribution to total U.S. needs. In the first three weeks of October, consequently, unloadings averaged less than 25,000 tons per day against an estimated requirement of about 40,000.[1] The theoretical discharge capacity of ports then in operation was only 28,000 tons.[2]

Although the final solution did not come until the opening of Antwerp, the entire port situation took an encouraging turn for the better in the first week of November. The improvement was attributable in part to the progress in overcoming the clearance problem at Cherbourg. More important, however, was the opening of two new ports, Le Havre and Rouen, which lay at least 100 miles nearer the front lines than did Cherbourg. Both ports were in operation by mid-October, making it possible to close all the minor ports in both Normandy and Brittany except Granville and Morlaix by the end of that month.

U.S. requirements went beyond the need for additional deep-draft facilities for direct ship-to-shore discharge. New coaster capacity was needed to replace existing facilities, most of which were located on lines of communication already too congested and too far to the rear. Coaster ports such as St. Vaast, Barfleur, and Isigny, for example, were

[1] Exclusive of wheeled vehicles and POL. Adm Appreciation, SHAEF G–4 Post–NEPTUNE Opns, 17 Jun 44, SHAEF 12 A Gp 370.2.

[2] SHAEF G–4 War Diary/Jnl, Basic Statistical Rpt 2, 14 Oct 44, App. 4 of Exec Br SHAEF G–4 War Diary/Jnl.

cleared by the same road and rail net which served the port of Cherbourg and the Normandy beaches. There was already a large enough accumulation of supplies in the Normandy depots to saturate the carrying capacity of those lines for months to come.[3]

There was also an urgent need for developing new bulk POL intake facilities. As of mid-October Cherbourg was virtually the only bulk POL port operating on the Continent. It possessed only one tanker berth, and that was located along the Digue de Querqueville in the Grande Rade, where stormy weather frequently interfered with berthing and discharge. A foretaste of future difficulties was given on the night of 4 October, when a storm destroyed eight of the ten unloading lines, completely shutting down intake for eight hours and materially reducing it for another twenty-four. A fairly heavy import of gasoline had been achieved thus far only by the use of large 15,000-ton tankers, but the advent of bad weather made it extremely doubtful that such vessels could continue to be handled at the Querqueville Digue. Early in October operating difficulties caused by bad weather, plus shutdowns occasioned by the failure of tankers to arrive, resulted in POL being withdrawn from the Cherbourg tank farms faster than it could be replaced, with the result that stocks again became dangerously low. Small tankers were still bringing POL into Port-en-Bessin, but there was danger that bad weather would also force a stoppage of imports there. By mid-October, then, it had become essential that POL ports be developed which would afford safe berthing in adverse weather.[4]

Finally, French officials were appealing for the use of some of the available port capacity to meet civil import requirements. There was small hope of meeting such demands in view of the urgent requirements for military purposes. Military needs of all kinds—railway rolling stock and shipping, as well as port capacity—so far exceeded the available means in October that the allocation of port capacity for French civil needs had to be postponed.[5] The opening of Le Havre and Rouen nevertheless aided greatly in halting the deteriorating port situation in October by more than compensating for the diminishing returns at the beaches and by making it possible to end the uneconomic use of most of the small ports in the Normandy and Brittany areas.

The Seine ports had figured in Allied planning from the start. Plans for the period after D plus 90 had assumed, in fact, that logistic considerations would require crossing the lower Seine and taking Le Havre and Rouen as a first priority operation after the capture of the lodgment area. Le Havre was particularly valuable. With its fourteen

[3] Ltr, Vissering, Chief of Mov by Land, Mov and Tn Br G–4 SHAEF, to Deputy Chief Mov and Tn Br, sub: Suggested Plan for Mov of U.S. Cargo Through Calais, 16 Oct 44, SHAEF G–4 825.1 Piers, Wharves, Docks, and Berths, III.

[4] Memo, Brig D. H. Bond, Chief G–4 Petroleum, SHAEF, for Crawford, sub: Dev of the Port of Le Havre, Seine River, and Rouen for Bulk POL, 27 Oct 44; Ltr, Normandy Base Sec to CG COMZ, sub: POL Situation, Normandy Base Sec, 17 Oct 44, EUCOM 463.7 Gasoline and Motor Oil IIB; CAO Mtg, 27 Oct 44, SHAEF AG 337–14.

[5] Memo, Maj Gen Charles S. Napier for CAO, sub: French Ports, 27 Oct 44, SHAEF G–4 825.1 Piers . . . , IV, COMZ G–4 Plant and Communications Diary/Jnl, 23 Oct 44, ETO Adm 145C.

basins and eight miles of quays, including facilities for the reception of large oil tankers and a pipeline to Paris developed as the result of a tremendous improvement program a few years before World War I, Le Havre had become the second port of France. Both Le Havre and Rouen, however, were intended for British rather than American use. They were expected to relieve British forces from dependence on their original Normandy landing beaches at about D plus 120, and were not to be turned over to U.S. forces until about D plus 210. By that date British lines of communication were to be based on ports farther up the coast.[6]

The rapid developments during the pursuit largely invalidated these plans. As early as 3 September the SHAEF Logistical Plans Branch recommended that the Seine ports be turned over to U.S. forces in view of the more than adequate facilities the British would soon have in such ports as Dieppe, Calais, and Boulogne, which were already uncovered, and in view of the imminent fall of Antwerp (which occurred the following day).[7] The 21 Army Group did not favor the release of Le Havre until it could be certain that Antwerp and Rotterdam would be available,[8] and for several days there was no decision on the matter. On 11 September, however, Maj. Gen. Charles S. Napier, the SHAEF Deputy G–4 for Movements and Transportation, recommended to the G–4 that Le Havre be assigned to the Communications Zone. General Crawford immediately asked the COMZ commander whether he was prepared to undertake the development of the port on the assumption that a portion of the port's capacity might initially have to be allocated to the British, and on the additional assumption that the Communications Zone in all probability might still have to open the port of Brest.[9] Supreme Headquarters did not wait for a formal reply to its query, but on General Lord's statement that the Communications Zone could assume the responsibility, notified General Lee on the 13th that it had decided to assign Le Havre to the Communications Zone, and informed him that the Allied Naval Commander was prepared to send two Royal Marine Engineer companies to assist in the work of rehabilitation.[10]

The COMZ commander was still dubious about the value of Le Havre. In his analysis of the port problem he noted that the port would contribute in only a limited degree to the shortening of the lines of communication. Only Antwerp and the other northern ports, in his opinion, could satisfy that need and provide the capacity required. He therefore thought it advisable that Le Havre be developed solely as an interim port with a capacity of between 8,000 and 10,000 tons, and with a minimum expenditure of effort.[11] Theater officials

[6] SHAEF Plng Study, sub: POST-NEPTUNE, Course of Action After Capture of the Lodgment Area, 30 May 44, SHAEF G–3 War Diary.

[7] Memo, Whipple for CAO, sub: Port Dev, 3 Sep 44, SHAEF G–4 825.1 Piers . . ., III; SHAEF G–4 War Diary/Jnl, 3 Sep 44.

[8] Cbl, 21 A Gp to ANCXF, 7 Sep 44, MA9, SHAEF G–4 825.1 Piers . . ., III.

[9] Memo, Napier for G–4, 11 Sep 44, and Cbl FWD–14635, Crawford to Lee, 12 Sep 44, SHAEF G–4 825.1 Piers . . ., III.

[10] Cbl FWD–14711, SHAEF to Lee, 13 Sep 44, SHAEF G–4 825.1 Piers . . ., III.

[11] Ltr, Lee to SHAEF, 14 Sep 44, COMZ G–4 Plans and Communications Diary/Jnl, ETO Adm 145C.

concurred in this view, and embodied the COMZ commander's recommendation regarding Le Havre in the port development plan which finally crystallized toward the end of the month.[12]

General Lee's misgivings as to the probability of developing a large discharge capacity at Le Havre in the near future stemmed in part from his knowledge of the port's condition. The enemy garrison at Le Havre had resisted to the bitter end, and had forced the Allies to subject the city to heavy bombardments from the sea, land, and air for a full week before it capitulated. When First Canadian Army forces entered Le Havre on 12 September, therefore, they found one of the most thoroughly demolished ports captured thus far. Port facilities had been destroyed with characteristic thoroughness. In addition, the repeated bombings had destroyed approximately two thirds of the city's business and residential sections and caused an estimated 6,000 civilian casualties. The bombings had also created an understandable resentment among the city's inhabitants toward their liberators, and at best an indifference to the activities of the units which shortly arrived to rebuild the port.[13]

The damage to Le Havre's port facilities followed much the same pattern as at Cherbourg and Granville. In one sense the destruction was actually more serious at Le Havre, for most of the port's facilities had centered around the wet basins, and a quick survey on 13–14 September disclosed that all the lock gates had been damaged and the basins rendered inoperable. This had the further result of subjecting all the basins to tidal action, and the hydrostatic pressure caused by the tides in turn caused many of the quay walls to fail.[14] In addition, there were the usual obstructions—the many sunken craft in the various channels and basins, demolished cargo-handling equipment, and bombed-out warehousing.[15]

COMZ officials initially established two general priorities for the rehabilitation of the port: the immediate development of tonnage reception of 1,500 tons per day from Liberty ships by the use of dukws and lighters, and then an increase in the port's discharge capacity to 7,000 tons, but without a major reconstruction effort.[16]

The principal engineer units dispatched to Le Havre to accomplish this mission consisted of two general service regiments, two PC&R groups, a port repair ship crew, a gas generating unit, a maintenance company, a dump truck company, and two Royal Marine Engineer companies. All were placed under the operational control of the 373d Engineer General Service Regiment, commanded by Col. Frank F. Bell. Work began on 20 September under a three-phase program drawn up by the commanders of the 373d Regiment and the other engineer units, and by the Channel Base Section Engineer.[17]

The first-phase program consisted

[12] Ltr, Hq ETO to CG COMZ, sub: Alloc of Ports, 19 Sep 44, SHAEF G–4 825.1 Piers . . ., III.

[13] Channel Base Section History, I, 207, ETO Adm 588; History of TC ETO, V, 16th Port, 40–41.

[14] *Final Report of the Chief Engineer, ETO,* I, 275.

[15] Port Construction and Repair, Hist Rpt 11, OCE ETO, 26ff.

[16] Ltr, Hq COMZ to Chiefs of Svs and Stf Secs, sub: Dev of Continental Ports, 27 Sep 44, EUCOM 400 Supplies, Svs, and Equipment, V.

[17] History of TC ETO, V, 16th Port, 8–9.

mainly of clearing the beaches of mines, wire, and tetrahedra; preparing landing sites for dukws and various types of landing craft; opening access roads connected with the inland highway net; and providing storage areas near the beaches. The second-phase program, which got under way in October, involved continuing work on roads and railways, repairing damaged quays, removing sunken vessels, repairing existing POL facilities and lock gates, and reconstructing storage. In some cases the removal of debris and the restoration alongside quays would have involved too time-consuming an effort. Instead, it was decided to provide entirely new berthing by the construction of a floating ponton pier and piers made with caissons diverted from the MULBERRY winterization program.[18] Third-phase work, which carried on into December, in many cases consisted simply of the continuation of earlier projects and the provision of various complementary facilities such as lighting and refrigeration.

Clearance of the beach areas had progressed sufficiently to permit the entry of a few vessels on 2 October. These were unloaded within the next few days by the 16th Port (Brig. Gen. William M. Hoge), which had arrived from Brittany to take over the operation of Le Havre.[19] The discharge rate was negligible in the first two weeks, for the presence of mines in the harbor temporarily limited the port's use to LCT's and coasters. As at Cherbourg, this resulted in delays in bringing some of the port equipment ashore, for portions of it were aboard Liberty ships and could not be transferred to dukws.[20] Lack of proper cargo-handling and clearance facilities contributed to the awkwardness of operations at first, necessitating multiple handlings of supplies.[21] On 13 October three Liberties were finally ordered forward, after which discharge improved steadily. In its first full week of operations Le Havre discharged about 2,000 tons per day. In the second week the average rose to 3,650, double the tonnage expected of it at that date.[22]

Progress in rehabilitation was excellent, and the encouraging discharge performance in the first weeks quickly dispelled earlier doubts concerning the port's value. As early as 23 October the plan of operations which the port commander proposed to the base section commander raised Le Havre's target to 9,100 tons (exclusive of POL and coal), which was to be achieved within thirty days.[23] Within another week the port was averaging about 5,000 tons, and on the last day of the month it fell just short of the 6,000-ton mark. By that time the value of Le Havre and Rouen was clearly recognized, not only because of their advantages over Cherbourg, but because they served as insurance against the possible neutralization of Antwerp by mine laying and by the enemy's use of V–1's and V–2's. The two ports took on a significance beyond all expectation as the opening date of Antwerp receded

[18] *Ibid.*, pp. 9–18; *Final Report of the Chief Engineer, ETO*, I, 275.

[19] General Hoge commanded the port at Le Havre only until 21 October, when he went to a new assignment. After an interim period of ten days, command of the 16th Port went to Col. Thomas J. Weed. History of TC ETO, V, 16th Port, 6.

[20] History of Channel Base Section, I, 207.

[21] History of TC ETO, V, 16th Port, 33, 36.

[22] CAO Mtg, 27 Oct 44, SHAEF AG 337–14.

[23] History of TC ETO, V, 16th Port, 30.

DUKWS TRANSFERRING CARGO *from ships to boxcars, Le Havre, 15 November 1944.*

further and further into the future. Le Havre and Rouen had proved to be the biggest nest egg since the capture of Cherbourg.[24] Early in November, therefore, the Communications Zone directed the Channel Base Section to develop still greater capacity at Le Havre, raising the target to 9,500 tons by 1 December.[25]

Le Havre's discharge capacity actually exceeded this target by the end of December, although full advantage could never be taken of its capacity. Port officials had recognized that the crux of the problem at Le Havre, as at Cherbourg, would be the ability to clear. Limitation on clearance resulted from the fact that all traffic from Le Havre had to cross British lines of communication, from the usual shortage in rail cars and locomotives, and from the lack of alongside berths where cargo could be transferred directly to rail. Despite these handicaps Le Havre continued to make a handsome contribution to the total tonnages discharged on the Continent, averaging more than 5,000 tons per day throughout November, and 5,400 tons in December. In January the port bettered this record with a daily average of 6,470 tons.

Le Havre never developed the number of alongside Liberty berths planned, for the use of caisson piers proved unsuccessful. An abnormally large percentage of the tonnage therefore continued to be brought ashore by lighters

[24] Memo, Col James McCormack, Chief of Mov Br G–4 12 A Gp, to Barriger, 31 Oct 44, 12 A Gp Tonnage 137.

[25] Ltr, Hq COMZ to CO Channel Base Sec, sub: Dev of Port Facilities and Port Clearance—Le Havre and Rouen, 6 Nov 44, EUCOM 825 Tunnels and Pipelines, Docks, Piers, Jetties.

and dukws, the latter alone accounting for 33 percent.[26] The extensive use of lighters had its disadvantages, of course, including multiple handling and interruptions from bad weather. The continued use of dukws brought its inevitable maintenance problems. One amphibian company estimated that its vehicles had operated the equivalent of 70,000 miles. Inability to replace worn-out vehicles and the lack of spare parts led to widespread cannibalism and other expedients, such as the manufacture of propeller strut bearings from applewood and rudders from scrap steel. At times in November the amphibian companies were operating with 76 percent of their vehicles deadlined.[27] In mid-January the port organization was strengthened by the arrival of the 52d Medium Port. Le Havre then reached a peak strength of about 20,000 men, of whom 4,000 were French civilians and an undetermined number were prisoners of war.[28]

Meanwhile, the port of Rouen, the ancient Norman capital lying seventy-five miles up the Seine River, had also helped relieve the deficit in port capacity by developing a discharge of several thousand tons. Rouen had been captured on 30 August, but obviously could not be utilized until Le Havre had also been taken and the Seine estuary cleared. Damage to this port hardly compared with that at Le Havre. Fortunately its quays were largely intact. But its cargo-handling equipment, such as cranes, was completely demolished and many vessels had been sunk, both along the quays and in the Seine channel between Rouen and Le Havre. The principal rehabilitation task consisted of removing these sunken craft and cranes and fell mainly to the U.S. Navy, aided by French civilians. Storage space at Rouen, both covered and open, was excellent, and the major engineer task there proved to be clearing debris, filling bomb craters, erecting cranes, and reconstructing railways.[29]

The COMZ directive of 27 September had established a target of 3,000 tons per day for Rouen. All cargo had to be discharged from coasters, since the port was accessible only to ships with a maximum draft of from nineteen to twenty-five feet. Although the rehabilitation of Rouen did not get under way until the beginning of October, the port was ready to receive cargo on the 13th, the same day on which Le Havre took its first Liberties. The port discharged its first supplies three days later. For the first few days the operation of the port was carried out by a detachment of the 16th Port, sent over from Le Havre. On 20 October the 11th Port, which had operated the minor Normandy ports until a few days before, arrived to take over Rouen.[30]

By the end of the month Rouen was handling well over 2,000 tons per day, and in the first week of November averaged more than 4,000 tons. The port's encouraging development soon made it possible for Le Havre to cease discharging coasters. On 8 November the Communications Zone went a step further

[26] History of TC ETO, VI, 16th Port, 119.

[27] *Ibid.*, V, 16th Port, 27–28.

[28] *Ibid.*, VI, 16th Port, 119.

[29] *Ibid.*, V, 16th Port, 119; *Final Report of the Chief Engineer, ETO*, I, 275–76; Engr Hist Rpt 11, pp. 37ff.

[30] The 11th Port relinquished control of Granville on 24 October, when the 4th Port of Cherbourg took charge there.

QUAY AT ROUEN LOADED WITH INCOMING SUPPLIES, *4 December 1944.*

and ordered all coasters except those carrying coal sent to Rouen,[31] making it possible to close all the shallow ports in Normandy and Brittany. At times Rouen took Liberty ships after they had been lightened to the proper draft at Le Havre.[32] Meanwhile the Communications Zone raised Rouen's discharge target to 7,500 tons.[33] Rouen did not meet the new target, but did perform very creditably in the next few months, averaging 4,200 tons per day in November, 4,138 in December, and 5,072 tons in January. Early in 1945 the port's operating strength rose to 9,000 Army personnel, 9,000 prisoners of war, and 5,000 civilians.[34]

[31] [Cutts] American Port Plans, II, 34-35.
[32] History of TC ETO, V, 11th Port, 2-3.
[33] Ltr, COMZ to Channel Base Sec, 6 Nov 44.
[34] History of TC ETO, VI, 11th Port, 58.

Le Havre and Rouen together averaged approximately 8,500 tons per day in November, the month preceding the opening of Antwerp, and accounted for approximately one third of all tonnage discharged in that period. The total daily discharge on the Continent had risen from approximately 25,000 tons per day to 27,300 tons since mid-October.[35] Le Havre and Rouen did not provide the final solution to the port problem, therefore, but they more than made up for the loss of the beaches, and placed a substantial portion of the total U.S. discharge capacity a sizable distance farther forward on the line of communications, relieving the desperate shortage in transport and serving as an important stopgap pending the opening of Antwerp.

[35] See Table 4, p. 124, below.

(2) Antwerp and the Other Northern Ports

Although it was not to receive cargo for nearly three months, Antwerp became the master key to all port policy and plans after its capture early in September. The advantages which Antwerp possessed over other ports had long been appreciated, and once the port was in Allied hands both the 12th Army Group and the Communications Zone lost little time in urging SHAEF to allocate a portion of the port's capacity to U.S. forces.[36] General Bradley strongly hoped that the early opening of the port would assure a constant source of supply for the armies and thus obviate the necessity for a long build-up period and postponement of the offensive.[37]

Small wonder that Antwerp's capture raised hopes of solving the Allies' long-standing logistic problem. Antwerp ranked with Hamburg, Rotterdam, and New York as one of the world's great ports, even though it did not approach the size of those cities in population. In 1938 alone it had registered 12,000 vessels and handled almost 60,000,000 tons of freight.

Antwerp is an inland port, situated on the right bank of the Schelde estuary, fifty-five miles from the sea. Unlike other ports on tidal streams, it could receive large seagoing vessels at all stages of the tide, for even the minimum depth along the quays in the river was twenty-seven feet. Furthermore, the Schelde was more than 500 yards wide at Antwerp and thus permitted easy maneuvering of the largest ships.

Antwerp's port facilities were located partly along the river itself and partly in a complex of wet basins built off one side of the river. Approximately three and one half miles of quays lined the right bank of the Schelde, forming the western limits of the city. The greater portion of the port lay to the north of the city and consisted of eighteen basins, to which access was obtained through four locks. These basins provided nearly twenty-six miles of quays. The port therefore offered more than twenty-nine miles of quays, and these were equipped with more than 600 hydraulic and electric cranes, plus numerous floating cranes, loading bridges, and floating grain elevators.

Storage accommodations were commensurate with these modern discharge facilities. There were nearly 900 warehouses, plus a granary with a capacity of almost a million bushels, and cold storage chambers with about 750,000 cubic feet of capacity. Petroleum intake and storage installations were also on a grand scale. Pipelines ran directly from tanker berths to the 498 storage tanks, which occupied 208 acres and could hold 124,000,000 gallons of POL.

Equally important were the excellent clearance facilities. Antwerp alone possessed more than 500 miles of rails, plus ample marshaling yards, and was well tied in with a Belgian transportation network consisting of 3,250 miles of railways and 1,370 miles of navigable waterways, including the Albert Canal, which

[36] Memo, Moses for Crawford, 10 Sep 44, SHAEF AF 323.3-2 Ports (Captured Ports); COMZ G-4 History, I, 45; Ltr, Lee to SAC, 14 Sep 44, EUCOM 800 Rivers, Harbors, and Waterways.

[37] Conf at COMZ, 18 Sep 44, G-4 History, I, 46.

connected Antwerp with the Meuse River.[38]

Antwerp therefore had a potential capacity which completely dwarfed that of all the other ports the Allies thus far had captured and put to use. Furthermore, Antwerp's appearance upon capture was in startling contrast to that of Cherbourg and Le Havre, for the port had suffered only minor damage. The swiftness of the British advance had allowed the enemy little time to prepare demolitions, and the port had been saved from complete destruction largely through the gallant action of a Belgian reserve lieutenant employed in the port administration who had worked out a tactical plan which effectively frustrated the enemy's attempted demolitions.[39]

Nor had Antwerp sustained the damage which Le Havre had suffered from Allied bombardment. The only demolitions of any importance had been carried out on the locks, thus preventing immediate use of the wet basins. But the quays lining the river were in good condition and, subject to the removal of two small coasters which had been sunk in the estuary, could accommodate twenty-three Liberty ships. Practically all of the port's 625 cranes and other unloading machinery were found in working order, and warehouses and sheds were also largely intact. Damage to the rail lines was limited to three demolished bridges, and there was a large quantity of immediately usable rolling stock in the port. In addition, local authorities reported an adequate supply of trained boat crews, crane operators, mechanics, and dock labor to man the port.[40]

Supreme Headquarters concluded that Antwerp's potentialities were ample to meet both U.S. and British needs. On 19 September SHAEF instructed General Lee to send a senior planner to 21 Army Group headquarters immediately to work out plans for the base layout and for sharing the port's facilities.[41] General Lee arranged a conference between COMZ and 21 Army Group officials, and representatives of the two headquarters met at Antwerp between 24 and 26 September. Tentative agreements were reached on the allocation of tonnage capacity and storage facilities, the use of rail lines, and on the port's command and administrative organization. A temporary division of the port between U.S. and British forces was also made, pending its complete rehabilitation, and agreement was reached as to the responsibility of the respective forces for undertaking various repair and reconstruction projects.[42] There was still considerable optimism over the prospect that Antwerp might soon be in operation, even though the approaches to the port were still in enemy hands.

The capture of Antwerp had not ruled out the use of other ports which had been uncovered along the Channel and

[38] History of TC ETO, VI, 13th Port, 74–77; Hist Rpt 11, OCE ETO, pp. 40–42.

[39] Maj Gen Sir Colin Gubbins, "Resistance Movements in the War," *Journal Royal United Service Institution*, XCIII (May, 1948), 219.

[40] Engr Hist Rpt 11, p. 43.

[41] Ltr, SHAEF TO CG COMZ, sub: Alloc of Ports, 19 Sep 44, EUCOM 800 Rivers, Harbors, and Waterways, I.

[42] Ports—Development of Antwerp, prep by Lt A. Gratiot, G–4 Sec COMZ, Jan 45, MS, ETO Adm 244; Ltr, Lt Col E. W. Weber, Roads, Railroads Br G–4 COMZ, to Potter, sub: Alloc of Facilities of Port of Antwerp, 27 Sep 44, ETO Adm 145C.

North Sea coast during the September drive. The Communications Zone had not lost sight of their possible value, and in the port plan which it submitted to SHAEF on 14 September recommended that all requirements be reviewed when the other north coast ports were captured so that an equitable apportionment of their facilities could be made.[43] At the end of the month the COMZ directive on port development specifically mentioned Calais, Boulogne, and Ostend as ports which might either be assigned to U.S. forces or shared with the British, and announced that the Channel Base Section and Navy would reconnoiter the entire coast east of Le Havre, particularly to find suitable beaching sites for LST's.[44] The latter were urgently needed to replace the OMAHA and UTAH Beaches as points of entry for tracked vehicles. Channel Base Section reconnoitered four of the ports—Dieppe, Le Tréport, Boulogne, and Ostend—within the next few days.

On 5 October U.S. and British officials met again, this time at Brussels, to review the entire port situation in the northeast and specifically to consider how the smaller ports could be utilized pending the opening of Antwerp. By this time the optimism over Antwerp had declined noticeably, and a target date of 15 November was accepted for the opening of the port. Le Havre and Rouen were still unknown quantities at this date. On the basis of either intelligence reports or reconnaissances already carried out, therefore, about ten of the smaller north coast ports were carefully studied during the four-day conference. A few of the ports, such as Dunkerque and Zeebrugge, were still in enemy hands; several others, like Nieuwpoort, Gravelines, Blankenberge, and the Calais–Ostend reach, which was considered as a possible landing site for vehicle-carrying LST's, were ruled out for one reason or another.

Only four ports were accepted as having potential value. These were Boulogne, Dieppe, and Ostend, which were assigned to the British with the reservation that U.S. forces should share in POL reception at Ostend, and Calais, which was assigned to the Americans. The Communications Zone concluded on the basis of a reconnaissance that Calais could provide the needed LST berths for vehicle discharge, and Channel Base Section made preparations for certain improvements there, including plans for the construction of an unloading ramp. But it failed to carry through on these plans, and when 21 Army Group requested permission to construct a train ferry terminal there, and later a hard for a vehicle discharge, SHAEF promptly approved. On 23 October it designated 21 Army Group as the agency henceforth responsible for developing, operating, and administering Calais, although it was understood that the Communications Zone would share in the use of the port.[45]

None of the ports mentioned above had anything but limited value and they were obviously no substitute for Ant-

[43] Ltr, Lee to SAC, 14 Sep 44.

[44] Ltr, Lord to Chiefs of Gen and Special Stfs and Base Sec Comdrs, 26 Sep 44.

[45] [Cutts] American Port Plan, II, 37–45; Cbls, 21 A Gp to Gale, 18 and 21 Oct 44, SHAEF AG 323.3–2 Ports (Captured Ports); Cbl, S–63627, G–4 SHAEF to 21 A Gp, 23 Oct 44, SHAEF G–4 825.1 Piers . . ., IV.

werp. Small wonder that the Supreme Commander and his logistic planners became more and more anxious, therefore, as the supply situation deteriorated in October without any sign that the great port might soon be placed in operation. The current concept of Antwerp's importance was well expressed by Colonel Whipple, the chief of the SHAEF Logistical Plans Branch, early in October. "The failure to open Antwerp," he wrote, "is jeopardizing the administrative soundness of our entire winter campaign. The placing of this operation as second priority within 21 Army Group had temporary justification while the Northern salient was being reinforced, but I see no excuse for it now. The present lack of support of troops of [the] US 3rd and 9th Armies and minimum support of 1st Army cannot be rectified until Antwerp is opened. Fifteen divisions are held impotent for lack of success in this relatively small operation, and this weakness may involve us in winter weather to such an extent that our advance into Germany may be delayed until spring." [46] In Colonel Whipple's view it was imperative that "21 Army Group be directed to place the clearing of Antwerp as highest priority and to make such other adjustments as are essential to insure it will not be further delayed."

General Eisenhower needed no convincing in this matter. He had assessed the true value of Antwerp some time before, and had repeatedly impressed on the 21 Army Group commander the urgency of getting the port into operation. By early October the Supreme Commander had become impatient and alarmed over the protracted attention which Field Marshal Montgomery was giving the Nijmegen bridgehead at the expense of the Schelde operation. Clearing the Antwerp approaches simply could not be postponed any longer, and in the second week of October General Eisenhower insisted that Montgomery give unequivocal first priority to that operation. Operations designed to clear the mouth of the Schelde were initiated by the First Canadian Army a week later, and were completed in the first week of November.[47]

It was with an obvious sense of relief that the Supreme Commander saw the 21 Army Group finally turn to the Schelde operation. Eisenhower had predicted that operations would come to a standstill if the port were not in operation by mid-November. In fact, he had emphasized the importance of Antwerp so frequently that General Marshall expressed fear that the ETOUSA commander was putting all his eggs in one basket. Late in October he cautioned General Eisenhower against relying too heavily on a single port, particularly in view of its vulnerability to rocket attacks. With the large concentration of shipping and supplies there, particularly ammunition, Antwerp offered a lucrative target, and enemy attacks would do tremendous harm to the war effort if

[46] Memo, Whipple for G–4, sub: Capture of Approaches to Antwerp, 8 Oct 44, SHAEF G–4 825.1 Piers . . ., III; SHAEF G–4 Log Plans Br Diary/Jnl, 8 Oct 44.

[47] Cbls S–61466 and S–61621 Eisenhower to Montgomery, 9 and 10 Oct 44, Eyes Only Cbls, Smith Papers; Ltr, Eisenhower to Marshall, 15 Oct 44, OPD Exec Office File 9, Bk 23.

offensive operations depended on this one vital but vulnerable asset.[48]

General Eisenhower immediately assured the Chief of Staff that, far from relying solely on that port, SHAEF was taking steps to develop every other port to its maximum capacity.[49] Le Havre and Rouen had in fact just come into operation; the bottleneck was soon to be broken at Cherbourg; and there even were hopes that before long Marseille might develop capacity in excess of the 6th Army Group's needs and help support the Third Army. It was expected, however, that Antwerp's opening would provide the additional capacity required to build forward reserves and to receive and maintain additional divisions in the line.[50] It would also make possible the discharge of ships carrying engineer and quartermaster Class II supplies and vehicles, all of which had had low priority in the preceding months because of the emphasis which had unavoidably been placed on the discharge of rations and ammunition. Finally, the opening of Antwerp would make it possible to eliminate the tremendous backlog of shipping waiting to be accepted in European ports.[51]

In the meantime U.S. and British forces had proceeded with the necessary rehabilitation of the port, and COMZ and 21 Army Group officials worked out detailed plans for administering the port and sharing its facilities. SHAEF had settled the issue of control of the port at the time it agreed that Antwerp should be developed for both U.S. and British use. General Lee had suggested joint control.[52] But the Supreme Commander had decided that the port would be opened under British control, since experience had shown joint operation of a port to be unsatisfactory.[53]

The manner in which this control was to be exercised was first outlined at the Antwerp meeting of 24–26 September. Additional matters, such as the labor situation at Antwerp, were taken up at the Brussels Conference on 5 October. The Communications Zone immediately thereafter sent qualified personnel to the various Anglo-American committees which were established to work out the detailed plans for the use of the port. Planning the clearance alone of a port the size of Antwerp was a tremendous undertaking, for it required close coordination to make the most efficient use of rolling stock and railway running rights, especially in view of the fact that the port was to be used by both the Americans and British.

The operational plan which American and British experts worked out in these weeks was finally formalized on 18 October in a "Memorandum of Agreement" signed by Miles H. Graham, major general in charge of administration, 21 Army Group, and by Colonel Jacobs, the Channel Base Section commander, who had been the chief COMZ representative in the negotiations with the British. The stated purpose of the agreement was to establish the basic plan and procedure for the development of the

[48] Cbl W–51862, Marshall to Eisenhower, 25 Oct 44, Eyes Only Cbls, Smith Papers.

[49] Cbl S–64077, Eisenhower to Marshall, 26 Oct 44, OPD Cbl Files.

[50] *Ibid.*

[51] COMZ G–4 History, I, 66.

[52] Ltr, Lee to Eisenhower, 14 Sep 44.

[53] Ltr, SHAEF to Lee, 19 Sep 44.

maximum capacity of the port, and to provide the necessary facilities and controls for road, rail, and inland water transport in order that the requirements of both forces might be met with minimum cross-haul and interference. A target date of 15 November was established for the completion of all work necessary for the opening and operation of the port.

Naval command of Antwerp, which included the control of shipping within the port, was vested in the Royal Navy; and the Naval Officer in Charge, or NOIC, was initially designated as the Chairman, Port Executive Committee. A British base sub area commander was made responsible for the local administration of the Antwerp area, and the defense of the port by air, land, and sea, was also a British responsibility. So far as berthing facilities were concerned, the port's inner basins were simply geographically divided, the northern section of the port being allocated to the Americans, and the southern to the British. It so happened that each of these areas also had a marshaling yard. The river berths were unassigned, their allocation being left up to the Port Executive Committee depending on current needs and on the basis of tonnage allocations established by SHAEF. At the first Antwerp meeting at the end of September a tentative division of tonnage capacity had allocated 25,000 tons to the Americans and 15,000 to the British. This was now changed to 22,500 and 17,500 tons respectively, exclusive of POL.

Joint use of Antwerp's facilities was thus avoided so far as possible, and in that part of the port specifically allocated to U.S. forces there was to be a U.S. Army officer designated as port commander. Above him was the Channel Base Section commander, who was responsible for the "coordination, control, and the administration" of all U.S. forces in the area. There were many "common use" facilities and installations, of course, such as POL, coal, grain, cold storage, signal, and repair, and where such facilities were jointly used the commandant of the area specifically allocated to the British was to co-ordinate British and U.S. activities in consultation with the U.S. port commander, both of whom were members of the Port Executive Committee.

The great importance of port clearance was recognized in the provision of a joint British-U.S. Movements and Transportation Committee, which was to plan and co-ordinate all traffic by road, rail, and inland waterway, and handle all dealings with Belgian transport organizations. The agreement of 18 October made an initial allocation of transportation facilities to the two forces, however. It gave U.S. forces primary rights over highways southeastward to Liège and Namur, and the British primary rights to Brussels; assigned to the Americans control over railways south and southeastward to Liège and (via Brussels) to Namur and Luxembourg, and to the British control over lines running to the north and northeast.[54]

The rehabilitation of facilities required for maximum operation of the

---

[54] Memorandum of Agreement on the Operation of the Port and the Clearance Therefrom for the Maintenance of British and U.S. Armies, signed by Jacobs and Graham, 18 Oct 44, in History of TC ETO, V, 13th Port, 1–5.

port was the responsibility of the 21 Army Group, although it was agreed that U.S. forces could be called on for whatever assistance was necessary to meet the 15 November deadline. U.S. forces undertook several projects, including the repair of the vital Kruisschans Lock, which led most directly into the American portion of the port, mine clearance in the inner basins, clearance and minor repairs to the quays and transit storage sheds, and repair and reconstruction of road and rail facilities assigned to the Americans.[55]

These projects were not fully completed by 15 November, but it was not for this reason that the port could not open on that date. Clearing the mines from the Schelde proved a time-consuming task, as at Cherbourg, and was not completed until 26 November. At that time much still remained to be done in the port itself, but of the 242 berths in the port 219 were completely cleared, all of the 600 cranes were in operating order, and all bridges needed for operations had been repaired.[56]

The Communications Zone had nominated seventy-odd ships for entry into Antwerp in the first ten days, almost all of them commodity-loaded—that is, loaded with a single type of supply, such as engineer supplies.[57] The long-awaited opening of the port finally took place on 28 November, when the *James B. Weaver*, a Liberty ship carrying personnel and organizational equipment for the port headquarters and a party of war correspondents, was berthed. Thirteen vessels entered the port on the following day, and seven more on the 30th.[58]

U.S. operations at Antwerp were organized and controlled by the 13th Major Port, which had operated briefly at Plymouth and Falmouth before moving to the Continent in October. When operations actually got under way the 13th was reinforced by the 5th Port, which began arriving from Brittany at the end of November.[59] The entire U.S. organization was commanded by Col. Doswell Gullatt, who had already had wide experience in both marine construction and port operations. Colonel Gullatt had commanded the 5th Engineer Special Brigade at OMAHA Beach, and earlier in his Army career, as District Engineer at Mobile, Alabama, had had varied experience in the construction of piers and docks, in canal dredging operations, and in general construction work.

Logistic planners estimated that U.S. discharge at Antwerp should reach 15,000 tons per day in December, 21,500 tons in January, and finally achieve the full tonnage allocation of 22,500 in March. Unloadings built up to the planned rate very rapidly. By the end of the first week the 13th Port had reached the 10,000-ton mark, and in the second week of December the port was already averaging 19,000 tons per day and accounting for approximately 48 percent of all the U.S. tonnage discharged on the Continent (exclusive of Marseille on the southern line of communications).[60]

The port had hardly achieved this per-

[55] Gratiot, Ports—Development of Antwerp.

[56] Final Report of the Port Executive Committee, 26 Nov 44, in History of TC ETO, V, 13th Port, 7–8.

[57] Cbl EX–65464, Lee to Gross, 20 Nov 44, in History of TC ETO, V, 13th Port, 6.

[58] History of TC ETO, V, 13th Port, 11.

[59] *Ibid.*, V, 13th Port, 1, and V, 5th Port, 6; VI, 13th Port, 77.

[60] See Table 4, p. 124, below.

formance when clearance became a bottleneck just as it had at Cherbourg. Clearance had not become a limiting factor through oversight. Mainly because of it, in fact, logistic planners had planned a maximum combined import at Antwerp of only 40,000 tons per day, knowing that the port possessed capacity far in excess of this target. The principal limitation so far as U.S. operations were concerned was in storage. Antwerp, with all its magnificent facilities, lacked sufficient warehousing to permit any sizable backlogging of cargo in the port itself, for it had been the practice in peacetime to clear incoming cargo via rail, highway, and canal immediately after it was unloaded.

It was evident after the first reconnaissance that there would not be adequate covered or open storage in Antwerp to satisfy both American and British needs. Antwerp, lying in the British zone, was a logical base for the support of 21 Army Group, and British officials, realizing the inadequacy of storage facilities in the area had opposed the establishment of U.S. base installations there.[61] Only a small amount of storage space, all of it uncovered, was allocated for American use, therefore, purely for intransit purposes, on the theory that all U.S. cargo would be promptly dispatched to depots in Liège and Namur or as near that area as possible.

COMZ supply planners were under no illusions as to the probable implications of this deficiency. Colonel Potter, the G–4 plans chief, estimated that an accumulation of more than 15,000 tons (less than a day's intake) would create a serious obstacle to further unloading and outshipment. Since there appeared to be no immediate solution to the problem, however, it was decided to hope for the best, and if it proved impossible to phase in certain types of cargo (particularly engineer supplies) at the rate at which forward depots could receive it, to "pile the stuff on the ground, and brace ourselves for the repercussions to come." [62]

Within two weeks of the port's opening about 85,000 tons of cargo had already accumulated in sheds and under tarpaulins back of the quays, threatening to hamper unloading operations. Storage space for 100,000 tons of supplies was being utilized in the U.S. section of the port, and space for an additional 50,000 tons was granted in the British area in December. It was estimated that in another ten days operations would be entirely dependent on the ability of the port to clear tonnage as discharged.[63]

The difficulty was attributable in part to the shortage of railway rolling stock, particularly in the first days after the opening of the port. Clearance by rail improved after the middle of the month, and eventually accounted for approximately 45 percent of the 313,500 tons cleared in the first month.[64] Clearance by barge, however, fell considerably below expectations. It had been hoped that fully a third of the port's intake could

[61] Summary of Events and Decisions 20 Sep to 30 Sep 44, in COMZ G–4 Plant and Communications Diary/Jnl, ETO Adm 145C; Gratiot, Ports—Development of Antwerp.

[62] Memo, Potter for Stratton, 18 Nov 44, COMZ G–4 Plant and Communications Diary/Jnl.

[63] Ltr, Lt Col R. W. Reisner, SHAEF G–4 Mov and Tn Br, to G–4, sub: Rpt of Trip to Antwerp, 20 Dec 44, SHAEF G–4 825.1 Piers . . ., V.

[64] Channel Base Section History, I, 225; History of TC ETO, V, 13th Port, 13.

Ships Discharging Cargo for Clearance by Rail, *Antwerp, 22 December 1944.*

be cleared in this manner, principally via the Albert Canal, which was to have opened by 15 December.[65] Delays in the removal of obstructions, particularly the wrecked Yserburg Bridge at the entrance, postponed the opening of the canal until 23 December, by which date 198 loaded barges had accumulated. There was some movement via canal after that date, but the total shipments by inland waterway totaled only 48,000 tons in December, equal to 15 percent of the total tonnage cleared.[66]

The clearance problem had barely shown signs of improving when it was aggravated afresh. Late in December the enemy counteroffensive, which threatened to overrun the advance U.S. supply installations in Belgium, caused the Communications Zone to place an embargo on all shipments to ADSEC depots. The embargo applied to barge as well as rail shipments, with the result that large numbers of loaded barges and rail cars began to accumulate in the port area. By 4 January nearly 3,500 loaded freight cars were awaiting dispatch, and the entire port had become seriously congested. At the ADSEC depots, meanwhile, thousands of cars were being held under load so that forward stocks could be kept mobile.[67]

To relieve the pressure on both Antwerp and the ADSEC depots General Plank, among others, had advocated the acquisition of overflow storage facilities in other Belgian cities as early as November, but without success. In January the backlog at Antwerp was finally relieved by opening such facilities in the Lille area, and by lifting the embargo. But the congestion was not easily cleared up, and the number of rail cars ordered for loading at the port consistently fell short of the number required.[68]

Antwerp's discharge rate inevitably reflected these difficulties. After attaining an average of 19,000 tons in the second week of December the port's intake fell to 13,700 tons per day for the remainder of the month, and to approximately 10,500 tons in the first half of January. It mattered little, therefore, that the port had a *discharge capacity* of between 80,000 and 100,000 tons per day as long as inadequate transportation and depot facilities limited clearance.[69]

Antwerp operated under another handicap which precluded its providing an ideal solution to the port problem. Long before it actually began to accept cargo the port came under attack from the enemy's vaunted secret weapons—pilotless aircraft and rockets. The Nazis had begun to employ the V–1's and V–2's in mid-October, and warned the people of Antwerp that they would send 3,000 planes over their city on the day the first Allied ship entered the port.[70] This threat was not fulfilled, but the enemy did maintain an almost constant rain of the dreaded missiles on Antwerp and other cities of Belgium for more than five months, terrorizing the population

---

[65] Gratiot, Ports—Development of Antwerp.

[66] Channel Base Section History, I, 225; History of TC ETO, V, 13th Port, 13–14.

[67] Channel Base Section History, I, 42–43, 239.

[68] Cbl E–88536, ETO to Channel Base Sec, 20 Jan 45; Cbl CBSG–4–1900, Channel Base Sec to TC ETO, 21 Jan 45; Cbl E–89714, G–4 ETO to Channel Base Sec, 23 Jan 45; Cbl 2078, Channel Base Sec to G–4 ETO, 25 Jan 45, Hq ETO Cbls, ETO Adm 405.

[69] Memo, Deep-Draught Ports on Continent, 26 Jan 45 (no signature) SHAEF G–4 825.1 Piers . . ., I.

[70] COMZ Comd and Stf Conf, 24 Nov 44, SHAEF G–4 337 Command and Staff Confs, I.

and causing many people to move into the country.

The V-bombs had surprisingly little effect on port operations, although their potential destructiveness forced the Allies to take special precautions in handling their most sensitive commodity, ammunition. Early in November Supreme Headquarters raised the question of admitting ammunition ships to Antwerp and requested the various interested headquarters to present their views on the matter. The Communications Zone recommended that ammunition be excluded entirely, and proposed that all Class V supply continue to be handled at Cherbourg and Le Havre. Neither the Allied Naval Commander nor the 21 Army Group considered it necessary to exclude ammunition from Antwerp, but did advocate that certain precautions be taken, including a restriction in the number of ammunition ships permitted in the port at one time, the dispersion of such vessels, and the prompt clearance of ammunition from the port so that there would be no accumulation at quayside.[71]

The policy which the chief administrative officer laid down a few days later generally followed these recommendations. It did not forbid the acceptance of ammunition at Antwerp, but restricted the quantity to the operational requirements at the discretion of the Communications Zone and the 21 Army Group, and specified that it be handled in a separate and remote section of the port, that no dumps be permitted even for sorting, and that special fire-fighting preparations be made.[72]

When Antwerp actually opened at the end of November the Port Executive Committee asked that all ammunition be excluded temporarily, and SHAEF approved this request, specifying that no ammunition would be unloaded at Antwerp for fourteen days except in emergency.[73] An exception was immediately made, however, in granting a request of 21 Army Group to admit certain British vessels,[74] and a few weeks later ammunition began to be received regularly, subject to the restrictions laid down earlier.

In mid-January the rules for ammunition acceptance required reconsideration. The scale of attacks by V-weapons had showed no signs of slackening. The main area of impact had in fact shifted to the docks, resulting in greater damage and increased casualties. In the opinion of the Port Executive Committee the current policy simply courted disaster. It therefore recommended much more stringent regulations. The problem was not serious for U.S. forces, since all ammunition on American account could easily be handled at other ports; the 21 Army Group readily agreed to have more ammunition discharged at Ostend and Ghent. Both the amounts and types of ammunition to be brought in via Antwerp accordingly were reduced to a min-

[71] Cbl S–65373, SHAEF to Major Commands, 4 Nov 44; Cbl Ex–61214, Lee to SHAEF, 7 Nov 44; Cble 060910A, ANCXF to SHAEF, 6 Nov 44; Cbl QM–1756, 21 A Gp to SHAEF, 8 Nov 44, SHAEF AG 323.3–3 (Ports) Port Capacities.

[72] Ltr, Crawford to Smith, sub: Acceptance of Ammo Ships to Antwerp, 10 Nov 44, and Cable to Major Commands, SHAEF SGS 800 Antwerp.

[73] Cbl 301504A, NOIC Antwerp to ANCXF, 30 Nov 44, and Cbl S–69457, SHAEF to COMZ *et al.*, 4 Dec 44, SHAEF SGS 800 Antwerp.

[74] Cbl QM–2261, 21 A Gp to SHAEF G–4, 4 Dec 44, and Cbl S–69623, SHAEF to 21 A Gp, 5 Dec 44, SHAEF SGS 800 Antwerp.

imum.[75] These restrictions were not relaxed until late in April, barely two weeks before the end of hostilities. Even then the number of vessels which could discharge at one time was limited, and no stacking of ammunition on the quays was permitted.[76]

In addition to imposing a handicap on discharge operations, the V-weapon attacks contributed greatly to the distress of Antwerp's inhabitants. Antwerp came under intensified attack by the dreaded "vengeance" weapons during the enemy counteroffensive in the last half of December. One of the most disastrous attacks occurred on the afternoon of 16 December, when a direct hit on the crowded Rex Theater killed 567 soldiers and civilians and seriously injured another 291. U.S. engineers, aided by other port units, worked nearly a week recovering bodies from the debris.[77]

Antwerp sustained nearly 4,000 hits before the attacks finally ceased at the end of March 1945. The city suffered heavy material damage, and sustained more than 10,000 casualties, two thirds of all those caused by the V-weapons on the Continent. Of these, about 82 percent were civilian, the remainder military.[78]

Under these attacks living conditions, already very bad, became worse. Shortages of food, clothing, and coal created great distress during the winter, and finally led to a strike among dock workers on 16 January. The strike was not directed against the Allied port organization, but was intended rather as a protest against the arduous working conditions and terrible economic conditions which forced people to resort to the black market for the barest essentials. The demonstration lasted only one day, and workers returned to their work on assurance from the burgomaster that more food and coal would be made available at regulation prices.[79]

Civilian labor was plentiful for the most part, although the movement of many workers out of the city produced a transportation problem. There were the usual difficulties over language, and over reading the complicated markings on U.S. cargo. But Belgian labor was both co-operative and industrious, and U.S. forces made the maximum use of the local manpower resources, reserving their own port battalions for supervisory jobs. In December an average of 9,000 civilian workers were employed in the U.S. section of the port, and on one shift a record 13,125 men.[80]

Antwerp recovered from the worst effect of the embargo and clearance handicaps in the second half of January and

[75] Cbl 131642A, NOIC Antwerp to 21 A Gp, 13 Jan 45, and Cbl MGA–5, 21 A Gp to SHAEF, 15 Jan 45, ETO Cbls, ETO Adm 404 and SHAEF SGS 323.3 Ports—Allocation and Development; Cbl S–75911, SHAEF to G–4 ETO, 21 Jan 45, ETO Cbls, ETO Adm 405.

[76] Cbl S–86010, SHAEF to TC ETO, 24 Apr 45, ETO Cbls, ETO Adm 415.

[77] History of TC ETO, V, 13th Port, 15; VI, 13th Port, 73–74.

[78] [Royce L. Thompson] Military Impact of the German V-Weapons, 1943–1945, 31 Jul 53, pp. 9–12, 51–53, MS, OCMH.

[79] Cbl 131642A, NOIC Antwerp to 21 A Gp, 13 Jan 45, EUCOM 471/1 Ammunition Policy; Ltr, R. S. MacTier, Chief Representative British Ministry of War Transport to Gale, 17 Jan 45, SHAEF SGS 323.3 Ports—Allocation and Development; Cbl, War Shipping Adm Antwerp to WSA, 17 Jan 45, ETO Cbls, ETO Adm 405; Cbl 171916, NOIC to SHAEF *et al.*, 17 Jan 45, SHAEF Cable Log IN, Smith Papers.

[80] History of TC ETO, V, 13th Port, 10–11; VI, 13th Port, 80–81.

REMAINS OF DECANTING SITE, *POL Depot, Antwerp, hit by a German V-weapon.*

by the end of the month was again discharging 18,000 tons per day. With the help of Le Havre and Rouen, which were contributing 12,000 to 13,000 tons, U.S. discharge by that time was averaging between 40,000 and 50,000 tons per day, double the intake in October. Antwerp did not provide the immediate solution to all logistic difficulties, but its opening at least eliminated *port capacity* as a limiting factor in the support of the U.S. armies. For the first time the Communications Zone enjoyed a surplus in discharge capacity, which permitted some choice in the use of ports and a more economic use of shipping and inland transport. It also opened up the prospect of relaxing the prohibition on civil imports. Whatever strain still remained on the logistic structure now centered on the transportation system, and even this was greatly relieved by the shortening of the lines of communication.

(3) *Southern France*

Providing the necessary port capacity for the support of Allied forces in southern France proved a far less protracted and less worrisome problem than in the north. Requirements were considerably smaller, for one thing, since plans initially called for the support of a force of only ten divisions via the southern line of communications. The problem was further simplified by the fact that there existed in southern France a major port, Marseille, the capacity of which was known to be ample to meet all Al-

SEARCHING FOR CASUALTIES IN WRECKAGE *caused by V-bomb hits in Antwerp.*

lied discharge requirements in that area.

Port discharge facilities in southern France had had an important bearing on the final decision to launch the DRAGOON operation. The southern France landings had long been planned as an operation closely linked to OVERLORD, and as an operation best calculated to utilize Allied forces in the Mediterranean, particularly French divisions. Early in 1944 the decision to strengthen the cross-Channel operation, which necessitated the withdrawal of landing craft from the Mediterranean, jeopardized the future of ANVIL, as the southern France operation was then called. The decision to strengthen OVERLORD at the expense of ANVIL led first to the realization that the southern operation could not be launched simultaneously with the Normandy invasion, and then to a prolonged argument between the British and U.S. Chiefs of Staff as to the best way in which to employ Allied forces in the Mediterranean. The issue still remained unresolved at the time OVERLORD was launched.

ANVIL had been envisaged as aiding OVERLORD by both drawing enemy forces away from the northern bridgehead and providing a way by which additional Allied forces could be committed against the enemy in France. At the end of June, with Allied forces bogged down in Normandy hedgerows and bottomlands, the desirability of a diversion in southern France designed to forestall enemy reinforcement of the northern defense line assumed added importance. The desirability of having another major port through which additional forces could be supported became an even more compelling argument with the acknowledgement by the OVERLORD planners that there were more divisions available for the European theater than could be supported through the northern ports.

At the end of June the shortage of shipping, which had been the major factor in the postponement of DRAGOON, no longer obtained. General Eisenhower at that time stated the case for the southern operation in the strongest possible terms, arguing that France was the decisive theater in Europe, and that a rapid concentration of the maximum forces against the enemy there could be achieved only by seizing another major port.

On 1 July, following an appeal from President Roosevelt, the British finally gave their consent to the DRAGOON operation, although even at this date they did not abandon their efforts to have the resources of the Mediterranean used elsewhere. In the first days of August, when it appeared that the Brittany ports would soon fall into Allied hands, they advocated that the DRAGOON forces be brought into France via Brittany, thus obviating the necessity for an assault of defended shores.

The proposal appeared to miss the whole point of the Supreme Commander's argument—namely, that logistic factors, particularly port capacity, limited the number of divisions which could be received in the OVERLORD lodgment. Moreover, it was impossible to foresee how soon the Brittany ports could be put to use. A long delay in employing the DRAGOON forces was unacceptable to General Eisenhower, who steadfastly insisted that the interests of OVERLORD would be served best by carrying out the southern France landings as planned. Loading for DRAGOON had already begun, and on 10 August the signal finally went to General Wilson, the Supreme Commander in the Mediterranean, instructing him to proceed with the assault as planned.[81]

As in OVERLORD, logistic plans for the southern France operation provided for support over open beaches in the first stages. But reliance on the beaches was to be of short duration, and no plans were made for elaborate artificial harbors as in the north.[82] Instead, there was to be an early shift to existing ports, first to Toulon, and eventually to Marseille.

Toulon, like Cherbourg, was primarily a naval base, although it had a considerably greater freight-handling capacity, estimated at 10,000 tons per day. But it normally handled relatively small quantities of freight, and was deficient in clearance facilities. Nevertheless, the planners concluded that Toulon could well serve as an interim port, which,

[81] See Pogue, *The Supreme Command*, pp. 108–17, 218–27, and Gordon A. Harrison, *Cross-Channel Attack*, UNITED STATES ARMY IN WORLD WAR II (Washington, 1951), pp. 164–73, for the detailed story of the ANVIL-DRAGOON controversy, and the following documents in particular: Cbl IZ–4741, JSM to SHAEF, 26 Jun 44, Cbl OZ–4191, Br COS to Wilson, 5 Aug 44, Cbl WX–76552, U.S. CofS to Br COS, 6 Aug 44, and Cbl FX–79418, Wilson to Br COS, 6 Aug 44, all in Cbl Log (in) SHAEF–Smith Papers; Cbl FWD–12704, SAC to CCS, 8 Aug 44, in Cbl Log (out) SHAEF, Smith Papers, and Memo, Smith for Hist Sec ETO, 22 Feb 45, in Smith Papers.

[82] In part because of the lateness of the season. DRAGOON was launched a full ten weeks later than OVERLORD.

along with the beaches, could easily handle the maintenance and build-up required for the consolidation of the bridgehead and the advance on Marseille. They hoped that the port would be captured by D plus 20 and that it would have an initial discharge capacity of about 2,000 tons per day.[83]

While Toulon was expected to have only passing importance in the maintenance of the DRAGOON forces, Marseille was planned to become the main gateway through which the southern forces eventually would be sustained. It was, in fact, essential for a build-up prerequisite to an exploitation up the Rhône valley. Marseille, with a population of nearly a million, was the second city of France and its foremost port. It had long played an important role in the commercial life of the Mediterranean, serving as early as the sixth century B.C. as an outpost of the Greek trading complex and coming into great prominence with the development of the north African colonies and the opening of the Suez Canal. It is largely an artificial port, consisting of ten basins with approximately thirteen miles of quays, almost all of which are served by rail. The port could accommodate all types of shipping, and possessed ample facilities for the transfer of all types of cargo between ships, barge-lines, trucks, rail cars, sheds, and warehouses. Clearance facilities were equally good, for hard-surfaced highways and standard-gauge railways linked the port with the major cities of France. In addition, the Marseille–Rhône Canal, extending fifty-seven miles to the northwest, connected Marseille with the satellite Port du Bouc, which had important POL receiving and storage facilities, and the Rhône River. Marseille did not equal Antwerp in size or facilities, but it was known to have a peacetime discharge capacity of 20,000 tons, which was ample to meet the estimated daily requirement of about 15,000 tons for the DRAGOON forces.[84]

Allied planners in the Mediterranean were under no illusions about the chances of capturing either Toulon or Marseille intact. They had had sufficient opportunity to observe the enemy's destructive ability in that theater, having witnessed one of the best examples of it at Naples. The Seventh Army fully expected that it would have to rehabilitate both Toulon and Marseille and scheduled the introduction of both engineer troops and equipment for the reconstruction task which they envisaged.[85]

The choice of Toulon and Marseille as the first major objectives dictated that the assault area be within easy striking distance of these targets. The sites finally selected for the landings lay in the general vicinity of St. Tropez–St. Raphael, between thirty and fifty miles northeast of Toulon. Landings were successfully carried out in that area on 15 August as scheduled, and engineer shore groups quickly organized the three beaches,

[83] Hamilton and Smith, Southern France and Alsace, Ch. II, pp. 20–27; ANVIL Appreciation and Outline Plans, Sec. I and Annex A, 22 Dec 43, AFHQ RG 727, Drawer 1104, G–3 OPs P–122.

[84] Preliminary Rpt, Strategic Engineering Study 84, Port and Terminal Facilities, Vol. III of Mediterranean France, prep by Bd of Engrs for Rivers and Harbors, Intel Br OCOE, WD, Oct 43, pp. 107–83, 249–87; CONAD History, 35 (HD); Report of Operations, Seventh U.S. Army, I, 327 (HD).

[85] Seventh Army Engineer Staff Section Reports, 1 January–30 September 1944, p. 10.

GENERAL DESTRUCTION AT MARSEILLE. *Note ships waiting in outer harbor.*

known as ALPHA, DELTA, and CAMEL, for the build-up of supplies and troops. The small neighboring ports of St. Raphael, St. Maxime, and St. Tropez were also captured and cleared for the use of small craft.

For the first two weeks the bulk of all maintenance supplies for the Seventh Army was brought ashore via the beaches, which averaged more than 10,000 tons per day. Logistic planners had calculated that the support of the DRAGOON force would require that about 278,000 tons of supplies be passed over the beaches in the first thirty days.[86] In the first month the beaches easily met this goal, handling 280,000 tons.[87]

Meanwhile both Toulon and Marseille were captured earlier than expected, providing a welcome bonus in port discharge capacity. The Seventh Army's operational plans had assigned to French forces the mission of capturing the two ports, giving first priority to the seizure of Toulon. The unexpected ease with which the landings were accomplished made it possible to improve on this plan. By 20 August French forces had already driven past Toulon to a point midway between that port and

[86] Hamilton and Smith, Southern France and Alsace, Ch. V, pp. 37–38.

[87] Seventh Army AAR, Annex 287, TC Sec—Beach Opns Summary, 15 Aug–19 Sep 44, Opns Rpts, L–1139 (141).

CLOSE-UP OF DAMAGED DOCK FACILITIES *and sunken craft, 1 September 1944.*

Marseille, creating the inviting opportunity of striking at both ports simultaneously. Such attacks were immediately ordered, and after a week of savage fighting Toulon and Marseille capitulated on the same day, 28 August. Toulon was captured a full week ahead of schedule, and Marseille almost four weeks earlier than expected.[88]

The condition of the two ports attested once more to the enemy's appreciation of their value for the Allies if surrendered intact. The demolitions and blocking do not appear to have equaled those at Cherbourg, where the enemy's destructive art apparently reached its height. Nevertheless, both Toulon and Marseille were useless for the moment. The first reconnaissance at Marseille indicated that all channels and entrances were completely blocked by sunken ships. Both the inner harbor and the waters beyond the outer mole were sown with hundreds of marine mines; jetties, quays, and cranes had been blasted; and the entire port area was mined and booby-trapped. Toulon had suffered even more severely, Allied naval and air bombardment having made a contribution to the damage.

Since the two ports had been captured at the same time there was little point in following the original plan of giving

[88] Seventh Army Report of Operations, I, 151, 167; Hamilton and Smith, Southern France and Alsace, Chs. XIII and XV.

priority to the restoration of Toulon. Marseille possessed better facilities for both discharge and clearance, and was better located to support the advance northward. After the preliminary reconnaissance, therefore, it was decided to concentrate most personnel and equipment resources at Marseille and to bring that port to its fullest development as quickly as possible.[89]

In accordance with this decision three Liberty ships, standing by with port construction personnel and equipment, including the 6th Port headquarters which was to operate the port, were immediately called forward and were unloaded from anchorage by dukws. The first survey had revealed that the southern end of the port could be restored quickly for lighterage operations. Construction was initiated first in that area, therefore, to provide hards and ramps for LCT's and dukws. Meanwhile U.S. naval units started on the task of clearing the harbor of mines and obstacles. They declared one basin free of mines as early as 3 September, and port ships began discharging shortly thereafter. Although seventy-five vessels had been scuttled in the harbor, a hydrographic survey revealed that channels and entrances fortunately were not as completely blocked as first appeared, and it was found that Liberty ships could safely pass around sunken ships and into the harbor.

Onshore rehabilitation was accomplished entirely by the 1051st Port Construction and Repair Group, supplemented by an engineer general service regiment and a dump truck company. It was unnecessary for the most part to construct timber-pile wharves or quays, as at Cherbourg, since sufficient berthage was made available by removing debris and patching quay walls. Rehabilitation was rapid, therefore, and on 15 September Marseille received its first Liberty ship for direct ship-to-shore discharge. Within another ten days sixteen alongside berths and twenty-three offshore berths were in use.[90] Toulon had been brought into use on 20 September, and Port du Bouc, mainly for bulk POL reception, on the 9th.

The rapid recovery of the ports made it possible to abandon the beaches somewhat earlier than anticipated. ALPHA Beach was closed out as early as 9 September; unloading at DELTA ceased on the 16th; and the closing of CAMEL Beach on the 28th of the month brought all movement over the beaches to an end.[91] During the period of their operation the beaches handled well over 300,000 tons of supply, and movements through the ports in September brought the cumulative cargo receipts, exclusive of POL and vehicles, to approximately 500,000 tons by the end of that month. In October the southern French ports discharged a total of nearly 400,000 tons, averaging about 13,000 tons per day.[92]

[89] Seventh Army Report of Operations, I, 328–29; CONAD History, pp. 41–42.

[90] Seventh Army Engineer Staff Section Reports, 1 Jan–30 Sep 44, p. 10 (Opn Rpts); History of TC ETO, V, Ch. VII (SOLOC), 3; CONAD History, pp. 40–41; Seventh Army Rpt of Opns, I, 330–31; Delta Base Section History, III, App. I, 6th Port Headquarters History, 13; *Final Report of the Chief Engineer, ETO,* I, 276–77.

[91] Delta Base Section History, Vol. III, App. I, 6th Port Headquarters History, 304; Cbl B–16776, Devers to Eisenhower, 28 Sep. 44, SHAEF Cable Log (in) 1944–45, Smith Papers.

[92] Delta Base Section History, 6th Port History, Nov 44–Mar 45, Exhibit B–8. The above is a minimum figure. There is wide disagreement in the discharge statistics for the southern France ports,

With discharges of this magnitude no difficulties in the support of the 6th Army Group were foreseen. At the end of October it was decided to turn the port of Toulon back to the French.[93] It had been used almost exclusively for the import of civil affairs supplies. Port du Bouc's chief function was to receive bulk POL; relatively minor tonnages of general cargo passed through that port. Marseille consequently accounted for the great bulk of all other classes of supply discharges in southern France, as was originally intended. At the end of October fifty-four berths were in use in the port, thirty-two of which could accommodate Liberty ships.[94]

While port discharge thus posed no serious problem, the southern line of communications at times suffered from the same limiting factor which plagued operation in the north—port clearance. A bottleneck first developed early in October, when clearance failed to keep pace with the rising discharge rate. By the middle of the month more than 40,000 tons had accumulated on the quays, forcing a slowdown in unloadings of all cargo except ammunition and items which could be cleared most easily. The principal cause, as in the north, was inadequate rail clearance capacity. For several days in October it was necessary to press every available vehicle into service, including horse-drawn wagons, to move the backlog of supplies.[95] Clearance, and therefore discharge, was again affected in January, February, and March 1945, when snowbound railways and shortages of rail cars, locomotives, and engine crews limited the rail traffic forward of the Delta Base Section.[96] For the most part, therefore, the southern line of communications was unable to provide the surplus capacity which, it had been hoped, would enable it to aid in the support of the 12th Army Group, although several divisions intended for movement through the northern ports were routed through Marseille in the fall of 1944.

In November, December, and January Marseille and its satellite, Port du Bouc, handled approximately 1,270,000 tons of general cargo, averaging 13,800 tons per day, nearly 90 percent of it at Marseille. (*Table 4*)[97] Work continued in these months to increase Marseille's discharge and clearance facilities, and by the end of January the number of available berths had risen to seventy-two, of which forty-five were suitable for Liberties.

The port was operated from the beginning by the 6th Port headquarters (Col. R. Hunter Clarkson), an organization with long experience in North Africa and Italy. By January the port employed an average of 18,000 men, of which nearly 6,000 were U.S. military

another tabulation in the 6th Port History (Exhibits F, F–2, and F–5) giving the total of 505,584 tons for October (of which 404,365 tons are attributed to Marseille), and the History of TC ETO (Chart 8A of Monthly Progress Reports, App. 7, in Pt. III of Vol. VII) giving the total of 524,894 tons, with no breakdown as to individual port performance.

[93] 6th Port Headquarters' History, Jul–Oct 44, App. I of Vol. III, Delta Base Section History, 19.

[94] *Ibid.*, Nov 44–Mar 45, Exhibit I.

[95] *Ibid.*, Jul–Oct 44, 14; History of TC ETO, V, Ch. VII (SOLOC), 9–10; Memo, Gilland, Deputy CG COMZ NATO to Adcock, G–4 6th A Gp, 15 Oct 44, ETO, Memos for Larkin.

[96] 6th Port History, Nov 44–Mar 45, 12–13.

[97] *Ibid.*, Exhibit B, B–8, and D. Transportation Corps figures are higher, totaling 15,600 tons per day for this period. History of TC ETO, App. 7, in Pt. III of Vol. VII.

TABLE 4—TONNAGES DISCHARGED AT CONTINENTAL PORTS: JUNE 1944–APRIL 1945

[Long Tons [a]]

Year and Month	Total	OMAHA Beach	UTAH Beach	Cherbourg	Normandy Minor Ports[b]	Brittany Ports	Le Havre	Rouen	Antwerp	Ghent	Southern France
1944											
June	291,333	182,199	109,134	—	—	—	—	—	—	—	—
July	621,322	356,219	193,154	31,658	40,291	—	—	—	—	—	—
August	1,112,771	348,820	187,955	266,644	125,353	9,499	—	—	—	—	174,500
September	1,210,290	243,564	150,158	314,431	100,126	75,198	—	—	—	—	326,813
October	1,309,184	120,786	72,728	365,603	58,816	77,735	61,731	26,891	—	—	524,894
November	1,402,080	13,411	12,885	433,301	48,707	64,078	148,654	127,569	5,873	—	547,602
December	1,555,819	—	—	250,112	50,749	27,327	166,038	132,433	427,592	—	501,568
1945											
January	1,501,269	—	—	262,423	47,773	—	198,768	157,709	433,094	15,742	385,760
February	1,735,502	—	—	286,591	41,836	—	195,332	173,016	473,463	69,698	495,566
March	2,039,778	—	—	261,492	39,691	—	192,593	268,174	558,066	172,259	547,503
April	2,025,142	—	—	181,043	47,542	—	165,438	240,708	628,227	277,553	484,631

[a] Exclusive of bulk POL and vehicles.

[b] Including Granville.

Source: Historical Report of the Transportation Corps, ETO, Vol. VII, April–June 1945, App. 7, Table 8A.

personnel (chiefly port battalions), the remainder consisting of French civilian labor, prisoners of war, and a small number of French Indochinese troops.[98]

(4) *The Shipping Tie-up*

In the first six months of operations inadequate port discharge capacity on the northern lines of communication had inevitable repercussions on other parts of the logistic structure. Its effect on the use of shipping—both coasters and ocean-going ships—was particularly serious.

The Combined Chiefs of Staff, recognizing the great need for shallow-draft shipping in the early stages of OVERLORD, had allocated 625,000 dead-weight tons of coasters for the first weeks of the operation. They had intended that this shipping should gradually be replaced by ocean-going vessels, leaving about 100,000 tons of coasters in cross-Channel service after D plus 42. Shortly before D Day, OVERLORD logistic planners, following a re-examination of shipping requirements in light of the enlargement of the operation and the planned acceleration in the troop build-up, got permission to retain an additional 150,000 tons of coaster shipping after D plus 42. They had argued that additional shallow-draft shipping would be needed for use inside the MULBERRIES, where ocean-going ships could not be accepted; for

[98] 6th Port History, Nov 44–Mar 45, Exhibit A.

the various special express services, in which rapid loading and discharging were required; and for taking fuller advantage of the capacity of the small ports. Modifications in the invasion plan also led to an increase in the allocation of MT shipping.[99]

Providing the needed coaster shipping for OVERLORD placed an additional strain on the economy of the United Kingdom, which relied heavily on coastal shipping for the movement of iron, coal, steel, and other commodities. The allocation of 625,000 tons to OVERLORD, representing about two thirds of the entire British coaster fleet, caused a drastic curtailment in movements and, in turn, a temporary shutdown of about one fourth of the United Kingdom's blast furnaces. British authorities naturally desired that coaster shipping be released from military use as early as possible.[100]

From the very start of the invasion there were shortages in practically every category of shipping, and hope quickly faded that shallow-draft vessels could be returned to coastal service as originally scheduled. Before the end of June U.S. authorities took steps to have additional LST's and MT shipping made available from the United States, and to have the release of coasters postponed.[101] The shortage remained critical throughout the summer and fall. In September, contrary to plans, 560,000 tons of coaster shipping were still engaged in cross-Channel service, and in November the total exceeded 600,000.[102]

Poor turnaround performance was the initial cause for the shortage, resulting in part from piecemeal or selective discharge on the far shore, in part to interruptions from bad weather, diversions from one port or beach to another, and unexpectedly long deadlining of vessels for repairs. An analysis of the worst period, late October and early November, revealed that sixty-three round trip voyages had required 1,422 ship-days instead of the planned 606, representing a turnaround time 135 percent greater than expected. The repair problem became particularly acute in November and December when 20 to 25 percent of the total coaster fleet was immobilized.[103]

At the root of these difficulties lay the shortage of deepwater berths on the Continent, which necessitated the extended use of shallow-draft facilities. Not until December, after Antwerp came into operation, was it possible to release 50,000 tons of coaster tonnage, and then only by withdrawals from support of 21 Army Group. The U.S. allocation was actually increased during the month.[104]

The need for a large coaster fleet continued to the very end of the war. Early in 1945 SHAEF refused to accept a reduction of the coaster allocation below 500,000 tons, which it insisted was the absolute minimum to meet operational needs. It actually desired a larger allocation, in part to obtain more flexibility

[99] Ltr, Gale to PAOs Committee, sub: OVERLORD Shipping Requirements, 17 Mar 44, and Cbl W–36141, CCS to SHAEF, 13 May 44, SHAEF AG 400.22–1 Shipments and Tonnage Allocations 1944.

[100] Cbl 7367, U.S. Mil Attaché, London, to U.S. Secy of State, 8 Sep 44, OPD Cbl Files.

[101] See *Logistical Support I*, Ch. X, Sec. 9.

[102] Cbl, U.S. Mil Attaché to Secy of State, 8 Sep 44; Mil Shipments Priority Mtg, 24 Nov 44, SHAEF AG 337–18.

[103] Mil Shipments Priority Mtgs, 5 Aug, 3, 10, and 24 Nov, and 9 Dec 44.

[104] *Ibid.*, Mtgs of 2, 9, and 30 Dec 44.

in the use of ports, and in part because coasters were the most economical means for coal shipments. At the end of February SHAEF again asked the Combined Chiefs of Staff for additional tonnage, requesting that all of the twenty-six Baltic type of coasters then under construction in the United States be allotted to the European theater to augment the available fleet.[105] In March the Combined Chiefs promised that all but three would be sent to Europe.[106]

The dearth of deepwater berths on the Continent had an even more far-reaching impact on ocean shipping. The theater's inability to berth and discharge all the deep-draft ships arriving from the United States inevitably led to an accumulation of shipping in both U.K. and continental waters—shipping which could be ill-spared from the world shipping pool.

The War Department first called the theater's attention to the problem in mid-July 1944, pointing to the many commodity-loaders and pre-stowed ships being held at anchor in the United Kingdom. Brig. Gen. Robert H. Wylie, the Deputy Chief of Transportation for Operations, presented figures showing that the European theater was getting more ships than it could handle, and questioned its justification for requesting 285 ships for August loading. He reminded General Ross and other theater officials that the shipping situation was critical, and that the retention of ships under load for excessive periods only aggravated the world-wide shortage by lengthening the turnaround time. General Ross could not agree that the theater had been wasteful in the use of shipping, and argued that the pipeline must be kept full, particularly for the eventuality of a breakout from Normandy. "I'm in the habit," he said, "of rather seeing a fellow spill a little over the side of the bucket and get his feet wet, than to worry about his brocade shoes." [107]

Before the end of the month the theater modified its requests somewhat, but stated that its "irreducible" minimum for August loading (September delivery) was 250 ships, of which 175 were intended for continental discharge, the other 75 for the United Kingdom. It estimated its needs for succeeding months as averaging about 265 ships, an increasing percentage of which it planned to send directly to continental ports. These requirements were based on estimated discharge capacities of 27,000 tons per day in September, and 40,000 tons thereafter, estimates which proved far too optimistic.[108]

Early in August the War Department presented additional figures to show that the theater was building up an excessively large bank of ships and that its discharge performance did not justify the

[105] A Baltic-type coaster is a vessel of about 1,800 gross tons designed to operate on short sea routes, modeled on coasters in use in northern European coastal traffic.

[106] Cbl FWD-17429, SHAEF to CCS, 28 Feb 45, and Cbl W-54347, CCS to SHAEF, 17 Mar 45, SHAEF AG 400.22-1 No. 1, 1945 Allocation of Tonnages—Policy.

[107] Tel Conf between Ross, Eyster, *et al.*, and Wylie *et al.*, 16 Jul 44, SHAEF G-4 Troops Flow 121/1 GDP-1, 69.

[108] Cbl EX-40282, Lee to Gross, 29 Jul 44, SHAEF AG 400-22-1 Shipments and Tonnage Allocations; Mil Shipments Priority Mtg, 29 Jul 44, SHAEF AG 337-18.

estimates of its needs. It pointed out that of 41 prestowed ships called forward from the United Kingdom, 10 had been held at anchor from 23 to 30 days, and that of 61 commodity-loaders called forward, 31 had been held at anchor more than 7 days and 9 of them from 23 to 30 days. Prestowed vessels thus far had averaged 46½ days in European waters between arrival in the United Kingdom and return sailing to the United States following continental discharge.

The War Department argued that this immobilization of shipping hazarded the support of operations in other parts of the world, and was unjustifiable in view of the frequency of convoys and the lessened submarine menace. By the end of September, it pointed out, the theater would have received 219 ships for continental discharge in that month alone, 44 more than ETOUSA had stated as its requirements. The theater, it stated, was unduly concerned about meeting future requirements, for sufficient shipping was available to support all operations in progress or planned if it were properly used. Once again the War Department assured ETOUSA that the current program of sailings from the United States would more than match the theater's most optimistic estimates of discharge capabilities, and that loading could easily be stepped up whenever the discharge rate indicated that the backlog of shipping was being reduced. But it warned that the theater must accept responsibility for deferring shipments from the United States when it became apparent that congestion would develop. It specifically asked that commodity-loaded vessels be phased so that they were not held at anchor longer than one convoy interval, or approximately seven days.[109]

The theater defended its shipping policy by noting that it was simply impossible to predict the progress of tactical operations accurately enough to schedule the arrival of shipping in exact consonance with needs. Weather conditions were also unpredictable, often interrupting discharge and delaying the towing of floating equipment to France. Finally, it argued that in the absence of suitable quays, discharge in French ports was still essentially a lighterage operation, which accounted for some of the slowness in unloading.[110]

Theater shipping officials were well aware of the serious backlog which was forming off the beaches and ports of Normandy, but did not feel justified in "turning off the tap" in view of the planned acceleration in the movement of divisions to the Continent in September. The Communications Zone objected strongly to any reduction in the program, arguing that priority demands for supplies could be met only if sufficient stocks were held offshore.[111]

The current practice was nevertheless recognized as both dangerous and wasteful. The Allied Naval Commander was concerned that congestion of shipping would not only invite attack from the sea and air, but that shipping in open

[109] Cbls W-76034, AGWAR to ETO, 5 Aug 44, W-81787, Gross to Somervell, 16 Aug 44, and W-81853, Lutes to Lee, 17 Aug 44, SHAEF AG 400.22-1 Shipments and Tonnage Allocations.

[110] Cbl EX-42044, Lee to Somervell, 8 Aug 44, SHAEF AG 400.22-1.

[111] Shipping Note for CAO, G-4 Mov and Tn, 19 Aug 44, SHAEF G-4 Mov and Tn War Diary/Jnl 3014/22 Mov, Folders 13-22; Min, Continental Mov and Shipping Committee, Mtg of 16 Aug 44, ETO 001 Meetings.

anchorage would soon suffer from equinoctial gales. He recommended limiting the number of vessels authorized to be in the U.S. area at any one time. One of the major causes for the congestion, selective unloading, was finally recognized as a bad practice by both SHAEF and COMZ officials. Late in August they ordered the practice stopped.[112] Apparently encouraged by tactical developments, the theater at the same time accepted without protest a proposed cut of approximately ten ships from each of six convoys sailing from the United States in the four weeks between 12 September and 10 October. In the end the theater canceled a total of 600,000 tons of supplies of all classes which had been scheduled for delivery by the end of November.[113]

These measures had no effect in September, and shipping continued to pile up in European waters. Late in the month it was estimated that only 90 ships would be discharged that month, and the prediction was made that even if unloadings were increased to 150 there would still be 271 on hand at the end of September.[114]

On 6 October the War Department announced a further cut of forty ships from European sailings. This time the theater protested vigorously. Rather than attempt to justify its earlier requests entirely on the basis of predicted port performance, however, it now offered a truer explanation for its exorbitant demands. Admittedly, the discharge rate had failed to come anywhere near the 40,000-ton figure which had been predicted a month before. The shortage of inland transportation had aggravated the problem by making it impossible to move the major continental supply reserves out of the Cotentin. Only rations, gasoline, and ammunition, plus a small tonnage of highly selective items, were being moved forward to the armies. In the absence of adequate discharge capacity, General Lee explained, the large backlog of ships had been the theater's only salvation by making it possible to meet high priority demands for specific items of supply. In other words, the large bank of ships in European waters had been the necessary substitute for supply depots which should have been established on the Continent. The loaded ships were in effect serving as floating warehouses.

While port capacity remained almost stationary, the theater's supply needs continued to grow with its troop strength, and the COMZ commander now urgently appealed to the War Department to meet what he referred to as "conservative estimates" of the theater's needs. These amounted to 139 more ships than the War Department offered to dispatch for October, November, and December. Lee estimated that the theater's discharge capabilities would be more than doubled by December, basing his prediction on the expected performance of Le Havre, Rouen, and Antwerp,

[112] Ltr, COMZ to Normandy Base Sec, sub: Discharge of Liberty Ships, 20 Aug 44, EUCOM 560 AT, Transports, Vessels, and Boats in General II; TWX, ANCXF to SHAEF, 31 Aug 44, SHAEF AG 400.22–1; Mil Shipments Priority Mtg, 26 Aug 44, SHAEF AG 337–18; Shipping Note for CAO, 19 Aug 44; Memo, Stratton for CofT, sub: Discharge of Ocean-Going Vessels on Continent, 6 Sep 44, EUCOM 400.22 Shipments General.

[113] COMZ G–4 History, I, 81.

[114] Ltr, Philip D. Reed, Mission for Economic Affairs, to Lee, 21 Sep 44, ETO 381/560 Shipping (OVERLORD).

and on the transfer of the Arromanches MULBERRY to American use. He predicted that the backlog would be virtually eliminated by that time and that some discharge capacity would actually be wasted by mid-December if the War Department carried out its proposed reductions.[115] In effect, however, the theater was demanding a bank of shipping not only large enough to insure full utilization of all continental berths which it forecast would become available, but, failing the development of adequate discharge capacity, one which would hold a large portion of the theater's reserves at anchorage off the ports and beaches and thus permit flexibility in the selection of items for which emergency need might develop.[116]

This time the War Department could not be shaken. The retention of shipping in the European theater, it claimed, was already threatening to strangle operations in other parts of the world. The War Department could not view with composure theater practices which, as related by General Lee himself, had recently made it necessary to call forward nineteen commodity ships loaded with engineer supplies to obtain an average of only 150 tons of priority cargo from each vessel.[117] On 9 October it notified the theater that the new sailing schedules would be followed until ETOUSA had demonstrated a capacity to unload at a faster rate and until the bank of ships had been substantially reduced.[118]

The War Department's action appeared amply warranted by the theater's performance, for the backlog of shipping continued to grow. By 20 October there were at least 240 ships in European waters, of which about 140, mostly commodity-loaders, consisted of Liberties under load in the United Kingdom.[119] In view of this obviously deteriorating situation the War Department imposed additional restrictions. On 20 October General Somervell told the theater that it would get no more commodity-loaded ships with rations, vehicles, or, with certain exceptions, ammunition until it had reduced its bank of such ships to a reasonable level.[120] In reply General Lee made his most urgent appeal yet, citing figures to show that practically all ammunition, ration, and vehicle ships had been called forward for discharge, and repeating the argument that the War Department's cuts would not only result in a loss of the selectivity which had made supply of the armies possible thus far, but result in serious over-all shortages.[121]

General Somervell was unmoved. The shipping situation, he said, simply would not permit the use of ships as base depot storage to the extent of nearly 200 commodity-loaders, some of which had been "on the hook" more than 60 days. He agreed that the theater should have a safe working margin of ships on hand at all times—perhaps 75 or 80—to pro-

[115] Some of Antwerp's capacity actually went unused at precisely that time, though not for lack of shipping.

[116] Cbl EX-53219, Lee to Somervell, 8 Oct 44, OPD Cbl Files; Ltr, Lee to Reed, 26 Sep 44, ETO 381/560 Shipping (OVERLORD); COMZ G-4 History, I, 81.

[117] Cbl, Lee to Somervell, 8 Oct 44.

[118] Cbl WARX-43793, CofT WD to ETO, 9 Oct 44, OPD Cbl Files.

[119] Mil Shipments Priority Mtgs, 7, 13, and 21 Oct 44, SHAEF AG 337-18; COMZ G-4 History, I, 86.

[120] COMZ G-4 History, I, 88.

[121] Cbl E-57846, Lee to Somervell, 26 Oct 44, OPD Cbl Files.

vide a measure of selectivity. But he rejected as entirely misleading the theater's figures on the number of ships which had been called forward to the continental ports for discharge. The fact remained that projected discharges in terms of completely unloaded vessels had not been realized; the theater had only released about 70 ships for return to the United States in October. In the War Department's view the theater had consistently overestimated its capabilities to receive supplies. The War Department, already under attack from shipping authorities for inefficient use of shipping, refused to accept the theater's assurances of improved discharge, and saw no alternative but to deny the theater's requests until it had demonstrated its ability to unload ships and restore them to useful service.

Two weeks later the War Department once more pointed out how unreasonable the theater's requests had been in view of its reception capacity. It cited the monthly shipping requests since July, which had they been granted, would have resulted in a bank of 500 idle ships in European waters. In his strongest reproof to date General Somervell wrote: "It is necessary . . . that your headquarters cease repeating by rote figures previously arrived at and take a realistic view of the situation. To do otherwise," he observed, "destroys confidence in the estimates, and delays our supply of the equipment which you can actually handle." [122] Meanwhile he sent Brig. Gen. John M. Franklin, the Chief of Transportation's director of water transportation and a former president of the United States Lines, to France to aid theater authorities in improving the turnaround of shipping and to gather more realistic estimates on the theater's future discharge capabilities.[123]

The War Department's concern over shipping was by no means confined to the European theater. The Pacific areas, particularly the Southwest Pacific, had also consistently overestimated their discharge capacity and had followed the same wasteful practice of holding loaded vessels at anchor. Meanwhile, Mediterranean requirements had unexpectedly risen sharply as civil relief needs claimed attention, contributing to the worldwide shortage.

By the fall of 1944 shipping agencies in the United States saw the need for drastic measures to impress upon area commanders the necessity of releasing shipping and to enforce a more rigid accountability on the handling of vessels. On 14 November the War Shipping Administration brought the problem before the Joint Military Transportation Committee, and a few days later the Joint Chiefs of Staff, acting on a proposal of General Somervell, approved and sent to the President their recommendations on the steps to be taken. Included were proposed reductions in allocations of vessels to the U.K. import program, British and Russian lend-lease, and civil relief.[124]

[122] Ltr, Somervell to Lee, 18 Nov 44, Hq ASF, European Theater—last half 1944.

[123] Cbls WAR-53834, Somervell to Lee, 28 Oct 44, and WARX-56447, 2 Nov 44; Cbl 49595, Marshall to Eisenhower, 20 Oct 44, OPD Cbl Files; Memo, Gross for OPD, sub: Cargo Shipping for the ETO, 11 Nov 44, OPD 400 ETO No. 113, 15 Nov 44 Cargo Shipping for ETO.

[124] Memo, Marshall for JCS, sub: Remedies for Existing and Prospective Shortages in Cargo Shipping, 17 Nov 44, with Incl, Study by Gen Somervell, same sub, OPD 560 Section VI, Cases 207-25.

President Roosevelt disapproved of most of the reductions and insisted that additional efforts be made to improve the use of shipping and to meet requirements from the United States, thus throwing the problem back into the laps of the military shipping agencies. This left them with little choice but to correct some of the abuses of current shipping practices.

Early in December the Joint Chiefs of Staff approved a memorandum defining policy on the use of shipping. It specifically prohibited the use of ocean shipping for storage purposes, ordered that shipping needs henceforth be based on a realistic appraisal of discharge capacities, and ordered a scaling down of supply levels to bring them within the capabilities of shipping. Selective unloading was prohibited save in the early phases of amphibious operations, as was also the use of ocean-going vessels for local, short hauls, a practice which the European theater had followed in moving supplies from the United Kingdom to the Continent.

Shortly thereafter the Joint Military Transportation Committee laid down a detailed reporting system by which theater commanders were thereafter required to report periodically on the status of all ships. Its obvious purpose was to disclose flagrant cases of ship retentions, to keep area commanders constantly conscious of the turnaround problem, and to provide planners with more reliable data on which to allot shipping. Within another few weeks this requirement was followed by instructions to the theaters that they establish shipping control agencies to co-ordinate the use of shipping within the theaters and to enforce strict economy in its use.[125] Such a watchdog agency, known as the Shipping Control Board, had already been established in the European theater in December.[126]

The War Department's several actions of November and December left the theater with no choice but to release ships at a faster rate. One of its first measures following General Somervell's firm stand on allocations was to consider the return to the United States of twenty-five of thirty-five Liberties which it had been using to transfer supplies from the United Kingdom to the Continent. Another step was to release ships still partially loaded with supplies not urgently needed. In this way about 35,000 tons of pierced-steel landing mats and other airfield runway surfacing were returned to the United States in twenty-one vessels, only to be shipped back to the theater on the next sailing. The New York Port naturally frowned on the practice, pointing out that cargo capacity equivalent to about six ships had been wasted. In any case, these measures actually brought little immediate improvement. The Liberties used in the cross-Channel service, for example, were not released until December.

Meanwhile the theater was given ample evidence that the War Department was keeping critical watch over its handling of shipping. Late in November, taking note of the idle vessels off south-

[125] Cbls WX-74985, JCS to Theater Commanders, 9 Dec 44, and WX-20236, CCS to SHAEF, 13 Jan 45, SHAEF SGS 540 Shipping Problems; D. S. Ballentine, *U.S. Naval Logistics in the Second World War* (Princeton, 1947), pp. 239-43.

[126] Cbl S-76892, SHAEF to CCS, 28 Jan 45, SHAEF AG 400.22-1 No. 1 1945 Allocations of Tonnages—Policy.

ern France, the War Department cut nine ships from the December sailings to Marseille. At the same time it called attention to the fact that of the last two convoys arriving in the theater, both of them loaded with supplies of the highest priority, only two vessels had been called forward for discharge.[127]

A few weeks later, when, despite the recent opening of Antwerp, it appeared that the northern ports would not achieve the target of 200 discharges for that month, the theater accepted without protest a proposed diversion of sixteen January sailings to Marseille, with the understanding that it might later request an increase if warranted by need and discharge capabilities. This left the northern ports with 175 instead of the scheduled 191 for that month.[128]

The long-awaited opening of Antwerp at the end of November promised to solve not only the discharge problem but the closely related shipping tie-up. Meanwhile the theater had also had some success in raising tonnage receipts by increasing the hatch rates. In mid-November, faced with what amounted to an ultimatum from the War Department on the use of shipping, port commanders instituted competition between unloading crews, offering special incentives in the form of awards and pass privileges to winners. These measures, in addition to more efficient port organization and operating procedures, brought about a rise in the hatch rate from 327 to 457 tons per ship per day by December.[129]

These improvements had hardly begun to have their effect when they were partially canceled by tactical developments. The German counteroffensive of mid-December resulted in a sudden embargo on shipments into the forward areas. Since few base or intransit storage facilities were available to U.S. forces in the Antwerp area, this stoppage in movements had its inevitable result in a saturated port and a partial shutdown in unloading operations. This again threatened to aggravate the shipping backlog. Foreseeing the chain of effects, the theater on 23 December voluntarily requested an immediate reduction of an additional twenty-four sailings in the next convoys.[130]

The month of December, which had opened so hopefully, had a daily discharge record of only slightly above 30,000 tons, and January's record was hardly better. Consequently the theater's earlier requests for 240 commodity-loaded vessels per month were certain to exceed its capabilities. Early in January, therefore, on the basis of more conservative estimates of discharge provided by General Franklin, the War Department cut the allocation for January and February to 175 per month.[131]

The bank of shipping shrank somewhat in January—to 116 idle vessels—

[127] COMZ G–4 History, I, 95.

[128] Cbl E–75769, ETO to AGWAR, 18 Dec 44, ETO In and Out Cbls, ETO Adm 397; COMZ G–4 History, I, 105.

[129] COMZ G–4 History, I, 89–93.

[130] Cbl EX–77774, ETO to AGWAR, 23 Dec 44, ETO In and Out Cbls, Adm 397.

[131] The theater had discharged 115 in November and 130 in December. Memo, Gross for Somervell, sub: ETO Shipping Situation, 4 Jan 45, Hq ASF Shipping 1945—Somervell file; TWX E–85355, COMZ to SHAEF, 12 Jan 45, SHAEF AG 400.22–1 Allocation of Tonnages—Policy, I, 1945.

but limited discharges that month led to the realization that the requested schedules would result in added accumulations of shipping in European waters. On 3 February the theater therefore asked for a cutback of thirty vessels scheduled for March arrival, and certain diversions to Marseille.[132] Almost simultaneously the War Department announced that it was cutting the March allocation by 61 ships, reducing the sailings from 233 to 172.[133]

ETOUSA objected to this drastic reduction, partly on the basis of anticipated improvements in rail clearance from the port of Marseille, which had been the biggest bottleneck on the southern line of communications,[134] and succeeded in getting sixteen ships restored to the March allocation. In mid-February, therefore, the sailing schedules called for 253 arrivals in that month (177 in the north, 76 in the south), and 306 in March (218 in the north and 88 in the south).[135] By that time—mid-February—steady improvement in both discharge and forward movement was evident in the European theater. Unloadings that month were to exceed 50,000 tons per day, a jump of about 20,000 over the record of December and January. With this improvement, less than three months before the end of hostilities, both the discharge and shipping problems, which had plagued logistic support of U.S. forces since D Day, finally appeared resolved.

[132] Cbl EX–74290, G–4 ETO to AGWAR, 3 Feb 45, ETO Cbls, Adm 400.

[133] Cbl WARX–30713, AGWAR to ETO, 2 Feb 45, Adm 407.

[134] Cbl EX–97331, ETO to AGWAR, 11 Feb 45, Adm 400.

[135] Cbl WARX–37577, AGWAR to ETO, 15 Feb 45, Adm 408; Cbl E–10453, ETO to AGWAR, 19 Feb 45, Adm 400.

CHAPTER V

Transportation Developments

(1) *Motor Transport: The Color Routes*

Of the various results of the pursuit the deficit in transportation had both an immediate impact on Allied capabilities and far-reaching effects on the workings of the entire logistic organization. It not only brought the pursuit to a halt, but deranged the entire logistic structure by forcing both combat and service echelons to abandon carefully worked out supply procedures in favor of "expedients" which upset the systematic and businesslike growth of the theater logistic organization.

Halting the pursuit brought no diminution in requirements for transportation. Fresh demands, such as the movement forward of newly arrived divisions, the redeployment of the Ninth Army from Brittany, and the winterization program, added to the standing requirements for the building of forward reserves, promised to absorb all the theater's transportation resources for weeks to come. The nature of operations in the fall made it necessary that cargo vehicles which had been withdrawn from tactical formations—particularly field artillery and antiaircraft units—be returned for use in their normal role.

Rail transportation showed steady improvement in September and October, but for a long time was unable to meet the full requirement for long-distance hauling. Motor transport consequently continued for another two months to carry large tonnages all the way from the beaches and ports to the army areas over lines of communication that stretched between three and four hundred miles. This it accomplished largely via the Red Ball Express, which had started operating on 25 August, and via additional express routes organized for similar missions.

The Red Ball Express had completed its original mission—the delivery of 75,000 tons of supplies to the Chartres–La Loupe–Dreux triangle—by 5 September. But there was no thought of discontinuing the operation on that date, and the Communications Zone gave it a new lease to operate indefinitely. On 10 September the route was altered and extended, the outgoing route diverging at Versailles, one branch bypassing Paris to the north and continuing to Soissons to serve First Army, the other continuing in an easterly direction via Melun to Sommesous in support of the Third. (*Map 5*) A week later the completion of a highway bridge at Chennevières, southeast of Paris, permitted a more direct routing and a saving of sixteen miles on the southern branch. Additional extensions to Liège in the north and to Saarbruecken in the south were considered, but the final extension, made on 20 September, affected only the route serving First Army, which was extended from Soissons to Hirson. This brought the total mileage of the Red Ball route to 924, the round trip on the northern route totaling 686 miles, and on the southern route, 590.[1]

The Red Ball Express continued to operate for another two months and chalked up the best ton-mileage records of its second phase operations in the last week of September. In that period, with approximately 5,400 trucks assigned, it averaged 8,209 tons dispatched in 1,542 trucks each day, the average load per truck totaling 5.3 tons, and the average round trip mileage totaling 714. The average trip required 71.2 hours to complete. For the entire month an average of 6,891 tons passed through the traffic control regulating point at St. Lô every day.[2]

TRUCKS LOADED WITH SUPPLIES *waiting to be unloaded at Soissons, terminal of the Red Ball Express route serving First Army.*

At the end of September the strain on motor transport eased somewhat as a result of the establishment of rail transfer points in the Paris area. Both the extent and the better condition of the rail net northeast of Paris made it feasible to handle substantially greater tonnages by rail in that area than could be moved west of the Seine. Plans were accordingly made to transfer a minimum of 4,000 tons of supplies per day from trucks to rail cars in the Paris area for movement to army railheads, and to decrease Red Ball hauling beyond the

[1] Later alterations had little effect on the total route mileage. On 20 September the portion of the return route between Courville and Alençon reached such a state of disrepair that further maintenance became impracticable, and an alternate route was substituted. On 5 October the return route from Sommesous via Arcis and Nogent-sur-Seine to Fontainebleau was eliminated and the portion between Sommesous and Rozay-en-Brie opened to two-way traffic. Similarly on 11 October the return from Hirson to Château-Thierry via Reims and Epernay was also eliminated and the section between Hirson and Soissons made a two-way highway. G–4 Periodic Rpt for Quarter Ending 30 Sep 44, Tn Sec ADSEC 319.1 Supplement to G–4 Periodic Rpt; Ltr, Lt Col M. N. Drake, Stat Br OCofT, to G–4 COMZ, 18 Sep 44, sub: Supply Info, Incl to Ltr, Stratton to G–4 12 A Gp, 19 Sep 44, sub: Supply Info, 12 A Gp 400.291 Supply Information, II.

[2] Memo, Maj Jerry House, ADSEC MTB Ln Off, for ADSEC Tn Off, 3 Oct 44, ADSEC 523.091 Red Ball–XYZ Routes.

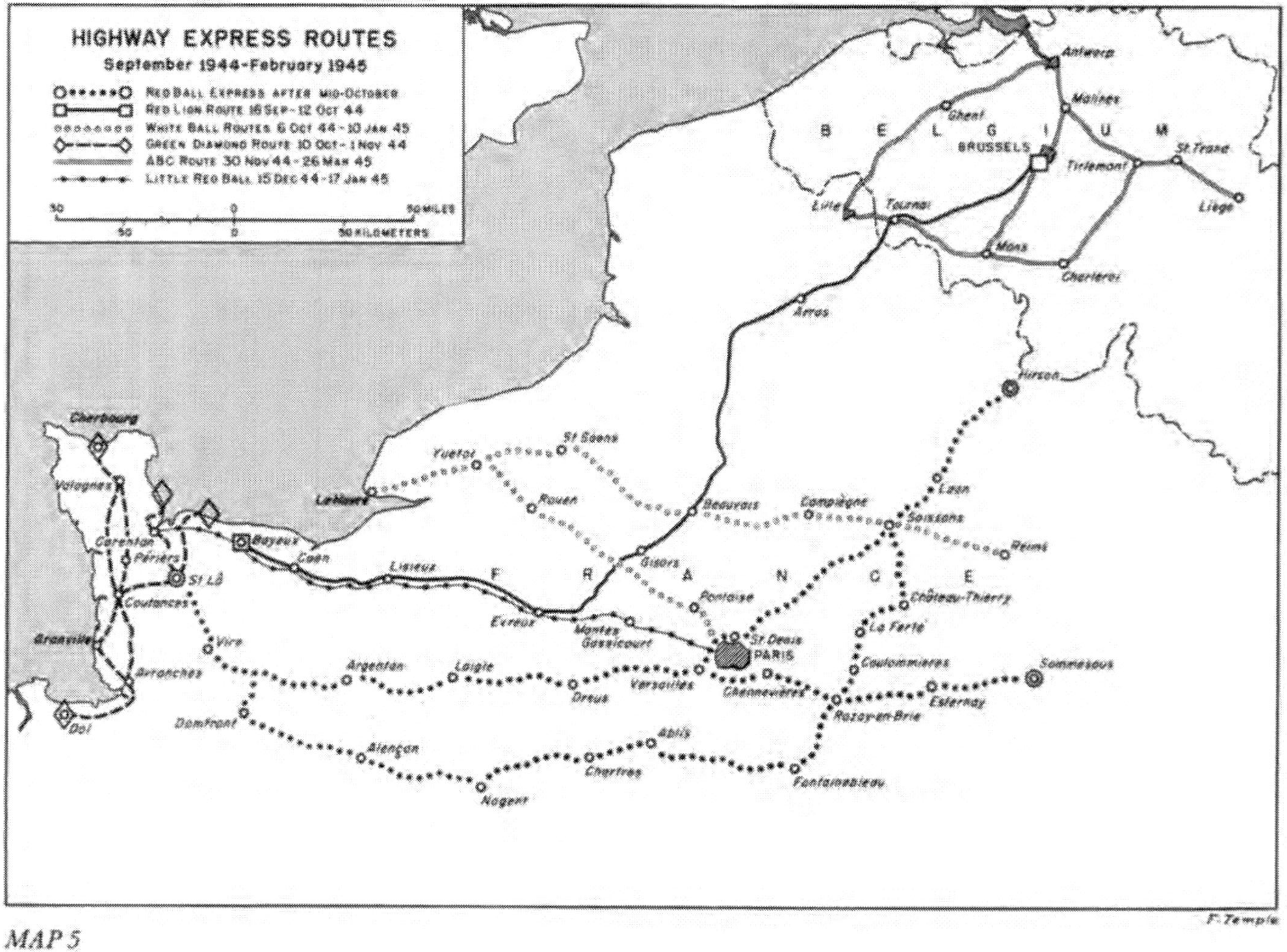
HIGHWAY EXPRESS ROUTES
September 1944–February 1945
Red Ball Express after mid-October
Red Lion Route 16 Sep–12 Oct 44
White Ball Routes 6 Oct 44–10 Jan 45
Green Diamond Route 10 Oct–1 Nov 44
ABC Route 30 Nov 44–26 Mar 45
Little Red Ball 15 Dec 44–17 Jan 45
50 0 50 MILES
50 0 50 KILOMETERS
BELGIUM
FRANCE
BRUSSELS
PARIS
Antwerp
Malines
Ghent
St Trond
Tirlemont
Liège
Lille
Tournai
Mons
Charleroi
Arras
Hirson
Cherbourg
Valognes
Carentan
Périers
St Lô
Coutances
Granville
Avranches
Dol
Bayeux
Caen
Lisieux
Le Havre
Yvetot
St Saens
Rouen
Beauvais
Compiègne
Soissons
Laon
Reims
Gisors
Pontoise
St Denis
Evreux
Mantes Gassicourt
Château-Thierry
La Ferté
Coulommiers
Sommesous
Esternay
Rozay-en-Brie
Chennevières
Versailles
Dreux
Laigle
Argentan
Vire
Domfront
Alençon
Nogent
Chartres
Ablis
Fontainebleau
F. Temple

MAP 5

Seine a corresponding amount. While this expedient entailed additional handling of supplies, it promised to increase the capacity of the available trucks by shortening the haul and decreasing the turnaround time.

Two transfer points were initially set up in the Paris area, one at Aubervillers la Courneuve (for First Army supplies), and one at Vincennes-Fontenay (for the Third Army). Both were well-developed yards where trucks could discharge their loads directly to empty freight cars. Aubervillers la Courneuve could accommodate 225 cars and work 20 at a time; Vincennes-Fontenay could hold 400 cars and work 115. Operations at both transfer points were organized and supervised by small detachments of officers and enlisted men working in two twelve-hour shifts, the actual transfer of cargo being carried out by a French labor force of between 300 and 350 men working in three eight-hour shifts. A third transfer point was established at Ruilly early in October to handle Ninth Army supplies. With the exception of a single crane at Vincennes, none of the yards was equipped to handle awkward or heavy lifts, hence only supplies that could be manhandled were accepted. With the inauguration of this plan a diversion point was established on the Red Ball route at Trappes, about twenty miles southwest of Paris, where manifests were examined to determine whether convoys should be routed to the transfer points or to the regulating stations serving the respective armies.

Truck-to-rail transfer was an immediate success, making it possible to discontinue much of the long-distance motor transport hauling beyond the Seine. In fact, tentative plans were made to discontinue Red Ball trucking beyond Paris entirely by 20 October. This goal was not achieved, for engineer supplies and equipment which were too heavy for transfer at Paris continued to be trucked the entire distance from the ports to the army depots beyond that date. Early in November the policy was laid down that all nontransferable items would be shipped straight through by rail and that trucks were to be used only for supplies that could be worked by hand and therefore transferred to rail at the Seine. In other words, there was to be no more long-distance hauling by truck beyond Paris.[3]

Less than two weeks later—on 16 November—the Red Ball Express ceased operations, its demise occurring on the same date the Normandy beaches closed down. In the course of its eighty-one days of operations the express service carried a total of 412,193 tons of supplies, some of them initially to the Chartres depot area, some directly to the armies, and in the last stages to the rail transfer points at Paris. In delivering an average of 5,088 tons per day its total ton-mileage came to nearly 122,000,000.[4]

Lacking precedent and experience,

[3] History of TC ETO, IV, Jul–Sep 44, OCofT, p. 9; V, Oct–Dec 44, 2d MRS, p. 14, and OCofT, p. 30; Memo, ADSEC Tn Off for ADSEC G–4, 9 Oct 44; sub: Weekly Transportation Narrative, ADSEC 319.1 Weekly Narratives, Tn Sec; COMZ G–4 History, III, 31–33; Ltr, Lt Col Harold L. Mack, ACofT for Movs TC, to CofT COMZ, 1 Nov 44, sub: Semimonthly Activity Rpt–Movs Div, EUCOM (TC) Semimonthly Reports to CofT Washington; G–4 COMZ Plant and Communications Diary/Jnl, Summary of Br Chiefs Mtg, 27 Oct 44, ETO Adm 145C.

[4] Hist Rpt of TC ETO, V, MTS, 12.

RED BALL EXPRESS TRUCKS *leaving a traffic regulating control point, September 1944.*

the Red Ball Express was plagued by problems of control and operational procedures through most of its history. The problem of control was inevitable in an organization operated by one COMZ section whose functioning involved the crossing of sectional boundaries. The Motor Transport Brigade, which operated the Red Ball, was an ADSEC organization. But all the COMZ sections traversed by the Red Ball routes had responsibilities affecting the efficient operation of the express service, such as road maintenance, traffic control, and signal communications, and these responsibilities were not carried out uniformly by the various section commanders. Reconciling the Advance Section's authority to operate the route with that of other section commanders proved an important stumbling block and evidenced a major defect in an organization which had a COMZ-wide function to perform. Confusion and disagreement inevitably arose between the Advance Section, Normandy Base Section, Seine Section, and Loire Section over maintenance of portions of the route, and late in September both the Chief of Transportation and Seine Section complained of unauthorized diversions and of changes in consignments made by the Advance Section at truck-to-rail transfer points.

Repeated attempts to delineate the responsibilities of the sections fell short of the goal. The essential defect of the system—the anomalous position of the Motor Transport Brigade—was finally recognized early in October, when the problem was resolved by transferring the control of motor transport operating intersectionally to a higher echelon. On 5 October the Motor Transport Brigade, a provisional organization in the first place, was dissolved and its personnel consolidated with the Motor Transport Service. The latter, operating under the

Chief of Transportation at the level of the Communications Zone, then assumed the duties of the Motor Transport Brigade. Truck units of the latter were attached to the base sections, but technical supervision and operational control of intersectional hauling was thereafter exercised by the Motor Transport Service. By coincidence the dissolution of the Motor Transport Brigade came shortly after the establishment of the truck-to-rail transfer points and therefore at a time when long-distance hauling east of Paris was already on the wane. With Paris becoming the main terminus of the Red Ball convoys it had become even more illogical that the Advance Section, already operating far forward, should control motor transport operating all the way back to the Normandy base.[5]

Meanwhile, through trial and error, Red Ball's operating procedures were also gradually improved. On 1 October a new standing operating procedure (SOP) on convoy make-up and control was put into effect, and a few days later a more clearly defined documentation procedure was adopted to correct earlier difficulties over marking and identification. Finally, on 2 December, in anticipation of possible future express systems, responsibilities of the COMZ sections were further amended to conform with the centralized control of intersectional hauling which had been instituted.[6]

[5] Daily Jnl, G-4 Plans and G-4 Plant and Communications, entries by Col Hansen and Col Potter, 18 and 29 Sep and 3 Oct 44, ETO Adm 145C, G-4 Plant and Communications Diary/Jnl; Hist Rpt of TC ETO, V, Oct-Dec 44, I, OCofT, 67; COMZ G-4 History, III, Ch. III.

[6] COMZ G-4 History, Chs. V and VI, and pp. 47-50.

While Red Ball was the first of the big express systems to be organized, and the most publicized, several other "color routes" on the model of Red Ball were established to meet specific needs in the fall of 1944. The first of these was the Red Lion Route, organized in support of the joint U.S.-British airborne operation carried out by the 21 Army Group in Holland. Red Lion's mission was to haul 500 tons of supplies per day (largely POL) from Bayeux to Brussels for a period of thirty days.

While U.S. forces furnished and operated the trucks—eight companies for most of the period—almost all other administrative services were provided by the British.[7] These included the loading and unloading of supplies, maintenance of the routes, the provision and staffing of camp sites and marshaling and control points, and the provision of medical facilities and water and rations. Vehicle maintenance was handled by two medium automotive maintenance companies, one stationed at each traffic regulating control point, where road patrols were based and repair work was performed. Each company established a small pool of 2½-ton trucks from which it could issue replacements in cases where repairs could not be carried out promptly.

[7] The eight U.S. companies—six of them equipped with 2½-ton 6x6's and two with 10-ton semitrailers—were withdrawn from the Red Ball Express and replaced with provisional companies formed with vehicles from the 26th, 95th, and 104th Divisions. The provisional companies, despite the lack of training, performed well, and their convoy discipline was reported to be generally superior to that of the regular truck companies. COMZ G-4 History, III, 53; Opn, Orgn, Supply, and Svs of TC ETO, Gen Bd Rpt 122, p. 44.

Red Lion convoys began their 300-mile runs to Brussels on 16 September and continued operations until 12 October. (*See Map 5.*) Normandy Base Section organized and operated the service. The route turned in a somewhat better performance than the Red Ball, profiting from earlier experience and enjoying certain advantages. Trucks carried a high average load of 5.9 tons, partly because of the density of the cargo. In addition, the operation benefited from the fact that all cargo was assembled at one dump in the Caen–Bayeux area, eliminating delays in pick-up and loading, and all trucks were unloaded at a single dump at the terminus of the route.

Red Lion convoys exceeded their target, delivering an average of 650 tons per day instead of 500, and handled a total of about 18,000 tons. Almost half of this consisted of supplies for the two U.S. airborne divisions participating in the Holland operation, a statistic often ignored by the partisans who so heatedly criticized this "diversion" of U.S. resources. Furthermore, the operation took place after the pursuit had definitely been halted and both the First and Third U.S. Armies had come up against the prepared defenses of the West Wall.[8]

Two other express services, one known as the White Ball Route, the other as the Green Diamond Route, were organized and placed in operation early in October. Both were relatively short hauls compared with the Red Ball and Red Lion systems. The White Ball Route was organized to take advantage of the shorter lines of communications from the newly opened ports of Le Havre and Rouen, its mission being to clear those ports, hauling supplies either directly to the armies or to rail transfer points at Paris and Reims. A quartermaster group headquarters (Transportation Corps) exercised operational control of the route, and Channel Base Section was made responsible for movement control. The White Ball Route started operating on 6 October 1944 and continued until 10 January 1945, with an average of twenty-nine truck companies participating. It handled a total of 134,067 tons of supplies on an average forward run of 113 miles.

The White Ball Route was modeled on the Red Ball Express, but performed rather poorly. Co-ordination and planning were noticeably deficient: depots were unaware of planned movements, labor was not provided at unloading points, both loading and unloading time was excessive, and neither line maintenance nor traffic control regulating points were provided until late in October. Lack of maintenance was reflected in the low rate of truck availability, only 32 of the 48 vehicles per company normally being fit for use as against the 40 expected under proper operating conditions. To make matters worse, trucks were frequently diverted for local use.[9]

The Green Diamond Route was organized by Normandy Base Section to

[8] Memo, Ravenhill for Vissering, 9 Oct 44, sub: POL Tonnage Lift, and reply, Vissering for Log Plans Br SHAEF, 12 Oct 44, SHAEF G–4 Tonnages —Cl III, POL 137/5/GDP—1; Hist Rpt of TC ETO, V, MTS, 15–18; COMZ G–4 History, III, 52–55.

[9] COMZ G–4 History, III, 55–56; Opn, Orgn, Supply, and Svs of TC ETO, Gen Bd Rpt 122, p. 44; Hist Rpt of TC ETO, V, MTS, 14–15.

move supplies from the ports and base depots in the Cotentin area to rail transfer points at Granville and Dol, near the base of the Cherbourg Peninsula. Although plans called for the use of forty truck companies, an average of only fifteen participated. The route was in operation only three weeks, from 10 October to 1 November, and delivered a total of only 15,600 tons of supplies. The Green Diamond Route was not a model of efficiency from the point of view of either planning or command supervision. In addition, it operated under a severe handicap imposed by mud, which made it almost impossible to handle the larger tractor-trailer combinations in the depots. They could be employed only by having cargo picked up by the smaller 2½-ton trucks and then transferred to the larger vehicles, an operation which proved highly uneconomical. In a sense the Green Diamond operation hardly belongs in the category of the special express routes, for it was more of a routine trucking operation and did not adopt most of the special operational procedures which characterized the larger express services.[10]

One of the most highly organized and efficient motor transport express systems came into being at the end of November. The ABC Haul was organized specifically to supplement rail and water facilities in clearing the port of Antwerp and moving the supplies discharged there directly from quayside to advance depots in the Liège–Mons–Charleroi area. Antwerp lay in the British zone, and the prompt clearance of the port was imperative because of the limited storage space available to U.S. forces there.

The ABC Haul derived its name from the fact that three nationalities—American, British, and Canadian—shared many facilities in the Antwerp area. Planning the operation involved the highest degree of co-ordination with the British on such matters as highway rights-of-way, restrictions on civilian traffic, and circulation routes through cities. The basis for this co-ordination was laid in the Memorandum of Agreement of 18 October, by which British and American officials had agreed on the use of the port. In all other respects the ABC Haul was strictly American in operation.

In addition to a high degree of co-ordination and organization, two features characterized the operation of the ABC route: the exclusive use of 4/5-ton truck-tractors with 10-ton semitrailers, one of the most efficient combinations for long-distance hauling; and the use of a marshaling yard or "surge pool." An average of sixteen companies of the big truck-tractor-semitrailer combinations was assigned to the ABC Haul, with a ratio of approximately two trailers for every power unit. Two of the truck-tractor companies did nothing but shuttle between the Antwerp surge pool and the quays, moving empties to the piers for direct loading from either ships or warehouses, and bringing loaded trailers back. At the surge pool fourteen companies assigned to over-the-road hauling picked up loaded trailers which

[10] Hist Rpt of TC ETO, V, Normandy Base Sec, 13, and V, MTS, 22; COMZ G–4 History, III, 52; Opn, Orgn, Supply, and Svs of TC ETO, Gen Bd Rpt 122, pp. 44–45.

TRUCK-TRACTOR AND SEMITRAILER STUCK IN THICK MUD *are pulled out by a D–7 tractor, Cherbourg area.*

had been formed into convoys and made the trip to the forward depots, a run which averaged ninety miles. A bivouac was established at Tirlemont, the halfway mark, where drivers were changed.

All the facilities which earlier experience had shown were necessary for efficient line-of-communications hauling were provided, including ordnance maintenance installations and road patrols, signal communications, aid stations, and so on. Control of truck movements was exercised from the Antwerp surge pool and the halfway point, and centralized control of the entire operation was achieved by having the route operated by a single base section commander—General Jacobs of Channel Base—although the route actually extended into ADSEC territory. The conflict of authority and responsibility which had plagued the Red Ball Express was therefore avoided, and long-distance truck transportation brought to a high degree of efficiency.

In January the operation was improved further by the establishment of surge pools at Liège, Mons, and Charleroi, where loaded convoys were received and then directed to forward dumps and depots in accordance with unloading capacities. This permitted a close control of movements at the receiving as well as the dispatching end, and made possible better line maintenance of equipment, a more efficient use of trucks, and a consequent saving in turnaround time.

The ABC Haul started operations on 30 November 1944 and continued until 26 March 1945. In that four-month period it cleared 245,000 tons from Antwerp, the average load per truck totaling 8.7 tons and the average round trip requiring twenty hours. The ABC Haul, like the use of Antwerp, afforded an excellent example of a highly organized and tightly controlled operation involving the crossing of national lines of communications and the joint use of logistic facilities.[11]

All the trucking systems described above were organized to handle large bulk shipments of supplies. In December 1944 still another motor transport service, known as the Little Red Ball, was inaugurated to meet a very special need—the fast delivery of small quantities of items urgently needed at the front. The requirement for a motor transport organization to fill this need arose from the fact that normal rail movements from Cherbourg to Paris required three days, while trucks could make deliveries in a single day.

The Little Red Ball route ran from Carentan, at the base of the Cotentin Peninsula, to Paris, following highway N–13 all the way. The service was designed to deliver only 100 tons per day—mostly medical, signal, chemical and quartermaster Class II items—and was operated by one truck company equipped with 10-ton semitrailers. It operated slightly less than five weeks, from 15 December 1944 to 17 January 1945, delivering an average of 106 tons per day.[12] A fast rail express service took its place a few days after it ceased operations.[13]

Independently of these various specially organized and for the most part short-lived color routes the Motor Transport Service had carried on another hauling operation of vital and continuing importance—the transportation of gasoline in bulk. The so-called POL hauls had started shortly after the Normandy landings, and were to continue without interruption till the end of hostilities. They were carried out by a special fleet of tank vehicles, consisting in the main of nine companies of 2,000-gallon semitrailers and five companies of 750-gallon tank trucks. From the beginning the Motor Transport Service, through a special POL Section, had exercised a highly centralized control over these operations, units of the tanker fleet being attached to the COMZ sections for administration only.

Most of the POL hauls were rather routine, did not involve an elaborate operating procedure, and were not favored by a special name or by publicity. Nevertheless the transportation of gasoline in bulk accounted for a high percentage of the total tonnages handled by truck. In the three-month period from October through December alone approximately 240,000 tons were forwarded from ports and pipeheads in this manner, the daily haul frequently exceeding 4,000 tons. In addition, substantial quantities of packaged gasoline and other POL products were transported

[11] COMZ G–4 History, III, 56–59; Hist Rpt of TC ETO, V, MTS, 18–21; Opn, Orgn, Supply, and Svs of TC ETO, Gen Bd Rpt 122, p. 45.

[12] COMZ G–4 History, III, 59; Hist Rpt of TC ETO, V, MTS, 13.

[13] The special rail service, known as the "Toot Sweet Express," is described in Chapter XV, Section 1, below.

LOADED 10-TON SEMITRAILERS, *Antwerp Surge Pool, waiting for the haul to forward depots, January 1945.*

by ordinary cargo trucks. In March 1945 the capacity of the POL tanker fleet was augmented by the addition of three companies of 10-ton semitrailers, each trailer mounting four 750-gallon skid tanks.[14]

The unexpectedly heavy burden which truck transport was forced to shoulder in the summer and fall of 1944 had its inevitable consequences in the attrition of the theater's motor transport resources. Vehicles underwent rapid deterioration in the fall of 1944 as the result of the grueling pace set by such expedients as the Red Ball Express. Truck units in the combat zone shared this to some extent, for they had joined in the long-distance work. Excessive speed, overloading, reckless driving, poor discipline and control, and the sacrifice of adequate maintenance in favor of short-term gains all contributed to the tendency. The result was a tremendous increase in repairs, which had already risen to 1,500 per day at the end of September.

The root of the trouble lay in poor maintenance, particularly first and second echelon, which suffered from both driver fatigue and lack of discipline. Plans for a "service station" type of line maintenance were not put into effect, partly because Red Ball was looked upon as a temporary expedient. Accidents, rather than mechanical failures, necessitated approximately one third of all vehicle replacement issues, reflecting both reckless driving and driver fatigue.

[14] Opn, Orgn, Supply, and Svs of TC ETO, Gen Bd Rpt 122, p. 44; Hist Rpt of TC ETO, V, MTS, 22–25, App. 7 to Ch. V, and Table 25.

TRUCK TRACTORS, 5-TON, HAULING 10-TON SEMITRAILERS *to Liège over the ABC Route, March 1945.*

In the fall General Mud stepped in to create additional hazards and hindrances in the forward areas and in many depots, causing overheated engines and shortening the life of brake systems. The available maintenance equipment was unequal to the suddenly magnified repair task. Four- and ten-ton wreckers in the heavy and medium automotive maintenance companies, for example, had to be augmented by 4-ton Diamond-T's normally used for towing.

Finally, lack of spare parts often kept vehicles deadlined. This was particularly true of tires. Overloading and lack of preventive maintenance took a heavy toll, and heavy damage was also caused by the sharp edges of the cut-off tops of C ration cans which were strewn along the roads. Theater reserves were quickly exhausted and repair facilities could not match the new demand. This deficiency was further aggravated by the lack of such supplies as camelback, an item essential in recapping.

Intensive efforts to improve maintenance were made in late 1944, particularly in the services needed along the lines of communications. These efforts eventually bore fruit in better vehicle availability, which rose from thirty per company at the end of November to thirty-five within the next month. Nevertheless, the replacement factor for cargo trucks, as in many other items of equipment, proved entirely too low for the conditions under which motor transport was used in the European theater. At the end of the year theater officials recommended that the original 2 percent factor be raised to 8 in the case of 2½-ton trucks, and 6 for the 10-ton semitrailer combinations.[15]

The Transportation Corps had continued its efforts throughout the summer and fall to acquire more of the heavier cargo units which had long since proved themselves much more efficient than the 2½-ton truck in the over-the-road hauling. The shortage of such units in the United States had forced the theater to accept substitutes, including many 2½-ton trucks, before D Day. By the end of December the Transportation Corps had succeeded in re-equipping thirty companies with the much-desired truck-tractor-semitrailer combinations, giving the theater a total of fifty companies equipped with the big 10-tonners.

Much of the new equipment was discharged at Marseille, where ordnance teams assembled the units and where companies sent down from the Communications Zone picked them up. Most of the drivers had had no experience in handling heavy equipment, and took a short training course while awaiting the assembly of their vehicles. When the trucks were ready they were driven to the quays at Marseille and loaded with supplies so that their initial lift capacity should not be wasted on the northward run. The newly equipped companies delivered these supplies to the army areas, and then proceeded to their assigned stations. Much of the equipment arriving in this manner was immediately committed to the ABC Haul out of Antwerp.[16]

The re-equipping program had already added materially to the capacity of the Motor Transport Service by the end of 1944. At the close of the year there were approximately 200 truck companies of all types from 2½-ton up, including tankers, refrigerator trucks, and even 2 companies of 45-ton trailers, in the Communications Zone, 84 of which came under the direct operational control of the Motor Transport Service. Under the Tables of Equipment this would have permitted an authorized strength of more than 10,000 vehicles, with about 8,300 in operation under optimum conditions. In actual practice, only about 75 percent of the authorized number was available, and a still smaller number in operable condition at any one time.

But the theater's motor transport resources were augmented steadily. By the close of the year seventy-five additional companies had been authorized for 1945, and the re-equipping program was also to continue. Early in the new year fourteen veteran companies arrived from Iran and were equipped with 10-ton diesel cargo trucks. Meanwhile other companies were shipped to the United Kingdom, where sufficient 10-ton semitrailer units were initially provided to equip and train five companies. On the Continent the Transportation Corps established a transfer pool at Chartres, where 2½-ton 6x6 companies could exchange their old equipment for the larger units.[17]

Approximately 30,000 men were employed in Motor Transport Service operations at the end of 1944, about three

[15] COMZ G–4 History III, 33, 38–39, 40–43; Hist Rpt of TC ETO, V, MTS, 6; History of the Normandy Base Section, D Day to V-E Day, 36, 38, ETO Adm 595; COMZ G–4 Plant and Communications Diary/Jnl, 26 Sep and 16 Oct 44, ETO Adm 145C.

[16] Hist Rpt of TC ETO, V, MTS, 3–4.

[17] *Ibid.*, 4, and App. 3 to Ch. V.

FOUR 750-GALLON SKID TANKS *mounted on 10-ton semitrailers used for transporting gasoline in bulk.*

fourths of them consisting of Negro enlisted personnel. The Transportation Corps had persisted in its efforts to get additional drivers so that truck companies could operate round the clock, but was only partially successful. Some overstrengthening had been carried out before the invasion. But the shortage of drivers remained acute throughout the summer, and in the fall the Transportation Corps was still recommending a substantial augmentation in strength of all truck units.[18]

Beginning in November, despite the serious maintenance and replacement problem, the Transportation Corps was in a much better position to meet sudden demands for motor transport occasioned by tactical developments, largely because of better planning for such eventualities. With the experience of the pursuit in mind, General Ross's staff made detailed plans to support a breakthrough should the November offensive produce one. Truck company bivouacs were reconnoitered and selected in the Namur and Verdun areas, and eighty companies with a lift capacity equivalent to 110 of the 2½-ton type were earmarked for rapid marshaling in support of any one or all of the armies in the

[18] In November the Transportation Corps asked the War Department to increase the strength of all truck companies to 140 enlisted men. Hist Rpt of TC ETO, V, MTS, 7, 28.

12th Army Group. The plan also provided for centralized operation of truck units through a specially organized forward echelon of the Motor Transport Service headquarters, operating under the control of either the main headquarters or the Advance Section.

While the tactical successes of November did not warrant the implementation of these plans, the preparations nevertheless stood the theater in good stead at the time of the German counteroffensive a month later, and were quickly adapted to that situation. Port clearance operations at Rouen and Antwerp were immediately curtailed and motor transport was released to meet more urgent demands. On 18 December, two days after the start of the attacks, the equivalent of 274 2½-ton trucks were taken off the White Ball Route and another 258 from Seine Section to rush combat formations—principally airborne units—forward from the Reims area. On the following day the White Ball Route released another 347 trucks for use in the redeployment of Third Army units. On 20 December additional diversions, including 10-ton semitrailers from the ABC Haul, were made to the Reims area. By the end of the month more than 2,500 trucks had been temporarily withdrawn from port clearance and static operations to handle emergency troop and supply movements. In some cases they evacuated forward dumps in danger of capture, as for example at Liège, from which tank trucks removed 400,000 gallons of aviation gasoline. Throughout these operations the Motor Transport Service ensured an efficient use of truck units by exercising centralized control over all movements in co-ordination with the COMZ G–4.[19]

(2) *The Railways*

While motor transport operated with greater and greater efficiency and gave a much-desired flexibility to the theater's transportation system, it constituted no substitute for the railways in the sustained movement of large tonnages over great distances. In the long run the railroad was the main workhorse of the transportation system, handling the great bulk of the tonnages.

In mid-September the railways had not yet assumed a large portion of the transport burden, although the Allied advance had uncovered almost the entire rail system of France, Belgium, and Luxembourg. Forward of St. Lô the lines then in operation had been rehabilitated in great haste, and only a few lines tentatively reached forward from the Seine, handling but a few thousand tons per day.[20] Very little additional mileage was captured in the next few months, and railway development was therefore confined almost strictly to the rehabilitation of the extensive network already in Allied hands.

At the end of September the 2d Military Railway Service, which operated the railways in the north, had under its jurisdiction approximately 2,000 miles of single track and 2,775 miles of double track lines. East of Paris three main lines of communications had been opened: one in the north to Liège via Compiègne, Cambrai, Valenciennes, Mons,

[19] Hist Rpt of TC ETO, V, MTS, 8–9, and V, OCofT, 31; COMZ G–4 History, III, 47, 59.

[20] G–4 report for quarter ending 3 Sep 44, Tn Sec ADSEC, 26 Oct 44, ADSEC 319.1 Supplement to G–4 Periodic Rpt.

Charleroi, and Namur; one via Soissons, Laon, and Hirson to Charleroi, where it tied in with the northern line; and a third extending directly east of Paris via Sezanne and Sommesous to the Nancy–Metz–Verdun triangle. By the end of December additional lines had been opened to raise the mileage to 3,500 of single track and 5,000 of double track, which was equal to about one third of all rail mileage in France.[21] *(See Map 9 below.)*

The condition of the lines varied greatly. By the end of December the lines in the original lodgment area and in northern Brittany were in fairly good repair. Northwest of Paris, however, where both Allied bombings and enemy demolitions had been heavy, much remained to be done. In the Paris area itself most of the bridges on either side of the city had been destroyed, forcing all traffic for some time to use the so-called "inner circle" route and the passenger stations in the very heart of the city. East of the Seine the railways had been left relatively intact, but damage was again heavy in the area where the enemy offered more determined resistance. In the north one of the two double-track lines which converged at Charleroi had several single-track sections, and beyond Charleroi only one double-track line was available to serve the railheads of the Advance Section, First Army, and Third Army at Huy, Liège, and Vise. At Liège itself only one bridge was left standing, creating a bottleneck which was not relieved until late in January. In the Metz area alone the enemy destroyed sixteen highway and five railway bridges and systematically demolished short stretches of track, making extensive use of the mechanical "track router."[22]

Bridge repair and reconstruction constituted by far the major task in railway rehabilitation, even though the footage destroyed averaged only 25 percent instead of the 50 percent expected. Nevertheless, by mid-November 180 damaged bridges had been repaired, and 125 had been rebuilt. Practically all this work was carried out by the Advance Section, whose engineer units were organized into groups, each with an allotted territory. Base and intermediate sections normally only carried out track work and completed marshaling yard rehabilitation.

U.S. engineers were initially dependent on British-designed bridging, produced in both the United States and the United Kingdom, for the U.S. Army had developed no military railway bridging. Shortly after the landings, however, American engineer units began to use captured rolled steel beams which the Germans had manufactured at the Hadir Steel Works in Differdange, Luxembourg, which greatly speeded bridge reconstruction. As soon as the Differdange plant was captured it began producing for the Allies and eventually provided beams for about 90 percent of all the bridges reconstructed on the Continent, including those for the Rhine, Elbe, and Danube crossings.[23]

Rehabilitation was only one of several

[21] Hist Rpt of TC ETO, V, MRS, 8–10; Teletype Conf, Gross *et al.* in Washington with Ross in ETO, 10 Oct 44, ETO Tn, 337 ETO, Teletype Confs.

[22] Hist Rpt of TC ETO, V, MRS, 9–10; Opn, Orgn, Supply, and Svs of TC ETO, Gen Bd Rpt 122, p. 56.

[23] *Final Report of the Chief Engineer, ETO,* I, 282–83.

problems which had to be met in bringing the railways to maximum usefulness. The Military Railway Service was short of experienced operating and supervisory personnel to begin with because of the inability of the zone of interior to release men with such training. Less than 15 percent of the strength of the operating battalions consequently consisted of men of "operating caliber" judged by normal standards. This deficiency was later aggravated by the course of tactical developments on the Continent. The Allies had never contemplated attempting to operate the railways of liberated countries exclusively with military personnel for an indefinite period. They had planned to return facilities to the respective countries in easy stages as areas were made secure from enemy attack, as near-normal operating conditions were restored, and as civilian organizations were reconstituted. Under these plans operations in the earliest stages during which the lines were being rehabilitated, known as Phase I, were to be handled by military units, with assistance from civilians wherever possible. Phase II operations were to begin once normal operating conditions were restored in a particular area or over a given line. Military officials were to supervise operations and retain complete control in that period, but trains were to be operated by civilian crews. Complete operational responsibility was to be restored to civilian agencies within a particular area in Phase III, although military personnel were to maintain close liaison with civilian agencies, and U.S. military requirements were to have movement priority.[24]

The rapid advance beyond the Seine had much the same effect on rail plans as on other features of the logistic plan. Units of the Military Railway Service were suddenly extended to a much greater degree than expected. In addition, difficulties in language and in documentation of shipments under French operation, plus the fact that the Cherbourg area for a long time remained the main base of supply, made it infeasible to return the railways to French control as rapidly as planned. The northern lines into Paris, in addition to the lines to the north, east, and southeast, consequently remained under Phase I operation, spreading the available operating personnel much more thinly than planned.[25]

Far more serious a limiting factor was the shortage of equipment, notably in locomotives and rolling stock. Captured equipment could not be counted on to meet all Allied requirements in view of the systematic destruction to which the French railways had been subjected by the Allied air forces and in view of expected German demolitions. The Allies had therefore planned to ferry substantial quantities of motive power and rolling stock to the Continent. Allied requirements for locomotives were originally estimated at 2,724 of the 2–8–0 type, 1800 of which were intended for American use, and 680 of the 0–6–0 type, 470 of them for U.S. use, plus a smaller number of diesels of various types. By

[24] Hist Rpt of TC ETO, V, MRS, 10–11; Military Railway Service, Gen Bd Rpt 123, pp. 11–12.

[25] Opn, Orgn, Supply and Svs of TC ETO, Gen Bd Rpt 122, pp. 53, 56. At the end of 1944 the 2d Military Railway Service consisted of 18 operating battalions, 4 shop battalions, 5 mobile workshop units, and 10 hospital train maintenance crews, organized into 5 grand divisions, and had a strength of 17,526 men. Hist Rpt of TC ETO, V, MRS, 35.

the end of June, 1,358 of the first type and 362 of the second had been made available in the United Kingdom, leaving a requirement of about 2,000 to be met from U.S. production. Four hundred and fifty of the locomotives shipped to the United Kingdom had been temporarily loaned to the British for use in the United Kingdom with the understanding that they would be released and shipped to the Continent as they were needed there.

Shortly after the landings in Normandy General Ross, alarmed over reported cutbacks in the production of locomotives in the United States, and over the failure of the British either to release engines on loan as scheduled or to deliver locomotives promised from U.K. production, informed the War Department that the theater would need all locomotives originally requested. The War Department, reluctant to undertake additional commitments because of interference with new tank production in the locomotive shops, asked the theater to exert all possible pressure on British officials not only to release the 450 engines loaned them, but also to make an all-out effort to meet earlier production commitments. The Army Service Forces meanwhile made similar representations to the British Ministry of Supply Mission in Washington.[26]

Late in August, Allied tactical successes, engendering hopes for an early victory, tended to relieve the anxieties of theater Transportation Corps officials over the adequacy of locomotives.[27] But this optimism was short-lived. The rapid advance across France only aggravated the shortage of motive power. Of the locomotives found on the French railways west of the Seine only about fifty could be placed in immediate use, and it was estimated that 80 to 85 percent of those recovered would be found inoperable.[28]

Lack of power was partially attributable to the uneconomic use of the available locomotives which resulted from poor management at the boundaries between operating battalions in Normandy, although this was eventually corrected. For a time it was necessary to double-head some trains. This not only doubled the requirement for motive power but for engine crews as well, creating a shortage of both personnel and power. The solution to this problem was found in employing French engine crews to man the second engine; 124 crews were recruited by the end of October.[29]

The urgency of increasing motive power on the Continent meanwhile found expression in plans to move 500 locomotives across the Channel in September, and at the rate of 20 per day thereafter. Arrangements were made several weeks later for the gradual release and shipment of the 450 U.S. locomo-

[26] Ltr, Lt Col Frank R. Crom, SHAEF, to Gen Maxwell, WD G–4, 31 Jul 44, sub: Locomotives for U.S. Troops in ETO, 1st Ind, CG ASF to G–4, 29 Aug 44, and Cbl WAR–83364, ASF to CG ETO, 19 Aug 44, all in WDGDS 453.3, WDGDS A47–2, Railroad Locomotives and Rolling Stock; Ltr, Lee to Somervell, 24 Aug 44, sub: Locomotive Requirements for ETOUSA, EUCOM 453 Railway Equipment 1944.

[27] Ltr, Michael Keith to WD G–4, 31 Aug 44, sub: Locomotives for U.S. Forces in ETO, WDGDS 453.3 WDGDS A47–2.

[28] Mil Shipments Priority Mtg, SHAEF, 2 Sep 44, SHAEF AG 337–18.

[29] G–4 COMZ Plant and Communications Diary/Jnl, Br Chiefs Mtg, 28 Oct 44, ETO Adm 145C.

AMERICAN LOCOMOTIVE LOWERED BY CRANE *from a seatrain to the rails at Cherbourg.*

tives then in use on the British railways.[30] These schedules proved unattainable, partly because of movement difficulties, and had to be revised.[31] Even more important, however, was the fact that the release of the 450 "Boleros" was closely linked with the problem of Britain's coaster fleet, the loan of the locomotives compensating in part for the extraordinarily extended retention of coaster tonnage in cross-Channel service.[32]

Late in November SHAEF pressed for the immediate shipment of 50 engines in the second half of December, followed by 100 per month to the end of March, holding out the hope that at least a portion of the coaster fleet might be released with the opening of Antwerp. The British agreed to release 150 locomotives by the end of December, and to prepare another 100 for dispatch in January. But they strongly indicated that the release of these and subsequent quotas might be conditioned on the return of a portion of the coaster fleet.[33]

At the very end of December the return of 100,000 tons of coaster shipping

[30] Mil Shipments Priority Mtgs, 2 Sep 44, SHAEF AG 337-18; Ltr, Keith to WD G-4, 18 Oct 44, sub: Locomotives for U.S. Forces in ETO, WDGDS 453.3 WDGDS A47-2.

[31] Mil Shipment Priority Mtgs, 9 Sep and 18 Nov 44.

[32] See above, Ch. IV, Sec. 4.

[33] Cbl S-68318, CAO to Br COS, 25 Nov 44, and Cbl 7016, Br COS to SHAEF, 30 Nov 44, SHAEF G-4 Rail Transportation 117/3 GDP-1.

led the British Chiefs of Staff to offer to dispatch 100 locomotives to the Continent during January, February, and March. Together with previous shipments, this promised to complete the release of the 450 engines which had been loaned for service on the British railways.[34] At the end of the year a total of 1,500 locomotives had been moved to the Continent, and an additional 800 captured engines—French, German, and Italian—had been repaired by French and American mechanics and placed in service.[35]

The same considerations which had led to the planned shipment of locomotives to the Continent had also led to planning substantial importations of rolling stock. More than 57,000 cars of various types, including box, tank, refrigerator, and flat cars, and cabooses were scheduled for shipment to the Continent, approximately 20,000 of which were shipped knocked down from the United States and assembled in British shops before D Day.[36]

Movement difficulties—notably the lack of reception facilities on the far shore—prevented shipment of much of this rolling stock to France in the first few months. Although a substantial amount was captured, shortages began to develop as soon as the breakout from Normandy imposed the long-distance hauling mission on the railways. Neither the condition of the French railways nor military requirements lent themselves to an efficient and economical use of even the limited rolling stock available. As early as mid-August the Communications Zone took measures to eliminate an evil which plagued the operation of the railways to the end of the war—namely, the tendency to hold loaded freight cars in the forward areas, thus removing them from circulation or at best lengthening the turnaround time. The Communications Zone impressed upon all the section commanders the necessity of prompt unloading of cars in view of the shortage of rolling stock and the limited sidings available. At the same time it authorized the chief of transportation to impose embargoes and divert shipments elsewhere if congestion developed at unloading points.[37]

By November an average of approximately 23,000 tons of supplies was being forwarded by rail east of the Seine each day, attesting to a tremendous increase in rail hauling capacity since mid-September.[38] But the very substantial increase in the number of freight cars available to the Allies by the end of November brought no final solution to the rolling stock shortage.[39] In mid-November, a serious jamming up of trains began to develop in the forward area, and

[34] Cbl 7635, Br COS to SHAEF, 30 Dec 44, SHAEF AG 617 Railway Case A, I.

[35] Hist Rpt of TC ETO, V, MRS, 30–31, and App. 7.

[36] *Ibid.*, V, MRS, 26–27, and App. 7.

[37] Ltr, Hq COMZ to Secs, 14 Aug 44, sub: Unloading of Rolling Stock, EUCOM 453 Railway Equipment.

[38] Hist Rpt of TC ETO, V, MRS, App. 4.

[39] On the basis of French claims, total rolling stock available to the Allies, French and Belgian combined plus 20,000 cars ferried over from the United Kingdom, should have totaled about 234,000. U.S. officials suspected that the French inventory included much rolling stock that wouldn't roll. It was advantageous to the French for purposes of compensation to include everything on wheels. Memo, Mil Railways Br G–4 SHAEF for Chief Mov and Tn G–4 SHAEF, 30 Nov 44, sub: Rail Mov of Freight and Pers to Support A Gp, SHAEF G–4, Rail Tn 117/3 GDP–1.

quickly extended back from the railheads. By 20 November eastbound trains occupied every block from the Belgian border to Namur. Within another few days the entire Belgian rail system became so choked with traffic that it was necessary to clear selected trains from the main track at stations near the French border and release their crews and engines. Similar developments in the Verdun area, the center of a big advance depot complex, necessitated the sidetracking of cars to rear areas, resulting in congestion, delays in the spotting and unloading of cars, and the tie-up of precious rolling stock. Meanwhile rail operations at Liège, already a bottleneck because of destroyed bridges, were partially disrupted by heavy V-bomb attacks carried out in the last ten days of the month.[40]

Part of the freight car tie-up difficulty stemmed from the tendency of the armies to keep as high a percentage of their reserves as possible on wheels, with the result that loaded cars accumulated on available sidings and rolling stock was immobilized. The Communications Zone had repeatedly called for "drastic action" to eliminate this costly practice. A more immediate cause was the decision to ship large tonnages of supplies in bulk directly from the ports to advance depots, contrary to established doctrine. Neither ADSEC nor army installations were prepared to accept the avalanche of supplies which began to descend on them toward the end of November. Unloading and storage facilities in the forward areas, including depot personnel, sidings, and switching facilities, were always at a premium. Moreover, they were intended for the primary function of *issuing* supplies to using units (which entailed the storage of a relatively small portion of the theater's reserves and presumed a fairly steady flow of maintenance needs), not that of the classification and segregation of supplies shipped in bulk, a mission normally assigned to intermediate or base depots.

The opening of Antwerp at the end of November did not help matters, despite the prospect of quicker turnaround on the shorter lines of communications, for the lack of storage space in the port necessitated the prompt forwarding of supplies. On the last day of the month there were already 11,000 loaded freight cars on the rails east of Paris. Within ten days the number had risen to 14,000, which was estimated to be more than double the normal operational needs. Turnaround time for some cars was between twenty and forty days.[41]

The Communications Zone, recognizing the seriousness of the tie-up, on 12 December outlined a specific program to relieve the congestion. Included in the instructions to the chief of transportation and section commanders were orders to unload commodity-loaded cars

[40] Hist Rpt of TC ETO, V, MRS, 14–15; Opn, Orgn, Supply, and Svs of TC ETO, Gen Bd Rpt 122, p. 56; Sgt. Joseph R. Ives, "Buzz Bombs and the Raids," *Army Transportation Journal*, I (September, 1945), 6–8.

[41] SHAEF estimated that the total loads east of the Seine should never be greater than three times the daily unloading capacity. Under existing conditions, with an estimated unloading capacity of 2,000 cars per day, the total cars under load should not have exceeded 6,000. Memo Concerning Transportation, unsigned, 13 Dec 44, Hq ASF Official Rpt Mission to ETO 4 Dec 44–13 Jan 45, by Gen Lutes, Tab M; Min, CAO Mtg, 1 Dec 44, App. C, SHAEF AG 337.14.

held as a rolling reserve, to reduce outloadings at Cherbourg, to build additional sidings in the ADSEC area, to stop and hold at Paris all ration trains consigned to Verdun, and to make substantial cuts in the over-all shipment of rations, which had averaged nearly 6,000 tons per day in the first days of December. To prevent this program from interfering with ship unloadings, Channel Base Section was instructed, as a temporary expedient, to continue the unloading of supplies from ships then working at Antwerp and to place them in temporary quayside storage or any suitable nearby area. Since this would in turn produce a backlog of supply stocks west of Paris and south of Antwerp, the chief of transportation was instructed to initiate a call-up system once normal shipments were resumed so that depot commanders would have some control over the daily load imposed on them.[42]

But the relief promised by this program was postponed. Within a few days the enemy counteroffensive in the Ardennes intervened and further aggravated congestion on the rails. Approximately 35,000 rail cars were allowed to accumulate in the forward areas and were held there against the possibility of large-scale troop and supply evacuations, with the inevitable effect of restricting port discharge and rail movements in the rear. The Communications Zone met the problem of the backlog and pressure on Antwerp partially by acquiring additional storage space from the British, partially by establishing inland holding and reconsignment points, particularly at Lille and Cambrai, which served as a cushion between the port and the depots, permitting clearance of the port to continue.[43] But General Ross later estimated that the counteroffensive had had the effect of setting back car unloadings and the movement of freight by nearly 14,000 cars or thirty-five shiploads of supplies.[44]

Rail transportation loosened up once more with the turning back of the German drive, and by the end of January the worst effect of the setback had been overcome.[45] But the control of movements and the unloading and release of rolling stock remained a thorny problem until the end of hostilities.

Damage to rail lines in the area of the December counteroffensive took the form mainly of destroyed bridges, caused by both Allied and enemy demolitions and Allied air attacks. At St. Vith the yards were completely destroyed. Outside the main battle area one major rail bridge over the Meuse had been destroyed at Namur. During an enemy air raid on 24 December a lucky hit had set off charges which Allied forces had placed in preparation for possible demolition. The bridge was immediately re-

[42] Ltr, Lord to CofT and Sec Comdrs, 12 Dec 44, sub: Reduction of Number of Loaded Cars Between Paris and Forward Areas, EUCOM 453 Railway Equipment, 1944; COMZ G–4 History, II, 97–102; Cbl EX–75445, Lee to Base Secs, 17 Dec 44, ETO Adm 397.

[43] Ltr, Lt Col R. W. Reisner of SHAEF G–4 Mov and Tn Br, to G–4, 20 Dec 44, sub: Rpt of Trip to Antwerp, SHAEF G–4 825.1 Piers . . ., 1944, V; TWXs S–73064, SHAEF to COMZ and 21 A Gp, 30 Dec 44, and EX–80904, COMZ to SHAEF, 31 Dec 44, SHAEF SGS 400.3/1 Supply Problems of Allied Advance; Channel Base Section History, I, 239; Hist Rpt of TC ETO, VI, Jan–Mar 45, Base Secs, 15.

[44] COMZ Stf and Comd Conf, 5 Jan 45, EUCOM 337/3 Conferences, Staff Weekly, I.

[45] Opn, Orgn, Supply, and Svs of TC ETO, Gen Bd Rpt 122, p. 56.

built, and was opened to traffic on 5 January.[46]

Problems of rail development and operation in the north were substantially duplicated in the south. A good rail net existed in southern France, double-track lines running north from Marseille along both banks of the Rhône, supplemented by another double-track route branching off at Valence and extending northward via Grenoble and Besançon. From these main lines an extensive network reached into Alsace via the Belfort gap and into Lorraine.

Rail operations in the south began on a very limited scale as early as 17 August (D plus 2), when small quantities of supplies were moved inland from the St. Tropez beaches on a narrow-gauge line. The first standard-gauge line was placed in operation shortly thereafter between St. Raphael and Aix-en-Provence. From the latter the single-track line north to Grenoble was then opened despite destroyed bridges across the Durance River at Meyrargues and the Buesch River at Sisteron. For a short time supplies were hauled by rail to Meyrargues and then trucked to Sisteron, where they were transferred back to rail. By mid-September temporary bridges strong enough to carry loaded rail cars, but not engines, had been completed at both points, and the line was open as far north as Bourg, 220 miles from the landing beaches. For a few weeks this route served as the principal rail line of communications, with a capacity of 1,500 tons per day.

In the meantime, the two main lines straddling the Rhône had been reconnoitered as far north as Lyon, where the headquarters of the 1st Military Railway Service, commanded by Brig. Gen. Carl R. Gray, was established on 14 September. Restoration of the line on the west bank of the river was initially ruled out, for all bridges but one had been demolished. The east bank line was in better condition, and steps were immediately taken to repair it and use it as the main supply route of the southern armies. Plans were initially made to locate the main advance depots in the Dijon area, and to extend the railways eastward from that city to Besancon and Belfort.

By 25 September the line was open as far north as Lyon with a capacity of 3,000 tons per day. By the end of the month the line was in operation north to Dijon and eastward to Besançon. Rehabilitation of the parallel eastern route north from Bourg continued, meanwhile, and was completed on 5 October, when the eastern line joined the other at Dol, halfway between Dijon and Besançon. At that time SOS Advance Headquarters at Dijon was accepting bids for the movement of 8,350 tons per day via rail. Within another week, with the acquisition of additional rolling stock and motive power, the capacity of the railways along the southern lines of communication had risen to 12,000 tons.[47]

By early October operational plans and the deployment of the Seventh U.S. and First French Armies dictated the

[46] Railroad Construction and Bridging, Hist Rpt 12, CE ETO, pp. 73–74.

[47] 6th A Gp G–4 AAR, Sep–Oct 44, p. 3; Hist Rpt of TC ETO, VI, MRS, 13–15; Opn, Orgn, Supply, and Svs of TC ETO, Gen Bd Rpt 122, pp. 59–60.

MAJ. GEN. FRANK S. ROSS, *Brig. Gen. Carl R. Gray, and Brig. Gen. Clarence L. Burpee walking alongside an American locomotive.*

development of rail lines northward as well as eastward. Epinal was selected as the main depot area for the support of the Seventh Army, and the rail line north from Dijon to that area was restored by the middle of the month. This gave the 6th Army Group two rail lines of communications beyond Dijon, one eastward via Besançon toward the Belfort gap, and one northeastward via Langres to Epinal in the direction of the Saverne gap. Railheads for the two armies remained in the vicinity of Besancon and Epinal for most of the following month. In mid-November the completion of a bridge over the Moselle made it possible to extend the northern line an additional thirty miles to Luneville. Meanwhile the line supporting the First French Army was extended to Clerval, approximately half the distance from Besancon to Belfort.

In December the rapid advance of the Seventh Army through the Saverne gap and the subsequent reversion to the defensive in the Colmar area had their effect on rail reconstruction priorities. The Seventh Army's capture of Strasbourg placed railheads out of comfortable reach of motor transport, creating a demand for rail support east of Luneville. The removal of track between Luneville and Sarrebourg and damage to a tunnel east of Sarrebourg presented a formidable reconstruction task. But by extra exertions the line was restored through Sarrebourg and Saverne all the way to Strasbourg and Haguenau and opened to traffic on 21 December. At the same time engineers of the 1st Military Railway Service extended the network supporting the First French Army by restoring the line north of Besançon via Vesoul and Lure to Champagney. Progress in that area was painfully slow because of heavy demolitions on bridges and tunnels and on the rails themselves. At the end of December the passing to the defensive in the Colmar region revived the importance of the Epinal–St. Die–Strasbourg line, which was given first priority so that units fighting on the northern side of the pocket could be supported.

As in northern France, rail rehabilitation was largely a matter of reconstructing bridges, of which forty-two were rebuilt by the end of the year, although the southern forces also had to contend with blown tunnels.[48]

At the end of December the 1st Military Railway Service had approximately 4,000 miles of track under its jurisdiction, and this mileage did not change appreciably in the next few months. The southern lines at that time had a rated capacity of 14,000 tons per day. But this was rarely if ever realized in performance. Like the 12th Army Group in the north, the 6th Army Group allocated the available tonnage between its armies.

Railway operations in southern France were conducted under Phase II conditions almost from the start, and very little mileage was ever operated entirely by military units. As in the north, shortages of personnel, motive power, and rolling stock plagued operations. The shortage of cars was aggravated, as on the northern lines, by the failure to unload cars. Beginning in October hundreds of cars were consistently held under load as mobile reserves.

Late in December the problem began to reach critical proportions as rail operations in the south suffered additional hazards and handicaps imposed by a severe winter in mountainous terrain. Extreme cold and drifting snow, which caused power failures, maintenance difficulties, and disruption of communications, plus sickness among the French crews and a shortage of coal, seriously hampered rail operations, resulting in the further piling up of loaded cars. At its worst, eight days of supplies accumulated and awaited unloading or movement into railheads. An attempt to break the log-jam by placing a forty-eight-hour embargo on all loadings at Marseille brought only passing relief, and the pile-up continued. The worst occurred at Is-sur-Tille, the Seventh Army regulating station, where 2,000 cars accumulated. Inclement weather also held up further rail reconstruction, with the result that the blown tunnel at Champagney continued to block the Lure–Belfort line, and the damaged Dannemarie viaduct the line between Belfort and Mulhouse.

Operating conditions worsened in January, when both forward deliveries and

---

[48] A Gp AARs, Sep–Nov, pp. 3–4, Dec, pp. 2–3; Record of Conversation, Maj. R. D. Hollis, SHAEF G–4 Sec, with Little and Lewis of G–4 Plans, 26 Oct 44, 6 A Gp, 27 Oct 44, SHAEF G–4 Diversion of Service Troops DRAGOON; Hist Rpt of TC ETO, VI, MRS, 14.

CONVOY OF TRUCKS CARRYING ESSENTIAL SUPPLIES *for Seventh Army travels over snow-covered winding roads in the Vosges mountains.*

unloadings dropped to an all-time low. At the end of the month 5,000 cars were still under load in yards and railheads of the two armies and of the Continental Advance Section. The supply of both armies reached a precarious state at that time, and led officials of the 6th Army Group and SOLOC to institute an even more rigid control over all transportation in order to ensure delivery of the minimum tonnages of essential maintenance items.

CONAD now began to allocate trains on a daily basis, specifying the exact number which were to go forward with Class I, III, and V supplies. Preparations were made to handle Class II and IV supplies almost exclusively by motor transport. For this purpose another special trucking operation, known, like that which operated briefly between Cherbourg and Dol in November, as the Green Diamond Route, was organized to operate between the CONAD depots and the armies. One thousand vehicles were drawn from Delta Base and other COMZ sections for the job, including 200 from CONAD itself. On 1 February sixteen companies, a majority of them of the 10-ton semitrailer type, and organized along the lines of the special motor transport services in the north, began shuttling critically needed supplies from Dijon, Langres, and Is-sur-Tille to Seventh Army depots and supply points.[49]

[49] 6 A Gp G–4 AAR, Jan 45; CONAD History, pp. 145, 164, 167; History of SOLOC, II, TC, 21: Ltrs,

These measures, plus a sudden improvement in rail operations resulting from better weather and an increase in motive power, quickly relieved the critical supply situation. Within six days the backlog of rail cars at Is-sur-Tille alone was reduced from 2,242 to 928 despite unrestricted movements into the yards. In the second week of February deliveries to the armies reached an all-time record—52,034 tons by rail and 7,853 tons by truck. By mid-February the rebuilding of forward reserves had progressed so well that the special trucking operation could be terminated.[50]

Both the 1st and 2d Military Railway Services had experienced trying times in December and January, having been put to severe tests by both winter weather and the demands occasioned by enemy offensives. Experience in both the north and south had underscored the importance of one of the most vital aspects of military transportation—movement control. Railway operations in a theater of war can rarely be conducted on a "scheduled" basis. Requirements for transport, reflecting the requirement for supplies, are subject to frequent changes. Rail lines, yards, depots, and handling equipment are often inadequate or destroyed. It is difficult to impose penalties for hoarding loaded rail cars. Existing facilities must therefore be used to the best possible advantage if the needs of the combat elements are to be met. This means the allocation of rolling stock and the matching of facilities—loading, movement, and unloading—in such a way as to avoid waste. Movement control was far from perfect in both the north and south in the fall of 1944, as attested by the wasteful idleness of rolling stock and the choking up of rail lines.

The organization and operation of the two services differed in at least one major respect. The 2d Military Railway Service in the north was operated by the Chief of Transportation, Communications Zone, although the various operating units of which it was made up were attached to the COMZ section in which they operated. By the end of 1944 movement control had become fairly centralized. The chief of transportation normally received movement requirements each month from the COMZ G–4. In conference with the various divisions of the Office of the Chief of Transportation, a decision was made as to how to fill the various movement needs, whether by rail, water, or truck. On the basis of this decision a movement plan was then drawn up, showing the ports at which certain tonnages would originate and the depots to which they were to be shipped. The 2d Military Railway Service, having been informed of the rail requirements for the plan, then prepared a detailed movement plan which was to be carried out through the railway operating battalions in co-operation with the COMZ sections involved.

The 1st Military Railway Service, unlike the 2d, had been established as a separate subcommand of the Mediterranean theater, and enjoyed an unusual degree of autonomy. As such, it possessed

Devers to Larkin, 2 Feb 45, and Larkin to Devers, 4 Feb 45, ETO, Memos for Gen Larkin; Ltr, Brig Gen S. L. Scott, Dir Planning Div ASF, to Lutes in the ETO, 12 Dec 44, sub: Shipment of Locomotive Coal from U.S. to Marseille, ASF, Official Rpt Mission to ETO, 4 Dec 44, 13 Jan 45 by Lutes.

[50] CONAD History, pp. 167–68.

its own engineers and carried out its own rail reconstruction, procured its own supplies and equipment, had its own operating units, and had assigned to it the necessary military police units to guard and protect supplies in transit. Movement control was not centralized for all modes of transportation as in the north. Once the tonnage bids were allocated, the 1st Military Railway Service assumed control of its own movements, which it handled quite independently of the chief of transportation.

Opinions differ as to whether this type organization had any advantages over that adopted in the north. Car shortages developed just as they did in the north; backlogs developed through failure to call cars forward; and the armies held on to loaded cars as rolling reserves. Clearing up the traffic backlog in February was in fact largely attributable to the fact that the transportation officer of Continental Advance Section for the first time assumed the function of traffic regulator.[51] Some of the features of the southern system nevertheless were adopted for the Military Railway System as a whole in the spring of 1945.

On 12 February 1945, when the separate Southern Line of Communications was dissolved and the northern and southern lines were finally integrated under one command, the two railway services were also brought under a central supervisory control—General Headquarters, Military Railway Service, established in Paris under General Gray, who became subordinate to General Ross as theater chief of transportation.

(3) Air Transport

No unusual developments took place in the field of air transportation in the weeks immediately after the pursuit. After achieving an average delivery of slightly more than 1,000 tons per day in the second week of September, when troop-carrier and cargo aircraft of the First Allied Airborne Army and converted B–24 bombers from U.S. Strategic Air Force plus a small contingent from the RAF combined in a final effort, supply by air fell off abruptly when planes of the First Allied Airborne Army were finally withdrawn for the Holland airborne operation.[52]

For about a week only B–24's were available for air supply in support of the 12th Army Group. These were used to good advantage in the transportation of gasoline in bulk from the United Kingdom. The bombers had begun to haul limited amounts of bulk POL on 9 September. Beginning on the 18th a major emergency effort was made to haul POL to three fields which the Ninth Air Force made available for the purpose on the Continent—St. Dizier, Clastres, and Florennes. The bombers could carry only 1,600 to 1,800 gallons of gas in four bomb bay tanks, and on the first day of the new effort only 11 planes, carrying 17,580 gallons, made deliveries. The next day 77 planes were dispatched to the Continent with 123,414 gallons. The lift continued until 30 September. The largest delivery was made on the 29th, when 197 planes moved 301,376 gallons. In the thirteen days of operations a total of 1,601 sorties were flown and 2,589,065 gallons of gas-

[51] Opn, Orgn, Supply, and Svs of TC ETO, Gen Br Rpt 122, pp. 60–61; Hist Rpt of TC ETO, V, MRS, 15, and VI, MRS, 1–2.

[52] See *Logistical Support I*, 578–83.

oline were delivered to the Continent. About three fourths of the POL was flown to the three U.S. fields, the remainder to a British field for 21 Army Group.[53]

Meanwhile, on 22 September the cancellation of a portion of the planned resupply missions in Holland made an additional 30 troop-carrier aircraft available, and the First Allied Airborne Army promised to increase the transport fleet to 600 craft as quickly as they became available. During the last week of September total cargo deliveries rose to a record 1,525 tons per day, partly through the return of troop-carrier aircraft, partly through the continued use of B–24's and some British Halifaxes. Despite the constant uncertainty over the availability of transport aircraft which characterized operations throughout the month, Allied planes flew 11,000 sorties and delivered more than 30,000 tons in September. This represented 60 percent of all the tonnage transported by air since D Day.[54]

The Communications Zone, faced with a sizable deficit in transportation, would have liked to continue utilizing the airlift to the fullest possible extent. At the beginning of October road and rail transportation were inadequate to meet even the daily maintenance requirements of the armies, Advance Section, and the Ninth Air Force, to say nothing of moving forward the 150,000 tons of supplies required to reconstitute the seven-day supply level authorized the armies. Between 6,000 and 7,000 tons of additional lift were needed if both maintenance and reserve goals were to be met by the end of the month. General Stratton, fully aware that this requirement was beyond the capacity of the available transport aircraft, nevertheless asked the Army Group G–4, General Moses, to seek the maximum possible allocation from the SHAEF Air Priorities Board.[55]

Because of its relative extravagance as a means of transport, the continued large-scale employment of aircraft for supply movement was not approved. The use of bombers for this purpose had been uneconomical from the start, and at the end of September the B–24's were withdrawn completely from their supply mission. Henceforth all air supply was to be carried out by aircraft of the IX Troop Carrier Command (U.S.) and the 46 Group (British). SHAEF immediately reduced the allocation of planes for that purpose and attempted to limit their use strictly to meeting emergency needs as originally intended. Bad weather and lack of forward fields further affected the scale of air deliveries during the fall, and for about two and a half months deliveries averaged only about 675 tons per day. By allocation,

[53] Min, Mtg, Area Petroleum Office, 14 Sep 44, USFET 463.72 Ground Force Gasoline, II; file USFET Petroleum Office, Airlift of Gasoline by Plane, Cabinet 3007, Drawer 3.

[54] Ltr, Wing Comdr G. G. Cradock-Watson to Distrib, 17 Oct 44, sub: Emergency Air Lift—September 1944, SHAEF G–4 581.2 Transportation by Air of Supplies and Equipment II; TWX VX–25382, First Allied Airborne Army to SHAEF, 22 Sep 44, SHAEF G–3 Resupply by Air 24518/Ops; CATOR Weekly Load Summaries, Jun–Oct 44, AEAF A–3, 505.27–30A, Air University Library, Maxwell Airfield Base.

[55] Memo, Stratton for Moses, 4 Oct 44, sub: Airlift Requirements for Supply of First, Third, and Ninth Armies for Period 9–29 October Inclusive, 12 A Gp Supply by Air, No. 133.

approximately two thirds of this tonnage went to the 12th Army Group.[56]

The administrative procedure for arranging supply by air was improved and simplified from time to time during the fall, but responsibility and authority of the various commands had to be clarified again and again. Supply SOP's specified that the armies were to state their needs and priorities and that the Army Group would establish tonnage allocations. Within those allocations the Communications Zone determined the means of delivery. Throughout the fall, however, both the armies and the Communications Zone frequently violated the intent of air supply policy. The armies requested air delivery of specific items, which was forbidden except in combat emergency, and the Communications Zone utilized air transportation for supplies no longer in critical shortage. Early in December Third Army complained that the airlift was being wasted in the shipment of gasoline, of which there was no longer a shortage, while its requests for the shipment of critical Class II and IV supplies, which it claimed were available in both the United Kingdom and Normandy, particularly signal items, had remained unfilled. Of a total of 3,227 tons of supplies received during November, it pointed out, 2,143 tons consisted of gasoline and only 250 of Class II and IV supplies, 55 of which consisted of signal items. Gasoline, it suspected, was being shipped because it happened to be on hand at U.K. airfields and was the most convenient cargo to handle.[57]

Early in December SHAEF further reduced the allocation of aircraft for supply and evacuation—to 150 planes from the IX Troop Carrier Command and 40 from 46 Group—the intention being to keep only a standby organization in operation at a reduced scale, but capable of immediate expansion. The major air supply effort of the month was made only a few days later, when it became necessary to resupply units isolated by the enemy break-through in the Ardennes. Between 23 and 27 December 850 planes were dispatched to Bastogne to parachute urgently needed supplies to the besieged 101st Airborne Division. Another sixty-one craft were dispatched with gliders, one of them bearing surgical teams. Although some planes were lost to enemy fire, and some supplies could not be recovered because of errors in dropping, the deliveries, totaling 850 tons and estimated to be about 95 percent effective, were considered the most successful ever made to the 101st Division.[58]

---

[56] Supply and Evacuation by Air, Gen Br Rpt 26, p. 30; Memo, Whipple for G–4 SHAEF, 22 Nov 44, sub: Air Lift, SHAEF G–4 581.2 Transportation by Air of Supplies and Equipment II; 12 A Gp AAR 3, 5 Nov 44, 12 A Gp 107A Rpt General Information. On several occasions small planes were used to fly emergency supply missions to isolated units. Late in October P–47 fighter aircraft were employed in the Seventh Army area to deliver ammunition, rations, medical supplies, and signal batteries to a battalion of the 36th Division isolated in the Foret de Champ. Belly tanks were used to carry rations. In November the 95th Division used artillery liaison planes several times to evacuate wounded and to supply units cut off by the swollen Moselle and isolated during the siege of Metz. Supply and Evacuation by Air, Gen Bd Rpt 26, p. 31.

[57] Memo, Muller, TUSA G–4, for Stratton, 12 Dec 44, with 14 Incls, attached to TUSA G–4 Periodic Rpt 19, 2–9 Dec 44, 12 A Gp 319.1 G–4 Rpts, I; Supply and Evacuation by Air, Gen Bd Rpt 26, pp. 15–17, 19–20.

[58] About fifty tons of this total were air landed.

C–47's Airdropping Supplies by Parachute *to an isolated unit of the 101st Airborne Division, Bastogne, 26 December 1944.*

A similar attempt to supply isolated elements of the 3d Armored Division in Belgium miscarried. Of twenty-nine aircraft dispatched to the division with gasoline and medical supplies on 23 December, twenty-three dropped their loads in enemy territory as the result of a misreading of map co-ordinates, and the remainder were either diverted or lost to enemy action. Bad weather frustrated a second attempt on 24 December.[59]

While the resupply of airborne units at Bastogne was regarded as a very creditable performance, few of the agencies involved were properly prepared for the emergency, and sharp words were exchanged between the 12th Army Group and COMZ staffs before the operation got under way.[60] After the emergency SHAEF ordered the Communications Zone to maintain balanced stocks of supplies required to support type units of airborne, armored, and infantry divisions in sufficient quantity to load all troop-carrier aircraft for a maximum two-day lift.[61] Shortly thereafter the Communications Zone prepared sixteen "bricks" of supplies at various airfields in the United Kingdom and on the Continent, each containing one day of supply for a division. The basic brick was designed to meet the needs of an airborne division and weighed 270.5 tons, but could be readily augmented with prepacked 76-mm. gun and 155-mm. howitzer ammunition to meet the needs of infantry and armored divisions. One additional brick was designed and packed for a regimental combat team.[62]

Supply by air continued at a relatively small scale throughout the period of the Ardennes battle, partly because of persistently bad weather. Deliveries averaged only 185 tons per day in January, the tonnage being divided about equally between 12th and 21 Army Groups.[63]

(4) Inland Waterways

The OVERLORD planners had not considered the inland waterways on the Continent of sufficient military value to warrant a large-scale rehabilitation effort. The policy was laid down of restoring waterways only in cases where minor repairs were required and where a clear military necessity existed. Any opportunity for the advantageous utilization of inland waterways for military transportation was counted as a bonus.

By September 1944 a clear-cut need had arisen to restore certain waterways in northern France to relieve the hard-pressed railways from the burden of coal movements into the Paris area. The Communications Zone at that time appointed an Inland Waterways Committee to recommend priorities for inland waterway restoration, estimate the manpower and equipment requirements, and

---

[59] Operation REPULSE, Rpt of the IX TCC, dated 3 Jan 45, SHAEF AG 581.2-1 Supply by Air; Ltr, Hq 101st Airborne Div to CG TUSA, 11 Jan 45, sub: Rpt of Air Resupply to 101st Airborne Div at Bastogne, SHAEF G-3 24518/Ops Resupply by Air; Supply and Evacuation by Air, Gen Bd Rpt 26, pp. 32-33. The 101st Airborne Division listed only 820 planes and 42 gliders as having delivered supplies to the Bastogne area.

[60] Memo, McCormack for Moses, 1 Jan 45, sub: Air Resupply of Isolated Units (location unknown).

[61] Cbl S-73225, SHAEF G-4 to COMZ, 31 Dec 44, SHAEF G-3 24518/Ops Resupply by Air.

[62] Cbl EX-91351, COMZ to SHAEF, 27 Jan 45, SHAEF G-3 23518/Ops.

[63] See 12 A Gp Supply by Air file for statistics on deliveries; also CATOR Weekly Load Summaries AEAF A-3 505.46-7, Air University Library, Maxwell Airfield Base.

act as the U.S. agent in all dealings with the French on these matters. Early in November, when the inland waterways began to assume importance in the movement of military supplies, an Inland Waterways Division was established in the Office of the Chief of Transportation.[64]

Four main waterways were eventually rehabilitated—the Oise, Seine, and Rhône–Saône Rivers, and the Albert Canal. None of these had been seriously damaged, with the exception of some of the locks. But all were clogged with demolished bridges, which not only obstructed barge navigation, but prevented the movement of floating equipment. French and Belgian authorities tried to salvage as many of the demolished bridges as possible, raising them and eventually restoring them to rail or highway use. Civilians carried out the bulk of the work, although U.S. troop units usually provided the skilled supervisory personnel and much of the heavy construction equipment. Practically no prisoner of war labor was employed.

First priority was given to the clearance of that portion of the Oise River system extending from the Chauny and Valenciennes coal fields to Conflans on the Seine, about thirty-five miles below Paris. Several locks, including the large one at Creil, had to be repaired, and thirty-four obstructions, mostly blown bridges, had to be removed. The 1057th Port Construction and Repair Group provided the skilled labor for this project, and also supplied the French—who performed about 60 percent of the labor—with fuel, equipment, and other supplies. The first objective, cutting a single 40-foot wide channel the entire way, was completed early in November and the first coal barges arrived in Paris on the 18th.

The Seine River, while part of the French coal distribution system, was restored primarily to facilitate the transportation of civil imports from Le Havre and Rouen. Demolished bridges and damaged locks and dams blocked traffic on the Seine as on the other waterways. The biggest single task was the repair of the locks of the Tancarville Canal, which connected Le Havre with the Seine at Tancarville, about fifteen miles up the river. The canal had been built to permit barges, loaded directly from ocean-going ships in a basin at Le Havre, to reach the Seine without traversing the mouth of the estuary, where strong tidal currents made navigation difficult. U.S. engineer units initially were too occupied with the reconstruction of Le Havre to assist in the repair of the locks, and the French port authorities therefore undertook the task unassisted. Lacking adequate salvage and engineer equipment, they made little progress. U.S. naval salvage equipment also failed to raise the sunken ebb gates, and the job was finally assigned to the 1055th Port Construction and Repair Group, which aided French contractors by furnishing the necessary equipment and trained operators. Repair of the lock was not completed until mid-March 1945.

French civilian organizations, assisted by U.S. and British construction units, meanwhile had restored the Seine to navigation for both civil and military traffic. But that traffic for a long time was

[64] *Final Report of the Chief Engineer, ETO,* I, 277; COMZ G–4 History, III, p. 69.

BARGE CONVOY ON ALBERT CANAL *carrying lumber from Antwerp to Liège, February 1945.*

plagued by one bottleneck or another. Barge traffic was obstructed at first by two ponton bridges, which, when they were opened to clear barge traffic, interrupted vehicular traffic across the river. A more serious stricture developed at Le Manoir, where a temporary railway bridge built by British forces left insufficient clearance for barges. Raising the center span several feet in October provided no permanent solution, for the Seine reached flood stage in November and again reduced clearance below the required minimum. At the end of November flood conditions threatened to knock out the bridge, and forced a temporary stoppage of barge traffic on the river because of the danger from swift currents. A decision to remove the bridge depended on whether British supplies forwarded from the Caen-Bayeux area could be handled via Paris. The issue was finally settled on 25 December, when a tug struck the Le Manoir bridge and put it out of commission. The bridge was then removed.[65]

Although the restoration of the Oise and Seine Rivers and the canal system was initially undertaken to meet civilian needs, it promised important benefits to the military forces by relieving the railways, particularly in the distribution of

[65] Port Construction and Repair, Hist Rpt 11, CE ETO, pp. 90–102; *Final Report of the Chief Engineer ETO,* I, 277; Ltr DWP to Potter, 4 Nov 44, sub: Opening of Seine to Navigation, and Summary of Br Chiefs' Mtg, 4, 16, 17, and 30 Nov 44, G–4 Plant and Communications Diary/Jnl, ETO Adm 145C.

coal. The Albert Canal in Belgium had a more direct military value, and its rehabilitation was the principal inland waterway project jointly undertaken by U.S. and British forces. The canal, built between 1933 and 1939, connected Liège with Antwerp, a distance of eighty miles. Eight groups of locks, each group of three built to accommodate one 600-ton and two 2,000-ton barges, reduced the water level from sixty meters at Liège (on the Meuse) to sea level at Antwerp. U.S. forces were assigned responsibility for restoring the approximately fifty miles of the canal between Liège and Kwaadesmechelen, British forces the remainder.

The Advance Section initially laid out twenty-four work projects on its portion of the canal, most of them involving bridge removal or repair of locks, which would open the canal to 600-ton barges. These tasks were carried out under the supervision of the 1056th PC&R Group, with the assistance of the 355th and 332d Engineer General Service Regiments and Belgian civilian contractors, and were completed early in December, as scheduled. Delays in removing the Yserberg Bridge at Antwerp at first prevented full use of the canal, but barges could be loaded just east of that point by trucking directly from the port area. In any event, transportation on the Albert Canal got off to an unspectacular start. Ice and flood conditions created operational hazards as on the Seine, and the enemy counteroffensive completely upset transportation plans in the forward areas, forcing an embargo on barge traffic. Once these difficulties were overcome, the canal played an important role in the clearance of Antwerp, eventually handling about 50 percent of the tonnage discharged there.[66]

The Rhône–Saône waterway was rehabilitated almost entirely by the French. Except for some local clearance at Marseille, and a limited traffic in POL, however, the Rhône had practically no military value, largely because of the lack of high-powered tugs required for operation on the swift waters of the system.[67]

The four months' period after the pursuit represented a transition so far as transportation developments were concerned. Its most obvious feature was the gradual assumption of the bulk of the long-distance hauling by the railways. Motor transport, particularly in the form of the color routes, continued to provide a degree of flexibility in performing special missions. But the cutback in motor as well as air transport signaled the end of expedients necessitated by the emergency conditions in the summer of 1944 and an eventual return to more conventional means of transport.

---

[66] Port Construction and Repair, pp. 83–90; *Final Report of the Chief Engineer, ETO,* I, 278; ADSEC Operational History, p. 108.

[67] COMZ G–4 History, III, p. 71; *Final Report of the Chief Engineer, ETO,* I, 278.

CHAPTER VI

Forward Movements

(1) The Tonnage Allocations System

For nearly two months after the end of the pursuit in mid-September 1944 the inadequacy of transportation lay at the root of most of the Allies' immediate logistic difficulties. The lack of port discharge capacity, while a potentially ominous limiting factor, did not as yet directly affect the Allies' ability to continue large-scale offensive operations. Additional port capacity in mid-September would have relieved the taut logistic situation only if it had resulted in shortened lines of communications—in other words, if ports farther up the Channel could have been brought into use. And while there were shortages of certain items of supply, substantial reserve stocks in most categories lay in the Normandy depots and in the ships offshore. The main problem was their movement forward. For the time being, therefore, the deficiency in transport was the common denominator of most of the Allies' supply troubles.

The prospects of meeting the growing needs of U.S. forces in mid-September were dim indeed. Minimum maintenance requirements in the combat zone already came to more than 13,000 tons per day,[1] and would rise as new divisions were brought forward. In addition, between 150,000 and 180,000 tons of supplies were needed in the forward areas for repairing or replacing Table of Equipment matérial, replenishing basic loads, building army and ADSEC reserves, and providing winter clothing.[2]

Against these requirements the Communications Zone was delivering only about 11,000 tons per day in mid-September, 7,000 of which were earmarked for the two armies, the remainder for the Ninth Air Force, the Advance Section, and various special demands. Only 40,000 tons of reserves, representing but a few days of supply, had been moved forward of St. Lô, and 75 percent of that still lay in dumps west of Paris.[3] These hard facts led inescapably to the conclusion that, temporarily at least, U.S. forces could not be supported at the desired scales. This meant rationing the available support in accord with operational priorities.

---

[1] On the basis of 550 tons per division slice, and including Ninth Air Force needs.

[2] Ltr, 12 A Gp to CG COMZ, 16 Sep 44, sub: Tonnage Rqmts of the Armies, and Memo, M. F. Hass for G–4 12 A Gp, 21 Sep 44, sub: Estimate of Tonnage Rqmts, both in 12 A Gp Tonnages 137; COMZ G–4 Plant and Communications Diary/Jnl, 19 Sep 44, ETO Adm 145C.

[3] Ltr, Vissering, Deputy G–4 Mov and Tn Br SHAEF, 14 Sep 44, sub: Observations on Trip to Paris, SHAEF SGS 617 Railroads; Memo for Record, Col T. F. Taylor, 12 A Gp G–4 Sec, 6 Sep 44, Memos—Moses, SHAEF 12 A Gp G–4, Folder 86.

A rationing system had in effect been instituted on the last day of August when the 12th Army Group commander divided the available tonnage, initially allotting a larger share to First Army than to Third.[4] On 5 September the army group commander made the first change in the allocation, dividing the first 7,000 tons equally between the two armies. By using their own transportation both armies were able to supplement their meager rations somewhat, but there was no prospect that the Communications Zone, unable for the moment to deliver even minimum maintenance requirements, might begin to forward reserves.[5]

A new allocation issued on 14 September renewed the equal sharing of the first 7,000 tons, but reflected a slightly greater optimism in allocating army tonnage in excess of 7,000 to First Army up to 1,500 tons and any additional tonnage to the Third Army. The Communications Zone was authorized at this time to use the trucks of the 104th, 95th, and 26th Divisions for two weeks to haul supplies.[6]

Further changes in allocations were made in the next two weeks in accordance with plans to shift the weight of operations to the north and in recognition of changes in the order of battle and in the strength of the armies. On 21 September General Bradley approved a new allocation giving 3,500 tons per day to Third Army and 700 tons to the Ninth Army, which was scheduled to take over a sector between the Third and First Armies, and assigning the remainder to First Army, but with the understanding that it would get a minimum of 5,000 tons.[7]

Before the new allocation could become effective, the transfer of an armored division from Third Army to First resulted in a corresponding shift of 400 tons from General Patton's forces, giving the First Army 5,400 tons per day and the Third 3,100.[8] The new apportionment went into effect on 27 September, at which time the First Army had ten divisions and the Third Army eight.[9] The Ninth Army was then in process of movement from Brittany, and within the next weeks placed a corps (the VIII) of two divisions in the line between the First and Third Armies.[10]

Forward deliveries roughly approximated the tonnage allocation in the next few weeks. Since the allocation constituted a starvation diet, the armies consumed virtually everything they received. Under these circumstances they were forced to confine their requisitions

[4] See *Logistical Support I*, 491.

[5] Memo, G–4 Movs 12 A Gp for G–4, 13 Sep 44, sub: Transportation for Replacements, 12 A Gp Rolling Stock 106; Memo, Moses for Bradley, 16 Sep 44, sub: Supply Situation, 12 A Gp Supplies, Misc, 126.

[6] Memo for Record, Moses, 14 Sep 44, 12 A Gp Supplies, Misc, 126.

[7] Memo for Record, Moses, 21 Sep 44, Memos—Moses, SHAEF 12 A Gp G–4, Folder 86; Diary of the 12 A Gp G–4, 21 Sep 44.

[8] A further reduction in Third Army's share was contemplated in the event of XV Corps' transfer to the Seventh Army, although Bradley hoped to keep the corps in the 12th Army Group and support it via the southern lines of communications.

[9] Memo for Record, Moses, 23 Sep 44, sub: Conf Held by Gen Bradley This Date, Memos—Moses, SHAEF 12 A Gp G–4, Folder 86; 12 A Gp G–4 Diary, 23 Sep 44.

[10] In view of its limited tonnage allocation, the Third Army requested authority to operate a train back to the beaches, offering to provide its own loading details and guards, but the request was denied. TUSA AAR, II, G–4, p. 18.

to absolute essentials, for all requirements without exception were charged against the assigned allocations. This meant that even mail absorbed tonnage whenever the armies desired its delivery. The 12th Army Group refused to make any exceptions for special requirements, as, for example, when Third Army requested that transportation exclusive of the daily allocation be provided for the movement of winterization supplies and equipment.[11]

In this situation it was naturally of the highest concern to the armies that they get the maximum tonnage allotted them, and that it consist only of useful items. In the frantic scramble of late September, however, the unavailability of supplies at precisely the time they were requisitioned, and the day-to-day uncertainty over transportation resulted in something less than perfect functioning of the allocations system. In addition, the modification of carefully worked out supply SOP's in the name of expediency, the hauling of supplies in army transportation, and the misdirection and misappropriation of supplies, the receipt of which was never acknowledged by anyone,[12] produced conflicting claims as to what was actually delivered.

Late in September General Lord reported to SHAEF that in the nine-day period from 16 to 24 September the Communications Zone had far exceeded its commitments to the armies, delivering 127,000 tons of supplies against a guarantee of only 63,000 tons and against requisitions totaling only about 66,000 tons.[13] General Moses, the 12th Army Group G-4, pointed out that these figures were inaccurate and misleading. The armies, he noted, had reported receipt of only 84,000 tons, including 23,000 brought forward in their own trucks, indicating that the Communications Zone had delivered only the tonnage it was committed to move under the current allocation. Furthermore, according to army and army group figures, 40,000 tons of the supplies delivered had come from stocks which had been accumulated in army service areas west of the Seine and which the armies had been forced to leave behind in their rapid advance. Credit for the delivery of this tonnage had therefore been given to the Communications Zone once before.

The 66,000-ton requisition was explained by the simple fact that the armies had been compelled to keep requests within the bounds of the tonnage allocations. They would have liked to requisition about 120,000 tons for maintenance alone. "We rationed tonnage," General Moses pointed out, "to distribute a scarcity, not to comfort CZ [Communications Zone] in continuing a famine." According to General Bradley's G-4, the armies needed 650 tons per division slice to fight effectively. Instead, they had received 550 tons per division, 400 of which were delivered by the Communi-

[11] Memo, Wilson for Hodges, 19 Sep 44, and Ltr, 12 A Gp to TUSA, late Sep, sub: Rqmts for Winterization of TUSA, both in 12 A Gp Tonnage 137.

[12] Notes on the Supply Situation of 12 A Gp, Col Charles W. McCarthy, n.d. [late Sep], SHAEF G-4 400.192 Supply Report I.

[13] Ltr, Lord to Smith, 26 Sep 44, SHAEF SGS 400.3/1 Supply Problems of Allied Advance; Ltr, Lord to Lutes, 27 Sep 44, EUCOM 400 Supplies, Services, and Equipment, General, V.

cations Zone, 150 by their own transportation.[14]

The statistics offered by General Lord were hardly calculated to inspire greater confidence in the Communications Zone, either as to claims of past performance or predictions of future capabilities. General Moses put it mildly when, in expressing his doubts on the latter, he said, "We do not know what we will actually receive but we feel fairly certain that there will be a number of things requisitioned that will not materialize."[15]

Even more exasperating and inexplicable to the armies was the Communications Zone's practice of shipping them supplies which they had not requested and for which they had no immediate need. To the armies, always suspicious of the Communications Zone's mysterious ways and jealous of their meager ration of tonnage, this was an inexcusable waste of the limited lift available to them. But requisitioned items were not always within reach of the technical services for loading on a particular day, and because the Communications Zone did not want to see the available outloading capacity go unused it frequently shipped substitute items, for which the armies admittedly might not have immediate use, as fillers.

The armies were not inclined to accept such explanations. In fact, they objected to dealing with supply allocations and movements in terms of *tonnages* at all. They suspected that the Communications Zone, in its enthusiasm to register high daily tonnage records, tended to overlook the importance which particular *items* might have for the armies, and to ship supplies most readily accessible, although they might have little or no local value. The armies desired that greater efforts be made to provide the specific items they requested. Under current procedures the Communications Zone simply canceled requisitions or the portions thereof which could not be filled within two days. The result was that the armies were inadequately informed as to what supplies they could expect to receive. No other aspect of supply was quite so frustrating to operational planning or more conducive to the padding of requisitions and hoarding.[16]

The problem of co-ordinating supply distribution between the Communications Zone, the army group, and the air forces, and the need to apportion the available resources in accord with operational priorities determined at the Allied level made it inevitable that Supreme Headquarters should eventually become involved in the allocation business. SHAEF prepared to take over the control of priorities early in October when the supply outlook was still extremely dismal. At the time practically no progress had been made in overcoming the worst effects of the pursuit. Neither First nor Third Army had been able to accumulate more than two days of supply and two units of fire, and Ninth Army was getting nothing beyond current maintenance needs. Motor transport was

[14] Ltr, Moses to Crawford, 2 Oct 44, SHAEF SGS 400.3/1 Supply Problems of Allied Advance; Memo, Moses for Bradley, 1 Oct 44, 12 A Gp Supplies, Misc, 126.

[15] Ltr, Moses to Crawford, 2 Oct 44.

[16] Notes on the Supply Situation of 12 A Gp, McCarthy; Memo, Col T. F. Taylor, Supply Br G–4 12 A Gp for G–4, 19 Sep 44, 12 A Gp Tonnage 137.

still being operated to the maximum of its capabilities, and higher echelon repair had not been performed since the early days of the beachhead. Heavy clothing and other winterization equipment and materials needed for the expected bridging operations were being moved up, but in nothing like adequate quantities.

Allocating the available supplies was essentially the responsibility of the SHAEF G–4, who announced that allocations would be determined on the basis of estimates of movement capabilities provided by the commanders of the three lines of communications (that is, the Communications Zone, the 21 Army Group, and SOLOC), on the tonnage bids submitted by the army groups, and on the projected scale, nature, and relative priority of tactical operations as outlined by the G–3. With this data the Logistical Plans Branch was to prepare a logistical study and recommend overall tonnage allocations. The concurrence of the G–3, as well as that of other interested parties, such as Movements and Transportation Branch, was of course necessary before the recommended allocation was finally sent to the G–4 for approval.[17]

The Logistical Plans Branch submitted its first allocations study on 8 October, covering the two-week period from the 15th to the 28th. Projected tactical operations in that period called for continued battering along the entire Allied line, with the greatest weight behind the attacks on the north. The 12th Army Group at this time consisted of twenty divisions, supported by an equal number of fighter squadrons. But five additional divisions were available for commitment, subject only to the provision of the necessary logistic support. Logistic planners calculated that the commitment of all twenty-five divisions and the building of minimum reserves for the three armies would require the delivery of 22,320 tons per day to the 12th Army Group, based on requirements of 840 tons per division (560 for maintenance [18] and 280 for reserves), plus 1,320 tons for headquarters and special troops, coal, and civil affairs supplies.

The 12th Army Group had actually submitted a bid for 18,000–20,000 tons per day for the allocation period in question, and the Communications Zone had estimated its average delivery capabilities at 15,000 tons, exclusive of Ninth Air Force and ADSEC maintenance stocks. While forward deliveries promised to be much improved over September, therefore, they were certain to fall far short of the needs of an enlarged combat force. It was clear that there would have to be either a reduction in the projected scale of support, a limitation in the number of divisions em-

[17] Memo for Record, Moses, 9 Oct 44, sub: Conf Conducted by Gen Bradley This Date, 12 A Gp Supplies, Misc, 126; Ltr, Whipple to Stf Secs, 10 Oct 44, sub: Alloc of Tonnage, SHAEF G–4 400 Supplies, General, III; Ltr, SHAEF to Major Comds, 16 Oct 44, sub: Alloc of Tonnages, SHAEF G–4 563.59 Tonnages and Estimates of, I.

[18] Maintenance requirements per division slice in the army group were divided as follows:

Class	*Tons*
I	78
II	75
III	144
IV	110
V	153
	560

ployed, or a combination of the two.

Within the limitations imposed by the Communications Zone's estimated delivery capabilities the first SHAEF allocations study concluded that twenty divisions could be supported if certain adjustments were made in the projected scales of maintenance and reserves. Only First Army, which had first priority in its effort, could be permitted to accumulate reserves and thus get the full 840 tons per division. Third Army's allowance of POL was to be somewhat less than normal but its ration of ammunition was to be doubled (300 tons per division as against 150 for divisions in the other armies) because of the heavy concentration of artillery on its front. This would give Third Army 610 tons of maintenance supplies per division as against 560 in the other armies. The allocation of the entire 15,500 tons on this basis would thus permit the maintenance of twenty divisions in combat at acceptable maintenance scales and an accumulation of reserves for half that force.[19]

The SHAEF G-4 decided to postpone implementation of the proposed allocation. Instead, on 9 October General Crawford assigned to the Communications Zone as a first priority commitment the delivery of 12,500 tons of supplies to the 12th Army Group each day plus 2,000 tons to the Ninth Air Force until further notice, without specifying the scales of maintenance or proportions for the various armies. He asked the Communications Zone to make every effort not only to meet these targets but to forward only the items requested.[20]

The decision to postpone implementation of the proposed allocation may have been inspired partly by the knowledge that Ninth Army's movement to a sector north of First Army was then being considered. More probably it resulted from the widely held doubts concerning the likelihood of achieving the delivery figures which the Communications Zone had submitted and which the SHAEF logistical planners had accepted as a basis for their recommended allocations. Colonel Whipple, chief of the Logistical Plans Branch, himself was extremely pessimistic in mid-October, asserting that with the scale of support then being provided the 12th Army Group only thirteen divisions were properly maintainable, although twenty-three were actually in line. Others, concluding that the Communications Zone had overestimated its own movement capabilities, expressed the the view that only maintenance requirements could be met until about mid-November, and that the reserves needed for a sustained offensive could not be established before the end of the month.[21] That some confusion and misunderstanding existed is indicated by the fact that on 22 October the 12th Army Group issued a new allocation,

[19] Alloc of Tonnages 1 to 12 A Gp for Period 15-28 Oct 44, Log Plans Br G-4 SHAEF, 8 Oct 44, SHAEF G-4 400 Supplies, General, III; also Ltr, Whipple to G-4, 8 Oct 44, sub: Tonnage Alloc with Incl, Logistical Study of Tonnage Rqmts, SHAEF G-4 563.59 Tonnages and Estimates of, I, 44.

[20] TWX S-61549, SHAEF G-4 to COMZ and 12 A Gp, 9 Oct 44, SHAEF G-4 563.59 Tonnages and Estimates of, I, 44.

[21] Memo, Whipple for Orgn and Equipment Opns A Plans G-3, 21 Oct 44, sub: Maintenance of 12 A Gp Divs, and Statement by Whipple, Divs Which Can Be Maintained on Adequate Scales, 20 Oct 44, SHAEF G-4, Logistical Forecasts; COMZ G-4 Plant and Communications Diary/Jnl, 21 Oct 44, ETO Adm 145C; Memo, Hass for G-4 12 A Gp, 18 Oct 44, sub: Daily Tonnages, 12 A Gp Tonnage 137.

dividing 15,000 tons between its three armies,[22] only to be told that the Communications Zone had been given a commitment of merely 12,500.[23]

SHAEF finally placed its new rationing procedure in operation for the period 5–18 November, the first allocation being based on a fresh logistic survey made late in October. By 5 November the regrouping of the 12th Army Group was expected to be completed, with the Ninth Army, consisting of three corps, deployed north of Aachen, on the First Army's left. By the end of the allocation period twenty-seven divisions were expected to be operational in the 12th Army Group. Tactical operations contemplated for the period included an advance to the Rhine by all four U.S. armies (including the Seventh) and a bridgehead for First Army at Cologne.

With these considerations in mind the 12th Army Group placed a bid for 33,430 tons per day—28,333 for the ground forces and 5,097 for the Ninth Air Force. SHAEF logistical planners considered some of the requests unreasonable and recommended a cut to 27,000 tons. Adding Advance Section's requirements to this figure brought the total demand for forward deliveries—that is, beyond Paris—to 31,200 tons.

The Communications Zone, meanwhile, had estimated its movement capability at more than 30,000 tons. But SHAEF planners considered this figure unrealistic, partly because it involved relying on fairly heavy movements out of Le Havre before the bridge across the Tancarville Canal was repaired, and partly because it was believed to include overoptimistic estimates on water transport from Rouen and on rail clearance out of Cherbourg. They did not consider the Communications Zone capable of delivering more than 21,900 to the combat zone, or 24,400 beyond Paris. This would leave a deficit of at least 5,000 tons.

In the end the estimate of the Communications Zone's movement capabilities was reduced still further, and on the G–4's recommendation Supreme Headquarters finally set the Communications Zone a first priority commitment to deliver 20,000 tons per day to the 12th Army Group—16,700 tons to the armies and 3,300 to the Ninth Air Force.[24] The 12th Army Group in turn suballocated this tonnage to its three armies in accord with operational priorities, the tonnages differing for "active" and "quiet" divisions and for armored and infantry units.[25]

The disparity between the army group

[22] Memo, Moses for Bradley, 22 Oct 44, Memos—Moses, SHAEF 12 A Gp G–4, Folder 86.

[23] Memo, Col Edwin N. Clark for G–4, 24 Oct 44, sub: Allocs of Tonnage by 12 A Gp.

[24] Ltr, Whipple to Current Opns Br G–4 *et al.*, 27 Oct 44, sub: SHAEF Tonnage Alloc 1, and Ltr, Crawford to CofS, 31 Oct 44, sub: Tonnage Allocs, SHAEF G–4 563.59 Tonnages and Estimates of, I, 44. The CAO, General Gale, disapproved the allocations study, insisting that the logistic planners were in no position to make cuts in the various bids on the basis of comparison with October consumption figures. General Crawford thereupon resubmitted the study without any analysis of the "reasonableness" of the various bids. But this did not alter the amount of tonnage available or its allocation. The only result was an increase in the deficit—on paper, at least—from about 6,500 tons per day to 13,500. Ltr, Gale to Crawford, 3 Nov 44, sub: Tonnage Allocs, and Ltr, Crawford to Gale, 4 Nov 44, same sub, SHAEF G–4 Tonnage Allocations, Log Plans Br 153/5, Folder 48.

[25] Breakdown of Current Allocs, G–4 12 A Gp, 12 A Gp Tonnage 137; TWX QX–23075, 12 A Gp to COMZ, 3 Nov 44, SHAEF G–4 563.59 Tonnages and Estimates of, I, 44.

requests and the tonnage targets which SHAEF regarded as attainable appeared to reveal a more glaring transportation deficit than ever, underscoring the need to develop the railways at a faster rate, to open Antwerp, to utilize inland waterways and local resources, particularly coal, more fully, and, if possible, to make joint use of the Rhône valley line of communications. Moreover, supply policy had favored the First Army, particularly in the matter of reserves, and had resulted in inequities in distribution out of proportion to the tonnage allocations.

Tonnage movements in the period 5–18 November fell short of the targets established by the first SHAEF allocation, averaging 12,400 tons to the armies and air force units in the 12th Army Group against an allocation target of 20,000, although the Communications Zone began to lay down increasing tonnages in the Advance Section.[26]

By this time, SHAEF Logistical Plans Branch completed its allocations study for the period 19 November–2 December, for the first time including all three army groups in its survey. By the end of that period 12th Army Group was expected to have a strength of thirty divisions,[27] and the 6th Army Group a strength of sixteen.[28] Operational plans for U.S. forces called for continued attacks eastward, the Ninth, First, and Seventh Armies pushing toward the Rhine, and the Third aiming at capture of the Saar. Tactical priorities went to the operations of the First and Ninth Armies, then to Third U.S. and Second British Armies, and then to the First Canadian. Tonnage bids for the period totaled 30,000 for the 12th Army Group and 11,660 tons for the 6th Army Group, including requirements for supporting air units. After an analysis of the bids the SHAEF planners again concluded that cuts could be made—the 12th Army Group's combined bid from 30,000 to 26,300 and the 6th Army Group's from 11,660 to 11,450. Requirements for the Advance Section brought the total forward needs on the central (12th Army Group) line of communications to about 30,000 tons.[29]

The Communications Zone estimated its delivery capabilities for this period at 22,200 tons per day, a figure which the SHAEF planners now considered conservative. Several factors, including the opening of the Seine for the clearance of Rouen, the reduction of British rail traffic from the rear maintenance area, the increased availability of POL at Rouen, the general improvement in rail transportation, and the expectation that Antwerp would soon open, led them to raise the COMZ estimate to 24,400 tons. Even assuming that deliveries could reach this figure there would be a deficit of about 5,000 tons. The Southern Line of Communications, on the other hand, estimated its capacity as 12,500 tons, giving it a surplus of about 1,000.

SHAEF planners recommended that

[26] Delivery statistics are from COMZ G–4 Weekly Reports on Status of Supply, from Stratton to 12 A Gp G–4, 12 A Gp 400.291 Supply Information; ADSEC Operations History, App. D; and TWXs, Lee to SHAEF, reporting on movements to armies and Ninth Air Force, SHAEF G–4 563.59 Tonnages and Estimates of, I, 44, and are not always in exact agreement.

[27] Including one at Lorient.

[28] 21 Army Group's strength was to remain unchanged at eighteen and one-third divisions.

[29] The total requirements of the three army groups, as approved by the planners, came close to 60,000 tons per day.

the Communications Zone be given a first priority mission to deliver 21,000 tons per day to the army service areas of the 12th Army Group and the Ninth Air Force (respectively 17,500 and 3,500). ADSEC requirements were not included in these figures. The current allocation already gave 80 percent (about 500 tons) of the available airlift to 12th Army Group and the remainder to the 21 Army Group, and the planners recommended that this apportionment be continued. Since operational priority was to shift to the 12th Army Group in the next allocations period, they also proposed that General Bradley's forces be favored to whatever extent was possible in the allocation of transportation facilities on the neighboring lines of communications. They accordingly recommended that in the allocation of railway facilities the necessary priority be given the Communications Zone for clearing Antwerp and that the forward movement of reserves planned by the 21 Army Group be postponed. If practicable, the Communications Zone was to begin the delivery of at least 1,000 tons per day to the Third Army or Advance Section via the southern lines.[30] The G–4 and CAO both approved the recommendations, and the new allocations went into effect as planned.[31]

A third allocation, prepared at the end of November and covering the period 3–16 December, did not differ substantially from the second. Operational plans again gave priority to the 12th Army Group, which was scheduled to have a strength of thirty-two divisions by mid-December. The army group's bid was not quite as high for the next two-week period, possibly reflecting a shade greater confidence in the Communications Zone's ability to fulfill its promises, but the SHAEF planners again reduced the requests—from 25,000 to 18,700 in the case of 12th Army Group, and from 10,200 to 8,300 in the case of 6th. The Advance Section was allowed 3,600 tons. The Communications Zone's prediction that it could deliver 24,000 tons per day was accepted as a reasonable estimate. The SOLOC had not met its earlier target, however, owing partly to the fact that many trains carried bulky Class IV supplies and equipment for the Rhine crossings, some of them averaging only 250 tons. Its estimate of 12,500 was reduced to 10,000. In any event, it appeared that all three lines of communications for the first time would easily meet the approved requirements of their army groups, including the associated air forces and advance sections. In fact, the planners foresaw a surplus of about 5,400 tons per day, and recommended that it be allocated to the two advance sections for the build-up of additional reserves behind the respective army groups which they supported. General Crawford approved the recommendations, setting a first priority commitment to the Communications Zone to deliver 20,000 tons per day to the army service areas of the 12th Army Group—15,400 and 4,600 tons

---

[30] SHAEF Tonnage Alloc 2, to cover period 19 Nov–2 Dec, SHAEF G–4 Log Plans Br, 12 Nov 44, SHAEF G–4 563.59 Tonnages and Estimates of, I, 44.

[31] Ltr, Crawford to CofS, 12 Nov 44, sub: Tonnage Allocs, 19 Nov–2 Dec, SHAEF SGS 400.3/1 Supply Problems of Allied Advance; TWX S–66813, SHAEF G–4 to Major Comds, 15 Nov 44, SHAEF G–4 563.59 Tonnages and Estimates of, I, 44.

respectively for the ground forces and the Ninth Air Force.[32]

The third allocation never ran its full course. Within a week the entire rationing system, under attack from several quarters and rendered largely unnecessary by the general improvement of supply, was discontinued. The black cloud which had hung depressingly over the logistic horizon since mid-September had actually begun to lift at the end of October, at the very time that logistic planners were determining how the meager resources should be apportioned. Total forward deliveries did not match the Communications Zone's predictions, nor were the gains necessarily reflected in the receipts in the army areas. In the last ten days of October, in fact, the Communications Zone claimed daily deliveries of only about 10,000 tons to the armies in 12th Army Group, and something less than 1,000 tons to the Ninth Air Force.[33] Even with these tonnages the armies managed to improve their reserve positions. October was a quiet month operationally, and by stringent control over expenditures the armies were able to accumulate savings. By the end of the month stocks in the combat zone of the 12th Army Group totaled more than 155,000 tons.[34] In some categories, such as Class I, levels were actually excessive, and the Army Group took steps to halt the flow into the army areas.

The Communications Zone, meanwhile, by mutual agreement with 12th Army Group began laying down a larger percentage of total tonnages in the Advance Section, although this was contrary to the allocations laid down by SHAEF.[35] By the time of the fall offensive, which began on 8 November, the Communications Zone was laying down about 8,000 tons per day in the Advance Section, where stocks had risen to nearly 100,000 tons. Army reserves by that time totaled about 180,000, bringing the stocks in the forward areas to a record 280,000 tons.[36]

A similar improvement had taken place on the southern lines of communication, where forward reserves, practically nonexistent early in October,[37] had risen to ten days for Classes I–IV, and to twenty-five days for Class V by mid-November.[38]

Serious shortages of many items continued to hamper operations and to worry all echelons. Nevertheless, tonnage movements were definitely on the rise, and the supply picture in both the 6th and 12th Army Groups was much brighter by mid-November. With cer-

[32] SHAEF Tonnage Alloc 3, to cover period 3–16 Dec, Log Plans Br G–4, 25 Nov 44, Ltr, Crawford to CofS, 27 Nov 44, sub: Tonnage Allocs, and TWX S–68947, SHAEF G–4 to Major Comds, 30 Nov 44, all in SHAEF G–4 563.59 Tonnages and Estimates of, I, 44. See also allocations papers in SHAEF AG 400.22–1, No. 2 Shipments and Tonnages Allocations 1944, and in 12 A Gp Tonnages 137.

[33] See Weekly Reports of COMZ G–4, sub: Supply Info, 12 A Gp 400.291 Supply Information II, and ADSEC Operations History, App. D.

[34] Memo, Hass, Supply Br G–4 12 A Gp, for G–4, 30 Oct 44, sub: Tonnages in Forward Areas, 12 A Gp Tonnages 137.

[35] Memo, Moses for Bradley, 31 Oct 44, sub: Supply Outlook, 12 A Gp Tonnages 137; Ltr, Whipple to Barriger, 11 Nov 44, sub: Comments of Col Wilson on Tonnage Allocs, SHAEF G–4 Tonnage Allocations, Log Plans Br 153/5.

[36] The sources present conflicting statistics. See Memo, Hass for Moses, 5 Nov 44, sub: Info on Supply Situation, 12 A Gp Supplies, Misc, 126; COMZ G–4 Plant and Communications Diary/Jnl, Br Chiefs Mtg, 8 Nov 44, ETO Adm 145C; and ADSEC Operations History, App. E, p. 3.

[37] TWX BX–17225, Devers to Eisenhower, 5 Oct 44, SHAEF Cbl Log, Smith Papers.

[38] 6 A Gp G–4 AAR, Sep–Nov 44.

tain exceptions, even ammunition, which had been at dangerously low levels throughout October, had become a less critical item of supply with the build-up of stocks in the Advance Section. Deliveries to the armies in the second allocations period—19 November to 2 December—averaged about 12,000 tons on the central line of communications and between 4,000 and 5,000 tons on the southern, and supplies continued to build up in the Advance Section at the rate of about 8,000 tons per day.

The general improvement brought with it a natural desire to return to normal supply procedures. The allocations system was inflexible and restrictive at best, and subject to abuse. So long as supply deliveries were characterized by uncertainty it was natural for the armies to seek whatever security they could in hoarding and in requisitioning beyond their actual needs. As long as all available tonnage was allocated, the Communications Zone lacked control over the flow of supplies, particularly as to the establishment of balanced stocks in the forward areas. Moreover, the system did not ensure that the armies would actually receive the items they most needed, for it tended to give an exaggerated and even harmful emphasis to over-all *tonnages* rather than *items* of supply, resulting in the delivery of supplies for which there was no immediate need, and in the accumulation of unfilled requisitions for thousands of tons of items on the critical list.[39]

The fact that the armies could set aside supplies as reserves at a time when deliveries to their service areas averaged only 11,000 tons against stated requirements for 25,000 to 28,000 bore out the suspicion that they had been overzealous in their requisitioning. First Army, in particular, was suspected of taking for granted the advantage in supply which it had enjoyed, first in its preparations for the Normandy landings, and later in connection with its operational priority, and had acquired a reputation for "asking for the moon." Lack of confidence in the Communications Zone's ability to meet its requirements undoubtedly accounted for the heavy demands. On the other hand, it was hardly surprising that the Communications Zone should question the dire urgency of army demands which listed as "critically short" such items as barber kits and handkerchiefs. Nor was mutual trust likely to be promoted by the attitude expressed in one of First Army's G–4 periodic reports, which noted, "The operation will be a success providing the Communications Zone has the ability and is *willing* to support the combat forces."[40] The Third Army, accustomed from the beginning to operating on a shoestring, accepted the hardships of the fall months with more equanimity, sometimes even failing to ask for enough. The Ninth Army, a late-comer arriving in the midst of scarcity, had a reputation for consistently professional staff work where supply was concerned, and could usually

[39] Ltr, Whipple to Barriger, 11 Nov 44, sub: Comments of Col Wilson on Tonnage Allocs, SHAEF G–4 Tonnage Allocations, Log Plan Br 153/5; Mechanics of Supply in Fast Moving Situations, Gen Bd Rpt 27, pp. 80–81.

[40] COMZ G–4 History, II, 19. (Italics are the author's.)

be counted on to limit its requests to actual needs.[41]

At any rate, by late November the Communications Zone was demonstrating its ability to deliver supplies to the combat zone in considerably greater volume than was required for maintenance alone. The allocations system had therefore served its purpose, and both the Communications Zone and 12th Army Group began to urge its early abandonment, for there was every prospect now of meeting a larger and larger percentage of the armies' requests directly from ADSEC depots, resulting in a greater flexibility, more rapid response to army demands, and a greater ease in maintaining balanced reserves. The re-establishment of forward reserves, although they still lacked many of the items the armies needed, was an especially significant development, for it presaged the return to a more conventional supply procedure, in which the Advance Section might once again become the sole agency through which the armies arranged for their day-to-day needs, thus obviating their repeated reaching back to the base areas for supplies. There was every reason to believe, moreover, that the discovery by the armies that they could count on having their needs promptly met would remove the desire to establish excessive stocks in their own areas. This improvement in reserves consequently constituted an important milestone in the road to recovery.

The developments of November had thus opened the way for ending allocations. In fact, 12th Army Group had already removed Classes I and V from its suballocations to the armies before the end of the month.[42] Finally, SHAEF discontinued all allocations on 9 December, bringing to an official end the long famine which had set in during the pursuit.[43] By that date the stocks of Class I, III, and V supplies alone had risen to 222,000 tons in the Advance Section, and the levels in the army areas were also very satisfactory. First Army actually asked the Communications Zone to stop shipping gasoline and Third Army even turned back one million gallons.[44] Within another week stocks in the Advance Section had risen to 294,400 tons, and in the combat zone of the 12th Army Group to 222,000, bringing the reserves in the forward areas of the central line of communications to well over 500,000 tons.[45]

Improvements on the southern line of communications were less spectacular, but stocks had risen sharply by mid-December—to 53,000 tons in the Continental Advance Section and to 80,000 in the Seventh Army area, reaching levels in the army area in excess of authorized reserves in all classes except POL.[46]

---

[41] Interv with Moses, 13 Aug 51, and Plank, 28 Jul 50.

[42] 12 A Gp G-4 AAR 4 (Nov 44), 12 A Gp 107A Reports, General Information.

[43] Memo, Col James W. Younger, 12 A Gp QM, for G-4, 20 Nov 44, sub: Turnover of Army Tonnages to COMZ, 12 A Gp Tonnages 137; 12 A Gp G-4 AAR for Nov 44; TWX S-70142, SHAEF to Major Comds, 9 Dec 44, SHAEF AG 400.22-1, No. 2, Shipments and Tonnage Allocations 1944.

[44] Comd and Stf Conf, 8 Dec 44, SHAEF G-4 337.

[45] ADSEC figures are from ADSEC Operations History, App. 4, army group figures from Information on Supply Situation Rpt 11, 18 Dec 44, Supply Br G-4 12 A Gp, 12 A Gp Supply—Reports of Status of, No. 131. There are some discrepancies.

[46] Ltr, Col Carter Page, Plans Br G-4 SOLOC, to CG SOLOC, 20 Dec 44, sub: Estimate of Supply Situation (Rpt 2), SHAEF G-4 319.1 Supply Status (SOLOC), I, 1945.

(2) The Ardennes Counteroffensive and Its Effect on Movements

It is typical of logistic operations that the solution of one problem often creates a new one. The acceleration in forward deliveries soon brought a new limiting factor into play—the inability of the depots to unload, classify, and store the huge tonnages which were being forwarded in bulk. The result, as already noted, was a serious congestion on the rail lines in the forward areas, thousands of loaded cars piling up.[47] The Communications Zone had just begun to give serious attention to this problem when the German counteroffensive struck on 16 December, endangering the huge stocks which had been built up in the forward areas.

The German offensive had particularly serious implications for the logistical support of U.S. forces in the north, for it threatened to cut directly across the lines of communications based on the port of Antwerp, which had just been brought into operation. In most immediate danger, of course, were the supply points and depots in the area of the First Army. But most of the ADSEC logistic structure backing up the First and Ninth Armies also lay directly in the path of the German drive, heavily concentrated along the Meuse between Liège and Namur and extending westward to Charleroi. An advance to the Meuse alone promised to overrun many army and ADSEC depots and place most installations north of the river within the reach of enemy artillery; an advance beyond the Meuse would threaten to cut rail lines from the south or the lines of communications based on Antwerp, in either case making it virtually impossible to support the First and Ninth Armies in combat.[48]

The immediate danger to forward supply installations made it imperative that the First Army service area be moved out of reach of the enemy, north and west of the Meuse. This involved superimposing army installations on the already congested area of the Advance Section and a general telescoping of supply operations. It posed a special dilemma because of the unusual character of the logistic structure in the north. As previously observed, the U.S. communications zone lacked storage facilities at Antwerp except for small amounts for in-transit purposes. U.S. supplies consequently had to be shipped in bulk to installations in the vicinity of Liège and Namur, where the Advance Section was forced to improvise base depot operations as best it could to handle bulk cargo and at the same time conduct retail issues to meet the day-to-day needs of the armies. Superimposing the First Army's service installations on this already cramped area was certain to aggravate the existing congestion. On top of this, First Army now presented the Advance Section with a formidable list of major combat equipment losses which required immediate replacement. In the face of these demands it was clear that the continued large-scale shipment of supplies in bulk from Antwerp would place an unbearable burden on the Ad-

[47] See above, Ch. V, Sec. 2.

[48] Ltr, Whipple to G–4 SHAEF, 28 Dec 44, sub: Logistical Support of U.S. First and Ninth Armies, SHAEF G–4 400 Supplies General, V.

vance Section and threaten orderly supply operations.

The obvious key to the situation was the acquisition of depot space in the vicinity of Antwerp which would relieve some of the pressure building up from the forward areas. The assignment of such space to U.S. forces had been discussed as early as October, when American and British staffs agreed on the apportionment of Antwerp's facilities. British officials had been unwilling to release accommodations to U.S. forces in the north, in part because they considered facilities in that area vital to their own operations, in part because the establishment of U.S. depots there would further contravene the principle of keeping national lines of communications separate. The Advance Section thereafter had repeatedly tried to convince the Communications Zone of the dangers inherent in the situation, but the Communications Zone had not pressed the matter.

The counteroffensive of December finally forced the issue. Late in the month SHAEF ordered 21 Army Group and the Communications Zone to provide depot space for U.S. forces in areas that could be defended in the event of a further enemy thrust. Although 21 Army Group still opposed the move, British officials at SHAEF admitted the necessity of relieving the pressure on Antwerp and agreed to permit U.S. forces to establish depots in several locations which the Communications Zone had already reconnoitered. On 4 January 1945 21 Army Group agreed to turn over space in the Lille–Tournai area. While the facilities did not come into use until after the danger from the counteroffensive had passed, they were necessary in any event for the organization of a supply system in proper depth.[49]

In the meantime certain "immediate action" steps had been taken to provide prompt relief of congestion in the forward areas. The Advance Section, by agreement with First Army, at once reduced forward deliveries of major maintenance items, and the army began to sustain itself by drawing on its reserves. On 19 December the Communications Zone in turn placed a temporary embargo on all shipments of other than Class II and IV supplies to the Advance Section in order to facilitate the evacuation of First Army supplies via Liège in case that became necessary. Ammunition moving forward from Cherbourg, for example, was halted at Soissons; POL shipments out of Antwerp were routed to Charleroi; rations consigned to Liège and Luxembourg were redirected to Charleroi and Verdun respectively.[50]

Meanwhile First Army supply chiefs placed in operation a plan to move all major army service installations to the west of the Meuse. On 18 December the 25th Regulating Station withdrew from Spa and a detachment at Liège took direct control of the evacuation of First Army's supplies, reversing the entire sup-

[49] *Ibid.;* Cbl S–73064, SHAEF to 21 A Gp and COMZ, 30 Dec 44, Cbl EX–80904, COMZ to SHAEF, 31 Dec 44, and Cbl EX–81864, COMZ to Channel Base Sec, 4 Jan 45, all in SHAEF G–4 Maintenance of British and U.S. Forces, 153/2/GDP–1; Col William Whipple, "Logistical Problems During the German Ardennes Offensive," *Military Review,* XXVIII (May, 1948), 19, 23; ADSEC Operations History, pp. 105–06; Notes on Mtg at SHAEF attended by Gale, Lord, Crawford, *et al.,* 3 Jan 45, SHAEF CAO War Diary.

[50] COMZ G–4 History, II, 78–79.

ply procedure by directing empty cars into forward railheads and expediting the loading and evacuation of supplies to depots west of the Meuse. In the next week approximately 5,600 carloads of supplies were removed to safer locations. In addition, First Army used 37 truck companies in constant supply and troop movements, clearing 196 convoys in the redeployment of nearly 250,000 men.

The evacuation of supplies got under way when the army began emptying two of its largest POL depots near Malmedy. directly in the path of the German drive. The evacuation of one depot, containing about 1,115,000 gallons of MT80 gasoline and allied products, began on 17 December. Within forty-eight hours all stocks had been removed except for 124,000 gallons, which were destroyed by burning. On two occasions during the evacuation reconnaissance elements of a German panzer division advanced to within 1,000 yards of the depot. A second installation, containing 2,226,000 gallons, was evacuated beginning on the night of the 18th and was completely cleared without loss by the morning of the 22d.[51]

In a few cases losses could not be prevented. Two ammunition supply points, one holding 2,000 tons of ammunition and the other about 800 tons, were eventually overrun, although U.S. troops continued to draw on one of them after the enemy had already reached the opposite end of the dump. One Class I truckhead issued rations for three days under sporadic mortar and small arms fire, was abandoned, retaken, and drawn down, and the remaining supplies were then burned.[52] The Advance Section evacuated some supplies, and lost about 2,700 tons, largely through destruction by V–1 attacks, which forced the abandonment of several installations.[53]

The Advance Section was responsible for the support of Third Army as well as the First and Ninth, and the enemy thrust into Belgium cut the direct wire communications between Namur and the Verdun–Luxembourg area. Fortunately the ADSEC supply complex in the south, centering at Verdun, for some time had been operating fairly independently of that in the north. The resulting flexibility in the ADSEC organization, plus good working relationships with the 12th Army Group and Third Army G–4's, permitted this partially truncated portion of the Advance Section to carry out its mission during this critical period without direct wire communications with the main ADSEC headquarters in the north and without changes in control or responsibility.[54]

Third Army's supply installations had not been endangered, since the German counteroffensive was directed away from the Verdun concentration, and Third Army was now assigned a major role in countering the enemy threat. On 18 December General Patton was ordered to turn over his right flank corps to the Seventh Army, and to reorient the Third Army's operations from the east to the north. He was also required to take over the VIII Corps, which could no longer

[51] Hist Rpt, 25th Regulating Station, 5 Feb 44–12 May 45; FUSA Rpt of Opns, 1 Aug 44–22 Feb 45, Bk. I, p. 128, Bk. II, p. 120, and Bk. III, p. 60.

[52] FUSA Rpt of Opns, Bk. II, pp. 120–21, 123, and Bk. IV, p. 5.

[53] ADSEC Operational History, p. 111.

[54] Whipple, *op. cit.*, p. 21.

be supported or controlled by the First Army as a result of the enemy breakthrough. Third Army thereupon initiated a wholesale redeployment of both combat and service units, a task which called for the closest co-ordination of movements and the efficient use of transport. Within three days of the order Third Army had turned over twenty-five miles of front to the Seventh Army, withdrawn two corps, and completed administrative preparations for an offensive on the new axis, an accomplishment which Colonel Whipple, chief logistical planner at SHAEF, characterized as one of the most professional performances of the entire war, easily ranking with the more spectacular accomplishments of the preceding August. New supply installations were opened in the general vicinity of Luxembourg City, Longwy, and Esch. In an attempt to re-equip units of the VIII Corps which had lost heavily in the initial enemy onslaught Third Army exhausted its reserves of many items.[55]

Ninth Army supply installations were not immediately endangered by the counteroffensive, but the army made preparations to evacuate or destroy about 100,000 tons of supplies and equipment. As in First Army, deliveries were immediately halted, and levels in the forward areas were reduced by issue.[56]

Both the Communications Zone and SHAEF were concerned with the security of areas other than those under imminent danger of German attack, and the planning staffs of both headquarters immediately studied the possible effects which the offensive might have on the logistic structure as a whole. The Antwerp base and its line of communications to Liège naturally ranked high in importance because of their role in the support of U.S. forces in Holland and Belgium. Next in importance were the adjacent depot concentrations and communications networks, such as Ostend–Ghent with its POL facilities; the Valenciennes–Mons–Charleroi area with its POL depots and coal mines; Soissons, Reims, and Verdun with their large ammunition stocks; and the rail net north and east of the Seine. General Lee immediately ordered special defense measures for all vital COMZ installations and for rail bridges, defiles, and tunnels, and the chief engineer issued detailed instructions on security measures, including demolitions, for POL discharge facilities, pipelines, pumping stations, and tank farms.[57]

While the Ardennes counteroffensive never achieved sufficient success to endanger COMZ rear installations seriously, its effect was nevertheless felt in various ways. Embargoes on forward movements and the immobilization of rail cars soon resulted in backlogs in the ports. At Le Havre, for example, cargo held in intransit storage increased by

[55] Whipple, *op. cit.*, p. 24; TUSA AAR, II, G–4, 37–40, and II, Arty, 17; Memo, Barriger for Moses, 27 Dec 44, sub: Forward Reserves, 12 A Gp Supplies, Misc, 126.

[56] Conquer: The Story of Ninth Army, 1944–45, pp. 121–22.

[57] TWX EX–76867, Lee to Sec Comdrs, 21 Dec 44, and TWX EX–78041, Lee to Sec Comdrs, 23 Dec 44, SHAEF AG 370.2 OVERLORD, 1945; Note by Plng Stf G–3 SHAEF, 23 Dec 44, sub: Counteroffensive Measures (Final Draft), SHAEF G–3 18008/Plans, 1944; COMZ G–4 History, II, 71.

Supply Trucks Passing Through Bastogne, *Belgium, January 1945.*

133 percent by the first week in January, with the result that the port failed to meet its discharge goals. At Antwerp tonnages held at the port increased by 78 percent. Some installations came under attack from the air, notably in the Advance Section, where rail lines and supply depots were subjected to heavy attacks, particularly by V–1's.[58] On the whole, however, supply operations were not seriously disrupted, and the Communications Zone met the emergency demands well, marshaling transportation for evacuation and troop movements, and handling many special shipments via rail, truck, and air, some to meet replacement needs resulting from heavy losses suffered by units which bore the brunt of the attack, some to meet the sudden demands for defense materials, such as wire and mines.

The partial embargo on forward shipments imposed at the beginning of the Ardennes battle naturally caused deliveries to the forward areas to fall off some in the week of 17–23 December. But deliveries were normal again by the last week of December, and in January averaged about 15,000 tons per day on the northern line of communications.[59]

In the south, as in the north, a congestion of rail cars had developed in the forward areas early in December, caused in part by hazardous winter operating conditions. In an effort to reduce the backlog SOLOC early in January instructed Delta Base Section to reduce the tonnage allocation to Continental Advance Section by 25 percent. Forward reserves reached dangerously low levels during the month. By early February, however, partly as the result of better operating conditions and partly through the special trucking operation described in the preceding section, deliveries were reaching record totals.[60]

The oft-repeated hope of providing logistic support for the 12th Army Group via the southern line of communications was never fully realized, at least not as originally envisioned. SOLOC's "surplus" consisted almost exclusively of port discharge capacity; the shortage of transportation was as persistent a limiting factor in the south as in the north, and precluded the shipment of supplies in large volume over the Rhône line of communications for U.S. forces in the north. SOLOC provided substantial assistance, but it took the form mainly of direct logistic support, initially of units transferred from the 12th to the 6th Army Group—as, for example, the XV Corps, which was shifted at the end of September—and later of additional divisions brought in via Marseille rather than the northern ports as originally intended. A shortage of service troops in the south eventually limited the extent to which SOLOC could assume supply responsibility for additional combat forces. Meanwhile, advantage was taken of the surplus port capacity in the south to bring in vehicles, which were assembled at Marseille and then driven north with gasoline and other supplies which

[58] COMZ G–4 History, II, 66, 68.

[59] ADSEC Operations History, App. D; COMZ G–4 Weekly Reports on Status of Supply, 12 A Gp 400.291 Supply Information.

[60] SOLOC Supply Situation Rpt 3 (dated 3 Jan 45) and 4 (dated 18 Jan 45), SHAEF G–4 319.1 Supply Status (SOLOC), I 1945.

had also been discharged at Marseille for that specific purpose.[61]

The record of forward deliveries echoes the story of transportation difficulties told in Chapter V. The frustratingly restrictive rationing system which the transport deficiency imposed in the form of tonnage allocations obviously pleased no one. The solution of the problem suddenly demonstrated how the elimination of one limiting factor could create a new one and, incidentally, highlighted another basic deficiency in the theater's logistic structure—the lack of intermediate depots. Finally, the counteroffensive of December demonstrated with dramatic force the reverberating effect which a major tactical event could have on movements within the entire logistic organization.

Not all the supply shortages which developed in the fall of 1944 can be attributed to the transportation problem; many in fact persisted well beyond the period of transport difficulties. What some of these shortages were and how they affected operations is described in the next three chapters.

[61] See below, Ch. X, Sec. 1, for a discussion of the troop build-up via SOLOC. History of SOLOC, MS, II, QM, p. 3, and II, TC, p. 15; TWX BX–17993, 6 A Gp to COMZ, 17 Oct 44, SHAEF SGS 400.3/1 Supply Problems of Allied Advance; Memo, G–4 6 A Gp for Current Opns Br G–4, 27 Oct 44, sub: Mov From Southern France of 1,000 Tons Per Day to Third Army, 6 A Gp G–4 Transportation Section General, I; Ltr (draft), Lord to Larkin, 23 Nov 44, EUCOM 560 AT, Transport, Vessels, and Boats in General, II.

CHAPTER VII

Supplying the Armies: Rations, POL, and Coal

(1) Rations

Not all the supply shortages which hampered operations in the fall of 1944 can be attributed to the inadequacy of transportation. Some developed as the result of higher loss or expenditure rates than had been expected, and some reflected production difficulties at the very source—that is, the zone of interior. There were some items in which the shortages were never critical in the sense of jeopardizing the success of combat operations. Among the supplies in this category were rations.

Ration levels in the combat zone reached their lowest point in the second week of September, when First Army reported one and one-half days of supply on hand and the Third Army less than one.[1] Both armies staved off total depletion of their reserves only by using captured supplies. In a few cases German rations were a welcome relief from the monotony of the operational rations on which the combat forces had subsisted during most of the pursuit. Third Army had already captured some enemy stocks, including flour for bread, at Châlons and near Reims in the last days of August. On 9 September it made another welcome addition to its diet by the seizure of 1,300 tons of frozen beef and 250 tons of canned beef in refrigerated storage at Homecourt, northwest of Metz. A few days later First Army made a smaller haul of fresh beef in a plant at Namur.[2]

Ration stocks in both the combat zone and the Advance Section remained far below authorized levels throughout September and early October. Stocks in the Communications Zone had been maintained at a fairly satisfactory level, although the bulk of them had remained in the port and beach areas. In mid-October the Communications Zone held 18.6 days of supply, but there was some concern as to whether it could maintain that level because of the emphasis which was then being given to the offloading of the more critically short ammunition.[3] COMZ levels did recede somewhat in

[1] FUSA and TUSA G-4 Periodic Rpts for Period 10-16 Sep 44, 12 A Gp 319.1 G-4 Reports.

[2] TUSA AAR, II, QM, 6; FUSA Rpt of Opns, 1 Aug 44-22 Feb 45, Bk. IV, pp. 56-57.

[3] See below, Ch. IX, Sec. 1.

the next few weeks, falling to about ten days of supply. Meanwhile advantage was taken of the general improvement in transportation to rebuild forward stocks. Reserves were rapidly rebuilt in all three armies of the 12th Army Group, and by the first week in November actually exceeded authorized levels, the First Army having built up its Class I stocks to the unprecedented level of nineteen days. At the start of the November offensive First Army had 13.4 days of supply, Third Army 5.9, Ninth Army 9.8, the Advance Section 4.8, and the Communications Zone 10.6.[4]

Ration levels were always rather volatile, but with the exception of the temporary interruption during the Ardennes counteroffensive, when First Army reduced its reserves by making large issues and then drawing its requirements directly from the Liège depot, the flow of rations to the combat zone was relatively smooth after transportation had become adequate in November. After the Ardennes battle the policy was adopted of moving the maximum allowable reserves well forward preparatory to the resumption of the offensive. By the first week of February, therefore, First Army had 5.9 days of supply, Third Army had 4.9, Ninth Army 7.9, and the Advance Section 15.03, the armies' stock totaling 15,700 tons and the Advance Section's 106,720 tons. The level in the Communications Zone at that time stood at 23 days and represented 289,135 tons.[5]

Aside from the tight situation in mid-September the main problem of Class I supply was one of quality rather than quantity. During the pursuit tactical conditions dictated that operational rations—that is, C's, K's, and 10-in-1's—would be the principal types of rations consumed, particularly in the combat zone. Consumption of operational rations in August and September was actually about double the rate originally expected. Consequently theater stocks were being rapidly depleted as the pursuit came to an end, and the issuance of operational types to certain groups, such as prisoners of war, had to be prohibited. Quartermaster plans from the beginning had called for an early shift to nonoperational, or bulk rations—that is, the B and eventually the A ration. The shortage of operational types thus constituted an additional compelling reason for the rapid shift which the chief quartermaster ordered upon the reversion to a more static type of warfare in mid-September.[6]

[4] FUSA's 13.4 days was based on a ration strength of 400,000 men and represented 11,316 tons; corresponding figures for TUSA were 320,000 men and 4,353 tons, for NUSA 234,000 men and 4,107 tons, for ADSEC 1,050,000 men and 16,556 tons, and for the Communications Zone 2,057,000 and 49,483 tons exclusive of about 13,000 tons of perishables. The ADSEC level was based on Advance Section's own strength plus that of the armies, and the Communications Zone's in turn on total theater strength, so that the combat forces were actually backed by additional stocks equal to the levels in those organizations, at this time fifteen days. 12 A Gp G-4 Periodic Rpts, with attached 12 A Gp QM Rpts, 12 A Gp 319.1 G-4.

[5] Army levels were based on ration strengths of 400,000 in First, 450,000 in Third, and 320,000 in Ninth. The ADSEC and COMZ levels were based on strengths of 1,290,000 and 2,354,000 respectively. Total theater stocks came to 411,570 tons. 12 A Gp G-4 Periodic Rpts, with attached 12 A Gp QM Rpts, 12 A Gp 319.1 G-4.

[6] For the more detailed treatment of the subsistence story see Charles F. Romanus *et al.*, The Quartermaster Corps: Operations in the War Against Germany, a volume in preparation for this series.

The large-scale swing to bulk rations highlighted a problem in distribution with which the theater had already had some experience. Delivering a balanced B ration, which consisted of approximately 110 separate components, called for careful handling along the entire supply line from New York Port to the using unit. Experience in July had already revealed how the loss of one or more components could disrupt the balance and create difficulties for cooks trying to follow published menus. This problem had been anticipated in the zone of interior by prestowing and commodity-loading ships with balanced blocks of rations. But the New York Port was not always consistent, and often made substitutions, particularly when the theater failed to submit requisitions in time to allow ninety days for delivery. Even when shipments were balanced at the point of origin, the effort might often be nullified by improper unloading or reloading at continental ports, by the breaking up of balanced trains, by pilferage, or by indiscriminate bulk shipments in the attempt to register large tonnage deliveries. Inderdepot shipments designed to marry up scattered components were impossible at the height of the transportation shortage in mid-September. At one point early in October an embargo actually had to be placed on deliveries out of Le Havre so that quartermaster units in Channel Base Section could sort ration components. By that time stocks had become so unbalanced and dispersed that it proved necessary to set up intermediate collecting points at Paris and Sommesous, where rations could be sorted and balanced loads again made up for delivery to the armies.[7]

Eventually the worst defects of the rations-handling problem were overcome by giving more attention to such matters as unloading and the make-up of trains, aided by the general improvement in transportation and the accompanying return of emphasis on selectivity rather than tonnage in forward movements. Subsequently intermediate depots, which were nonexistent for several months after the breakout, were also established, where large bulk receipts could be handled. For some time, however, the imbalance of subsistence stocks often threatened to present the Quartermaster Corps with the paradox of scarcity in the midst of plenty—that is, of having ample Class I supplies but few rations.

The manner in which this imbalance could affect the over-all ration level was well illustrated early in February 1945. At that time the theater's level of balanced rations was determined by the supply of coffee, of which there were only 7.6 days of supply on hand, although there were much higher levels of all other components. If coffee was disregarded sugar became the determining item, of which there were 19.7 days on hand. If both coffee and sugar were disregarded yeast became the determining factor, of which there were 20.7 days on hand, and so on. In other words, an additional 12.1 days supply of coffee would

[7] Romanus *et al.,* The Quartermaster Corps: Operations in the War Against Germany, MS, Ch. VIII, pp. 10–11, 14–20; Quartermaster Supply in the European Theater of Operations in World War II, prep by QM School, Camp Lee, Va., II, Subsistence, 24–28, MS OCMH.

have raised the over-all level of balanced rations to 19.7 days and the actual number of rations from 17,850,000 (represented by 7.6 days) to 46,200,000 (19.7 days); an additional day's supply each of coffee and sugar would have brought the theater level up to 20.7, representing 48,560,000 rations.[8] At times this imbalance resulted in a drain on operational rations, which could be ill-spared. During the more fluid operations beginning in mid-December operational rations again were in great demand, and the shortage caused Third Army to conserve K rations for front-line troops, limiting issues to rear area units to 10 percent of the total number of operational rations requested. Third Army continued to restrict issues until late in January.[9]

Aside from the Ardennes interlude, steady progress was made during the fall and winter in providing bulk rations throughout the theater. By the end of January between 85 and 90 percent of all troops on the Continent were receiving either the B or A ration. The percentages were only slightly lower in the combat zone.[10]

As a part of the program to replace the operational ration with the bulk ration the Quartermaster Corps in the European theater also attempted to provide perishable items, including meat, dairy products, and fresh vegetables and fruits, with the intention of eventually converting the B ration into a type A ration. The chief quartermaster had approved plans to introduce perishables onto the Continent as soon as possible after the assault, beginning at approximately D plus 30 with issues to 40 percent of the troops, providing 50 percent of them with items in the third month, and 60 percent by D plus 90. Plans called for all troops to receive the A ration by D plus 240. Deliveries were to be made from the United Kingdom at first, but the chief quartermaster hoped to have ocean-going reefers discharge directly on the Continent by the end of the second month.

These plans proved far too ambitious. Planning for the introduction of perishables was dominated from the beginning by the problem of providing adequate cold storage. Providing perishables, like delivering POL via pipeline, required a co-operative effort. The Transportation Corps was to handle reefer shipments to the Continent and move stocks in the Communications Zone within the capabilities of the rail system, the Quartermaster Corps was to transport perishables in the Communications Zone and to the army areas in excess of rail capacities in refrigerated trucks, and receive, store, and issue them at refrigerated warehouses, and the Corps of Engineers was to construct or rehabilitate and maintain static cold-storage warehouses.

---

[8] Daily QM Situation Rpt, 11 Feb 45, attached to Weekly Report on Status of Supply (COMZ), 10 Feb 45, 12 A Gp 400.291 Supply Information. Another typical example is afforded by the reduction of Advance Section's level of rations by four days at the beginning of December because of the lack of the salt component. 12 A Gp G–4 Periodic Rpt 18, for Period 26 Nov–2 Dec 44.

[9] TUSA AAR, II, QM, 13–14, 17.

[10] TUSA reported 84.4 percent of its troops receiving the A or B ration in January. FUSA's consumption figures indicate that about 86 percent were getting the bulk ration. TUSA AAR, II, QM, 17; FUSA Rpt of Opns, 1 Aug 44–22 Feb 45, Bk. IV, pp. 112–13.

Before D Day neither the engineers, who had been requested to provide cold storage rising from 6,000 tons at D plus 90 to 44,000 tons at D plus 240, nor the Transportation Corps, which was asked to increase coastal reefer tonnage and to provide refrigerated rail cars, road vans, and barges for supplementary storage, could promise to meet the chief quartermaster's requirements. The only solution to the deficit appeared to be the use of ocean-going shipping for supplementary storage. But permission to use large reefers for that purpose, repeatedly urged by the chief quartermaster, was denied by the War Department.

In mid-July, a week after the first bulk rations were issued on the Continent, Maj. Gen. Robert M. Littlejohn, the theater chief quartermaster, announced plans to convert the B to an A ration. At that time less than 1,000 tons of cold storage were available in Normandy. In any event, the breakout only a few days later largely canceled these plans, at least so far as the combat elements were concerned, for the armies quickly reverted to operational rations for the period of mobile warfare which followed. Meanwhile, the tight supply situation forced plans for the construction and rehabilitation of static cold-storage plants to give way to more urgent needs, such as road, railway, and bridge reconstruction; and the desire to have reefer rail cars transferred from England to France likewise gave way to the need for items of greater tactical importance. Consequently the goal of providing fresh food items to 40 percent of the theater's troops by D plus 60, and 60 percent by D plus 90, was not realized.

Early in September refrigerated storage facilities on the Continent still totaled a mere 1,400 tons against the original requirement of 6,000. August requirements had been met only by using as a floating warehouse a 3,000-ton reefer loaned by the British. Despite the outlook during the pursuit, however, General Littlejohn had announced plans for providing the A ration to 85 percent of all troops on the Continent by 1 November, and expressed the hope of providing one pound of fresh meat and dairy products per man per day by February 1945. These objectives exceeded even the preinvasion plans, which had already failed to materialize. Haphazard distribution practices, in part the product of the serious reefer rail car shortage, continued to frustrate plans for realizing the perishables program. Again the chief quartermaster asked the War Department to permit large reefers to cross the Channel and discharge at continental ports rather than in the United Kingdom. This request was finally granted early in October.

By one expedient or another, and through the gradual accumulation of transport and storage facilities, the perishables program began to improve during the fall. Ocean-going reefer space was more economically used during the winter months by shipping cheese and meat products such as smoked ham, bacon, and salami in nonrefrigerated dry cargo space. In September cold storage was opened in Paris, and early in October storage captured earlier at Homecourt and Namur was turned over to the Advance Section.

While the cold storage capacity still

totaled only 23,600 tons at the end of January 1945, the deficit (amounting to 37,800 tons) was partially overcome by the better use of distribution facilities. Improved unloading techniques and more reliable transportation schedules shortened the turnaround time for rail cars. Local procurement of fresh vegetables and fruits helped relieve some of the burden on both transatlantic shipping and on continental storage. Finally, an unexpected development in the use of refrigerated vans helped mightily in offsetting the shortage of rail cars. Because the range of their efficiency was believed to be limited to about seventy-five miles, reefer vans were expected to be used only for short trips in delivering perishables from cold storage in forward areas to distribution points. Necessity eventually caused them to be pressed into service for long-distance hauls, however, and when their operational efficiency was proved through continued use on long hauls reaching all the way back to the ports, the decision was made to release all refrigerated vans for such hauling, and to use open, nonrefrigerated trucks for the shorter runs, such as those between Le Havre and Paris, and between Antwerp and Namur. But mid-December deliveries had improved to the point where 90 percent of all troops in the theater were receiving perishable items.

Unbalanced depot stocks continued to distort the Class I supply picture. The lack of only a few components, such as sugar, evaporated milk, or lard, often prevented the attainment of the desired levels. Nevertheless, by early 1945 distribution of fresh meats and vegetables was sufficiently good to warrant a change in the nomenclature of the bulk ration from B to A.[11]

Dehydrated foods—mostly in the form of powdered milk, eggs, and potatoes—were common components of both the A and B field ration, facilitating distribution and conserving shipping space for items which otherwise could not have been provided. In general, however, mess personnel were inadequately instructed in their use, and only cooks with imagination and an inclination to experiment discovered formulas to make them palatable.[12]

(2) POL

Motors as well as men had huge appetites in the type of war fought in 1944. The supply of POL, like that of rations, felt the effects of the pursuit for several weeks after the halt in mid-September. In general, improvement in the supply of POL had to await improvement in the means of distribution, which meant that recovery was postponed until November.

Deliveries of gasoline had risen slightly for a few days early in September, making it possible to restore unit reserves in some of the front-line formations.[13] But the improvement was deceptive. By the third week of September the Third Army, struggling to enlarge its bridgeheads over the Moselle, reported that it had less than a half day of supply on hand and that it was reinstituting

[11] Romanus *et al.*, Quartermaster Operations in the War Against Germany, Ch. VIII, pp. 20–32; Quartermaster Supply in the ETO, II, Subsistence, 20–35. In January and February First Army reported that fresh meat issues averaged 1.08 meals per man per day. FUSA Rpt of Opns, 1 Aug 44–22 Feb 45, Bk. IV, p. 69.

[12] QM Supply Opns, Gen Bd Rpt 109, p. 76.

[13] TUSA AAR, II, G–4, 6–7.

rationing.[14] First Army, having just breached the first line of the West Wall defenses south of Aachen, reported that it had no motor gasoline at all for issues and that its corps and divisions had no fuel other than that remaining in the tanks of their vehicles.[15]

First Army, taking advantage of its higher tonnage allocation, succeeded in rebuilding its reserve to about seven days by mid-October. But Third Army averaged less than two days of supply until the end of October. In mid-September it had rationed gasoline at the rate of 5,000 gallons per day for infantry divisions, and 25,000 gallons per day for armored divisions. But deliveries were highly unpredictable and continued to fall short of requirements. Late in October the ration was set at 6,500 gallons and 12,500 gallons respectively for infantry and armored divisions. Throughout these weeks issues within the army averaged only 235,000 gallons per day.[16]

The shortage of gasoline, unlike the situation in many other items of supply, had become more than a local matter and could not be attributed solely to the shortcomings of inland transportation. Stocks had also declined in the base areas. Reserves in the Communications Zone, which in mid-August had stood at fourteen days for all troops on the Continent, fell to about two and one-half days in the first week of October, causing grave apprehension in the 12th Army Group because of the potential threat to future operations.[17]

The alarming decline in continental stocks, first in the combat zone and then in the communizations zone, resulted from a combination of causes, including increased consumption, insufficient reception and storage capacity, and difficulties in distribution. In the first two months of continental operations the consumption of POL had been at a rate considerably below the preinvasion planning factor of 153 tons per division slice, and even in August rose to only 158.8 tons.[18] In September, in spite of the slowing up of operations, consumption rose to an unprecedented 248.3 tons per slice, reflecting the maximum sustained demand on all line-of-communications transportation.[19] Consumption dropped back to 197.2 tons in October, and to

[14] FUSA G–4 Periodic Rpt for Period 17–23 Sep 44, with attached diary, and TUSA AAR, II, G–4, 7.

[15] TWX, ADSEC to ETO G–4, 22 Sep 44, SHAEF SGS 400.3/1 Supply Problems of Allied Advance; FUSA G–4 Periodic Rpt for Period 17–23 Sep 44. Army and army group reports of stocks on hand varied somewhat, apparently because of different methods of computing levels. The reserve in days of supply varied, e.g., depending on whether assigned or ration strengths were used, whether requirements were computed on the basis of pounds per man per day or on the basis of miles per vehicle per day. Furthermore, levels might be expressed separately for each petroleum product or for Class III as a whole.

[16] TUSA G–4 Periodic Rpts; Cbl F–1503, TUSA to Crawford, 24 Oct 44, 12 A Gp 463.7 Gasoline and Motor Oil, I.

[17] Daily QM Situation Rpt, COMZ, 8 Oct 44, Incl to COMZ G–4 Weekly Report on Status of Supply, 7 Oct 44, 12 A Gp 400.291, Supply Information, II; 12 A Gp G–4 Periodic Rpt for Period 8–14 Oct 44.

[18] The figure was based on an average troop strength of 960,000 and twenty-four division slices. Daily consumption averaged 3,804 tons and totaled 117,923 tons for the entire month. Memo, Cummings for OCQM COMZ, 7 Nov 44, sub: Plng Estimates of POL Rqmts, U.K. and Continent, USFET 400.42 Petroleum, USFET Petroleum Office.

[19] This was based on an average strength of 1,250,000 and 31.25 slices. Daily consumption averaged 7,758 tons and totaled 232,728 tons for the month. Memo, Cummings for OCQM COMZ, 7 Nov 44.

164.2 by December.[20] Before these figures were available, however, Col. Elmer E. Barnes, the chief petroleum officer of the theater, on the assumption that September conditions would prevail for several months, recommended a substantial upward revision of the planning factor.

Late in November the theater began requisitioning its future needs on the basis of a new factor of 217 tons per division slice.[21] Revisions in the percentage breakdown of the various POL products were made at the same time. Preinvasion estimates had provided that 79 percent of the total POL tonnages should consist of MT80 gasoline (80-octane motor vehicle type). The experience of the first few months showed that motor transport gasoline accounted for fully 90 percent of all POL products consumed. The November revisions compromised with this statistic and raised the percentage of MT80 to 85.

While the consumption of gasoline in September, when it averaged 6,500 tons per day, proved abnormally high in terms of the number of divisions employed, there was not likely to be any reduction in the total tonnages required in succeeding months in view of the steady build-up. The fact that 90 rather than 79 percent of all POL would have to consist of MT80 gasoline alone made it certain that requirements would exceed the original estimates. In October the Communications Zone estimated that 8,600 tons of motor transport gasoline would be needed every day.[22]

Neither the distribution nor the reception facilities of late September offered any prospects of handling such tonnages. In mid-September the Major Pipeline System was still far from adequate to move bulk POL requirements forward. At that time the most advanced of the three pipelines was still twenty miles short of the Seine and was dispensing at Chartres; a second line was dispensing at Domfront; the third (for aviation gasoline) had reached Alençon. Further construction had been suspended temporarily, and the rail tonnage required for the movement of pipe and other construction materials was sacrificed to more urgent supply movements. This decision was almost immediately reconsidered, and two daily trains were allotted for the movement of construction materials so that the pipelines could be extended to Coubert, about ten miles beyond the Seine, where tank storage was available.

Progress thereafter was slow. One 6-inch line reached Coubert early in October, but the other two lines were not completed to that point until December. Plans to extend the Major System to Metz, and eventually to Mainz, did not begin to materialize until late in January 1945, when construction eastward toward Châlons-sur-Marne was finally

[20] Romanus *et al.*, Quartermaster Operations in the War Against Germany, p. 104; Min, Mtg, American Petroleum Products Comm, London, 30 Jan 45, dated 2 Feb 45, EUCOM 337 Confs General, III, 44.

[21] This was later reduced to 192 tons on the basis of data which included the fall and winter months. QM Supply Opns, Gen Bd Rpt 109, p. 152; Memo, Barnes for OCQM, 26 Oct 44, sub: Supply of MT80 and Derivatives to Far Shore, and Memo, Cummings for OCQM, 7 Nov 44, sub: Plng Estimates of POL Rqmts, USFET 400.12 Petroleum.

[22] Ltr, Hq COMZ to CGs Base Sec, 26 Oct 44, sub: Forward Shipments of MT80 Gasoline, ADSEC 463.7 Gasoline and Motor Oil.

PORTION OF THE MAJOR PIPELINE *passing over an improvised trestle bridge.*

undertaken. Coubert consequently remained the easternmost terminus of the Major System through most of the winter. Completion of the three lines to that point gave the Major System a total of 850 miles of pipeline. The Major and Minor Systems combined consisted of 950 miles of pipeline and 850,500 barrels of storage.[23]

An important change had meanwhile been made in the operational control of the pipelines. As intended, the pipelines were constructed by engineer units of the Advance Section, the lines later being turned over to other COMZ sections wherever they assumed area control. The resulting divided responsibility for a service which, like the Red Ball Express, had become intersectional in its operations, had the usual defects, and it soon became evident that centralized control was imperative if maximum efficiency was to be achieved. On 23 September, therefore, all construction and operation of the pipelines were brought under the control of a single agency of the Office of the Chief Engineer, ETO, known as the Military Pipeline Service. Thereafter all pipeline troop units were attached to the Military Pipeline Service for operational control, and to the various COMZ sections in which they happened to be located only for supply and administration. The new organization permitted

[23] Monthly Rpts Construction Div OCE ETO, ADSEC Files, 51–104; COMZ G–4 History, V, 14–15; TWX E–46157, COMZ to SHAEF, 11 Sep 44, SHAEF SGS 800 Harbors, Opening, Use, Construction; COMZ G–4 Plant and Communications Diary/Jnl, 4 Oct 44, ETO Adm G–4 145C; Petroleum, Oil, and Lubricants, Hist Rpt 13, CE ETO, pp. 94–95.

CAMOUFLAGED PUMPING STATION *along the pipeline from Cherbourg.*

the chief engineer to exercise effective unified control over all planning, construction methods, and operation.[24]

The inadequacy of the Major and Minor Pipeline Systems,[25] both as a means of inland transport and as an intake facility, became evident as the weather began to worsen toward the end of September. Port-en-Bessin, the terminus of the Minor System, was particularly vulnerable because of its dependence on the submarine TOMBOLAS for tanker discharge. The planners had foreseen this, and an additional overland pipeline had been constructed from Cherbourg to Port-en-Bessin to ensure the full use of the storage facilities of the Minor System, which served both U.S. and British forces. Port-en-Bessin possessed a maximum intake capacity of 6,000 tons per day, but its average performance could not be counted on to exceed 2,000 tons. Clearance difficulties at Cherbourg made it imperative that Port-en-Bessin be used to its maximum potential, and POL officials planned to keep the port in operation as long as possible. In mid-September U.S. and British officials allocated one third of the port's capacity to U.S. forces, which included continuing the transfer of up to 1,000 tons per day over the line from Cherbourg. In the first week of October,

[24] *Final Report of the Chief Engineer, ETO,* I, 309; Petroleum, Oil, and Lubricants, Hist Rpt 13, CE ETO, pp. 84–85; COMZ G–4 History, V, 19; ADSEC G–4 Periodic Rpt for Quarter Ending 30 Sep 44, ADSEC 319.1 G–4 Periodic Rpts; Report on POL Plans and Construction to 8 May 45, n.d., ADSEC Engineer Completion Rpts Bulk POL Installations.

[25] Beginning in January 1945 the combined facilities of the Major and Minor Systems were usually referred to as the Western System.

however, bad weather gave a foretaste of future difficulties when it caused two TOMBOLA berths to be put out of action and necessitated the diversion of tankers to Cherbourg. Brigadier D. H. Bond, chief of the G–4 Petroleum Branch at SHAEF, foresaw that difficulties in discharge via ship-to-shore lines might very soon sharply reduce the amounts of gasoline that could be made available at Port-en-Bessin and warned that U.S. forces should no longer expect a large portion of the Minor System's reduced output.[26]

Although it handled 80 percent of all the gasoline brought to the Continent, Cherbourg itself was proving unsatisfactory as a POL port, for its operations were frequently affected by bad weather, and its capacity was limited by inadequate intake and storage facilities. On the night of 4 October a small storm destroyed eight of the ten intake lines at the Digue de Querqueville, stopping all discharge for eight hours and materially limiting it for another twenty-four. Storage facilities, which totaled about 250,000 barrels for MT80 and at first appeared ample, proved inadequate. Pipeline breaks farther inland, such as occurred when a flood washed out a section of the line between La Haye-du-Puits and St. Lô, slowed clearance of the port and forced tankers to wait offshore.[27]

Discharge had been rather limited all along by the fact that only one berth had been provided, which, located along the Digue de Querqueville in the exposed Grande Rade, was unavoidably vacant during the periods between tankers. But the Communications Zone refused to authorize the expenditure required to provide a second berth and additional offloading lines which Normandy Base Section requested in mid-October.[28] By that time POL officials had taken steps

[26] Min, Mtg, 21 A Gp, SHAEF, and ETO POL Officials at 21 A Gp Rear Hq, 12 Sep 44, SHAEF G–4 463.7 Gasoline and Motor Oil (General) 1945; Memo, Bond for G–4 SHAEF, 22 Sep 44, sub: Protection of Port-en-Bessin for Winter Opns, Ltr, Col. H. Goodfellow, 21 A Gp, to G–4 Petrol SHAEF, 30 Sep 44, sub: Bulk Imports—MT80, Memo, Bond for G–4 POL COMZ, 8 Oct 44, sub: Bulk Imports MT80 Port-en-Bessin, and Memo, G–4 Petrol SHAEF for Bond, 8 Oct 44, sub: Gen Info Tanker Position Channel Ports, all in SHAEF G–4 463.7 Gas and Motor Oil.

[27] Notes on Conf, Office of the G–4 12 A Gp, 5 Nov 44, 12 A Gp 463.7 Gasoline and Motor Oil, I; Min, CAO Mtg, 10 Nov 44, SHAEF AG 337–14; Memo, Maj J. M. D. Heald, Chief Plng and Rqmts Br ETO G–4 POL, for Brig Gen Weaver, 23 Oct 44, USFET Petrol Officer 400.42B Capacity for Bulk POL on Continent.

[28] With two berths Cherbourg, according to Normandy Base, might have maintained an uninterrupted discharge at the maximum rate of 405 tons per hour. Another source states that the Cherbourg installation could handle 500 tons per hour. Ltr, Normandy Base Section to G–4 COMZ, 17 Oct 44, sub: POL Situation Normandy Base Sec, with Ind, 3 Nov 44, EUCOM 463.7 Gasoline and Motor Oil, 44, IIb, and Ltr, Lt Col H. C. Ferrell, Chief Stocks and Shipping Br G–4 Petrol Br COMZ, to ANPB, 30 Sep 44, sub: Shipping Facilities—Far Shore, USFET Petrol Office 400.42B Capacity for Bulk POL on Continent.

The PLUTO project, calling for underwater pipelines from the Isle of Wight to Cherbourg, was largely a failure. Two flexible 3-inch cables were laid in August, but extreme difficulties were encountered because of leaks and breaks, and no gasoline was ever pumped through them. A third line began delivering gasoline in mid-September at the rate of about 140 tons per day, but the total deliveries were insignificant. In November Brigadier Bond reported all lines broken down and inoperative, and recommended the abandonment of the project. Memo, Bond for G–4 SHAEF, 23 Nov 44, sub: Abandonment of PLUTO Project—Isle of Wight to Cherbourg, SHAEF G–4 463.7 Gas and Motor Oil; Memo, Col Dan Gilmer for Gen Hull, 2 Nov 44, sub: Cross-Channel Pipelines, OPD Exec Office File 9; SHAEF G–4 Weekly Logistical Summaries for 10–16 Sep, 17–23 Sep, and 8–14 Oct 44, SHAEF G–4 War Diary/Jnl.

to develop additional bulk intake capacity elsewhere. The first relief was afforded at Ostend, a British-operated port, where U.S. forces were initially authorized to draw 500 tons, later 1,000 tons, of MT80 gasoline per day. Discharge at Ostend began in the second week of October and was uneven because of the vagaries of the weather. Nevertheless, deliveries there relieved to some extent pressure on Cherbourg.[29]

Reception and distribution facilities were also substantially augmented by the development of what was known as the Seine River System, based on Le Havre and, farther up the river, Port Jerome and a satellite of Rouen, Petit Couronne. Both offloading and storage facilities were found relatively undamaged at Le Havre, where tankers could discharge either into smaller tanker vessels for transshipment to Rouen, or into shore storage, whence bulk POL could either be delivered to Port Jerome through an existing 10-inch pipeline, or shipped out via rail or truck.

Plans were immediately made to develop Le Havre's discharge capacity to 5,000 tons per day, 3,000 of which were to be shipped to Port Jerome, via the pipeline. Engineers of the Military Pipeline Service began rehabilitating existing facilities in mid-October, starting with unloading lines and storage at Le Havre, and clearing the pipeline to Port Jerome. The pipeline was not placed in operation until December, but Le Havre received its first tanker on 31 October, and unloading into storage and decanting began immediately. Facilities at the other two ports were gradually brought into use, and work on the various installations continued until February.

By the end of January 1945 the Seine River System, combining tanker berths and storage, decanting, and loading facilities at Le Havre, Petit Couronne, and Port Jerome, was virtually complete. Included were two 6-inch pipelines from Petit Couronne to Darnetal, across the Seine, taken over from the British.[30]

While the development of the Seine River System had high priority, Allied planners counted even more heavily on Antwerp to bolster continental reception capacity. The great Belgian port was known to have POL facilities matching those for general cargo reception. Storage capacity alone totaled 2,600,000 barrels and was captured virtually undamaged. A survey of the port in September revealed that only minor alterations would be required to provide the necessary tank truck and tank car loading facilities. Antwerp's value was to be further enhanced by the construction of pipelines which were eventually to deliver bulk gasoline across the Rhine. Plans which the Military Pipeline Service submitted to the COMZ G-4 in September called for laying one 6-inch and four 4-inch lines from Antwerp to Ko-

[29] SHAEF G-4 War Diary/Jnl, Petrol Br, 8 Oct 44; COMZ G-4 History, 22; Quartermaster Supply in the European Theater of Operations, QM School, Camp Lee, Va., IV, Fuels and Lubricants, 46.

[30] Monthly Reports Construction Div OCE ETO, Sep 44-Jan 45, ADSEC Records; Memo, Bond for G-4 SHAEF, 27 Oct 44, sub: Dev of Port of Le Havre, Seine River, and Rouen for Bulk POL, and Memo, Lord for Crawford, 29 Oct 44, sub: Dev of Bulk POL Import Facilities at Le Havre and Along the Seine River, EUCOM 463.7 Gasoline and Motor Oil, 1944, IIb; SHAEF G-4 War Diary/Jnl, 9, 17, and 27 Oct 44; History of the Channel Base Section, I, 148, ETO Adm 588.

POL STORAGE TANKS, *used by the Americans, Petit Couronne, France, December 1944.*

blenz. Subsequent alterations in the plan named Cologne as the eastern terminus, and finally Wesel, much farther north.

The COMZ G–4 approved the plans for the Antwerp, or Northern, System, as it was called, early in October and instructed the Military Pipeline Service to proceed with construction immediately, although there was as yet no prospect of receiving POL through Antwerp because the Schelde estuary had not yet been cleared. Construction was initially held up for lack of materials and equipment, but work finally got under way on loading facilities in Antwerp with the arrival of equipment via rail from Cherbourg and through the loan of additional equipment from 21 Army Group. The first tanker berthed at Antwerp on 3 December.

Work on the pipeline did not begin until 8 December, more than a week after the port opened. Construction then started simultaneously at several points along the route, and the lines were completed to Maastricht by the end of January, where dispensing of both MT80 and aviation gasoline began early the next month. Construction was suspended at that point, and Maastricht remained the eastern terminus of the Northern System until early in March, when the extension of the lines northeastward was undertaken. In a minor switch from original plans the 6-inch line was eventually used for MT80 and two of the 4-inch lines for aviation gasoline. POL facilities at Antwerp, like cargo-handling facilities, were used jointly by U.S. and British forces. The Communications

Zone had at first bid for 675,000 of the 2,600,000 barrels of storage. By January 1945 U.S. forces had been allocated a total of 950,000 barrels.[31] Intake of the system eventually averaged more than 30,000 barrels per day, some of which was pumped forward for decanting at Maastricht, and later at Wesel.[32]

The need for additional intake and storage facilities was paralleled by a similar requirement for more adequate means of distribution, of which the pipelines were only a part. The ideal method of distribution, as contemplated in POL plans, called for the reception and forwarding of gasoline in bulk to storage facilities in the Communications Zone, and retail distribution, particularly to combat elements, in 5-gallon cans.[33]

All decanting from bulk and packaging was intended to be carried out by the Communications Zone. But the speed of the pursuit, the lag in pipeline construction, the condition of the railways, and the shortage of 5-gallon cans all combined to upset these intentions.[34] In mid-September First and Third Armies were receiving gasoline mainly by trucks hauling from the pipeheads at Chartres and Alençon to Soissons and Sommesous. By the end of the month First Army was receiving a large portion of its requirements in bulk via tank trucks and tank cars and doing its own decanting into 5-gallon cans. Third Army for a while got most of its gasoline packaged and by rail, but eventually it also set up its own decanting points. Inadequate decanting facilities at the pipeheads and other shortcomings of the pipeline also made it necessary to ship gasoline, both in bulk and in cans, by rail from Cherbourg.[35]

The entire distribution problem was severely aggravated in October by a growing shortage of 5-gallon cans. The lowly "jerrican," so named by the British, who, followed by the Americans, had copied the German container after discovering its superior merits, had a role in gasoline supply hardly suggested by its size.[36] Gasoline might be shipped from the port via pipeline, tank car, or tank truck; but it had to be delivered in packaged form to the ultimate consumer. In the last analysis, therefore, the retail distribution

[31] Petroleum, Oil, and Lubricants, Hist Rpt 13, CE ETO, pp. 99–105; Monthly Rpts Construction Div OCE ETO, ADSEC Files.

[32] *Final Report of the Chief Engineer, ETO,* II, App. 33–C–4.

[33] The 55-gallon drum was never considered satisfactory for retail distribution because of its weight and bulk, and its use was therefore kept to a minimum on the northern lines of communications. Many cases of hernia were attributed to the attempts to manhandle the awkward 55-gallon drum. QM Supply Opns, Gen Bd Rpt 109, p. 139.

[34] QM Supply Opns, Gen Bd Rpt 109, pp. 139–40.

[35] Quartermaster Supply in the ETO, IV, Lubricants, 44–45; Study G–4 SHAEF, 28 Sep 44, sub: Mov of POL, SHAEF G–4 463.7 Gas and Motor Oil, III; History of G–4 POL, Normandy Base Section Hist Rpts, pp. 3–4, ETO Adm.

[36] The U.S.-developed container, sometimes referred to as the "ameri-can" was also essentially a copy of the German can except for the closure. The jerrican had a simple locked-cap, cam-operated spout; the ameri-can had a larger opening with a screw cap, to which a flexible nozzle was supposed to be attached for pouring. Quartermaster officials designed the American container in the belief that the gas tank openings on U.S. vehicles, which were either countersunk or flush with the body, could be reached only with a nozzle attachment. But combat experience actually proved this to be its greatest defect. Nozzles were easily lost and were difficult to replace. Lacking them, or preferring not to bother with them, drivers often wasted gasoline through spilling. Quartermaster Supply in the ETO, IV, 29.

THOUSANDS OF JERRICANS *are filled from railroad tank cars at a decanting area, Belgium, December 1944.*

of gasoline depended in large part on an adequate supply of 5-gallon cans.

U.S. forces had built up a stock of about 12,000,000 cans before the Normandy invasion. But this number was expected to suffice only for the initial stages of continental operations. Quartermaster planners subsequently concluded that about 800,000 new cans per month would be required to cover losses (estimated at 5 percent per month after D plus 60) and to maintain a can population commensurate with the troop build-up. The chief quartermaster accordingly placed an order with the British War Office for nearly 4,500,000 cans to be supplied from U.K. manufacture by the end of 1944. Nearly 2,000,000 of them were intended for the air forces with the understanding that they would be turned over to the ground forces after their first trip in accordance with the practice of using them only once for aviation fuel.[37] A large portion of U.S. can requirements had already been met by British production, in part through the shipment of an American plant to England early in 1943.[38]

Tactical developments in the first three months were largely responsible for up-

[37] Ltr, Cummings to Gen Peckham, 19 Sep 44, sub: Reqmts for 5-Gallon Blitz Cans, USFET 458.11 Cans 1943–45; Romanus *et al.,* Quartermaster Operations in the War Against Germany, Ch. VIII, 98–99.

[38] The plant had been set up in Middlesex and was operated by the firm of Magnatex Limited under the control of the Ministry of Supply. Quartermaster Supply in the ETO, IV, 30–31.

setting the chief quartermaster's plans for supplying an adequate number of jerricans. The rapid advance, in addition to increasing the consumption of POL, had, by placing Allied forces far beyond planned phase lines, resulted in a much longer turnaround time—that is, the time required to fill, forward, and return cans—than that on which the required supply of cans had been based.[39]

The loss of cans had also been much higher than expected. Retail distribution of gasoline in the early phases had been based on the principle of exchanging a full can for an empty one. Units were permitted to draw 100 full cans only by turning in an equal number of empties. This simple but essentially sound SOP was widely disregarded in the heat of the pursuit, resulting in a trail of abandoned or discarded jerricans stretching from Normandy to the West Wall. Hundreds of thousands lay in abandoned dumps and bivouacs; thousands more had been used to build sidewalks in the mud, or as chairs, and for hundreds of other purposes not intended; others had found their way into French homes. By mid-October the chief quartermaster noted that 3,500,000 could not be accounted for.[40]

Meanwhile two of the sources of supply showed signs of drying up. The air forces had given notice that they could not ensure the return of their quota of cans, stating that they were needed for static reserves because of the depletion of current working stocks. As for procurement in the United Kingdom, which U.S. forces had counted on as their main source after D Day, the British War Office first advised that it could allocate only 221,000 cans per month to the Americans against the request for 500,000, and subsequently expressed a desire to retrain the entire U.K. output for British forces.[41] In mid-September the chief quartermaster therefore reluctantly turned to the War Department to meet the theater's needs, placing a requisition for 7,000,000 cans. The War Department offered to provide only 5,400,000 of this number. All but two can-producing plants in the United States had been closed down, it explained, and it did not favor reopening idle plants and drawing labor away from other urgent production. It would be much more economical, the War Department suggested, to increase going production in the United Kingdom.[42]

Can production was one of the several fields in which the United States and Britain eventually found it necessary to collaborate closely. Early in the fall, when it became apparent that requirements were outrunning production facilities, British and U.S. officials in Washington agreed to set up an Allied Container Advisory Committee to co-ordinate more closely the collection of information on requirements and potential sources of supply and to allocate production. Late in November, apparently as a result of the committee's initial

[39] Original plans had been based on a seven-day turnaround. British planners had used a more realistic thirteen-day factor.

[40] Quartermaster Supply in the ETO, IV, 33.

[41] Ltr, Cummings to Peckham, 19 Sep 44.

[42] Cbl, ETO to AGWAR, 19 Sep 44 Cbls, ANPB to COMNAVEU, 19 and 23 Sep 44, Cbl, ETO to AGWAR, 30 Sep 44, and Cbl CR-1351, ANPB to ETO, 7 Oct 44, all in USFET 463.72 Ground Force Gasoline, II; Ltr, Barnes to War Office, 21 Oct 44, sub: Supply of Returnable Cans From U.S., USFET 458.1 Containers 1942-45.

deliberations, the British agreed to provide about 550,000 cans per month to American forces.

Also the chief quartermaster had begun to explore the possibilities of meeting a portion of U.S. needs from another source—local procurement on the Continent. Negotiations during the fall produced agreements with the French for the manufacture of 9,000,000 cans and with the Belgians for 2,000,000. Both programs were dependent on imports of sheet steel from the United States. Production was scheduled to get under way in February in Belgium and in April in France.

Early in January the chief quartermaster re-estimated U.S. requirements and, on the basis of maintenance and turnaround factors developed during operations thus far, concluded that U.S. forces would need about 1,300,000 new cans per month in 1945 to maintain a workable can population for the gradually increasing troop strength. U.S. and British officials agreed then that approximately 550,000 of these should be provided from British production, and that the rest should come from U.S. and continental production, the zone of interior contribution depending on progress in getting French and Belgian production under way.

The local procurement programs failed to make a significant contribution to U.S. requirements before the end of hostilities, largely because of difficulties in getting sheet steel from the United States. As of V-E Day the French had manufactured only a token number of the original commitment. Shortly before the end of hostilities the chief quartermaster estimated that U.S. forces needed a can population of 19,000,000 to support the current troop strength. But the target was not met. What the actual count was in the last month is not known.[43]

The theater also had initiated a vigorous campaign to recover some of the lost cans. With the help of the Allied and U.S. Information Services, and employing *The Stars and Stripes,* the French and Belgian press, and the radio and newsreels, it widely publicized the importance of the jerrican's role in winning the war, and made an unprecedented appeal to civilians and soldiers alike to search for the wayward containers and return them to the supply stream. Through the French Ministry of Education a special appeal was made to French children to round up cans, offering prizes and certificates for the best efforts. In this way approximately 1,000,000 cans were recovered. At the end of November 2,500,000 were still "AWOL." [44]

The first improvement in POL distribution had been realized in October, when the minimum requirements of MT80 for U.S. forces east of the Seine rose to 5,900 tons (1,616,600 gallons) per

[43] Romanus *et al.,* Quartermaster Operations in the War Against Germany, Ch. VIII, pp. 145–48; Quartermaster Supply in the ETO, IV, 32–35; Cbl, BAS (Washington) to War Office, 14 Nov 44, Ltr, Barnes to GPA, 2 Dec 44, sub: Availability of 5-Gallon Returnable Cans, Ltr, Lt. Col. J. R. Spencer to Chief G–4 Petroleum Br SHAEF, 24 Jan 45, with Min, Mtg, Allied Container Advisory Committee, Ltr, Col. J. B. Franks, DCQM ETO, to QM U.K. Base, 25 Jan 45, sub: Returnable Packages—Rqmts, Availability, and Procurement, and Ltr, Littlejohn to QMG, 2 Apr 45, sub: Jerrican Rqmts, all in USFET 458.11 Cans 1943–45.

[44] Ltr, Lee to Sec Comdrs, 21 Nov 44, sub: Supply of Gasoline on the Continent, EUCOM 463.7 Gasoline and Motor Oil, IIb: Quartermaster Supply in the ETO, IV, 33.

day.[45] Early in the month the Communications Zone specified that 600 tons of this requirement were henceforth to be drawn from Ostend for delivery to First Army by tank car, thus relieving to a small extent the strain on the Cherbourg line of communications. U.S. and British forces were also constructing three 6-inch pipelines to Ghent in order to reduce the road and rail haul from that port.[46] Large shipments, all packaged, still had to be made from Normandy Base Section, totaling 1,600 tons by truck and 800 tons by rail. The remainder, 2,900 tons, was scheduled to be decanted at the pipehead at Coubert.[47]

Rain and mud began to hinder operations at Coubert in October, and, in any case, only one line extended that far eastward. Fortunately the Military Pipeline Service had located an unused autodrome with fifteen miles of paved road at Linas, only a few miles east of the take-off point at Dourdan. Construction of storage tanks and eighty double risers at Linas, and the laying of two lines connecting it with the main pipeline turned this installation into the biggest decanting point on the Continent. The installation at Coubert, used mainly for rail shipments, was enlarged later, after an air attack on the pipehead during the Ardennes counteroffensive. Standby facilities, consisting of two storage tanks, the necessary rail-loading risers, and a connecting pipeline were constructed at Grisy Suisnes, on an alternate rail line a few miles to the north.[48]

First Army was the initial beneficiary of the slightly improved POL deliveries in October. By the middle of the month its reserve actually exceeded the authorized level, rising to 6.55 days of supply, and within another week it reached 10.4 days.[49] Late in the month it became apparent that this rebuilding of First Army's reserve had been carried out at the expense of other formations. By contrast, Third Army's situation had actually deteriorated. Late in the month it reported that receipts of gasoline in the preceding three weeks had fallen short of requests by more than 2,000,000 gallons. Its reserves had receded despite strict rationing, and in the last week of October, as it prepared for resumption of the offensive, amounted to less than one and one-half days of supply.[50]

Third Army had reported its critical situation in POL directly to Supreme Headquarters, which took immediate measures to have the imbalance between the armies righted. By the time of the November offensive the reserves were fairly well equalized: First Army then

[45] The total MT80 requirements for U.S. forces on the Continent exceeded 2,000,000 gallons per day, and of aviation gasoline, 700,000 gallons. Ltr, Barnes to ANPB in Washington, 3 Oct 44, sub: Bulk Distribution on Continent to Forward Areas, USFET Petrol Office 400.42B Capacity for Bulk POL on Continent.

[46] Ltr, Lord to 12 A Gp, 6 Oct 44, sub: Supply of POL on the Continent, 12 A Gp 463.7 Gasoline and Motor Oil, I.

[47] Ltr, Hq COMZ to Sec Comdrs, 14 Oct 44, sub: Forward Shipments of MT80 Gasoline, ADSEC 463.7 Gasoline and Motor Oil.

[48] Petroleum, Oil, and Lubricants, Hist Rpt 13, CE ETO, pp. 93–94; Quartermaster Supply in the ETO, IV, 47.

[49] FUSA and 12 A Gp G–4 Weekly Periodic Rpts for Oct 44.

[50] TWX F–1503, TUSA to G–4 SHAEF, 24 Oct 44, and Memo, G–4 Petroleum for G–4 SHAEF, 23 Oct 44, sub: MT80 Status—12 A Gp, SHAEF G–4 463.7 Gas and Motor Oil, 1944; Rapport in ETO Plant and Communications Diary/Jnl, 25 Oct 44, ETO Adm 145C; TUSA G–4 Periodic Rpts for Oct 44.

SIGN APPEALING FOR THE RETURN OF JERRICANS *posted along a street in Charleroi, Belgium.*

held 7.3 days of supply, Third Army 6.7, and Ninth Army 9.3.[51] Levels in the Advance Section and Communications Zone had also been precariously low throughout October, and rebuilding them was deliberately postponed until those in the combat zone had been re-established. At the start of the November offensive the Advance Section held stocks amounting to only .27 days, and the Communications Zone 6.5 days.[52]

The experience of October had demonstrated once more how, in a period of unpredictable deliveries, a loosely supervised allocations system could create inequities. First Army had naturally taken advantage of its favored position in tonnage allocations to provide its own insurance against the insecurity which had become so characteristic of operations since early September. The army group, aware of the temporary imbalance which had resulted, saw as the only solution to the problem the recovery of the logistic structure to the point where it could be depended on to meet the armies' requests more promptly. Under current conditions, in which reserves were meager or nonexistent, the long reaction time between the submission of requisitions and the receipt of supplies—as much as nine or ten days—constantly threatened interruptions in the flow of supplies which could seriously handicap the armies. The army group agreed that a more rigid control over the accumulation of reserves in the combat zone should be imposed, but only after a substantial buildup of supplies in the Communications Zone and more adequate transportation reduced the time required to fill requisitions.[53]

These conditions were gradually met

[51] 12 A Gp Periodic Rpt for Period 5–11 Nov 44. The respective Army G–4 Periodics reported 6.2, 5.34, and 7.9 days of supply on hand at that time.

[52] 12 A Gp G–4 Periodic Rpt for Period 5–11 Nov 44; TWX S–63946, Crawford to TUSA, 25 Oct 44, and TWX E–58485, Lee to TUSA, 29 Oct 44, SHAEF AG 463.7–1 POL Tankers and Targets, 1944.

[53] Cbl E–58380, COMZ to 12 A Gp, 28 Oct 44, and reply, Ltr, 5 Nov 44, sub: Distribution of Available Supplies of MT80, 12 A Gp 463.7 Gasoline and Motor Oil, I; Notes on POL Conf, Office of the G–4 12 A Gp, 5 Nov 44, dated 10 Nov 44, 12 A Gp 463.7 Gasoline and Motor Oil, I.

in November. The improvement in POL supply generally paralleled the brightening of the entire supply picture occasioned by the opening of Le Havre and Rouen and the improvement in transportation. Le Havre possessed four tanker berths, but work was required on storage and rail clearance facilities before the port's full potential could be realized. Gasoline was at first shuttled up the Seine to Rouen, where tank trucks were used for clearance.

Early in the month the Communications Zone stopped shipping jerricans all the way back to Cherbourg, thus cutting down on their turnaround time and alleviating the can shortage. Except for movements through the pipelines all shipments out of Cherbourg henceforth were made in bulk via rail tank car.

At the same time the Advance Section took steps to provide bulk storage in the forward areas. The largest depots were established at Liège, which served First Army, Luetterode, which served the Ninth, and in the vicinity of Verdun, which served the Third. The Advance Section eventually took over decanting from the armies. The attempts by the armies to fill their own cans had not proved entirely successful. Dispensers were initially lacking and often operated inefficiently, and the arrival of bulk trains was highly unpredictable. As an example, Col. Andrew T. McNamara, the First Army quartermaster, reported that on a single day 109 tank cars had arrived at the First Army railhead without warning. In the absence of adequate dispensing facilities, tank cars, always in short supply, consequently were immobilized. Once more the desirability of getting advance notice of shipments was underscored. One 12th Army Group staff officer expressed the sentiment of all army quartermasters when he noted that "if ever Com Z desired to endear itself with the Armies," one means of doing so would be to give advance information of shipments.[54]

The opening of Antwerp at the end of November finally gave the desired flexibility to the POL distribution system by providing additional intake capacity on short lines of communication. Although the projected pipelines from that port to Maastricht had not yet been constructed, both intake and inland transportation were now quite adequate to meet all Allied needs in the near future. With the opening of Antwerp to POL tankers, distribution of MT80 gasoline was accomplished roughly as follows: most of the gasoline discharged at Antwerp found its way either to First or Ninth Army, or to the Advance Section for its own use, a large portion of it being sent in bulk via rail tank car, the remainder going by way of the Liège storage depot, where it was first packaged. The U.S. allocation at Ostend had the same destination, being piped to Ghent, then forwarded in bulk by rail to Liège, where it was also packaged before delivery to using units. Most of the gasoline entering the Continent at Cherbourg eventually went to Third Army or to the Advance Section, going forward via pipeline to Coubert, and the remaining distance either in bulk by rail or to

[54] Memo, Col K. R. Bendetsen, 12 A Gp, to COMZ Ln Off, 8 Nov 44, sub: Advance TWX Notice of Rail Shipments, 12 A Gp 463.7 Gasoline and Motor Oil, I; TUSA G-4 Periodic Rpts for Nov; ADSEC Operations History, p. 121; Notes on POL Conf, Office of G-4 12 A Gp, 5 Nov 44, 12 A Gp 463.7 Gasoline and Motor Oil, I.

U.S. First Army's POL Reserves, *stored along the roadside near Spa, Belgium, 7 December 1944.*

the Verdun complex, where it was packaged. A portion of the Major System's output was drawn off at Chartres or Linas for local use in the Seine Section or shipped via rail to Brittany, and a small amount was decanted in Normandy for use there. Le Havre's intake was consumed largely in Channel Base Section or in the Oise Section. The total deliveries arriving via tanker and disposed of in this way came to about 7,000 tons. Much smaller tonnages continued to arrive on the Continent in 5-gallon cans. Aviation gasoline arrived largely via Antwerp and Cherbourg, and from the latter was forwarded via pipeline to Chartres.[55]

The improvement in POL intake and distribution facilities brought with it a reassuring situation with regard to continental reserves. Reserve levels in the combat zone remained fairly stable after they had been rebuilt in November. In mid-December, at the start of the enemy counteroffensive, First Army had 7 days of supply on hand, Third had 8.8 days, and Ninth Army 12.8. Meanwhile reserves in the Communications Zone had staged a remarkable recovery, rising to 6 days in the Advance Section and to 12.03 days in the Communications Zone.[56] The building of these levels removed much of the anxiety which the armies had felt only a few weeks earlier.

On 16 December the German counteroffensive suddenly threatened to destroy much of the work of the preceding month in the north. Attacks outside the immediate area of penetration took the form of air attacks on the pipehead at Coubert and on the Third Army decanting point at Mancieulles. But these proved to be only halfhearted attempts to disrupt POL supply and caused little serious damage. In much greater danger were those installations in the area of

[55] Quartermaster Supply in the ETO, IV, 49–50, and Apps. X–A and X–B.

[56] The levels represented tonnages as follows: FUSA, 14,220; TUSA, 14,351; NUSA, 15,474; ADSEC, 46,391; and COMZ, 68,771. As was explained in the case of rations, the ADSEC level represented reserves not only for its own units but for all the armies, and the Communications Zone's level in turn represented reserves for the entire theater, so that the combat forces were actually backed by additional stocks equal to the levels in those two organizations, at this date totaling eighteen days of supply. 12 A Gp G–4 Periodic Rpt for Period 10–16 Dec 44, with attached QM Rpts.

First Army and the Advance Section which lay directly in the path of the offensive. Fortunately the First Army's two main POL dumps in the vicinity of Spa and Stavelot, containing 12,300 tons of gasoline, were successfully evacuated, and about half of all the POL products in the ADSEC depot at Liège, under attack from V–1's, was also loaded and moved back. The Advance Section lost about 900,000 gallons of gasoline as the result of fires started by German planes on two successive nights, and First Army destroyed a small quantity to prevent its capture. But on the whole losses were small.

In view of the close proximity of reserves from which the armies could draw, no attempt was made to maintain authorized levels. First Army, which had reserves of nearly 3,500,000 gallons on the eve of the attack, allowed its on-hand stocks to drop to less than 400,000 gallons at the end of December, both through issues and evacuation. Meanwhile all forward shipments were stopped, and most trains of bulk products were halted and decanted at Charleroi. First Army continued to maintain a reserve of only about one day of supply in January as it continued to reduce the "bulge." But POL again began to flow into Liège during the month, and decanting was resumed there. By early February the distribution of POL had returned to normal.[57]

(3) Coal

The supply of solid fuels was probably more consistently plagued with difficulties than that of any other item. Coal was needed for a variety of purely military purposes, including space heating, cooking, and hot water, for coffee roasting, static bakeries, bath units, and laundries, and above all for hospitals and for the railways. In addition, it was needed to provide minimum essential public utilities for the civilian population.

The supply of coal, like that of POL, was a responsibility of the Quartermaster Corps, which procured, stored, and issued the fuel to using forces on the basis of established priorities and allowances. Like POL, coal was a "common-user" item, but its supply was handled somewhat differently because of the source of procurement. All coal used by the Allied forces had to come from Britain or be procured locally on the Continent. Supreme Headquarters therefore exercised a closer control over the use of coal, screening both military and civilian requests, and allocating fuel to using agencies on the basis of priorities and availability. For this purpose a Solid Fuels Section had been set up within the Petrol and Fuel Branch, G–4, of SHAEF in March 1944. The G–4 of the Communications Zone eventually also organized a separate Coal Section within its Movements Branch to assemble all U.S. coal requirements, to co-ordinate procurement, shipment, unloading, storage, and distribution, and to maintain liaison with Supreme Headquarters on coal matters.[58]

Supreme Headquarters estimated that

[57] ADSEC Operations History, pp. 121–22; COMZ G–4 History, V, 24–25; FUSA Rpt of Opns, 1 Aug 44–28 Feb 45, Bk. IV, pp. 67, 72.

[58] QM Supply Opns, Gen Bd Rpt 109, p. 163; COMZ G–4 History, V, 2; Col. W. R. Gordon, Deputy Chief, Solid Fuels Section, SHAEF, "Coal in War," *The Journal of the Royal United Service Institution*, XCI (November, 1946), 564.

the total Allied requirement for coal in the first three months on the Continent would be about 111,000 tons. Since the landings were to take place far from the French coal fields, this requirement had to be met entirely through imports from the United Kingdom. In fact, the pre-D-Day forecast of operations and the expectation that the Germans would destroy mine shafts as they had in World War I made it unlikely that the Allies would be able to draw on continental resources until much later. OVERLORD plans provided that until D plus 41 all coal (about 14,000 tons) would be shipped to the Continent in eighty-pound sacks. Thereafter shipments would be made in bulk, with Caen (in the British sector), Granville, and the minor Brittany ports handling most of the discharge.[59]

Receipts lagged from the very start. The first sacked coal was not unloaded at UTAH Beach until early in July, and Cherbourg did not begin to receive shipments until later in the month. Fortunately the need for coal was not great during the first months. By the end of August, however, three developments had made the supply of coal an increasingly urgent problem: a tremendous expansion in rail traffic had begun; Paris, captured earlier than planned, required coal for its utilities; and cold weather was approaching. By early September it became clear that Allied needs would not be met by shipments from the United Kingdom. Imports had already fallen far behind schedule, in part because of inadequate discharge capacity and a shortage of rail cars, but also as the result of bad weather. In the first five days of September Channel storms delayed shipping and no coal was available at all for discharge at Cherbourg. Shortly thereafter a shortage of rail cars became the bottleneck, holding discharge to a fraction of the 2,500 tons per day which Cherbourg was planned to handle. Port discharge capacity was generally unsatisfactory. Caen lacked adequate crane facilities; Granville, which was planned to serve almost exclusively as a coal port, was still not ready to receive shipping, and did not open until the end of the month; and the shortage of shallow-draft shipping limited the use of the smaller Brittany ports like St. Brieuc. The result was that only a fraction of the current import target of 6,000 tons was being achieved.[60]

The poor import prospects made it all the more imperative that local production of coal be restored as quickly as possible. Early in September members of the Solid Fuels Section of SHAEF made a reconnaissance of the recently uncovered Nord and Pas-de-Calais coal fields to survey the stock position of coal above ground and to determine the condition and the productive capacity of the mines, their supply requirements, and transportation factors in making coal available. They found approximately 1,000,000 metric tons of coal in ground stockpiles, another 15,000 tons in loaded cars, and about 100,000 tons on barges which were landlocked in various northern French canals as the result of destroyed bridges and other damage. Only about 100,000 tons of these stocks were

[59] Quartermaster Supply in the ETO, IV, Fuels and Lubricants, 70; COMZ G-4 History, V, 3.

[60] Min, Coal Conf Held at the Hotel Crillon, Paris, 16 Sep 44, SHAEF G-4 337 Conf, 1944; Gordon, *op. cit.*, p. 564; COMZ G-4 History, V, 3.

of the quality suitable for locomotives and gas plants, the types which were most urgently needed. Production, which before the war had achieved a maximum of 100,000 tons per day, had recently dropped to 30,000 tons owing to several causes, including a lack of pitwood, labor unrest, and inadequate transportation. But the condition of the mines was generally good, rail connections with Paris had been restored, there was an adequate supply of competent labor, and officials were optimistic about solving the labor difficulties. The big need was timber for pit props, of which there was only a ten-day supply on hand.[61]

At the conclusion of the survey SHAEF took immediate steps to promote the fullest possible exploitation of indigenous resources in the liberated countries and prepared to exercise complete control over the allocation of coal for both military and civilian needs. On 11 September General Eisenhower asked the SHAEF Mission to France to request the French Government to provide as much coal as necessary to meet the needs of the Allied forces in France on the basis of the recently negotiated reciprocal aid agreement, and to grant to the Supreme Commander full authority to establish priorities for the supply and transportation of coal during the period of military operations and for about three months after the cessation of hostilities. Similar requests were made to the Dutch and Belgian Missions within the next few days. In making the request the Supreme Commander noted that both port discharge capacity and shipping were urgently needed to handle other supplies. Moreover, U.K. production was already strained to the utmost, and, in any event, larger quantities of locomotive coal could not be made available from that source. Consequently the Allies were now forced to depend largely on continental resources. Coal and railway wagons had become munitions of war, and in view of their scarcity the provision and movement of coal for civilian purposes had to be kept to the barest minimum.[62]

The procurement and distribution of coal was rapidly assuming the proportions of a big business. To administer the program more adequately SHAEF reorganized and reinforced the staff dealing with the problem, combining G–5's coal section with the Solid Fuels Section of G–4 and thus adding qualified mining engineers to the staff. The headquarters was organized along functional lines to deal with production, requirements, shipping, internal transport, distribution, and statistics. Subsections were eventually created for each liberated country and for Germany. Within six months the section had a strength of over 400 British, American, French, Belgian, and Dutch officers and men. Its principal missions were to effect a centralized control over the procurement and distribution of solid fuels, both to military and civil agencies, to keep imports to a minimum, and to bring about the fullest possible exploitation of the

[61] Rpt of Reconnaissance of the Nord and Pas-de-Calais Coal fields by Members of the Solid Fuels Sec, Current Opns Br G–4 SHAEF, 7–11 Sep 44, SHAEF G–4 337 Conf 1944, III.

[62] Ltr, SAC to French Mission, 11 Sep 44, sub: French Coal Resources in Allied Expeditionary Force Zone, SHAEF G–4 Coal 137/10 GDP–1. The letters to the Belgian and Dutch Missions were dated 13 and 14 September respectively.

indigenous resources of the liberated and occupied areas. Fulfilling these missions required, among other things, that the section collect fuel requirements from all the military formations and national authorities, that it screen essential civil requirements for public utilities and essential industries, that it allocate fuels after evaluating these needs, and that it take whatever measures were necessary, such as requesting mine supplies and machinery, to increase the production of coal.[63]

Neither imports nor local production improved sufficiently to meet the rising requirements in the months which followed, and the entire coal picture therefore remained dark. In October unloadings came to only about one third of the required imports. Early in the month the primary cause was the lack of rail cars to clear the ports. A few weeks later bad weather was given as the main reason for discharge targets not being met, particularly at Granville and St. Brieuc.[64] In southern France, where logistic support depended almost exclusively on the military railways over a line of communications more than 400 miles in length, the shortage of fuel was especially critical. Stocks dwindled to an eight-day supply in November. Locomotive coal had always been a problem there; it had been imported from abroad and from the northern French fields even in peacetime. Damaged rail bridges and the lack of rail cars now made overland movements difficult, and Marseille lacked the proper discharge facilities. As in the north, there were plenty of coaster berths but no coasters. Late in November the British War Office agreed to meet at least a part of southern France's critical need by shipping 25,000 tons from the United Kingdom.[65]

Continental coal production made some gains during the fall, but was plagued by endless difficulties. In southern France production rose to nearly 70 percent of normal in November, and in an effort to achieve a higher measure of self-sufficiency Allied authorities rushed repairs on the Tarascon–Beaucaire bridge across the Rhône so that coal from the southern fields could be used on the Rhône line of communications. The southern mines used prisoners of war with satisfactory results.[66]

In both the south and north the most persistent bottleneck in mining operations was the shortage of pitwood, of which approximately one ton was needed for every thirty tons of coal produced. Early in the fall SHAEF gave the Communications Zone the responsibility for arranging an adequate supply of mine timber, and the Communications Zone in turn dealt directly with French regional authorities. Transportation, as

[63] Gordon, *op. cit.*, p. 565; SHAEF Stf Memo 119, 24 Oct 44, sub: Functions and Responsibilities of Solid Fuels Sec, SHAEF, SHAEF SGS 463.3 Coal Supply, I.

[64] TWX S–61191, SHAEF to COMZ, 6 Oct 44, SHAEF SGS 463.3 Coal Supply, I; Rpt of Allied-French Working Party, 20 Oct 44, SHAEF G–4 Coal 137/10 GDP–1; Min, 17th Mtg, Subcommittee at MOWT on Shipment of Bulk Coal to OVERLORD Area, 7 Nov 44, SHAEF G–4 337 Conferences, III.

[65] Min, Mtg at Versailles on problem of transfer of solid fuels responsibility for southern France from AFHQ to SHAEF, 25 Oct 44, SHAEF G–4 337 Conferences, III; Rpt on Survey of the Southern French Coal Situation, G–4 SHAEF, 2–10 Nov 44, SHAEF AG 463.3 Coal, Case A, I; Cbl S–66487, SHAEF to War Office, 12 Nov 44, TWX BWRT41, 21 A Gp to SHAEF, 21 Nov 44, and Cbl 95956Q (Ops) 2, War Office to SHAEF, 22 Nov 44, all in SHAEF SGS 463.3 Coal Supply, I.

[66] COMZ G–4 History, V, 6.

usual, was one of the major stumbling blocks, and in many cases the Communications Zone had to make trucks, gasoline, or tires available to pitwood contractors.[67]

During the winter coal production and, in turn, transportation were adversely affected by both bad weather and labor unrest. Lack of adequate foods and clothing led to widespread strikes, particularly in the Belgian mines, where production fell off 40 percent in January 1945.[68] Severe winter weather added to the difficulties that month. Coal deliveries to Paris, a large portion of which normally were made by water, fell to 12,000 tons per day against a minimum requirement of 20,000, in part because barge movements were blocked by ice, and in part because of the shortage of locomotives. In mid-January the director general of the French railways reported that 646 trains were delayed for lack of motive power.[69]

As early as the summer of 1944 the prospects of coal shortages had led General Littlejohn to urge the use of wood as fuel wherever it could be substituted for coal. Early in the fall the Procurement Division of the chief quartermaster's office made detailed arrangements with French, Belgian, and Luxembourg authorities for the production and delivery of fuel wood during the fall and winter months. Two logging camps began operating in the Foret de Cerisy in Normandy as early as September, employing prisoner of war labor, and additional camps were eventually established in the Brittany, Loire, and Oise Sections.

The project was not a spectacular success. The program was complicated by a lack of tools, equipment, and transportation, by the necessity to provide housing for prisoners, by the inaccessibility of many of the camps, and, in addition, by unsatisfactory co-operation from civilian authorities. At the end of January 1945 barely 36,000 cords of an original requirement of 1,000,000 had been produced. Considering the labor expended in the production of a cord of wood, its fuel value, and the transportation involved, as compared with the effort required in the production of coal, the endeavor was hardly economic. An attempt to supplement the meager supply of solid fuels by the production of peat was found to be even less worthwhile and was abandoned after a month of cutting in Normandy.[70]

Although coal imports more nearly equaled the targets in February 1945, combined imports and indigenous production never sufficed to meet the minimum essential civilian and military needs during the winter of 1944–45.[71] SHAEF normally allocated only 65 to 70 percent of the amounts requested by U.S. forces, and deliveries rarely exceeded 50 percent of the total needs. Conservation measures adopted in the

[67] *Ibid.*, p. 8; Quartermaster Supply in the ETO, IV, 76; TWX, SHAEF Mission to Belgium to SHAEF G–4, 6 Jan 45, SHAEF SGS 463.3 Coal Supply, I; Rpt of Mtg of Allied-French Working Party, 11 Oct 44, dated 14 Oct 44, SHAEF G–4 337 Conferences 1944.

[68] Cbl R–57441, ADSEC to G–4 COMZ, 31 Jan 45, ETO Adm 406, ETO Cbls; CAO Mtg, 2 Feb 45, SHAEF AG 337–2 CAO Mtgs, 1945.

[69] Hist Rpt of TC ETO, VI (Jan–Mar 45), OCofT, 142A; TWX S–75146, SHAEF to COMZ *et al.*, 15 Jan 45, SHAEF SGS 463.3 Coal Supply, I.

[70] Quartermaster Supply in the ETO, IV, 73, 85–88; QM Supply Opns, Gen Bd Rpt 109, p. 162.

[71] CAO Mtg, 9 Mar 45, SHAEF AG 337–2 CAO Mtgs, 1945.

fall of 1944 consequently had to be strictly enforced. Original allowances for space heating were cut in half, and the use of coal for utilities in cities such as Paris was rationed to permit only a few hours of gas and electricity each day.[72]

[72] The original cold-weather factor of eight pounds per man per day was reduced to four. Quartermaster Supply in the ETO, IV, 75; COMZ G-4 History, V, 7.

CHAPTER VIII

Supplying the Armies: Equipment

(1) Class II and IV Shortages in General

With the exception of a few critical items, the shortages in Class II and IV supplies in the combat zone, like the shortages in other classes, could initially be laid to the deficiencies of inland transportation. Class II and IV items were especially handicapped in this respect, for the staples of supply—gasoline, ammunition, and rations—had first call on available lift and replacement of destroyed or worn-out equipment had to be postponed as long as possible. In September, out of an average daily tonnage allocation to the First Army of 4,076 tons, only 442 tons could be assigned to Class II and IV items, against which only 322 tons were actually delivered. There was some improvement in October, but of an allocation of 5,880 tons, Class II and IV supplies were still assigned less than 1,000 tons, against which deliveries came to 637.[1] Even these figures exaggerate actual performance, for the receipts included many items which the Communications Zone added as "fillers" and which the armies had not requisitioned. Moreover, the true supply picture was distorted by the emphasis on tonnages at the expense of items. A ton of radio spare parts, as signal officers pointed out, could be worth 2,000 tons of pole line hardware.[2]

In September and October the armies' critical lists grew longer every week. Shortages initially resulting from inadequate transportation were in many cases aggravated and perpetuated by unexpected demands arising from the armies' advanced positions, by higher rates of attrition than originally expected, and finally by production shortfalls in the United States. Supply shortages were common to all the technical services, but the most serious ones were in the signal, engineer, quartermaster, and ordnance services.

In the signal service the most persistent shortages which affected operations were in radios, spare parts, batteries, and field wire. Field wire was used at the rate of about 66,000 miles per month when available, but shortages forced strict rationing and permitted allocations of barely half this amount. In November First Army reported 4,700 miles on hand, which it calculated was sufficient for only one and one-half days of large-scale operations, and reported shortages of between 10,000 and 17,000 miles. By contrast, Seventh Army listed normal requirements of only 250 miles

[1] TUSA AARs, Sep, p. 59, and Oct, p. 55–56.

[2] TUSA AAR, II, Sig, 10.

per day, which were being met to the extent of only 30 miles.[3]

In the engineer service the shortage of bridging provided an example of unexpected demand arising from the armies' advanced positions. Third Army successively crossed the Marne, Meuse, and Moselle in addition to smaller streams in September, and in that one month built 52 treadway, 6 heavy ponton, 2 infantry support, 170 timber trestle, and 67 Bailey bridges. Demolitions were more and more extensive as the advance slowed down, and in the area of the XII Corps, southeast of Châlons, all bridges over the main streams had to be reconstructed.

Shortages of tactical bridging made it imperative that Bailey bridges be replaced as rapidly as possible by more permanent structures so that the tactical bridging could be shipped forward and reused. Although relatively small amounts of tactical bridging were needed in October, when operations almost came to a standstill, both First and Third Armies took the opportunity to replace temporary bridges with more permanent timber structures. To meet the requirements for lumber for this purpose Third Army alone placed contracts with twenty-one French mills. Late in October the armies again called for large shipments of bridging and stream-crossing equipment in preparation for a possible break-through to the Rhine in the coming offensive. By that time the forward shipment of supplies had been complicated by an additional factor—mud—which made some of the stocks in the Normandy depots inaccessible.[4]

The shortage of paper also proved a major engineer supply problem, for the demand for maps exceeded all expectations. French paper stocks were far from adequate, and engineers eventually printed about 10,000,000 maps on the reverse side of captured German maps.[5]

In the quartermaster service clothing, tentage, and mess equipment, including stoves, were the most persistent shortages. The supply of quartermaster Class II and IV items, like that in the other services, first presented difficulties in August with the inability of transportation to meet the demands growing out of the unexpected tactical developments. Class II and IV supplies were relegated to positions low on the priority list as long as gasoline, rations, and ammunition remained the more urgent needs. In September, of 54,200 tons offloaded in the ports, only 15,400 were cleared, leaving a backlog of nearly 40,000 tons. By December the backlog had grown to 88,600 tons.[6] Of a total average lift of 4,076 tons allocated to the First Army in September, only 102 tons were earmarked for quartermaster Class II and IV supplies, and the average daily delivery during the month was only 39 tons.[7]

---

[3] TUSA AAR, II, Sig, 11–13; FUSA Rpt of Opns, 1 Aug 44–22 Feb 45, Bk. III, p. 178; SUSA Rpt of Opns, III, 878; COMZ G–4 History, I, 51–52.

[4] TUSA AAR, II, Engr, 6, 10; Ltr, NUSA to 12 A Gp, 4 Nov 44, sub: Supply of Stream-Crossing Equipage, with Inds, 12 A Gp to COMZ, 6 Nov 44, and COMZ to 12 A Gp, 24 Nov 44, EUCOM 475, Equipment of Troops, River Crossings.

[5] COMZ G–4, History, I, 51–52.

[6] QM Supply Opns on the Continent, Sec. IV of Study, Supply of Winter Clothing to ETO, prep for Col Charles Garside by the CQM ETO, 17 Apr 45, with supporting papers as exhibits, in ASF, Dir of Mat, Pur Div.

[7] FUSA AAR, 1–30 Sep 44, p. 59.

Meanwhile shortages originally caused by the inadequacy of inland transportation were made more accute by port discharge deficiencies and to a lesser extent by the inability to move stocks across the Channel from the United Kingdom. At the end of September there were seventy-five ships in the theater commodity-loaded with all types of quartermaster supplies, for which only fourteen berths were available. At the end of October the theater had eighty such ships and only eighteen berths in which they could be worked.[8] Backlogs of quartermaster supplies, both in loaded ships and stored in the port areas, continued high even after Antwerp's opening, and were not completely eliminated until February 1945.

The shortage of coaster shipping for cross-Channel movement meanwhile voided the planned reduction of U.K. stocks. Receipts from the United Kingdom in August totaled 29,000 tons, representing but 53 percent of the 55,000 tons allocated. General Littlejohn appreciated the reasons for deferring the movement of Class II and IV supplies in the first few months. But early in September he became concerned with the possible effects on winterization requirements. On 7 September he informed General Lee that the month's Class II and IV quartermaster lift requirements alone totaled 56,750 tons, 10,000 of which were needed for the movement of winter clothing, 10,350 tons for winter tentage, 29,500 for combat maintenance, and 900 tons of clothing for prisoners of war. The issue of winter clothing, he insisted, had to be completed by 1 October if the fighting efficiency of troops was to be maintained.

Littlejohn proposed to carry out the winterization program by a 6,000-ton airlift, partially from the United Kingdom and partially from Normandy Base Section, and by shipments via LST and small coasters to Brittany ports, to a discharge point in the Seine, and to a rail connection in the Calais area.[9] General Stratton immediately turned down the request for air transport and ruled out most of the other proposals as infeasible at the moment. Cross-Channel tonnage allocations at the time were dictated largely by what the 12th Army Group decided should be moved forward on the Continent, and in view of the current tactical situation and attendant optimism concerning a quick end to the fighting there was little likelihood that the field commands would favor a diversion of transportation, including airlift, from the movement of gasoline to the movement of clothing.[10]

General Littlejohn persisted in emphasizing the urgency of the winterization program, however, and shortly thereafter put the issue of the needed allocations directly up to General Bradley.[11] The change in the tactical situation in the next few weeks, while it brought no immediate improvement in transport, permitted a shift in emphasis. Late in the month the chief quartermaster reached an agreement with the the-

[8] Study of 17 Apr 45.

[9] Ltr, Littlejohn to Lee, 7 Sep 44, sub: Tonnage from U.K. to Continent, 12 A Gp Tonnage 137.

[10] Memo, Stratton for Littlejohn, 8 Sep 44, sub: Tonnage Lift From U.K., Sec. IV of Study of 17 Apr 45.

[11] Ltr, Littlejohn to Stratton, 9 Sep 44, sub: Transportation for QM Supplies, Exhibit G to Sec. IV of Study of 17 Apr 45.

ater G–4 on plans for shipping clothing from the United Kingdom to the Continent and for its movement to the forward areas. The operation finally got under way in the first week of October, and on the 13th General Littlejohn announced that the winter clothing and equipment then available in the theater had been delivered to the armies. Approximately 6,500 tons of clothing overshoes, blankets, and other equipment were moved forward, 41 percent of it by air.[12]

(2) *The Case of the Winter Uniform*

Transportation was only one aspect of the winterization problem. Behind this problem lay more basic shortcomings, particularly with respect to winter clothing. Viewed in retrospect, it is clear that planning and decision-making in both the War Department and the theater, as well as co-ordination between the two, left something to be desired.

A controversy eventually developed over winter clothing that involved questions of both quality and quantity. Inadequacies of the winter uniform in Europe on both counts were several times brought to the attention of the public via the newspapers. Critical articles appearing in January and February 1945—particularly one in the *Washington Post*—finally evoked an angry blast from General Littlejohn, who charged that a malicious campaign had been launched to discredit him. The articles precipitated one of the most acrimonious intraservice squabbles of the entire war, and finally led to an investigation.[13]

The *Washington Post* article had given due recognition to such factors as the abnormal severity of the 1944–45 winter in Europe, the unexpectedly high attrition of clothing during the summer and fall, and the unfortunate habits of American soldiers regarding the proper fitting of clothing. But the article struck a sensitive spot in implying that the theater had placed its orders too late to ensure adequate early winter protection against wet cold weather and that it had failed to adopt combat-tested items recommended by the War Department. Most of the controversy over winter clothing centered on these two closely related points.[14]

U.S. forces had already experienced the distress of operating without adequate clothing. In the winter of 1943–44 the Fifth Army in Italy had found the

[12] Outfitting the Soldier, Vol. III of Quartermaster Supply in the ETO in World War II, 280–84, and App. XXXIII, MS, OCMH; Sec. IV of Study of 17 Apr 45; SHAEF G–4 Basic Statistical Rpt 2, 14 Oct 44, in SHAEF G–4 War Diary/Jnl, Exec Br.

[13] The *Washington Post* article, written by George Connery and titled "U.S. Western Front Clothing: A Factual Report," appeared on 18 February 1945. Dispatches from other correspondents in Europe had appeared in various papers beginning in October, implying criticism of the War Department, and had caused General Somervell to ask the theater for explanations. Cbl 4102, Somervell to Lee, 14 Oct 44, CofS ASF, ETO 1944; Cbl WAR–24746, AGWAR to Somervell, 23 Jan 45, ETO Cbls, ETO Adm 405; Cbl WAR–38662, AGWAR to Lee, 17 Feb 45, ETO Adm 408; Ltr, Littlejohn to Somervell, 2 Mar 45, Hq ASF, Somervell file, ETO 1945.

[14] A follow-up article, written by Edward T. Folliard and titled "Trench Foot Scourge Ends, But Many Yanks Still Are Hospitalized," appeared in the *Washington Post* on 4 March 1945 and dealt primarily with the causes of trench foot and the problem of adequate footgear. It included replies by General Littlejohn to the claim that the theater had not requisitioned overshoes and shoepacs early enough, but did not retract statements made in the Connery article.

then standard field uniform inadequate to protect its troops fighting in the mountains around Naples.[15] Meanwhile, The Quartermaster General had developed a simplified uniform, based on the layering principle, and adaptable to combat wear in cold wet climates, which the Army Ground Forces had approved for standardization, "subject to minor modifications," as early as March 1943.[16] Late in February 1944, The Quartermaster General sent Capt. William F. Pounder, an officer in the Research and Development Branch, Military Planning Division, OQMG, to England to familiarize ETOUSA officials with the new items and explain their advantages. The items recommended had undergone tests in either the continental United States or Alaska, and they were now sent to the Mediterranean theater as well where they were tested on troops of the 3d Infantry Division in the Anzio beachhead. While the Italian test was not exhaustive, particularly from the point of view of performance in severe weather, the Mediterranean theater found the new uniform far superior to the combination then in use and eventually equipped the three divisions that it provided for the Seventh Army, which operated in France the following winter, as well as the Fifth Army, with the new clothing.[17]

The main items recommended which distinguished the proposed uniform from the one then in use were the M1943 sateen field jacket, the high-neck wool sweater, the combat service boot, the shoepac, and the leather glove with wool insert. General Littlejohn had already had a preview of the new items on a visit to the United States in November 1943, and they evoked considerable interest when they were shown to the chief quartermaster and his staff in the United Kingdom. The M1943 field jacket, a wind- and water-repellent garment with a pile liner which could be worn over a jacket or sweater in cold weather and which became the item of greatest controversy, was initially well received, since it was to replace the unsatisfactory 1941 Parsons jacket. But General Littlejohn was not satisfied with the production figures which Captain Pounder was able to furnish and, lacking assurance that the new jacket would be delivered in sufficient quantities to dress units uniformly, stated that he would make no special effort to procure it.[18] Pounder continued to press for decisions on various articles he had brought with him, and advised early requisition of accepted items. But

[15] Ltr, Brig Gen Joseph P. Sullivan, Army QM Fifth Army, to Littlejohn, 28 Jan 43, Littlejohn Reading File.

[16] Memo, McNair for CofS, 10 Mar 43, copy of this and other pertinent documents in Supply of Clothing and Equipment to ETO, 1944, Pt. 4—Documentation (Corresp and Special Rpts), dated 5 Apr 45; ASF OQMG File A45-280, Drawer 7. For a discussion of the new winter uniform proposed by TQMG, see Erna Risch, *The Quartermaster Corps: Organization, Supply, and Services, Volume I,* UNITED STATES ARMY IN WORLD WAR II (Washington, 1953), pp. 88–97.

[17] Rad CM-IN, 1 Jun 44, Protective Clothing, CG U.S. Army Forces in NATO to WD #F-53022, signed Devers, 31 May 44. Copies of this and other papers on supply of winter clothing to NATOUSA are in ASF OQMG Folder. Supply of Clothing and Equipment to ETO, 5 Apr 45, 205.03 A 45-280, Drawer G 1616.

[18] Ltr, Pounder to Col Georges F. Doriot, Chief Research and Dev Br, Mil Plng Div OCQM, 29 Mar 44, Study, Supply of Clothing and Equipment to the ETO, 5 Apr 45, QMG, ASF Dir of Mat, Pur Div.

GENERAL DWIGHT D. EISENHOWER *wearing the Eisenhower jacket, Paris, August 1944.*

he was unsuccessful, and he eventually returned to the United States.[19]

While the theater quartermaster initially rejected the new M1943 jacket on the ground of uncertainty as to its availability, other considerations appear to have influenced his decision. A new waist-length wool jacket, designed to replace the wool serge coat and to double for combat and dress, had been under development in both the United States and the theater for some time.[20] Theater commanders, including General Eisenhower, desired a wool jacket resembling the one that was part of the widely admired English "battle dress," and General Bradley, in March, expressed his opinion that such a jacket, made of rough wool, would be warm enough to protect a soldier in combat without an outer jacket or overcoat. In any case, it could be worn in cold wet weather under the loose-fitting M1943 jacket if this became available, and the theater now urged the War Department to adopt the type of wool jacket it desired.[21] In mid-March 1944 it asked for 4,259,000 of these jackets to be delivered by the end of 1944. Early the next month, on a visit to the United States, General Littlejohn obtained acceptance of the basic design of the new short wool jacket, and after his return to London the War Department notified the theater that it had settled on the design of the jacket and scheduled shipments of 2,600,000 in the last quarter of 1944.[22]

The War Department had never intended, however, that the "Eisenhower jacket," as the ETO model was later called, should replace the new M1943 jacket. In May it informed the theater that the latter, worn in combination with the high-neck wool sweater, had

[19] Rpt, Dev of Cold Climate Clothing, prep by Pounder for Doriot, 11 Oct 44, in Study of 5 Apr 45.

[20] See Erna Risch and Thomas M. Pitkin, *Clothing the Soldier of World War II,* QMC Historical Studies, 16 (September, 1946), pp. 54–58 for the history of the development of both the wool field jacket and the M1943 jacket.

[21] Ltr, Eisenhower to Marshall, forwarded to TQMG by Somervell; Extract from Incl 4, Temperate Zone Clothing: Jacket Field Wool, 2d Ind Brig Gen Herman Feldman (SPQR 400.34 T/E 21), ASF OQMG, 13 Mar 44, to Dir Mob Div ASF, photostats in Summary of Action Taken to Supply Winter Clothing to ETO; Gen Gregory's Personal File, Hist Br OQMG; Memo given Capt Pounder by Lt Col Cohen, ETO, 8 Mar 44; and Ltr, Pounder to Doriot, 29 Mar 44, photostats of both in Supply of Clothing and Equipment to ETO 1944.

[22] Risch and Pitkin, *Clothing the Soldier of World War II.*

been approved and was intended to replace the old Parsons jacket. Because of decreasing stocks and size shortages of the old 1941 jacket, shipments of the M1943 were already being set up against requisitions from the theater. But the theater now made it clear that it did not desire the M1943 jacket, except in limited numbers for parachutists, and it called attention to an agreement which General Littlejohn had reached with General Clay of the Requirements Division, ASF, during his visit to the United States in April under which the old-type field jackets were to be supplied pending the initial deliveries of the new Eisenhower jacket.[23]

Obviously concerned over the possible consequences of the theater's decisions, the War Department asked that the matter be checked with Supreme Headquarters. The War Department met General Littlejohn's original objection by assuring the theater that it could have the 1943 jacket, presumably in the needed quantities, if it so desired.[24] A few days later Maj. Gen. Edmund B. Gregory, The Quartermaster General, forwarded a study prepared by the Research and Development Branch summarizing all the data then available from field and laboratory tests on the merits of the various items. General Gregory left no doubt of his misgivings over the adequacy of the theater's proposed winter uniform, which consisted of the wool field jacket, the overcoat, and the raincoat: the raincoat, he said, could not be considered a combat garment; the overcoat provided entirely inadequate protection against rain and wind, as had been amply demonstrated in Italy; and the combined bulk and weight of the raincoat and overcoat seriously impaired the mobility of the soldier. ETOUSA's rejection of both the 1943 jacket and the high-neck sweater, in Gregory's words, "would leave troops [in the European theater] without any garment designed for efficiency in wet cold climates." [25]

The theater nevertheless on 1 June confirmed its earlier decision on the 1943 jacket, although it now decided to accept the high-neck sweater. Both Generals Bradley and Hodges, according to the theater quartermaster, had concluded that the overcoat was a necessary part of the winter uniform, and Generals Smith and Crawford at SHAEF had concurred in these recommendations. Two weeks later the theater placed a requisition for 2,250,000 of the sweaters, of which the New York Port agreed to deliver 1,750,000 by September.[26]

Officers in the Research and Development Branch of the OQMG were clearly disturbed over the theater's decision. Col. Georges F. Doriot, chief of the branch, pointing out that the winter climate of northeastern France and Belgium was similar to that of Italy, predicted a repetition of the experience of

[23] Study, Wind Resistant Jacket, Sec. VI of papers supporting Littlejohn letter of 2 Mar 45, ASF Div of Mat, Prod and Pur Div; Cbl E-28364, ETO to WD, 18 May 44, Study of 5 Apr 45, QMG ASF.

[24] Cbl WAR-39574, Clay to Lee, 20 May 44, Study of 5 Apr 45, QMG ASF.

[25] Ltr, Gregory to ACofS OPD WD, 25 May 44, sub: Clothing Efficiency, ETO, Study of 5 Apr 45, QMG ASF.

[26] Cbl E-30871, ETO to WD, 1 Jun 44, Pt. III (Documentation), Study of 5 Apr 45, QMG ASF; Study, Wind Resistant Jacket, Sec. VI of papers supporting the Littlejohn ltr of 2 Mar 45, ASF Dir of Mat, Prod and Pur Div; Cbl, EX 33226, CG ETOUSA to CG NYPE, 15 Jun 44, no sub; ASF Pur Div Files, Supply of Clothing and Equipment to ETO, DRB AGO.

American troops in Italy during the preceding winter, if ETOUSA persisted in adopting a uniform which, in his opinion, had already been proved inadequate. Captain Pounder likewise had emphasized to the theater quartermaster the point that the area in which U.S. troops could be expected to be operating fell into the wet cold classification and warned that U.S. troops would be improperly clothed unless such items as shoepacs, ski socks, and the woolen sleeping bag, in addition to the 1943 jacket, were adopted. In view of ETOUSA's recent communications, however, the OQMG had no choice but to eliminate requirements for the 1943 jacket for the European theater except for the limited needs for parachutists, and to divert production to other items.[27]

On 20 June the theater startled the OQMG with an urgent request for all available information on winter clothing for operations in cold wet climates, information which the War Department had sought with indifferent success to have the theater consider earlier. Colonel Doriot promptly forwarded the desired data and took the opportunity to urge the theater chief quartermaster again to accept the M1943 jacket and to issue it in addition to the high-neck sweater and wool field jacket. Without it, he warned, the front-line soldier would not be adequately protected against cold wet conditions. Both the overcoat and the raincoat, he pointed out, had failed to meet this requirement where mobility was desired. Time was already running desperately short for getting production under way which would meet ETOUSA's needs for the coming winter.[28] But nothing came of this exchange, at least for the moment, and a note of resentment at the War Department's repeated urgings was indicated in General Littlejohn's remark early in July that it was not his policy "to force these new items down the throat of troops." [29]

General Littlejohn held high hopes at first of getting the required quantities of the much desired Eisenhower jacket. But deliveries of both finished garments and cloth from the United States lagged from the start, and prospects of meeting the original commitment faded rapidly. Early in July the chief quartermaster concluded that it would be at least six months before a sizable number of troops could be supplied with the new jacket.[30] Nevertheless, he refrained from making any requests which might divert production from that program. Pending receipt of the new jacket, therefore, he preferred to take substitutes, such as the obsolescent Parsons jacket and even the wool serge blouse, rather than accept the newer 1943 jacket, receipts of which he claimed were already complicating his supply situation. Early in July he asked the War Department to ship the entire

[27] Ltr, Doriot to Rqmt Div ASF, 3 Jun 44, sub: Secret Radiogram CM157 dated 1 Jun 44, Protective Clothing; Memo for Record Only, Doriot, n.d.; and Ltr, Pounder to Doriot, 30 Jun 44, all in Study of 5 Apr 45, QMG ASF.

[28] Ltr, Doriot to Littlejohn, 24 Jun 44, sub: Confidential Radio E–39902, Study of 17 Apr 45; Dev of Cold Climate Clothing, 11 Oct 44, by Pounder, Study of 5 Apr 45; Study, The Supply of Quartermaster Clothing, Subsistence, and Equipment to the ETO, 5 Jan 45, prep in OQMG for Somervell, Hq ASF QMC Reference Book.

[29] Study, The Supply of Quartermaster Clothing . . . , 5 Jan 45.

[30] Memo, Littlejohn for G–1, 12 Jul 44, Littlejohn Reading File.

remaining stock of 479,000 of the old 1941 jackets to the European theater.[31]

The chief quartermaster had also placed great emphasis, as had General Eisenhower, on the desirability of having a dressy uniform, and was hopeful throughout the summer that this requirement would be met by the new wool jacket. His concern over appearance in fact led him to protest repeatedly against the shipment of trousers of a lighter shade which did not match the jacket.[32]

Littlejohn's determination on this point was encouraged by the course of tactical operations after the breakout at the end of July. The mounting optimism of the next few weeks was soon reflected in theater supply policy. The OCQM expressed its confidence as early as 15 August that the war would not go into another winter. On that date it submitted a requisition to the War Department for winter clothing specially designed for severe cold for one field army —353,000 men—but purely as a precautionary measure and not in anticipation of any need arising from tactical developments.[33]

The decision not to requisition special winter clothing earlier had been deliberate and understandable. The theater chief quartermaster had decided on the basis of an analysis of the climatic map of Europe that no special cold climate clothing would be needed. A comparison of the climatic map with the expected rate of the Allied advance showed that U.S. forces would not enter the "cold wet" area, beginning roughly with the Ardennes, until D plus 330, or May 1945. But the phase lines on which those plans had been based represented the course of operations as expected before D Day and hardly constituted a valid basis for planning in mid-August.[34]

The mid-August requisition, according to the OQMG in Washington, was already one month late in arriving, judged by the theater's own policy recommendation on the requisitioning of winter clothing.[35] More important than the tardiness of the order, however, was its size, which appeared far too small to the OQMG. But when the War Department queried the theater and pointed out that the size of the requisition would result in production cutbacks, the theater on 5 September confirmed the requisition.[36] General Littlejohn expressed his own optimism at this time in a personal letter to General Gregory, in which he wrote: "You and I know that

[31] Study, Wind Resistant Jacket, ASF Dir of Mat, Prod and Pur Div.

[32] Eisenhower ltr cited n. 21; Ltrs, Littlejohn to Gregory, 4 Sep 44, and Littlejohn to Feldman, 14 Sep 44, Littlejohn Reading File, Army War College Library.

[33] Cbl EX-43895, COMZ to AGWAR, 15 Aug 44, copy in Outfitting the Soldier, App. XXX; Secs. I (Planning) and III (Requisitions), Study of 17 Apr 45. The issue of "special winter clothing," designated as "arctic (cold-wet)," was based on an allowance of such clothing which a theater commander was authorized to draw in anticipation of exceptional climatic conditions.

[34] Secs. I (Planning) and III (Requisitions), Study of 17 Apr 45.

[35] Study, Cold Climate Clothing, Sec. IV of papers supporting Littlejohn ltr of 2 Mar 45, ASF Dir of Mat, Prod and Pur Div.

[36] Ltr, Feldman, DQMG, to CG SOS, 21 Aug 44, sub: Requisitions and Current Army Supply Program Anticipated Shipments, Study of 5 Apr 45; Study, Requisitions, Sec. I of papers supporting Littlejohn ltr of 2 Mar 45, ASF Dir of Mat, Prod and Pur Div.

SERVING A HOT MEAL TO COLD INFANTRYMEN, *Belgium, January 1945. Note cloth overshoes worn by the men.*

the serious fighting cannot long continue." [37]

The confidence which these messages reflected was not confined to the office of the ETOUSA quartermaster. Headquarters, Communications Zone, had asked all the supply services to review their requirements and prepare stop orders in anticipation of the expected end of hostilities. The chief quartermaster even took measures to control the issue of winter clothing to ensure that occupation troops would be the first to get it.

The halt of the pursuit in mid-September and the prospect of winter operations gradually dissipated the rampant optimism which had begun to influence supply policy. Moreover, the slowing down in operations brought to light a new factor to complicate the supply of clothing—the discovery that maintenance and replacement factors had been far from adequate. During the pursuit, General Littlejohn pointed out, it had been impossible to obtain accurate data on either stocks on hand or consumption. Now it was found that wear and tear had been much heavier than expected in units constantly on the move, and that men had lost or discarded large quantities of individual equipment despite attempts to enforce supply disci-

[37] Ltr, Littlejohn to Gregory, 4 Sep 44, Littlejohn Reading File.

pline. General Littlejohn estimated that the consumption of major items of clothing and equipment had been at a rate two and one half times that prescribed by War Department maintenance factors.[38]

With this additional argument the ETOUSA quartermaster on 18 September placed the first of several requests for large quantities of winter clothing and equipment, asserting that he was now confronted with the necessity of completely re-equipping a minimum of one million men, about 100,000 French territorials, and a large number of prisoners of war. He asked that the supplies be made available for distribution on the Continent not later than 10 October.[39] In the next two weeks the theater made an appeal for additional quantities of winter equipment, including blankets and sleeping bags. The shortage of blankets was especially critical, having been aggravated by the large number of prisoners (300,000 at the time). On 10 October the theater quartermaster indicated that an additional 500,000 men would have to be re-equipped within the next sixty days.[40]

Much as it deplored the theater's resort to emergency requisitioning, there was little the War Department could now do but attempt to meet what apparently were legitimate needs. General Gregory assured the theater quartermaster that his office would do everything possible to provide men in the European theater with a serviceable uniform from stocks available in the United States, although he could not promise to provide matched items for "promenades on the streets of Paris." He also took the occasion to remind the ETOUSA quartermaster of the efforts which the OQMG had made to supply the theater with adequate equipment. "As you know," he wrote, "my office has on several occasions made definite recommendations to you as to the proper uniform required for the climate in which you are now operating. In addition to these recommendations being made from a climatology point of view, they were also made from a production point of view, considering the over-all size and deployment of the United States Army." [41]

The reference to production suggested that the emergency requisitions for large amounts of winter clothing would not be easily met on such short notice. The Requirements Branch of OQMG pointed out that between 18 September and 1 October the theater had requested 850,000 overcoats after indicating in August that it needed only 45,000 in addition to those previously shipped. At the end of July the theater stated a need for 450,000 overshoes for the remainder of the year, which, with those already in stock would have permitted issue to 75 percent of the troops. But in the last week of September the theater had suddenly placed a demand for an additional 1,173,000 pairs. Calls

[38] Ltr, Littlejohn to Feldman, 4 Sep 44, Ltr, Littlejohn to Gregory, 18 Sep 44, sub: Rqmts of Winter Clothing for ETO, Ltr, Littlejohn to Col Ira K. Evans, 18 Sep 44, and Ltr, Maj Gen W. M. Goodman to Col H. A. Malin, 8 Oct 44, all in Study of 5 Apr 45.

[39] Memo, Littlejohn for Gregory, 18 Sep 44.

[40] Teletype Conf between London and NYPOE, 1 Oct 44, Study of 5 Apr 45; Cbl EX-53583, CQM to AGWAR, 10 Oct 44, EUCOM 400 Supplies, Services, and Equipment, IV.

[41] Ltr, Gregory to Littlejohn, 28 Sep 44, Study of 5 Apr 45.

were also made for 2,900,000 wool drawers and 2,500,000 wool undershirts over and above previous requests for 1944.[42]

The War Department was able to fill the 18 September requisition with little trouble, although it drained U.S. stocks in many items. Filling the subsequent requests required certain substitutions, including some used overshoes,[43] and led The Quartermaster General to ask the theater quartermaster to review all woolen clothing requirements and to make as many reductions as possible. He warned that any additional requests would have to be accompanied by detailed justification.[44]

In mid-October the theater submitted the first of several recommended revisions of replacement factors and asked for approval of requisitions based on the new tables. Most of the new factors exceeded those currently authorized by 100 to 150 percent and were greeted with a critical eye in the OQMG. Requests for such changes almost invariably precipitated a long argument between the theater and the War Department. The War Department's hesitancy about approving increases was inspired in part by the belief that a basic misunderstanding existed on the part of theater officials as to the purpose of replacement factors. The War Department had repeatedly contended that excessive consumption for short periods of time did not warrant radical revisions. Replacement factors, it emphasized, were not established to meet the needs arising from temporary fluctuations in consumption, but rather the average losses or maintenance needs over a long period of time. Replacement factors were used for procurement or production planning, especially for items having a long lead time, and there was an understandable reluctance to change them unless a definite long-term trend was indicated. Excessive losses resulting from unusual and nonrecurring situations, the War Department insisted, should be met through special requisitions. It was particularly hesitant to permit upward revisions for winter clothing items in which it suspected that initial issue requirements had entered into the theater's demands. Troops were only beginning to wear the overcoat, for example, for which a 100 percent increase in the replacement factor was requested. The War Department consequently refused to give blanket approval to the theater's requests for increases. Subsequent recommendations submitted by the theater in November and December were also partially rejected.[45]

Deliveries made against the September and October requisitions were largely completed by mid-December. Meanwhile, receipts of the wool jacket, which

[42] Memo, Chief Rqmts Br Mil Plng Div ASF for Dir of Mil Plng Div, 4 Oct 44, sub: Winter Clothing and Equipment for the ETO, Study of 5 Apr 45; Special Rpt prep by Tech Info Br, 11 Nov 44, Study of 5 Apr 45.

[43] Memo, Gregory for CofS, 3 Nov 44, sub: Requisitions for the ETO, OPD 400 ETO Sec. IV, Cases 109–22.

[44] Outfitting the Soldier, p. 285.

[45] Memo, Gilmer for Handy, 16 Oct 44, sub: Emergency Winter Clothing Rqmt in ETO, OPD 400 ETO Sec. III, Cases 109–22; Outfitting the Soldier, pp. 50–53, 286–87; Study, Maintenance Factors, Sec. III of papers supporting Littlejohn ltr of 2 Mar 45, with atchd documents, ASF Dir of Mat, Prod and Pur Div; Special Rpt prep for Gregory by Tech Info Br OQMG, 11 Nov 44, Study of 5 Apr 45.

the theater so ardently desired, had not risen above a trickle.[46] Production of the Eisenhower jacket had encountered one difficulty after another, and it became more and more evident in the fall that the War Department would not meet its commitments to deliver 2,600,000 by the end of the year. In mid-September only 14,000 had been shipped against a scheduled delivery of 500,000 in that month.[47] Despite this disappointing performance, General Littlejohn continued to omit the M1943 jacket from the list of acceptable substitutes. On 2 October he asked for an additional 1,500,000 of the old Parsons jackets, but had to accept some substitutes, including wool serge overcoats. Troops expressed a strong dislike for the overcoat, however, frequently discarding it in fast-moving situations, and at the end of the month the chief quartermaster acknowledged that it was unsatisfactory as a combat garment and canceled his earlier acceptance of it as a substitute for the 1941 jacket. Faced with shortages in the wool jacket and the rejection of the overcoat, the chief quartermaster now asked the New York Port for 800,000 M1943 jackets to meet deficiencies in all types of jackets to the end of 1944. The War Department immediately assured him that practically the entire requisition could be filled.[48]

PARSONS JACKET 1941 *is worn by Field Artillery men, Belgium, January 1945.*

Despite efforts to expedite the delivery of clothing called for in the September and October requisitions, front-line troops fought through a large part of the winter inadequately clothed. Third Army reported in November that 60 percent of its troops lacked sweaters, 50 percent lacked a fourth blanket, and 20 percent lacked overshoes in the proper size. Smaller percentages needed jackets and raincoats.[49] The problem became

[46] Outfitting the Soldier, p. 289.

[47] The jackets eventually delivered to the theater were held in the United Kingdom, and no wholesale issue was made until after V-E Day. Whether the jacket would have served its intended purpose in the European theater was never determined. Experience in the Mediterranean theater later revealed that troops never regarded the jacket as a component part of the field uniform. They preferred to regard it as a dress item, to be worn on furlough or in rest areas, and for that reason also tended to fit it too snugly to be worn over the sweater. Risch and Pitkin, *Clothing the Soldier of World War II*, p. 58.

[48] Ltr, Littlejohn to QMG ASF, 31 Oct 44, sub: Quartermaster Supply Situation in ETO, Study of 5 Apr 45; Study, Wind Resistant Jacket, ASF Dir of Mat, Prod and Pur Div; Cbl S-76466, SHAEF to AGWAR, 24 Jan 45, SHAEF G-4 War Diary/Jnl.

[49] Maj Paul A. Siple, Report on the Adequacy of Winter Clothing in the ETO, 1944-45, 12 May 45, DCofS file, 420 ETO; TUSA AAR II, QM, 10.

most acute in December when the weather turned bitterly cold and damp. Frantic efforts were made to supply clothing which would provide the necessary protection. Uniformity and standardization consequently went out the window, for troops wore what was available, including arctic and limited standard items. Lack of a suitable outer garment led them to don additional woolen undershirts and socks. Improvement finally came in January with the arrival and distribution of clothing from the United States.[50]

The story of the field jacket was closely paralleled in the case of the shoepac, one of the major items which differentiated the uniform recommended by the War Department from that initially adopted by the theater. The shoepac is essentially a combination rubber and leather boot which gives far better protection against water than either the leather boot or cloth overshoe. It was designed to fit over two pairs of socks, one of them a heavy ski sock, and had removable insoles. Later models of the shoepac gave the needed arch support which the combat boot had provided.

In some ways the footwear problem was more complex, and there was more room for legitimate differences of opinion as to what constituted adequate protection under winter conditions. In all its contacts with the theater on the subject of winter clothing, and during the visits of General Littlejohn to Washington and of Captain Pounder to London, the OQMG had consistently included either overshoes or shoepacs in its recommendations. On the basis of tests it had recommended the shoepac as the most suitable item for combat troops under the conditions expected on the Continent.[51]

Early in July 1944 General Littlejohn indicated his awareness of the problem when he wrote to the OQMG that he unquestionably would be called on to furnish overshoes or the equivalent thereof to all men in the theater for the coming winter and indicated that this would necessitate a substantial requisition at an early date.[52] The requisition which he submitted two weeks later, however, called for sufficient overshoes to equip only 75 percent of U.S. troops on the assumption that the combat boot, which was then beginning to replace the old service shoe with leggings, would suffice for a portion of the continental strength. The first request for shoepacs was made on 15 August as part of the requisition for special winter clothing and equipment for one field army.

With the onset of cold wet weather in September it was realized that the combat boot, although an excellent dry weather item, did not offer suitable protection against water and mud, and that 100 percent of the troops on the Continent would need overshoes. The combat boot, like the flesh-out service shoe, was not leakproof, and troops used both the authorized dubbing and the forbidden shoe polish in an attempt to water-

[50] 12th A Gp Rpt of Opns, XII, 201–04; Outfitting the Soldier, pp. 292–93.

[51] Special Rpt of 11 Nov 44; Study, Overshoes Arctic and Combat Boots, Sec. VII of papers supporting Littlejohn ltr of 2 Mar 45, ASF Dir of Mat, Prod and Pur Div.

[52] Study, Overshoes Arctic and Combat Boots.

proof them. Late in September the theater made the first of its supplementary requisitions, calling for 293,000 overshoes. Within two weeks it submitted an additional request for 1,300,000. Early in December it submitted its needs for the first three months of 1945, calling for 500,000 overshoes and an equal number of shoepacs.[53]

The shortage which the theater faced pending the receipt of these supplementary shipments was aggravated from another source. Shoes and boots which had been fitted during the summer, when men were wearing light woolen or cotton socks, became too tight when worn with two or more pairs of heavy woolen socks. The inevitable result was a demand for larger sizes. This requirement led to a demand for larger overshoes as well. Size tariffs did not allow for the needed high proportion of E, double-E, and triple-E widths. The OQMG's adoption of a special winter tariff which allowed for greater widths in all types of footgear did not meet the theater's immediate needs. Overshoes in the larger sizes were lacking well into January. To make matters worse, the cloth-type overshoe tore easily and leaked badly, and the first shoepacs were of an early model which lacked a raised heel and an arch support. Meanwhile many troops adopted the expedient of wearing overshoes without shoes or boots, using several pairs of socks and improvised cardboard insoles.

The lack of adequate footwear became inseparably associated with the precipitate rise in the incidence of trench foot which occurred in the second week of November. Trench foot eventually caused more than 46,000 men to be hospitalized and accounted for 9.25 percent of all the casualties suffered on the Continent. Trench foot is an injury, not an infection. Its cause is long exposure to cold and wet conditions which result in crippling injury to the blood vessels and muscle tissues of the feet. Trench foot is characterized by discoloration and painful swelling, and requires evacuation and prolonged hospital treatment. A large percentage of those affected were unable to return to combat duty; some could no longer perform any military service. The highest rates normally occurred among units (usually infantry divisions) living under wet and cold conditions in relatively static situations. Cold wet conditions, however, were only the most constant factor in the cause of the injury. Failure to rotate troops, improper foot care, and inadequate footgear and clothing, all contributed to the high incidence.

The European theater had been warned about trench foot. The experience of the previous winter in Italy had led the War Department to advise the theater in the summer of 1944 on methods of prevention and control. Theater headquarters in turn drew up directives which were duly passed down through the various echelons. But the seriousness of trench foot as a casualty producer was not widely appreciated outside of a few units which had already had experience with it (as in the 6th Army Group, for example), and the instructions, particularly regarding individual care of the feet, were poorly enforced. Numerous cases of trench foot were reported during the fall. But the problem suddenly be-

[53] *Ibid.*

came serious with the launching of Third Army's offensive in the second week of November, when 1,500 cases were hospitalized. In calling attention to this precipitate rise in trench foot casualties, Col. Alvin L. Gorby, the 12th Army Group surgeon, noted that the condition was largely preventable and called for a campaign to combat it.

The 12th Army Group shortly thereafter issued a circular directing its subordinate commands to enforce preventive measures and threatening disciplinary action for noncompliance.[54] But a vigorous theaterwide control program which emphasized command responsibility in enforcing foot care was not launched until the end of January 1945. Training directives, pamphlets, and various media of public communications, such as *The Stars and Stripes* and *Army Talks,* were then employed to give the widest possible publicity to the nature and seriousness of trench foot and to the measures by which it could be combated. More important, the major commands now formed trench foot control teams, usually consisting of a line officer and a quartermaster or S–4 officer, to work with unit surgeons and to assist in training and in the supervision of control measures. In addition, noncommissioned officers of demonstrated ability and experience were designated at the company level to supervise and check on foot discipline and to ensure that certain routine preventive measures were taken by individual soldiers, such as the proper wearing of clothing, keeping feet dry and avoiding constriction, and massaging the feet to improve circulation.[55] These measures, aided by more moderate weather, brought a distinct improvement in the next two months. By that time, however, the loss of personnel from trench foot and frostbite already approximated the strength of three divisions in the 12th Army Group.

The importance of effective indoctrination, discipline, and individual hygiene in the control of trench foot had been amply demonstrated. Incidence had varied greatly in units with the same type of footgear and living under substantially the same conditions. The effectiveness of control measures, moreover, was found to be directly related to the state of discipline of a unit. Poor discipline was reflected in a high venereal disease rate, a high court-martial rate, a high AWOL rate, and a high trench foot rate.[56]

Nevertheless, lack of adequate winter clothing and footgear was recognized as an important contributory cause of the casualties resulting from cold.[57] Because healthy feet depend in part on a warm body and hands, the War Department had from the beginning emphasized that its proposed winter uniform, whose individual items complemented each other, be considered as a whole. Lack of proper body clothing therefore contributed to foot troubles.

As to footgear itself, the OQMG apparently did not regard any single combination as completely satisfactory

[54] 12 A Gp Rpt of Opns, XIII, 49–51.

[55] Trench Foot, Gen Bd Rpt 94, pp. 6–7.

[56] *Ibid.*, p. 7; 12 A Gp Rpt of Opns, XIII, 157–63.

[57] Ltr, Lt Col Mason Ladd, Dir Legal Div SGO ETO, to CG COMZ, 25 May 45, sub: Rpt of Study of Records and Investigation Relative to Incidence of Trench Foot, Micro Reel 114, Hist Docs World War II, ASF, Item 1318.

for all situations. In defending the ETOUSA clothing record, the theater quartermaster later called attention to a War Department statement that the shoepac was not necessarily the answer to trench foot, and that the combination of service shoes with overshoes was probably the best combination under most conditions on the Continent. But the service shoe-overshoe combination was admittedly a heavy and awkward combination in any situation requiring mobility, and combat troops frequently discarded the overshoes. The War Department, having concluded that the shoepac was the best article for unusual wet and cold conditions in which men were compelled to stand in water for long periods, had recommended its adoption for combat troops in the spring of 1944.

That the shoepac was not the sole answer to trench foot was shown by experience in the Seventh Army, which was 90 percent equipped with the shoepac and still suffered a sizable number of casualties.[58] But the incidence had not been as high as in other armies. This could probably be attributed to the fact that the veteran Seventh Army was somewhat more trench foot conscious as a result of its earlier experiences. But the ETOUSA chief surgeon pointed out that a survey of one general hospital revealed that there were 29 percent fewer cases among troops who wore the shoepac. In many cases the shoepacs had failed to prevent trench foot only because of faulty instruction in their use and fitting.[59] The theater chief surgeon had reported as early as December 1944 that the shoepac had been found to be the only mechanical aid which contributed substantially to the prevention of trench foot.[60]

INFANTRYMAN WEARING A FIELD JACKET M–1943 *tries on a new pair of shoepacs with wool ski socks, January 1945.*

It appears to have been clearly established that trench foot would have been much less prevalent had combat troops been equipped with the shoepac, whatever the shortcomings of the earlier models. Beyond the requisition for one

[58] *Ibid.*

[59] Study, Overshoes Arctic and Combat Boots.

[60] Cbl EX–78065, ETO to WD, 23 Dec 44, in Study, Overshoes Arctic and Combat Boots.

field army, ETOUSA did not request additional quantities of the shoepacs until December. In the case of footwear, therefore, as in the case of other items of clothing, it was the War Department's view that the theater had not given full consideration to experiential data from other theaters and that the chief quartermaster had been slow to adopt winter clothing items which on the basis of both tests and combat experience had been proved superior to the uniform proposed by the ETO.

Responsibility for providing adequate clothing obviously was shared by the theater and the War Department, and responsibility for the shortcomings in this field must also be shared, although in precisely what degree it is difficult to say. Assigning blame for failures in the supply of adequate winter clothing in the winter of 1944–45 is not a simple matter, for some of the decisions on winter clothing had complex origins.

An investigation of the clothing controversy was carried out by a committee headed by Col. Charles Garside in the spring of 1945. The investigation dealt almost exclusively with the question of supply—that is, quantity—and not with the adequacy of the uniform from the point of view of protection. The results were inconclusive. In general, the committee found that both the theater quartermaster and the OQMG had planned and acted with intelligence and foresight to meet winter clothing problems, and explained the difficulties over clothing supply as stemming largely from unforeseeable circumstances such as production problems in the United States, transportation difficulties in the theater, extraordinarily high attrition rates, and misjudgment with regard to the end of the fighting in the theater.[61]

Maj. Gen. Clinton F. Robinson, chief of the Control Division in the ASF, reviewing the investigating committee's findings for General Somervell, did not accept them in their entirety. He concluded that aside from the unforeseeable difficulties both the War Department and the theater had been remiss in some respects, the War Department primarily for the lateness or inadequacy of research, the theater for improper requisitioning practices and failure to forward requisitions sufficiently far in advance.[62]

The controversy over the adequacy of various items of clothing—that is, the question of quality—is more complex. To begin with, the Army had been long in arriving at final decisions with respect to various development items in winter clothing. There were conflicting schools of thought within the QMC, and the merits of different principles or theories—the layering idea versus others, for example—were being debated at least as late as December 1943, when a new Table of Equipment was adopted. Personality conflicts clearly account for some of the controversy which developed between General Littlejohn and the OQMG, and these arose at least in part from the fact that the theater quartermaster had personally sponsored and

[61] Investigation, Supply of Clothing to ETO, 1 Jan 44 to 28 Feb 45, DCofS, 420 ETO; see also Ltr, Garside to Brig Gen Albert J. Browning, 12 May 45, same file.

[62] Memo, Robinson for Somervell, 21 Jun 45, sub: Investigation Supply of Clothing in ETO, DCofS files, 420 ETO.

promoted the ETO jacket in the theater.[63]

To what extent the problem of availability—that is, production—entered into the theater's initial rejection of certain items of the War Department's recommended winter uniform is hard to say. General Littlejohn made much of the lack of assurance on this point in explaining his original decision not to requisition some of the new items. The War Department tended to discount this argument. General Gregory later pointed out that his recommendations had been made with production capabilities in mind, and General Lutes also later claimed that the new 1943 jacket was being produced in ample quantities beginning late in 1943. The implication was that the War Department would not have offered the theater the new items if they could not have been made available in the required quantities, and that the theater's rejection was not justified on that count.[64]

It is clear in any case that the War Department and the theater had not come to an understanding as to what would be required by, and what should be supplied to, the theater, particularly with respect to outer garments of the winter uniform. It is perhaps surprising that there should have been room for debate about the make-up of the combat uniform as late as the spring of 1944. From the theater's point of view there apparently was enough uncertainty and indefiniteness about the question to permit it to plead its own interests and preferences. The War Department was obviously reluctant to impose its decisions and judgment on the theater in the matter. Its indulgence in this respect was not abnormal. The independence which the theater enjoyed in many matters was in line with traditional policy. It was one which the War Department had reason to regret on occasion, most notably, for example, in the handling of manpower resources.[65] Perhaps this is the most serious indictment that can be made in the controversy over winter clothing. Whatever the indictment, one incontrovertible fact stands out: the ETOUSA combat soldier wore a uniform that was deficient in proper protection against the cold wet conditions under which he had to fight in the winter of 1944–45.

By the late winter, as the result of substitutions and improvisation, the outstanding characteristic of the ETOUSA uniform was its lack of standardization and simplicity. By that time seventy different items had been issued, including six types of jackets and seven types of trousers, creating insurmountable supply problems.[66] In mid-January General Littlejohn summoned the quartermasters of all the major commands to a conference for the purpose of eliminating some of the unsatisfactory items and of reaching an agreement on a single winter combat uniform. Littlejohn had

[63] See, for example, Ltr, Littlejohn to Gen Maxwell, 17 Mar 44, and Ltr, Littlejohn to Feldman, 7 Jul 44, Littlejohn Reading File; also Ltr, Somervell to Robinson in ETO, 6 Mar 45, ASF, ETO 1945.

[64] Ltr, Gregory to Littlejohn, 28 Sep 44, Study of 5 Apr 45; Memo, Lutes for Dir of R&SC, 22 Jan 45, sub: Supply of Winter Clothing to ETO, Lutes file ETO 1945.

[65] See Ch. XI, below.

[66] Gen Bd Rpt 109, p. 127.

first planned to call such a meeting in December, but the Ardennes counteroffensive caused it to be postponed.

Little was accomplished at the conference which finally met at Paris on 29 January, for the quartermasters of the 12th Army Group saw no point in discussing items with which most U.S. forces had had no experience. To the First, Third, and Ninth Armies the M1943 combat uniform—consisting of the M1943 jacket with pile liner, the high-neck sweater, the ETO jacket, scarf, and woolen underwear and shirt—was largely an unknown quantity. Col. James W. Younger, the Army Group quartermaster, expressed astonishment that it had not even been made available for field tests.[67] Representatives of the 12th Army Group asked for an opportunity to test the 1943 uniform before attempting any decision, and the chief quartermaster agreed to make available small quantities of the complete uniform for tests in all three armies, in the Ninth Air Force, and in the XVIII Airborne Corps.

These units had hardly had sufficient opportunity to test the uniform when the second clothing conference met on 17 March, attended by representatives of the major commands, the chief quartermaster, the chief surgeon, and the OQMG, including a shoe expert from the War Production Board.[68] The conference disclosed a wide range of opinion among the armies on the various items, and there was complete accord on only a few items such as underwear and shirts, and on the demand that leather be reversed on the combat boot. The greatest controversy arose over the type of jacket to be adopted. Third Army, which had carried out tests in the 4th Armored and 26th Infantry Divisions, particularly favored the ensemble designed for armored units, which included a widely admired combat jacket. But the production of this ensemble had already been terminated in the United States.

The diversity of opinion on many items led General Littlejohn to appoint a committee headed by Colonel Younger to consolidate the many recommendations and summarize the consensus of the conference. On the most controversial item the tabulation of preferences was not conclusive, for four of the five armies voted for both the M1943 jacket and the armored combat jacket. But the final uniform recommendations of the committee closely resembled the M1943 uniform that the War Department had repeatedly proposed, which included the wool field jacket (for dress wear), the M1943 combat jacket (modified somewhat), shoepacs, service boots, a trenchcoat type of field coat, ponchos, and leather gloves with wool inserts.[69]

The conference made certain com-

[67] Min, Mtg at 12th A Gp Hq, 22 Feb 45, 12 A Gp QM No. 3 Clothing Conf.

[68] The WPB representative was Lawrence B. Sheppard, Assistant Director of the Leather and Shoe Division, and formerly of the Hanover Shoe Company, who had come to the theater in response to General Littlejohn's request for a shoe expert. For correspondence on this subject see file, Footwear and Socks for Use in the ETO, ASF Dir of Mat, Prod and Pur Div.

[69] Min, Second Clothing Conf, 17–19 Mar 45; Memo, Younger for 12 A Gp Stf Secs, 23 Mar 45, sub: Winter Clothing Conf; and Memo, Younger for CG 12 A Gp, 14 Mar 45, sub: Winter Combat Uniform, all in 12 A Gp QM No. 1 Winter Combat Uniform, and No. 3 Clothing Conf.

promises because of known production limitations, and was not completely successful in deciding on a simple, single uniform. But it proposed the elimination of 21 items then authorized for issue in the theater, the reduction in the number of sizes by 59, and a reduction in the number of basic fabrics from 10 to 4.[70]

General Littlejohn's personal report to The Quartermaster General on the results of the conference, in which he underscored the armies' preference for the armored combat jacket, only rekindled the old controversy between the OQMG and the office of the theater quartermaster. General Gregory's reply strongly suggested that the theater quartermaster's conclusions did not accurately represent those of the clothing conference. More important, he considered General Littlejohn's conclusions inadequately supported by experiential data, for they were based largely on the experience of men who had not had an adequate opportunity to test the complete 1943 uniform. Admitting the popularity of the armored combat jacket and trousers, General Gregory showed that the preponderance of evidence from those who had used both the M1943 uniform and the former combination had indicated a decided preference for the 1943 ensemble. Only the Fifth and Seventh Armies, he maintained, had had any substantial measure of experience with the items comprising the authorized uniform, and he found it highly significant that those experienced organizations had arrived at the same conclusions.[71] The whole argument had of course long since become academic so far as the ETOUSA soldier in the winter of 1944–45 was concerned.

(3) Weapons and Vehicles

Shortages of ordnance equipment were probably the most serious in the Class II and IV category because of the immediate and direct effect which the lack of both tactical and cargo vehicles and weapons could have on operations. Shortages ranged from major items such as tanks, trucks, and artillery pieces to tires and tire patches, trailers, automatic weapons, fire control equipment, and antifreeze.

While transportation affected deliveries in the late summer and fall, shortages of ordnance equipment were mainly the result of high attrition and inadequate receipts in the theater. Early in November SHAEF provided the War Department with statistics to illustrate the rate at which supplies were being

[70] Gen Bd Rpt 109, p. 128. Another survey of the adequacy of winter clothing in the ETO, made by Maj. Paul A. Siple, a QMC technical observer who had accompanied Admiral Richard E. Byrd on his antarctic expeditions, reached conclusions very similar to those of the armies on various individual items. Siple, in addition to pointing out the deficiencies of individual items, emphasized that U.S. troops lacked both discipline and training in the proper use of clothing. Many of the difficulties he traced to their insistence on choosing undersized, tight-fitting garments for winter wear in an effort to retain the close-fitting characteristics of civilian clothing. British troops were better prepared in this respect, particularly with regard to the care of the feet, having had a costly experience in World War I. Maj Paul A. Siple, Report on the Adequacy of Winter Clothing in the ETO, 1944–1945, 12 May 45, DCofS files, 420 ETO; Memo, Col Rodney H. Smith, Asst Deputy G–4, to DCofS, 30 Jan 45, EUCOM ETOUSA G–4 Planning Directives Series H OVERLORD.

[71] Memo, Littlejohn for Gregory, 9 May 45, sub: Winter uniform, and Ltr, Gregory to WDGS, 11 Jun 45, both in WDGS G–4 Winter Uniform for Use in the ETO, OCQM, ETO, WDGDS A47–2.

consumed or expended in the European theater. Every day, it reported, nearly 1,200 small arms weapons, 1,300 bayonets, and 5,000 tires were lost. Every month 700 mortars, 375 medium and 125 light tanks, 900 2½-ton trucks, 1,500 jeeps, 100 cannon of various calibers, and 150 tubes had to be replaced. These were total losses, and did not take into account unserviceable equipment which could be repaired. In the latter category, for example, were the 100 2½-ton trucks which had to be taken off the Red Ball route every day.[72] Lack of spare parts for these vehicles and of adequate maintenance and repair facilities resulted in a rising number of deadlined vehicles. These totaled 15,000 in November.[73]

The shortage of many Class II and IV items was attributed in part to War Department replacement or maintenance factors, which the theater claimed did not match monthly losses. The First Army showed that the loss rate for the 4.2-inch mortar, for example, was approximately double the authorized 12.5 percent per month, and stated that the consumption rates for all signal equipment were far above the maintenance rates established by the War Department.[74] The importance of having adequate replacement factors lay in the fact that it was on the basis of them that theater reserve levels were established. Consumption of supplies or losses of equipment which greatly exceeded the maintenance or replacement factors could result in a sudden reduction of reserves because of the normally long lag in delivery time.

The problem of replacement factors was nowhere better illustrated than in the case of the medium tank. Attempts to get the replacement factor for the M4 tank revised had a long history, dating back to preinvasion days when the theater had predicted that losses in the landings would not be covered by the currently authorized factor of 7 percent. As with other items of equipment, however, the War Department insisted that any requests for revision must be backed by experiential data from actual combat. In June it raised the factor to 9 percent, but, as before, mainly on the basis of experience in Italy, for no conclusive data were yet available from operations in France.

Losses in the first three months were considerably above the existing replacement factor, and thus tended to confirm the theater's earlier assertions. In mid-August ETOUSA reported that its reserves were exhausted; by mid-September it was finding it increasingly difficult to keep armored units at their authorized Table of Organization and Equipment (T/O&E) strength.[75] The War Department meanwhile had agreed

[72] Cbl S–65533, SHAEF to AGWAR, 5 Nov 44, SHAEF G–4 400 Supplies, General, IV; Memo, Somervell for Dir Office of War Mobilization and Reconversion, 7 Dec 44, Hq ASF–President–White House–Exec Office 1944, Somervell file.

[73] COMZ G–4 History, I, 130.

[74] FUSA Quarterly G–4 Rpt for Period Oct–Dec 44, FUSA 319.1 G–4 Periodic Rpts FUSA.

[75] The figures cited in *Logistical Support I*, 522–23—namely, 26.6 percent for June, 24.4 for July, and 25.3 for August—which are taken from the 12th Army Group Report of Operations (XI, 67), may be exaggerated. Another source indicates that the theater reported losses of 6.4 percent in June, 7.3 percent in July, and 20.6 in August, and an average cumulative loss rate of 14.7 for the first eleven weeks of operations. Memo, Maj Gen W. A. Wood, Actg Dir Plans and Opns ASF, for Dir Rqmts and

to expedite the shipment of tanks already released. But receipts did not meet requirements despite the lower losses which attended the revision to more static operations in the next few months. Losses in September came to 16.5 percent of the theater's T/O&E strength as compared with 25.3 percent in August. In October the rate fell to 9.8 percent. In November the rate advanced to 11.2 percent and in December shot up to 22.8, reflecting the greatly intensified combat activity.[76]

Early in October the War Department had announced an increase in the replacement factor for the medium tank from 9 to 11 percent. But this revision was made on the basis of the combined loss experience of 9.9 percent in the North African and European theaters in the month of July, and did not reflect the experience of August and September. By October the cumulative loss rate, according to the 12th Army Group, was nearer 20 percent in the European theater.[77] Revisions in the replacement factor consequently lagged far behind current experience.

Under these circumstances the theater found it impossible to maintain units at their authorized strength, to say nothing of reconstituting reserves. At the end of September First Army, having operated with approximately 85 percent of its authorized strength in tanks during the month, adopted the expedient of temporarily suspending the Tables of Organization and Equipment of armored units so far as medium tanks were concerned, and reducing the authorized strengths in order to effect an equitable distribution of the available tanks and to establish a small reserve. The new T/O&E's temporarily cut the authorized strengths in 75-mm. and 76-mm. gun tanks from 232 to 200 for armored divisions organized under the old T/O&E (the 2d and 3d Divisions only), from 168 to 150 for divisions organized under the latest T/O&E (all remaining armored divisions), and from 54 to 50 for separate tank battalions. The Ninth Army later adopted the same provisional T/O&E's for its armored formations.[78]

The situation saw no improvement during the fall months. By the end of November there were on hand in the theater only 3,344 tanks against a T/O&E requirement of 3,409 and an authorized on-hand reserve requirement of 937.[79] No true reserve existed, therefore. The 12th Army Group reported

Stock Control Div, n.d., sub: Replacement Factor for Certain Items in ETO, Incl III to Memo, Wood for Somervell, 4 Jan 45, sub: ETO Replacement Factors or Days of Supply for Tank, Medium; Mortars, Radio Sets, and Ammo, Hq ASF Notebook of Memos, Ltrs, etc., on Supply, Somervell file.

[76] The actual count in 75-mm. and 76-mm. tank losses in the 12th Army Group in these months was as follows: 318 against a T/O&E strength of 1,927 in September, 199 out of a T/O&E strength of 2,032 in October, 280 out of 2,509 in November, and 576 out of an authorized strength of 2,[illegible]25 in December. 12 A Gp Rpt of Opns, XI, 67.

[77] The records are conflicting. 12th Army Group figures are cited, except as otherwise noted, for the sake of consistency. They are believed to be high.

[78] Ltr, FUSA to Corps, Divs, and Tank Bns, 30 Sep 44, sub: Medium Tank Loss Replacement Policy, FUSA 470.8 Armored Cars and Tanks; Ltr, I. D. B. to Chief AFV&W Sec ETOUSA, 2 Dec 44, sub: Status of 75-mm. and 76-mm. Gun Medium Tanks, SHAEF G–3 O&E 470.8 Tanks, II.

[79] Memo, Brig Gen F. A. Heileman for Somervell, 30 Dec 44, sub: Medium Tank Situation in ETO, Hq ASF, European Theater—last half 1944. The figure 937 is based on a 75-day reserve. The theater had actually been authorized to requisition on the basis of a 75-day reserve plus a 60-day shipping factor.

that two of its tank battalions had fewer than ten serviceable tanks, and field commanders in general deplored the fact that armored units had to operate at from 10 to 25 percent below authorized strength. Furthermore, the current theater troop basis failed to provide the one tank battalion per infantry division which field commanders considered a necessary minimum. As a partial remedy steps were taken to convert two battalions of the 10th Armored Group to composite tank battalions for use with infantry divisions.[80]

The field commands had continued to urge the theater to obtain a higher replacement factor, arguing that a larger flow of replacement tanks was imperative if the habitual infantry-tank co-operation which had characterized all operations thus far was to continue. The 12th Army Group noted that at no time since the middle of August had the armies had their full T/O&E allowance of tanks, and that not since the early days of the Normandy beachhead had they possessed a reserve. It maintained that a 25 percent reserve in each army was an operational necessity.[81]

One factor which plagued all supply between the zone of interior and the theater, and which never seemed to get sufficient consideration, was the time lag between the submission of requisitions and the eventual delivery of replacements. Because of the handling problem involved in the case of tanks, this time lag had a marked bearing on the theater's supply position. Time required to submit loss reports to the War Department and to obtain releases of replacement tanks, time required to move tanks from factories to ports and to load them for shipment, time required for unloading in the theater in the absence of adequate discharge facilities, and delays in reporting losses, including tanks deadlined for lack of spare parts, all contributed to what must have seemed an interminable lag between requisition and delivery. In addition, the large number of tanks habitually under repair in the various echelons appeared to justify increasing the reserve factor. Theater officials were cognizant of these factors in the supply of tanks as well as other items, and it was for this reason that they had requested and been granted a sixty-day shipping factor in addition to the seventy-five-day reserve level. The low replacement factor, however, had become the more vital consideration, and it was the focal point of all attempts to remedy the situation.

Improvement finally came in December, the month of reckoning for several other major logistic problems, partly as the result of the first-hand investigation which Lt. Gen. LeRoy Lutes, Director of Operations, ASF, made during his visit to the theater. The theater presented the ASF official with an exhaustive analysis of its medium tank situation. It concluded that with currently scheduled receipts it would meet its T/O&E needs for the next two months, but predicted that it would be short approximately 1,100 tanks against its authorized reserves. It argued that a large portion of the reserve must be on hand to provide a cushion against

[80] Ltr, I.D.B. to Chief AFV&W Sec, 2 Dec 44, sub: Status of 75-mm. and 76-mm. Gun Medium Tanks, SHAEF G-3 O&E 470.8 Tanks, II.

[81] Ltr, 12 A Gp to CG ETO, 11 Dec 44, sub: Tank Rqmts, NUSA 451 Vehicles.

losses. Additional tanks totaling 22 percent of the T/O&E—the equivalent of the sixty-day shipping and order time at 11 percent per month—it stated, should be on release and in the pipeline to ensure adequate maintenance requirements. On this basis it asked for the release of 1,102 tanks for January lift in addition to those already scheduled. The analysis and request were also dispatched to the War Department.[82]

As was so often the case, the theater's data conflicted with the picture as seen in the War Department, partly because of the different replacement factors used and partly because of the War Department's habit of considering all tanks, whether actually on hand in the theater or just released, as part of the theater's assets. The theater and the War Department rarely saw eye to eye on this matter, as was also evidenced in the argument over ammunition.[83] On the other hand, the theater had used a replacement factor of only 13.1 percent in its computations, far below the exaggerated claims of 25 percent which had repeatedly been made by the armies.

In any case, General Lutes was now completely convinced that the theater's shortage of medium tanks was much more critical than indicated by the War Department studies, and he agreed that it was imperative to ship additional tanks to Europe as quickly as possible and to raise the replacement factor. On 17 December officials in Washington went over the entire problem with the Lutes mission in a transatlantic telephone conference. Washington now accepted the theater's computations and agreed to release 1,010 medium tanks in addition to the 250 scheduled for delivery to the port by the end of January.

Only two days before this the War Department had raised the replacement factor from 11 to 14 percent. On General Lutes' recommendations it now agreed to raise it still further, to 20 percent, but with the understanding that this was a temporary concession and was to apply only during the next critical months, or until 1 May 1945.[84] In a memorandum to General Somervell several days later Lutes admitted that a higher replacement factor should have been adopted earlier even though there were doubts as to whether U.S. production could have met the demand. The theater later reported a cumulative loss rate of only 12.8 percent through January 1945.[85]

While the War Department's action promised to place the theater in a healthy situation by the end of February, it provided no answer to the plight in which the theater found itself as a result of the Ardennes counteroffensive. First Army's medium tank losses in December were to total nearly 400, three times the casualties suffered in the preceding

[82] Cbl EX-84824, Lee to AGWAR, 15 Dec 44, Tab H to Official Rpt, Mission to ETO, 4 Dec 44–13 Jan 45, Hq ASF.

[83] See above, Ch. IX, Sec. 2.

[84] Memo, Col R. P. Hollis for Lutes, 10 Dec 44, sub: Critical Ord Items Listed in G–4 Rpt, 12 A Gp, Memo, Lutes for Lee, 11 Jan 45, sub: Rpt on the Activities of Lutes Party, and Record of Tel Conf Between Paris and Washington, 17 Dec 44, all in Official Rpt of Lutes Mission, Hq ASF.

[85] 12th Army Group claimed a cumulative loss factor of 16.4 percent. Memo, Lutes for Somervell, 23 Dec 44, sub: Medium Tanks, ETO, Official Rpt of Lutes Mission; Ltr, ETO to 12 A Gp, 22 Feb 44, Ind to Ltr, 12 A Gp to ETO, 11 Dec 44, sub: Tank Rqmts, EUCOM 470.1 Combat Armored Cars and Tanks; 12 A Gp Rpt of Opns, XI, 67.

MEDIUM TANKS *ready for shipment from Marseille to combat areas, 10 February 1945.*

month.[86] To meet its immediate replacement needs the theater had to seek relief closer at hand. On 19 December it appealed to its neighbor to the south, the Mediterranean theater, to release seventy-five tanks which had been consigned to U.S. forces in Italy but which for some reason had been unloaded at Marseille and were already in Delta Base Section. ETOUSA proposed to repay this loan by asking the War Department to divert to the Mediterranean an equal number then being loaded at New York for shipment to Europe. General McNarney, the MTOUSA commander, readily agreed. In fact, a few days later the Mediterranean theater announced that it was releasing a total of 150 tanks for transfer to France.[87]

ETOUSA next asked 21 Army Group to survey its resources to determine whether any number of tanks up to 500 could be made available, promising repayment in February. Montgomery responded by offering to release 351 tanks to U.S. forces, 254 of which were delivered to the First Army and 97 to the Third before the end of the month.[88]

[86] FUSA Rpt of Opns, 1 Aug 44–22 Feb 45, Bk. III, p. 64.

[87] Cbl S–71959, Smith to McNarney, 19 Dec 44, and Cbl FX–71545, MTOUSA to SHAEF, 20 Dec 44, SHAEF Cbl Logs 1944–45, Smith Papers; Cbl FX–73970, MTOUSA to SHAEF, 24 Dec 44, SHAEF G–3 O&E 470.8 Tanks, II.

[88] Cbl E–78893, COMZ to 21 A Gp, 26 Dec 44, and Cbl SD–8692, 21 A Gp to SHAEF, 26 Dec 44, both in SHAEF G–3 O&E 470.8 Tanks, II.

British forces could easily afford such a transfer, for they held disproportionately high reserves—totaling 1,900 Shermans—in the United Kingdom. At the end of the month General Somervell directed Maj. Gen. James K. Crain, U.S. executive of the London Munitions Assignments Board (LMAB), to bid for 1,000 of these tanks. Somervell pointed out that the establishment of a 35-percent reserve for both U.S. and British forces and the restoration of equality between the two would actually require the transfer of 1,147 tanks.

But these instructions were rescinded. In view of the measures recently taken to meet ETOUSA's shortages from the United States, it was decided instead to assign the entire output of U.S. production to U.S. forces until their reserves in Europe totaled 2,000, which was expected to require four months. Complete parity on a percentage basis was not expected to be achieved before June, since British forces were promised a hundred mediums per month beginning in April. Under these arrangements no attempt was to be made to repay the loan of 351 tanks made by 21 Army Group in December. In any event, action had finally been taken to correct the maldistribution of reserves between the two forces.[89]

Efforts to provide the theater with armor of better quality continued throughout the fall and winter, but with little tangible result for the theater in this period. Production, and consequently deliveries, of the 76-mm. gun tank fell behind schedule, in part because of design changes in suspension and tracks and because of tooling up for the newer 90-mm. gun tank.[90] The theater therefore took no action to request the complete elimination of the 75-mm. gun tank until January 1945, although priority was given to the loading of the 76-mm. gun model in the New York Port whenever possible. While the theater received increasing numbers of the 76's, the obsolescent short-barreled 75 continued to be the principal weapon of armored units throughout the fall and winter.[91]

Equipping 300 M4's with British 17-pounder guns, plans for which had been made in August 1944, was postponed again and again because of the shortage of reserves with which to make the conversion.[92] Improved ammunition, high-velocity, armor-piercing (HVAP), for the 76-mm. gun was shipped to the theater in this period, but receipts amounted to less than two rounds per gun per month until March 1945.[93]

The 105-mm. howitzer tank, first hailed as meeting the need for a tank with better high explosive ammunition,

[89] Cbl WX-84700, Somervell to LMAB, 30 Dec 44, Memo of Decision on Medium and Light Tanks by Somervell, Crain, Crawford, Lord, Saylor, *et al.*, 10 Jan 45, Cbl LM1-57455, Troopers to 21 A Gp, 5 Jan 45, Cbl UKX-22667, LMAB to Shingler in ASF, 21 Jan 45, and Cbl UK-19942, LMAB to SHAEF, 30 Dec 44, all in SHAEF G-3 O&E 470.8 Tanks, II; Memo, Lutes for Lee, 1 Jan 45, sub: Medium Tanks, EUCOM 470.8 Combat Armored Cars and Tanks, II; Record of Tel Conf Between Col Hollis (Paris) and H. A. Markle (Washington), 28 Dec 44, and Cbl WARX-83327, Somervell to ETO, 28 Dec 44, both in Official Rpt of Mission to ETO, Lutes, Hq ASF.

[90] Rpt, Hiland G. Batcheller, 14 Nov 44, sub: Critical Programs: A Rpt to the WPB, Hq ASF–President–White House–Exec Office 1944, Somervell file.

[91] Final Hist Rpt AFV&W Sec ETOUSA, D Day to Date of Inactivation, 6 Jun 44 to 24 May 45, p. 23, ETO Adm 540.

[92] 12 A Gp Rpt of Opns, XI, 48.

[93] Final Rpt, AFV&W Sec, p. 23.

arrived in satisfactory numbers and was never in critical supply. But this tank did not live up to expectations because of its lack of a powered turret traverse. A later model contained the desired power traverse mechanism, but tanks so equipped did not reach the New York Port until the final month of hostilities and never saw combat.

The 90-mm. gun tank, first known as the T26 and later as the M26 or Pershing, and long-awaited as the answer to the theater's need for armament that could match the Germans', did not become available for shipment until January 1945. A token number first saw action on 15 February. Theater officials had twice revised their recommendations as to the ratio in which they desired the 105-mm. howitzer tank and the newer 90-mm. gun tank supplied. Before D Day they had advised a ratio of one 90-mm. gun tank to three 105's. In October, on the basis of combat experience with the latter, they recommended a ratio of two to one. Early in January 1945 the ratio was further altered to four to one in favor of the 90-mm. gun tank.[94]

The shortage of trucks was evident early in the fall and extended well into the winter. All armies reported critical deficiencies and a high deadline rate, and in October representatives of SHAEF, ETOUSA, and 12th Army Group met to establish priorities for the issue of vehicles.[95] Operational losses had far exceeded War Department replacement factors and by the end of November had exhausted continental reserves. Lack of the desired percentage of heavy duty trucks (over 2½ tons) had accentuated the rapid deterioration, for the smaller 2½-ton general purpose vehicle was not well suited to long-distance hauling, which, along with overloading and improper maintenance, accounted for a high mortality rate. Trucks with larger capacity had been in production in the United States for more than a year, partly as a result of pressure applied since the North African campaign. But heavy duty vehicles had never been manufactured in large numbers in the United States, and production thus far had fallen short of the goals. Pressure to turn out heavy duty vehicles meanwhile had a detrimental effect on the production of the standard types, with resulting deficits in the output of 2½-ton trucks also. In both cases the foundry industry, which was unable to expand rapidly enough to supply the castings for the full complements of axles, transmissions, and engines, was the most persistent bottleneck.[96]

Like most deficiencies, the vehicle shortage had many ramifications. More vehicles were actually made available to the New York Port than could be shipped to the theater in the fall months of 1944. Lack of port capacity on the Continent, with its resultant accumulation of ships in European waters, had led the War

[94] *Ibid.*, pp. 9, 16; 12 A Gp Rpt of Opns, XI, 46; Cbl EX-98320, COMZ to Somervell, 14 Feb 45, Cbl WX-39078, AGWAR to Lee, 17 Feb 45, and Cbl, SHAEF to Somervell, 8 Mar 45, all in SHAEF G-3 O&E Tanks, II.

[95] Memo, Kibler for G-4 12 A Gp, 11 Oct 44, sub: Status of Gen Purpose Vehicles, 12 A Gp Vehicles (up to 13 Nov 44), No. 141; Ltr, Wilson to CG SOLOC, 4 Dec 44, sub: Informal Rpt for Nov 44, Hq SOLOC 319.1 CONAD Rpts.

[96] Rpt, Hiland G. Batcheller, 14 Nov 44, sub: Critical Programs: A Rpt to the WPB, Hq ASF-President-White House-Exec Office 1944, Somervell file.

Department to cut the number of sailings to the theater. The higher priority given to the shipment of weapons, combat vehicles, tires and tubes, antifreeze, and all types of spare parts for the equipment already in the theater left little space for vehicles on the allotted commodity-loaders, with the result that virtually no general purpose vehicles could be lifted in November and December. Paradoxically, the theater claimed late in December that, despite the over-all port deficiency, unused discharge capacity actually existed in both the U.K. ports and at Cherbourg which, though unsuitable for other types of supplies, could have accepted thirty vehicle ships and thus given the theater about 30,000 badly needed trucks.[97] But General Lutes, when told of this during his visit to the theater in December, found no evidence that theater officials had made a point of this in communications with the War Department.[98]

Early in January the theater claimed shortages of 33,000 vehicles and estimated that with currently scheduled shipments it would still be short about 30,000 on 1 February and 35,000 on 1 March. To meet at least part of this deficit it initially asked the War Department to dispatch twenty-five ships solidly loaded with vehicles for arrival in February. After a more searching review of the theater's future needs, which considered the requirements for operations beyond the Rhine and took into account the civil relief needs and the poor prospects for local procurement, General Somervell, who was then in the theater, increased the requisition to thirty ships for February and directed that an additional fifty be dispatched for March arrival. Maj. Gen. Charles P. Gross, the Chief of Transportation in Washington, at first questioned the theater's ability to receive such numbers in view of the continuing bank of idle ships in European waters. But Somervell was apparently convinced of the theater's discharge capacity and ordered the shipments set up without further discussion.[99] Shipments made against this directive brought substantial relief by the time of the Rhine crossing in March.

Tires were another item in which truly critical shortages developed in the fall of 1944. The War Department had actually warned all major commands of potential shortages as early as December 1943, and announced its allocations policy, just as it had for other supplies in which it was known that production would fail to meet demands. Headquarters, ETOUSA, passed this information on to its subordinate commands and directed them to enforce rigid compli-

[97] Vehicle assembly facilities were more than equal to discharge capacities and, with the completion of arrangements with the Ford Motor Company at Antwerp, reached a monthly capacity of 31,500 by the end of the year.

[98] Unsigned Memo, 21 Dec 44, Memo, Chief Ord Off ETO for G–4 COMZ, 26 Dec 44, sub: Supply of GPVs, Memo, Hollis for Lutes, 25 Dec 44, sub: Deficit of GPMV for Initial Equipment of ETO Units, and Memo, Lutes for SAC, 26 Dec 44, sub: Vehicles for Equipping Units, all part of Tab G to Official Rpt of Mission to ETO, 4 Dec 44, 13 Jan 45, Lutes, Hq ASF.

[99] Cbl E–84296, ETO to G–4 to SHAEF, 9 Jan 45, Cbl WARX–89701, AGWAR to ETO, 10 Jan 45, Cbl E–84558, ETO to AGWAR, 10 Jan 45, Cbl WARX–90495, AGWAR to ETO, 12 Jan 45, Cbl S–74716, SHAEF to CCS, 13 Jan 45, Cbl WARX–20354, Gross to ETO, 13 Jan 45, Cbl E–86337, Somervell to AGWAR, 14 Jan 45, and Cbl EX–86788, ETO to AGWAR, 15 Jan 45, all in ETO Adm 398 and 404, IN and OUT Cables.

ance with preventive maintenance and conservation standards.[100]

Enforcement of these well-conceived regulations was something less than ideal. This, plus the grueling conditions which accounted for a high mortality of vehicles, only accentuated the shortages already caused by production shortfalls in the United States. It was soon apparent that the War Department replacement factor of 7.5 percent was much too low in view of the constant use to which vehicles had been put. By early September every command was aware of the deterioration which was snowballing into critical proportions. An inspection of trucks in the Advance Section at that time revealed that 70 percent of its vehicles had already run an average of 10,000 miles. Even assuming 100 percent tread wear, which did not take into account tires replaced because of damage resulting from cuts, overload breaks, or accidents, the average mileage expectancy of tires was calculated to be about 12,000 miles. From this it was evident that several thousand tires would have to be replaced within the next few weeks. The same situation could be assumed to apply in other commands.[101]

Replacement of worn-out tires fell behind as early as September. By the end of November the 12th Army Group, then deeply engaged in its autumn offensive, claimed that the lack of tires was actually affecting operations.[102] No immediate and substantial relief was in sight. On the basis of available stocks and foreseeable replacements from the United States the Communications Zone in fact concluded that the theater faced an emergency of the gravest nature. Repair and retreading had fallen far behind for lack of materials, particularly camelback. Delays in the shipment of this vital commodity, combined with the small shipment of new tires, led the Communications Zone to predict a deficit of at least 250,000 tires by the end of January and the deadlining of 10 percent of the theater's vehicles. General Lee, while describing the dark outlook at a command and staff conference early in December, noted that the chief of transportation had already been forced to deadline all one-ton trailers and one thousand vehicles for lack of tires.[103]

Some measures had already been taken to meet the crisis. In November Brig. Gen. Hugh C. Minton, director of the Production Division, ASF, had arrived in the theater to survey local productive capacity, and after an inspection of eight plants in France and Belgium recommended the reactivation of the local industrial capacity. Late in the month representatives from SHAEF and the Communications Zone met and decided to form a rubber committee which was to promote the development of civilian resources for tire production and repair

[100] Ltr, WD to Major Cmds, 7 Dec 43, sub: Critical Tire Shortage, and Ltr, Hq ETO to FUSAG *et al.*, 8 Feb 44, sub: Tire and Tube Maintenance, 12 A Gp 451.92 Tires, I.

[101] Ltr, Capt Leland E. Copple, Chief Tire Sec ADSEC, to G–4 Motor Transport Brig ADSEC, 9 Sep 44, sub: Tire Replacement Survey, ADSEC 451.92 Tires and Tubes; Ltr, ETO to Maj Comds, 9 Apr 44, sub: Tire and Tube Maintenance, 12 A Gp 451.92 Tires, I; COMZ G–4 Hist, V, 40.

[102] TWX, FUSA to COMZ, 1 Oct 44, and TWX Q–24367, 12 A Gp to COMZ, 25 Nov 44, 12 A Gp 451.92 Tires, I.

[103] Comd and Stf Conf on Tire Conservation, Hq COMZ, 8 Dec 44, ADSEC 451 Vehicles; Cbls E–67929 and EX–68973, COMZ to ETO, 27 Nov and 1 Dec 44, respectively, 12 A Gp 451.92 Tires, I; COMZ G–4 Hist, V, 40.

on the Continent and allocate a fair share of the tires produced to civilian and military needs. Representatives from the G–4 and G–5 Divisions of both SHAEF and the Communications Zone, SHAEF Mission to France, the General Purchasing Agent, and the theater ordnance officer were appointed to the committee, which immediately took steps to restore the tire industry to production.

The program did not promise substantial relief in the immediate future. As was the case in several other attempts at local procurement, it created as many problems as it solved. Production in France and Belgium depended on the import of raw materials and on the allocation of transportation and power, all of which were critically short, and in addition entailed the division of the product between military and civilian needs. The program got under way in good time, however, and bore the first fruit on 4 January, when the first tire was turned out from American synthetics at the Goodrich plant in Paris. Before the end of the month the Goodrich plant was producing at a rate of 4,000 tires per month and the Michelin plant at the rate of 2,000. Before long six plants were in operation in France and Belgium. The War Department also considered the possibility of procurement in Spain, but the shortage of carbon black made this impossible. In fact, lack of this vital component forced cutbacks in U.S. production and threatened to force suspension of the program in Europe.[104]

The theater had also launched a vigorous campaign to enforce preventive maintenance. Tire inspections had shown that 40 percent of all replacement needs could be laid to underinflation, overloading, or other abuses. The widest possible publicity was therefore given to the importance of preventive maintenance, including such points as proper tire pressure, maximum speeds, rotation of tires, proper wheel alignment, avoidance of overloading, and daily inspections for cuts and bruises. Both *The Stars and Stripes* and the Armed Forces Network radio were enlisted in the campaign to make vehicle drivers conservation conscious, and a slogan contest was held, with prizes in the form of War Bonds and two-day passes to Paris.[105]

The War Department meanwhile had been bending every effort to increase its shipments to the theater. To meet emergency needs it combed the zone of interior for extra tires, collecting all stocks from posts, camps, and stations, removing tires from all unserviceable vehicles that were not immediately repairable, and stripping spares from all vehicles except for minimum emergency pools for convoy operations. Increasing production proved a thorny problem, partly because of material and manpower shortages, but also because of bad labor relations in the rubber industry. Labor and management eventually signed an agreement which brought about a moratorium in labor disputes affecting tire production. The War Department at the same

[104] Ltr, COMZ to SAC, 6 Jan 45, sub: Rubber Committee, and Cbl W–60623, AGWAR to COMZ, 29 Mar 45, SHAEF AG 451.92–1 Tires and Tubes; COMZ G–4 Hist, V, 45–47.

[105] Comd and Stf Conf, Hq CZ, 8 Dec 44; COMZ G–4 Hist, V, 46, 48–49.

time released skilled workers to tire plants to ease the manpower shortages.[106] By these expedients the desperate theater shortage was gradually eased.

Production shortfalls in vehicles and tires were fairly symptomatic of the difficulties which had plagued munitions output in general. In both cases the shortages had been foreseen for several months, for they had resulted in part from a general slackening of effort in the United States. The production of munitions had fallen off as early as November 1943 and had remained for several months at levels unequal to expected demands. The contagious optimism growing out of the midsummer pursuit in France encouraged this trend, causing workers to leave factories turning out badly needed munitions to seek other employment. These developments, combined with the increasing demands arising from unexpected consumption and attrition in the active theaters, inevitably led to critical deficits.

The War Department had taken measures to counteract this trend with a determined drive to stimulate production. In an attempt to ease the manpower shortage General Somervell in the summer of 1944 first secured the release of servicemen to foundries, and then secured authorization to furlough up to 2,500 men for ninety-day periods to aid in the manufacture of 105-mm. shells. At the same time the ASF enlisted the help of the War Production Board, the War Manpower Commission, and the Military Affairs (Mead) Committee of the Senate to impress upon labor the need for greater output. General Eisenhower had already made an appeal via the press in August for the maximum flow of munitions. Early in January General Somervell asked the Supreme Commander to send a message to both management and labor emphasizing the critical needs of the European theater.[107]

[106] Memo, Somervell for Dir Office of War Mobilization and Conservation, 7 Dec 44, Hq ASF–President–White House–Exec Office, 1944, Somervell file; Memo, no signature and n.d. [*c.* December], sub: Heavy Duty Truck and Bus Tires, Hq ASF Notebook of Memos, Ltrs, etc., re Supplies; Byron Fairchild and Jonathan Grossman, *The Army and Industrial Manpower,* UNITED STATES ARMY IN WORLD WAR II (Washington, 1959).

[107] Ltrs, Somervell to Lee, 4 Aug 44 and 1 Dec 44, Ltr, Somervell to Eisenhower, 12 Dec 44, and Memo, Lutes for Somervell, 17 Dec 44, all in Hq ASF European Theater–last half of 1944; Memo, Somervell for Clay, 7 Dec 44, Hq ASF–President–White House–Exec Office 1944, Somervell file; Cbl, Somervell to Eisenhower, 3 Jan 45, Cbl S–73675, SHAEF to WD, 4 Jan 44, and Cbl WAR–87134, Marshall to Eisenhower, 5 Jan 45, all in Hq ASF, ETO 1945, Somervell file; Cbl E–81755, COMZ to AGWAR, 5 Jan 45, and Cbl W–22616, AGWAR to COMZ, 19 Jan 45, all in SHAEF AG 451.92–1 Tires and Tubes, 1945.

CHAPTER IX

Supplying the Armies: Ammunition

(1) The October Crisis

In the entire eleven months of operations on the Continent no supply problem plagued U.S. forces more persistently or constricted their operations more seriously than the shortage of field artillery ammunition. Restrictions on expenditures were imposed shortly after the Normandy landings because of unloading difficulties at the beaches. Such restrictions continued with little relaxation until the end of hostilities because resupply from the United States was uncertain.[1] In well over half of all types of artillery ammunition the theater was able to maintain stocks in excess of the authorized level of seventy-five days of supply at War Department rates. But in the major items accounting for the great bulk of all expenditures the aggregate stocks on hand in the theater were almost without exception below the authorized levels throughout the eleven months of operations.[2]

Ammunition supply prospects appeared favorable for a short time early in September, and the 12th Army Group, although increasingly skeptical of the Communications Zone's optimistic forecasts, made relatively liberal allocations to the armies in the hope of crashing through the West Wall on the momentum of the pursuit. By the middle of the month this policy had left deep holes in the theater's reserves, reducing reserve levels in the major types by an average of twenty days of supply from the preceding month.[3] Exhaustion of some categories was expected within as little as two weeks. Inadequate discharge facilities continued to account for much of the delay in deliveries. But in the case of the heavier calibers the War Department simply was not releasing sufficient quantities. Now an Allied force of growing size stood at the German border and demanded huge quantities of field artillery ammunition to break through the steel and concrete of the West Wall.[4]

Increasing uncertainty over future

[1] See *Logistical Support I*, 445–48 and 525–43 for early difficulties.

[2] History of Planning Division ASF, App. 18–A, Ammunition Supply for the European and Mediterranean Theaters, 15 Aug 45, OCMH. This study contains tabulations of on-hand stocks, monthly expenditures, and authorized and actual day-of-supply levels for the major types of artillery ammunition.

[3] History of Planning Division ASF, pp. 132–70.

[4] Ltr, 12 A Gp to CG COMZ, 16 Sep 44, sub: Items of Ammo in Critical Short Supply, EUCOM 471 Allocation of Ammunition, II; 12 A Gp Ord Sec Jnl, 15 Sep 44.

ammunition availability characterized the last two weeks of September as the armies attempted to widen the breaches in the German defenses. Army group continued to allocate ammunition for eight-day periods, and the armies fired at substantially higher rates than in the preceding month. But the allocations reflected the hand-to-mouth supply situation and fell far short of the rates agreed to in the month before the invasion or desired by the field commands. The allocation for the period 27 September–5 October, for example, permitted daily expenditures of only 3.8 rounds per gun for the 240-mm. howitzer and 3.1 rounds for the 8-inch gun.[5] Apprehension over future deliveries led both First and Third Armies to impose a strict rationing of critical types. Subordinate units in turn exercised additional economies in an effort to build up forward stocks.[6]

At the periodic allocation meeting on 1 October the Communications Zone presented figures which indicated some improvement in supply, and the army group therefore granted somewhat larger expenditure rates for the next eight-day period, 5–13 October.[7] Two days later the full seriousness of the ammunition situation was finally brought to light. The First Army ammunition officer showed that the allocation was completely unrealistic, for the ammunition which the army had been authorized to expend did not exist in army depots and could not be obtained from the Communications Zone. Both the army group and the armies had long doubted the reliability of the Communications Zone's availability forecasts. Their suspicions now appeared confirmed.[8]

A full investigation of the situation in the next few days revealed that the ammunition shortage had reached truly critical proportions. At the next regular allocation meeting on 9 October, attended as usual by the G–3, G–4, ordnance, and artillery officers of the 12th Army Group, and a representative of the Communications Zone, plus the ammunition officers at First, Third, and Ninth Armies, it was revealed that reserve stocks of certain critical items were near exhaustion despite the fact that expenditures had been lower than predicted. Widely varying availability figures for the allocation periods since early September convinced the conferees of the "absolute unreliability" of such figures provided by the Communications Zone.[9]

There appeared to be two main causes for the alarming situation that had developed: inadequate discharge of ships, and a recent decision authorizing First Army to increase its reserves in noncritical items from three to five units of fire. The latter decision had served to

[5] Memo, Hinds for G–3, G–4, and Ord Sec 12 A Gp, 24 Sep 44, sub: Ammo Allocs 27 Sep–5 Oct 44, 12 A Gp 471/1 Ammunition Allocations.

[6] Ltr, Hodges to Bradley, 28 Sep 44, sub: Increased Alloc, 12 A Gp 471 Ammunition; TUSA AAR, II, Ord, 12; Rpt, Ammo, Western Europe, 1–30 Sep 44, Ammo Off FUSA to Ord Off FUSA, 25 Oct 44, Gen Bd file 471/1 Ammunition Supply for Field Artillery, entry 31, Box 43.

[7] 12 A Gp Ord Sec Jnl, 1 Oct 44.

[8] *Ibid.*, 3 Oct 44; Ltr, Lt Col J. P. Daley, Arty Sec 12 A Gp, 10 Sep 44, sub: Rpt of Visit to Eagle Tac, 9 Sep 44, 12 A Gp 319.1 Reports; Rpt, FUSA Ammo Off to FUSA Ord Off, 25 Oct 44, sub: Ammo Supply Rpt, Gen Bd file 471/1 Ammo Supply for Field Artillery.

[9] Ammunition Supply of Field Artillery, Gen Bd Rpt 58, p. 25, citing an inclosure to conference recommendations of 9 October.

draw most of the remaining ammunition into the area of the First Army.[10]

COMZ officials had recognized the seriousness of the unloading situation earlier, and in the last days of September the G-4 and ordnance officials had worked out a plan calling for the unloading of eight Liberties at a time, six of them at Cherbourg and the remainder at the beaches.[11] The Communications Zone had obviously counted on this program in presenting its availability figures on 1 October. But circumstances beyond its control intervened to upset its schedule and discredit its predictions. First, higher authority (presumably SHAEF) ordered berths at Cherbourg freed to give priority to troop debarkations. Then storms virtually stopped operations at the beaches, with the result that an average of only two ships had been worked at a time and barely a thousand tons of ammunition per day had been discharged in the first week of October. This precipitate drop in unloadings, combined with the generally poor discharge record for most of September, had had its inevitable effect on COMZ depot stocks, which were reduced practically to zero in all critical items through shipments to the armies.[12]

Drastic measures were obviously needed to forestall disaster. On 11 October, on the recommendation of the conferees, 12th Army Group forced the most stringent economy in ammunition expenditures yet imposed on U.S. forces by canceling the allocation already in effect for the period 5–13 October and reallocating the available ammunition to the armies with the warning that no additional amounts would be issued until 7 November.[13]

In some cases the amounts now authorized for the thirty-three day period were less than those originally authorized for the eight-day period; and since the allocation was retroactive to 5 October some of the ammunition had already been expended. In making the new allocation the army group included only unobligated balances in base depots, stocks in the continental pipeline, and stocks which were expected to be in field force depots. Because of the uncertainty of discharge it refrained from counting as assets all stocks afloat off the Normandy beaches and ports, quantities which the Communications Zone had ill-advisedly included as available for purposes of allocation.[14] First Army actually had to give up some of the stocks it had built up under the previous authoriza-

[10] Memo, Hinds for G-3 12 A Gp, 10 Oct 44, Gen Bd file 471/1 Ammunition Supply for Field Artillery.

[11] Memo, Clark for G-4 SHAEF, 11 Oct 44, sub: Ammo Allocs to Central Group of Armies for Period 13 Oct–7 Nov, SHAEF G-4 471 Ammunition.

[12] 12 A Gp Ord Sec Jnl, 9 Oct 44; Ltr, Capt W. E. Gunther, SHAEF G-4 Div, to Chief Current Opns Br G-4, 9 Oct 44, sub: Rpt on Ammo Mtg; Memo, Hinds for G-3 12 A Gp, 10 Oct 44, Gen Bd file 471/1 Ammunition Supply for Field Artillery; Memo, Clark for G-4 SHAEF, 9 Oct 44, and unsigned tabulation of statistics on ammunition ships working, discharging, etc., in SHAEF G-4 471 Ammunition.

[13] TWX's, 12 A Gp to Armies, 11 Oct 44, SHAEF G-3 O&E 471 Ammunition; Gen Bd Rpt 58, p. 25. The available ammunition was divided between the armies on the basis of activity factors which gave 55 percent to First Army, 35 percent to Third Army, and 10 percent to Ninth Army. Ltr, Gunther to Chief Current Opns Br G-4, 9 Oct 44, sub: Rpt on Ammo Mtg, Gen Bd file 471/1 Ammunition Supply for Field Artillery.

[14] Memo, Clark for G-4 SHAEF, 11 Oct 44, sub: Ammo Allo to Central Group of Armies for Period 13 Oct–7 Nov, SHAEF G-4 471 Ammunition; TUSA AAR, II, Ord, 14.

tion to accumulate five units of fire in order to bring about a more equitable distribution.[15]

Meanwhile the Communications Zone agreed to step up unloading at once, setting a target of 6,000 tons per day. To meet this goal General Lee informed General Bradley that he was taking action to ensure the working of at least 12 ships at all times, and promised a continuous flow of the required ammunition by not later than 24 October. At this time there were 35 loaded ammunition ships in European waters. The Communications Zone actually gave immediate priority to the discharge of 16 ships—6 at Cherbourg and 10 at the beaches. But bad weather was expected to hold down unloading at the beaches to the equivalent of 2 ships, so that current priorities actually assures an effective discharge of only 8. General Stratton proposed to better this program in the near future by assigning additional berths at Cherbourg, adding berths at Morlaix, working additional ships at the beaches, and transferring ammunition to LST's in the United Kingdom for discharge at Le Havre. These measures entailed a temporary sacrifice in the discharge of other supplies. They also necessitated forwarding ammunition to the armies in bulk, and this would result in receipts of some items not requested. But it was expected that the net effect would be to expedite the delivery of needed items too.[16] The selection of 7 November as the date until which the armies would have to get along on stocks already on the Continent was predicated on the successful achievement of the proposed schedule. Not until that time was it believed that the continental pipelines could be filled and a continuous flow of ammunition assured.[17]

Pending the acceleration of unloadings, the Communications Zone attempted to provide some relief by collecting scattered remnants of ammunition long since left behind by the armies in the base area. Approximately 4,000 tons of the types desired by the field commands, which had not been picked up on COMZ records, were recovered in this way in old dumps in the vicinity of St. Lô, Mortain, and Alençon. Additional quantities were collected and shipped forward from an ammunition supply point recently used by the VIII Corps in Brittany.[18]

In the south the 6th Army Group, which still depended on the North African theater for logistic support, faced similar difficulties as it approached the borders of Alsace. In mid-September the artillery officer of the Seventh Army reported that if, as expected, that command met stronger resistance in the near future it would be "in a hell of a fix for

[15] Memo, Hinds for G–3 12 A Gp, 10 Oct 44, Gen Bd file 471/1 Ammunition Supply for Field Artillery.

[16] 12 A Gp Ord Sec Jnl, 9 Oct 44, TWX E–53841, Lee to Bradley, 12 Oct 44, and Memo, Stratton for Crawford, 12 Oct 44, sub: Discharge of Ammo Ships, SHAEF G–4 471 Ammunition; Ltr, Col H. A. Nisley, 12 A Gp Ord, to G–4 12 A Gp, 30 Oct 44, sub: Estimate of Ord Class V Situation, ADSEC 381 Supply Plan; Col Rapport in COMZ G–4 Plant and Communications Diary/Jnl, 11 Oct 44, ETO Adm G–4 145C.

[17] Ltr, Gunther to Clark, 9 Oct 44, sub: Rpt on Ammo Mtg, SHAEF G–4 471 Ammunition.

[18] Ltr, Nisley to G–4, 30 Oct 44, sub: Estimate of Ord Class V Situation, ADSEC 381 Supply Plan; Memo, Clark for G–4 SHAEF, 9 Oct 44, SHAEF G–4 471 Ammunition.

ammunition." [19] As in the north, lack of adequate transportation and discharge were the most immediate causes for the shortages, and in mid-October the unloading of certain critical types of ammunition was given the highest priority at Marseille. The 6th Army Group, like the 12th, imposed severe limitations on expenditures in order to conserve stocks for the November offensive.[20] At the same time 6th Army Group attempted to obtain ammunition from ETOUSA for the recently transferred XV Corps on the ground that the ammunition requirements for that corps, until recently assigned to Third Army, must have been requisitioned from the War Department by the European theater. But the 12th Army Group insisted that it had agreed to the transfer of the XV Corps, including considerable artillery, only in the belief that the 6th Army Group was better able to support it. In view of ETOUSA's own desperate situation there was no prospect of its coming to the aid of the forces in the south.[21]

The October crisis had precipitated a long overdue reform in the system of control over ammunition issues and expenditures. The procedure in operation up to this time was essentially the one adopted by First Army in mid-June, when the first restrictions were imposed. Under this system First Army had prescribed, usually for four-day periods, the number of rounds of each type of ammunition that each corps could fire. Until the end of July First Army's control was all-embracing, for, as the highest U.S. command on the Continent, it enjoyed complete control over the entire supply machinery from the water's edge to the front line, determining priorities in discharge and the location and level of reserves. When the 12th Army Group became operational on 1 August it continued to allocate ammunition in essentially the same manner, although it increased the rationing period to eight days. Under the system an Ammunition Allocation Committee composed of representatives from the G–3, G–4, artillery, and ordnance sections of 12th Army Group (later augmented to include representatives from SHAEF and the Communications Zone), met periodically and, on the basis of availability forecasts presented by the Communications Zone, established expenditure rates which would provide ammunition for tactical missions and the build-up of the desired reserve. Army group then authorized the armies to draw whatever ammunition was required to build their reserves to the authorized level, but prohibited them from firing more than a specific number of rounds of each type during the allocation period.

This system, though a model of simplicity, had several weaknesses and did not conform to doctrine laid down in field service regulations. For one thing, with a Communications Zone established on the Continent, the army group, unlike the First Army in the first two months,

[19] Gen Bd Rpt 58, p. 23, citing Ltr to Arty Off 12 A Gp, 17 Sep 44.

[20] Memo, Lt Col G. E. Nichols, Arty S–4 SUSA, 22 Nov 44, in Sixth Army Group History (mimeo); History of Delta Base Section, III, App., p. 14; 6th A Gp G–3 Sec Final Rpt World War II, pp. 92–93.

[21] Ltrs, Hinds to Maj Gen John A. Crane of AFHQ, 13 Oct 44, Wilson to Hinds, 19 Oct 44, and Hinds to Crane, 30 Oct 44, all in 12 A Gp 471 Ammunition General; Cbl F–43310, McNarney to Eisenhower, 24 Oct 44, and Cbl S–63913, SHAEF to AFHQ, 25 Oct 44, SHAEF Cbl Logs, Smith Papers.

lacked control over the physical distribution of ammunition and the disposition of reserves. While it prescribed maximum expenditure rates and supply levels, it had to depend on the Communications Zone for figures on ammunition availability. If those figures proved unreliable and shortages developed, the army which first learned of the threatened deficit might monopolize most of the stock of a particular item by requisitioning the quantities needed to achieve its authorized reserve, as had actually occurred in the case of First Army. Secondly, the system made no provision for informing the armies on supply beyond the next allocation period, with the result that all long-range planning was clouded with uncertainty. Finally, the system provided no incentive for conserving ammunition during quiet periods. It thus encouraged wasteful firing, for all ammunition unexpended at the end of an allocation period reverted to army group control. Those that did not shoot up their ration felt justified in falsifying their expenditure reports in order to carry over their savings into the next period as a rainy day reserve.[22]

Twelfth Army Group took the first step in overhauling the entire control procedure on 11 October, when it allocated the existing stocks of critical items on the Continent and established credits for the three armies in the various depots. Ten days later it announced the adoption of a credit system on a permanent basis, to become operative as soon as stocks had again been rebuilt in the forward depots, a step expected to be completed early in November. Under this system the army group proposed to continue making allocations from time to time as before on the basis of activity factors determined by the G–3. Using these allocations it intended to establish credits for each command in designated COMZ depots, copies of the credits being furnished each command and the regulating station serving it. The system's main feature, in contrast with past practice, was that credits would henceforth be written only against ammunition physically present in forward depots. In other words, allocations would be based on stocks actually available for issue and not on quantities on manifest or on expected deliveries. For the next few weeks the armies' tonnage allocations were to be used to rebuild ADSEC ammunition depots—mainly at Liège, Verdun, and Soissons—from which the armies were then to draw on their accounts. The new system gave the army group a more complete control of the distribution of available stocks and gave reasonable assurance to the armies that ammunition allocated to them would actually be available. The new procedure also provided an incentive to save, and thus encouraged prudent shooting, for all ammunition allocated to the armies remained to their credit whether expended or not.[23]

The armies still regarded the system which the army group outlined on 21 October as defective in two respects: like the earlier rationing system, it did not provide the armies with information on future supply, so necessary for

[22] Gen Bd Rpt 58, pp. 54–56; Mechanics of Supply in Fast Moving Situations, Gen Bd Rpt 27, p. 77.

[23] 12 A Gp Rpt of Opns, XII, 141–442; Gen Bd Rpt 58, pp. 57–58; Gen Bd Rpt 27, pp. 77, 83.

planning purposes; and, because the army group still imposed maximum expenditure rates, it gave the armies little latitude in the use of their ammunition. Under the system just announced, rationing was to be retained as part of the credit system. This meant in effect that the armies would be given allocations of ammunition but told that they could only fire at certain rates.

The army group artillery section, headed by Brig. Gen. John H. Hinds, had repeatedly recommended that ammunition be allocated on a credit basis and without limitation on its use. General Hinds argued that rationing was contrary to the basic principle of giving a commander a mission and the means without dictating the details of method. The armies, he maintained, were closer in both time and space to the battle than the army group headquarters, whose only justification for continuing rationing was its knowledge of resupply prospects. He urged the elimination of expenditure limitations, therefore, and proposed instead that the armies be kept fully informed of the resupply situation, and that they be permitted to use their own judgment as to how to expend the ammunition made available to them.[24]

On 5 November the army group adopted this proposal and replaced the former limitation on expenditures with a periodic forecast of future supply. It now began issuing thirty-day forecasts to the armies, taking into account all ammunition on hand and becoming due. Each forecast was divided into three ten-day periods. The amounts shown for the first period constituted a firm commitment. This was divided between the three armies and credits for each were established in the forward depots. The amounts shown for the two succeeding periods constituted estimates of resupply issued for planning purposes alone.[25]

Ammunition could still be drawn only against established credits. Forecasts therefore had to be issued every ten days, consisting of an actual allocation in the form of credits, and revised estimates on future supply. In its first forecast, issued on 6 November, only a few days before the November offensive, the army group informed the armies that it did not plan to establish any reserves of its own and warned that they must now use their own discretion in determining the scale of firing and in establishing reserves to take care of fluctuations in the flow of supply and to meet emergencies.[26]

Neither the credit system nor its accompanying forecast procedure guaranteed an adequate supply of ammunition. But a true credit system had at last been worked out along the lines prescribed in field service regulations in which allocations were based on actual availability, in which the army group possessed adequate command control over the distribution of ammunition, and in which the

[24] Memo, Hinds for G–3, G–4, and Ord Sec 12 A Gp, 4 Nov 44, sub: Ammo Estimates, and Ltr, Hinds to Brig Gen C. E. Hart, FUSA Arty Off, 8 Nov 44, 12 A Gp 471/1 Ammunition Allocations; 12 A Gp Ord Sec Jnl, 4 Nov 44; Gen Bd Rpt 58, pp. 58–59.

[25] Credits and forecasts were based mainly on activity factors prescribed by the G–3 and by the distribution of weapons.

[26] Ltr, 12 A Gp to FUSA, 6 Nov 44, sub: Ammo Supply of FUSA, FUSA 471 Ammunition 1Ar–15, Drawer 2; Ltr, 12 A Gp to NUSA, 6 Nov 44, sub: Ammo Supply of NUSA, 12 A Gp 471 Ammunition, V; 12 A Gp Ord Sec Jnl, 5 Nov 44.

armies enjoyed the maximum freedom in employing the means placed at their disposal.[27] The armies considered the new system a godsend.[28] Whether it could be made to operate successfully and a return to rationing avoided depended, in part, on the prudence with which they exercised their newly won discretion.

The Communications Zone, in the meantime, had gone forward with its program of accelerating discharge and restocking forward depots. It initially estimated that its efforts would not be reflected at the guns until 7 November, although General Crawford thought this unduly pessimistic and believed the date could be advanced as much as two weeks. In any case General Bradley favored restricting expenditures until a flow of ammunition sufficient to support a sustained offensive was assured.[29] Achieving the target of 6,000 tons per day depended largely on what the weather would permit, particularly at UTAH and OMAHA, and when additional berths at Morlaix and Le Havre could be brought in. The necessity of having the equivalent of twelve ships discharging at all times was predicated on an unloading rate of 500 tons per ship per day.

Unloadings improved immediately after SHAEF ordered the step-up on 11 October, but the discharge target was not achieved until the 23d, when 7,617 tons were offloaded. The peak performance was registered on 4 and 5 November, when discharges exceeded 10,000 tons. By that time the Communications Zone had reduced the backlog of ammunition ships in the theater, and unloadings again fell off. Discharge thereafter was to depend on the rate of arrivals from the United States. In the twenty-five-day period from 19 October to 12 November discharges averaged 6,614 tons.[30]

Much of the tonnage discharged in October was shipped directly to the forward areas. To reduce movement time from port to advance depot the Communications Zone decided to bulk-load trains for dispatch to the armies and the Advance Section, thus bypassing the base depots where classification and segregation were normally carried out. While this practice created a new problem in the forward areas—the segregation of ammunition by lot number—it speeded the build-up of stocks in both army and ADSEC depots.[31] There was some fear at first that the emphasis on ammunition shipments might adversely affect the delivery of Class II and IV supplies. But transportation improved steadily in these weeks, and the supply of other items did not suffer.[32] By the time of the November offensive ADSEC depots contained nearly 60,000 tons of ammunition compared with 3,500 tons a month before, and the armies held five units of fire in

[27] See FM 100–10, Field Service Regulations, Administration, 15 November 1943, pars. 75–77.

[28] Ltr, Col Edward T. Williams, TUSA Arty Off, to Hinds, 11 Nov 44, 12 A Gp 471/1 Ammunition Allocations; Gen Bd Rpt 58, p. 63, citing a statement by the FUSA ammunition officer.

[29] 12 A Gp Ord Sec Jnl, 19 Oct 44.

[30] Memo, Clark for G–4 SHAEF, 15 Nov 44, Gen Bd file 471/1 Ammunition Supply for Field Artillery; 12 A Gp Ord Sec Jnl, 11 and 18 Nov 44.

[31] Ltr, Nisley to G–4 12 A Gp, 30 Oct 44, sub: Estimate of Ord Class V Situation, ADSEC 381 Supply Plan.

[32] Memo, Hass for G–4 12 A Gp, 14 Nov 44, sub: Supply of Class II and IV, 12 A Gp Supplies, Misc, 126.

STREET FIGHTING IN AACHEN, *15 October 1944*.

most of the major items of field artillery ammunition.[33]

The crisis in ammunition supply left its mark on the fighting in October. The shortage of ammunition, more than any other factor, determined the character of tactical operations that month. General Bradley had immediately recognized that major offensive operations were out of the question until minimum reserves were reconstituted and a steady flow of ammunition was assured. Except for the action leading to the capture of Aachen by the First Army, therefore, and minor probing attacks, activity was relatively light along the entire front occupied by U.S. forces.

While the ammunition shortage restricted the operations of all three armies in the 12th Army Group, expenditure reports show that Third Army operated under the severest handicap. In one of the worst weeks of the ammunition famine—from 15 to 21 October—firing by the 105-mm. howitzer, the main artillery support weapon of the division, was held to 1.1 rounds per gun per day of action against a desired expenditure rate of 60, and a total of only 3,401 rounds was fired. First Army, by comparison, fired at the rate of 30 rounds per weapon per day and expended a total of 109,469 rounds. Third Army's 155-mm. howitzers fired at the rate of .4 rounds per day, expending a total of 553 rounds, while First Army's fired at the rate of 15 rounds per gun per day, expending a total of 24,341 rounds. Much of First Army's heavier firing was done in the attacks on Aachen, and Third Army's only major action in October—the attack on Metz—

[33] ADSEC Operations History, App. E; Gen Bd Rpt 58, p. 29.

TABLE 5—ARTILLERY AMMUNITION EXPENDITURES, 15–21 OCTOBER 1944

Weapon	First Army		Third Army		Ninth Army	
	Total rounds fired	per day per gun in action	Total rounds fired	per day per gun in action	Total rounds fired	per day per gun in action
105-mm. howitzer M2	109,469	30.0	3,401	1.1	15,946	18.1
4.5-inch gun	2,940	12.0	172	.7	401	2.4
155-mm. howitzer	24,341	15.0	553	.4	2,171	3.6
155-mm. gun self-propelled	2,001	8.0	315	3.8	(a)	(a)
155-mm. gun M1	5,941	10.0	640	2.5	930	5.5
8-inch howitzer	3,819	15.0	391	1.1	284	1.7
8-inch gun	159	2.0	66	1.6	(a)	(a)
240-mm. howitzer	627	3.7	35	.3	(a)	(a)

[a] No data.

Source: Memo, Col T. B. Hedekin, 12 A Gp Arty Sec, for G–3 12 A Gp, 19 Nov 44, 12 A Gp 471 Ammunition General.

had to be called off for lack of ammunition.

Comparative expenditure figures for the major artillery weapons in the three armies for that week are given in Table 5. In the period from 11 October to 7 November Third Army's expenditure in all calibers, which totaled 76,325 rounds, barely equaled its expenditures on a single day at the height of the Ardennes battle in December.[34]

The small ration necessarily forced drastic restrictions on the employment of field artillery. In the XX Corps (Third Army), for example, the artillery commander issued instructions enjoining the use of artillery for anything but counterattacks endangering the battle position, counterbattery against active enemy guns, and observed fire on only the most lucrative targets.[35] Since 75-mm., 76-mm., 3-inch, and 90-mm. tank and antitank ammunition and 40-mm. antiaircraft ammunition were fairly plentiful, all armies turned to tank destroyers, tanks, and antiaircraft weapons for employment in their secondary role as artillery.[36]

Both First and Third Armies also made maximum use of captured enemy guns and ammunition, in some cases equipping American units with German weapons, like the 10.5-cm. howitzer, and firing captured ammunition, in others using enemy ammunition in American weapons, as was successfully done in the case of the 155-mm. howitzer and the 81-mm. mortar. In the last week of October 80 percent of all the ammunition fired by the XX Corps in Third Army consisted of captured ammunition. On 10 October XX Corps in the course of its attacks on Maizieres les Metz fired a time-on-target mission using German 88-mm. guns and 105-mm. howitzers, Rus-

[34] TUSA AAR, II, Arty, 32.

[35] Gen Bd Rpt 58, p. 28.

[36] *Ibid.;* Ltr, Kean to Corps Comdrs FUSA, 15 Oct 44, sub: Fld Arty Firing, FUSA 471 1Ar–20; TUSA AAR, II, Ord, 14; Ltr, Lt Col John Ray, FUSA Ammo Off, to FUSA Ord Off, 12 Nov 44, sub: Ammo Supply Rpt, Western Europe, 1–31 Oct 44, Gen Bd file 471/1 Arty Sec, Entry 39, Box 43.

sian 76.2-mm. guns, and French 155-mm. howitzers, in addition to American 155-mm. guns and tanks and tank destroyers.[37]

Despite these expedients the shortage was seriously felt all along the American front. Unit after unit reported its inability to take targets of opportunity under fire, and complained that the inability to use its artillery took the teeth out of its attacks. The artillery officer of the 35th Division (Third Army) reported that one of the division's regiments had been twice repulsed for lack of artillery support in attempts to take an objective.[38] The commanding general of the VI Corps (Seventh Army), operating in the St. Die area in northern Alsace, reported that he could provide adequate artillery support in the attack to only one division at a time.[39] Artillery support was particularly important during bad weather because of the absence of air support. Bad weather also resulted in more unobserved fire, which in turn involved greater expenditure.

In the entire period of operations on the Continent the month of October provided the clearest case of supply deficiency thwarting tactical operations. In one sense the ammunition shortage epitomized the dire effects of the pursuit, for its immediate causes were largely attributable to the inadequacy of transport and discharge capacity which had resulted from the dash across France.

(2) Contention With the War Department

Theater officials, while attempting to solve the immediate crisis by accelerating the unloading of ships, realized that the ammunition problem had another side. A more ominous shortage threatened, particularly in the heavier calibers and in mortar ammunition, because of inadequate shipments from the United States.

The theater had forewarned the War Department of rising requirements for ammunition as early as March 1944, when it raised its estimates of future needs substantially over the figures it had presented in January, and again in May, when the tactical commands had adopted new "agreed rates," which the theater thereafter used to substantiate its requests for future shipments. Just before D Day the War Department had given assurance that it would meet the theater's "initial requirements," although it predicted shortages in 60- and 81-mm. mortar, 105-mm. howitzer, 8-inch, 240-mm. howitzer, and 155-mm. gun ammunition at D plus 30 and continued shortages in certain categories at D plus 60. The Communications Zone in turn had informed the tactical commands that they could depend on receiving their requirements in virtually all types of ammunition through D plus 70.[40]

As early as July the theater had been forced to request additional shipments

[37] Ammunition Supply and Operations, European Campaign, Gen Bd Rpt 100, p. 23 and App. 7; TUSA AAR, II, Arty, 11; Gen Bd Rpt 58, p. 28.

[38] See observer reports submitted to Maj. Gen. Everett S. Hughes by Lt. Gen. Ben Lear on the adequacy of supply arrangements in the ETO. ETO 400 Supplies, Services, and Equipment.

[39] Rpt, Col Louis T. Heath, G-3 Arty Sec 6 A Gp, to G-3, 23 Nov 44, Sixth Army Group History.

[40] See *Logistical Support I*, 537-38, and Gen Bd Rpt 58, pp. 9-10.

of several types, and early in August the Supreme Commander made a personal appeal for additional releases of the two most critical types—155-mm. howitzer and 81-mm. mortar ammunition. The War Department was able to provide some relief in these categories, but by mid-September U.S. forces faced more serious shortages, mainly as the result of their tactical successes. Late in the month the theater, pointing out that U.S. forces had advanced much faster than expected and now faced the heavily fortified West Wall, reported an urgent need for ammunition in the heavier calibers—8-inch howitzer and gun and 240-mm. howitzer. All three types had already been rationed for several weeks and, according to the theater, if expended at the desired rates—that is, in concentrations required to break through the German defenses—would be exhausted in from fifteen to twenty-five days.[41] The theater concluded that only the immediate shipment of ammunition already set up for future loadings would alleviate the current shortage. It asked that approximately 90,000 rounds be dispatched on two fast freighters without delay.[42]

Within twenty-four hours of the request the War Department announced that it would meet the demand for 8-inch howitzer ammunition in full and would come within a few hundred rounds of filling the need for 240-mm. howitzer ammunition, but that it would fall far short of the requirement for 8-inch gun ammunition, in which production was very low.[43] With these releases the War Department left no doubt that it was according the highest operational priority to the European theater, for these shipments exhausted the stocks of these items in the zone of interior and entailed the diversion of all October and November commitments to other theaters and the suspension of training of newly formed heavy artillery units in the United States.[44]

The late September releases in the heavy calibers went far toward alleviating the shortage in the theater and, together with the speed-up in unloadings, put the theater in a much improved position for the November offensive.

Two weeks later—on 14 October—the theater submitted its requirements for November loading and thereby precipitated a new and more voluminous exchange with the War Department over the reasonableness of the requests. The theater, recalling an argument which had begun in August over ammunition requisitioning practices, and apparently anticipating difficulties, took pains to emphasize that its requirements had been carefully computed and that they should not

[41] The Twelfth Army Group had asked for a supply adequate to permit firing 8-inch howitzer, 8-inch gun, and 240-mm. howitzer ammunition at the rate of 25, 15, and 15 rounds per gun per day, respectively, as against the current ration of 13, 10, and 11 rounds.

[42] Cbl EX-49415, COMZ to AGWAR, 23 Sep 44, 12 A Gp Ammunition, Allocations and Credits, I.

[43] The War Department announced shipments of 56,250 rounds of 8-inch howitzer ammunition against the request for 55,511, 21,490 rounds of 240-mm. howitzer ammunition against the request for 22,130 (a shortage of 640 rounds), but only 3,848 rounds of 8-inch gun ammunition against the requirement of 11,023. Total production of the last in September came to only 5,000 rounds.

[44] Cbl, Marshall to Eisenhower, 24 Sep 44, Memo, Marshall for Byrnes, 25 Sep 44, and Memo for Record, 24 Sep 44, sub: Request for Heavy Arty Ammo (ETO), all in OPD 471 ETO Sec. I, Cases 1-16; Cbl WX-41189, AGWAR to Larkin, 4 Oct 44, SHAEF AG 471-5 Summary of Ammunition Position.

be judged on the basis of past expenditures, since firing had been restricted from almost the beginning because of the nonavailability of ammunition. It admitted that the shortages had resulted primarily from unloading and transportation difficulties within the theater. But it was confident that these problems would soon be overcome and predicted that expenditures would certainly increase.[45]

The War Department was not impressed with the theater's explanations and proceeded to deny a large portion of its requests on the ground of either unavailability or lack of justification for the demands. In the case of 155-mm. howitzer and gun ammunition the War Department stated that it was providing only 56 and 36 percent respectively of the theater's requests because they were in excess of both the authorized theater level based on the War Department day of supply and the War Department allocation of those two critically short items. For the same reasons it offered to ship only 26 and 51 percent respectively of the theater's requests for 81-mm. mortar and 105-mm. howitzer ammunition. In some cases the hard fact of nonavailability simply precluded shipments in the amounts desired. The War Department pointed out that the theater was already getting all the 8-inch gun and 240-mm. howitzer ammunition and almost all the 8-inch howitzer ammunition being produced, and that its request for 105-mm. howitzer ammunition—5,328,000 rounds—was two and one-half times the total October production. In other categories releases could not be increased substantially without serious detriment to other theaters. Finally, ignoring the theater's explanation of its unloading difficulties, it again pointed to the excessive number of vessels awaiting discharge in European waters.[46]

The theater promptly responded with a more detailed justification of its demands. Regarding the problem of idle ammunition ships, it noted that under the recently inaugurated speed-up program all vessels in European waters would be unloaded in twenty or thirty days. It explained, furthermore, that although quantities of certain items presently afloat—particularly 105-mm. howitzers, 155-mm. howitzer, 8-inch gun, and 240-mm. howitzer ammunition—were sufficient to meet current shortages, this was only because past expenditures had either been limited by rationing and inadequate transportation, or had been low because of the nature of operations during the pursuit. Ordnance officials presented figures to show that quantities afloat would not be sufficient to cover shortages had expenditures not been restricted, and that expenditures at the armies' desired rates, or even the authorized War Department day of supply rates, could not have been supported from War Department releases. In brief, shipments from the United States had fallen far short of the theater's requests.

The SHAEF G–4 estimated that all ammunition then afloat would be ashore by 3 November, but that there still would be shortages in all categories at that time. On that date, in other words,

[45] Cbl EX–54310, COMZ to AGWAR, 14 Oct 44, SHAEF AG 471–5 Summary of Ammunition Position.

[46] Cbl WX–48152, AGWAR to SHAEF, 18 Oct 44, SHAEF G–4 471 Ammunition.

the immediate cause of the deficit would shift from inadequate discharges to shortages in the theater. In a memorandum for record General Crawford noted that the theater had begun to warn the War Department of expected shortages in 81-mm. mortar and medium artillery ammunition as early as January 1944, and expressed the opinion that the War Department should by this time have taken action to increase production. Most exasperating of all from the point of view of the theater was the War Department's repeated reference to the fact that past expenditures had been below the day of supply rate, which ignored the theater's argument that past expenditures had been restricted and should not be used as a measure of future needs.[47]

The War Department's challenging questions were inspired in part by its knowledge of production shortages in the United States and in part by the suspicion that the theater's requests were not fully justified. The theater had repeatedly agitated for higher day of supply rates—the rates on which ammunition requisitions and the accounting of stocks in the theater were based. Both theater and War Department officials had recognized that operational experience on the Continent would probably dictate revisions for some items of ammunition. The theater, after consulting with the field forces, had recommended certain changes as early as 21 June despite its lack of conclusive experiential data at that time. The War Department granted some of the desired revisions in August, but it rejected others as unjustified either by expected combat activity or by expenditure reports from the Mediterranean area.[48]

In a sense all subsequent discussion of changes in the day of supply rates was academic, for dominating the War Department's attitude was the grim fact that current production in the United States simply could not meet the mounting demands from overseas theaters. In fact, the August revisions did not in reality go into effect. The War Department approved them with the warning that ammunition was not immediately available in sufficient amounts to permit shipments at the new rates, and cautioned the theater that the increases should in no way be considered as a basis for requests to achieve maximum authorized levels in the theater.[49] Moreover, there had been no real test of the adequacy of current rates, for ammunition was never supplied at the established rates. Late in September, when the 12th Army Group asked the armies for recommended changes in the supply rates based on their combat experience thus far, First Army replied that it did not feel competent to propose changes, for there was no way of knowing whether the authorized rates would have been adequate in view of the fact that ammu-

[47] Cbl S-63033, SHAEF to AGWAR, 19 Oct 44, Ltr, Crawford to Handy, 20 Oct 44, sub: Rpt on Ammo in Short Supply, and Memo for Record, Crawford, 19 Oct 44, all in SHAEF G-4 471 Ammunition; Report on Ammunition in Short Supply, prep in Office Chief Ord Off COMZ, 17 Oct 44, Gen Bd file 471/1 Ammunition Supply for Field Artillery.

[48] Ltr, AGWAR to CG ETO, 25 Aug 44, sub: Estimated Ammo Requirements, EUCOM 471/1 Ammunition Policy. The new rates were officially communicated to the theater on 1 September 1944. Ltr, Secy of War to CG ETO, 1 Sep 44, sub: Day of Supply for Ammo, ADSEC 471 Ammunition.

[49] Ltr, AGWAR to CG ETO, 25 Aug 44.

nition had never been supplied at those rates.[50]

Fundamental to the arguments over the adequacy of ammunition supply, and tending to confirm War Department doubts as to the legitimacy of the theater's requests were the differences in interpretation of the ground rules governing the calculation of requirements to cover shipping and distribution time and, to a lesser extent, the accounting of stocks actually in the pipeline. The same problem had arisen in connection with major Class II and IV items in which the loss rates were unpredictable, such as tanks.[51] The War Department had authorized ETOUSA to have between forty-five and seventy-five days of supply of ammunition on hand in the theater. Experience in the first two months, as the theater had pointed out as early as August, had shown that a minimum of fifteen days was required for discharge and shipment to a depot before ammunition could be considered available for issue. Interruptions in supply, losses from enemy action, and variations in expenditure further reduced the quantities available in the forward areas. Thus, although as many as seventy-five days of supply plus the number of days of supply representing the shipping time might be released to the port of embarkation, there was no assurance that the minimum of forty-five days of supply would actually be available for issue on the Continent. Ammunition in the zone of interior or on manifest, the theater argued, was not an asset until received. To achieve even the minimum level required for adequate support of the field forces necessitated that shipments be based on the seventy-five day authorized level plus anticipated expenditures through the date of arrival in the theater. Since requirements were placed three months in advance, therefore, they included not only quantities needed to achieve the authorized level, but expenditures expected from the time the requisition was placed until the time of delivery. This explained why requests exceeded the level based on the War Department day of supply rate.[52]

Furthermore, the theater had not counted basic loads of ammunition—the amounts which individuals and unit vehicles were allowed to carry—as part of the theater's assets. To include them, the theater argued, only distorted the true status of ammunition supply. Ammunition in the hands of troops, it maintained, was lost insofar as theater stockages were concerned, for it was a frozen asset which could not be moved from

[50] Ltr, 12 A Gp to Armies, 23 Sep 44, sub: Day of Supply for Ammo, with replies, 12 A Gp 471 Ammunition Day of Supply.

[51] See *Logistical Support I*, 522–23, and see above, Ch. VIII, Sec. 3.

[52] Ltr, ETO to ASF, 31 Aug 44, sub: Ammo Rqmts for October Loading, EUCOM 471 Allocations of Ammunition, II; Cbl WARX–26353, WD to ETO, 6 Sep 44, Cbls, SS&P Plng Div 201.02, A46–371; Cbl S–63033, SHAEF to AGWAR, 19 Oct 44. A later analysis of ammunition shipments covering the period July 1944–January 1945 showed that an average of 54 days elapsed between the submission of a requisition on the zone of interior and the arrival of the first rounds in theater waters, and that an additional 24 days elapsed before the ammunition actually became available for issue: 14 to discharge, 7 to move to an intermediate or advance section, and 3 to inventory and transmit records to the army group. An average of 78 days therefore elapsed before the first round was issued as the result of a theater request. Memo, Lt. Col. J. B. Goodell, Deputy Chief, Ammo Supply Div 12 A Gp Ord, for ExO, 29 Jan 45, sub: Levels of Supply in ETOUSA, 12 A Gp 319.1 Reports.

one unit to another to meet changing tactical conditions. Moreover, basic loads represented substantial quantities when compared with the seventy-five-day level, ranging from five to ten days in the case of artillery ammunition and from ten days upward in the case of small arms ammunition. To count these as part of the total theater supply level would limit the actual reserves to a dangerous level. The theater protested strongly against policy.[53]

These interpretations of the rules were unacceptable to the War Department. It agreed with the theater that ammunition afloat between the port of embarkation and ETOUSA ports should not be counted against the theater's authorized level. For accounting purposes it considered ammunition as part of the theater's stocks only after the vessels had come under theater control. But it did not agree with the theater's insistence on maintaining a level of forty-five to seventy-five days physically on the Continent, and it rejected the argument that basic loads should not be included as part of total theater stocks. The War Department had intended that the authorized maximum level—seventy-five days in ETOUSA's case—should represent all stocks under the theater's control whether afloat or in depots. It appeared that for each thirty-day period the theater was requisitioning three months' requirements without taking into account the War Department's own system for keeping in the pipeline the amounts necessary to offset time for processing orders and shipping. By its own calculation the War Department contended that there already was enough ammunition en route or scheduled for loading to maintain the theater's level until 1 December.[54]

For the moment, at least, this argument was unresolved. In the meantime General Eisenhower, faced with the demands for the November offensive, personally cabled General Marshall on 20 and 22 October in an attempt to convince the War Department of the gravity of the theater's ammunition situation. In the view of the Supreme Commander, three facts stood out: Certain types of ammunition were unavailable because of insufficient production in the United States; the War Department in the past had not released the quantities requested; and now the War Department was also cutting down on the lift allocated to the European theater.[55]

Once more General Eisenhower emphasized that the theater could not be assured of an adequate supply as long as shipments were based on past expenditures in view of the necessity to ration heavy calibers since shortly after D Day. It appeared clear to him that every expedient must be applied to step up production. Referring to the speed-up in unloadings, he stated that that program could be supported for the time being with the ships then on hand and the twenty arriving in November, but not beyond the end of the month in view of the cut in sailings. General Eisenhower stressed that the uncertainty of ammunition supply was worrying commanders

[53] Cbl EX-60321, Lee to AGWAR, 4 Nov 44, SHAEF AG 471-5 Summary of Ammunition Position; Cbl E-52905, Lee to AGWAR, 7 Oct 44, EUCOM 471/1 Ammunition Policy.

[54] Cbl W-49721, AGWAR to SHAEF, 20 Oct 44, SHAEF G-4 471 Ammunition.

[55] See above, Ch. IV, Sec. 4.

at all levels, and that tactical plans for the immediate future hinged on the assurance of an adequate supply. He urged the immediate shipment of 75,000 tons of mortar and artillery ammunition, promising the highest unloading priority and a speedy turnaround of shipping.[56]

Once more the Supreme Commander's personal appeal brought results. The Army Chief of Staff promptly assured General Eisenhower that everyone in the War Department was cognizant of the theater's problem and realized the need for "generous supply and firm commitments." General Marshall saw no prospect of increasing the October and early November loadings in the critical calibers—that is, 105-mm. and larger and 81-mm. mortar ammunition. But he believed that substantially all of the theater's needs to the end of December could be met, and he promised that ETOUSA would get the maximum quantities becoming available, subject only to meeting the minimum operational requirements of other theaters.

Two aspects of the problem still concerned him. He pointed to the fact that the nineteen ships already in European waters plus those en route would give the theater 3,000,000 rounds of 105-mm. ammunition and 1,200,000 rounds of 81-mm. mortar ammunition, implying that the situation was not as desperate as suggested. In addition, the Chief of Staff deplored the divided responsibility for supply of the northern and southern forces in the theater. Requisitioning by the North African theater for the 6th Army Group made duplication possible, and, in the case of critical items, might result in unnecessary deficiencies elsewhere. He asked that requisitions submitted by General Devers to the North African theater for the southern armies be cleared and approved by the European theater before submission to the War Department. Better still, he suggested that a unified supply system with undivided responsibility would be welcome.[57] Meanwhile he assured General Eisenhower that the top officials of the War Department from the Secretary of War down were personally working on the ammunition problem.[58]

(3) The November Offensive and the Bull Mission

In the meantime Generals Bradley and Devers launched the November offensive with the ammunition which the armies had so painstakingly accumulated and husbanded during October. Early in the month the 12th Army Group announced the activity factors which were to govern the distribution of all

[56] Cbls S-63259 and E-56546, Eisenhower to Marshall, 20 and 22 Oct 44, OPD Cbl Files.

[57] Special shipments had only recently been made to southern France for overland delivery to troops in the north, and additional shipments were made to that area for troops transferred from the 12th to the 6th Army Group. The War Department believed these to be duplicate requisitions. Responsibility for the supply of the 6th Army Group was actually assumed by the Communications Zone on 20 November, and requisitions from the two army groups were henceforth to be consolidated and submitted by the Communications Zone. Memo for Record, unsigned, 17 Oct 44, sub: Status of Arty Ammo, OPD 471 ETO Sec. II, Cases 17-30; Ltr, COMZ to SAC, 25 Nov 44, sub: 6th A Gp Ammo Supply, SHAEF AG 471-5 Summary of Ammunition Position.

[58] Cbl W-50677, Marshall to Eisenhower, 22 Oct 44, Eyes Only Cbls, Smith Papers; Cbl W-51013, Marshall to Eisenhower, 23 Oct 44, and Ltr, Lutes to Crawford, 25 Oct 44, in SHAEF G-4 471 Ammunition.

ammunition on hand on 7 November and becoming available in the next three weeks. For active divisions in the First and Ninth Armies the G–3 assigned a factor of .60, for active divisions in Third Army a factor of .40, and for all inactive divisions a factor of .25. On this basis the number of active and inactive divisions in each army determined that 50.1 percent of the ammunition would go to First Army, 22.8 percent to the Ninth, and 27.1 to the Third. This allocation did not take into account the variation in the number of guns in each army, and the rates of fire therefore bore little relationship to the activity factor. But since it did take into consideration the number of divisions, it ensured that an active division on the First Army front would be supported by about the same number of rounds of ammunition as an active division in the Ninth Army's sector despite variations in the rates of fire for individual weapons.[59]

All five armies in the 12th and 6th Army Groups, including the U.S.-supported First French Army, made fairly heavy expenditures of artillery ammunition, particularly in the initial attacks. In the Third Army the XII and XX Corps, with Metz, and eventually the Saar as their objectives, did their most active firing since their commitment in August. For several days Third Army's 540 105-mm. howitzers fired at the rate of about forty-five rounds per gun per day. Ninth Army's expenditures for the same weapon exceeded seventy rounds for a few days at the beginning of its attacks toward the Roer River.[60] Total expenditures of high explosive shells for the 105-mm. howitzer M2 in November came to 2,507,000 rounds, the highest expenditure of any month thus far.[61]

But ammunition was far from plentiful. In the north the knowledge that neighboring British units were better supplied at least partially explained the decision to place the American 84th Division under the control of the British XXX Corps, which was better able to give it adequate artillery support.[62] First Army complained that only its VII Corps, driving toward the Roer, could fire "reasonable amounts." Third Army reported that its expenditure record merely reflected ammunition availability, not the rates at which the army desired to fire. In the Seventh Army, whose 648 105-mm. howitzers fired at the highest rate thus far—forty-nine rounds per weapon per day—the artillery officer exercised a rigid control over expenditures, restricting firing immediately after the initial break-through in the Sarrebourg and Belfort areas in fear of future shortages at the Siegfried Line and Rhine.[63]

Doubts over future supply were not unfounded. Even before the launching of the November attacks the theater had had additional warning that the ammunition crisis would continue. On 30 October, only a week after its response to Gen-

[59] Memo, Hinds for G–3, 9 Nov 44, 12 A Gp 471/1 Ammunition Allocations; 12 A Gp Ord Sec Jnl, 10 Nov 44.

[60] Memo, Hedekin, Arty Sec 12 A Gp, for SHAEF G–4 Ln Off, 21 Nov 44, 12 A. Gp 471 Ammunition General.

[61] History of Planning Division ASF, App. 18–A, p. 152.

[62] *Conquer: The Story of Ninth Army, 1944–45*, p. 83.

[63] Gen Bd Rpt 58, pp. 29, 32–33; TUSA AAR, II, Ord, 16; Memo, Col C. W. Stewart, Deputy Chief Current Opns Br G–4 SHAEF, for G–4, 25 Jan 45, SHAEF G–4 471 Ammunition.

eral Eisenhower's appeal, the War Department notified the theater that the recent demands from the European as well as other theaters had completely drained zone of interior stocks of certain items. In some cases releases had actually obligated total anticipated production through 10 November because of unforeseen shortfalls in October production.[64]

A week later the War Department spelled out its warning in greater detail, informing the theater that despite increasing production, it had grave doubts about meeting ETOUSA's demands in the next three to six months. Current shipments, it noted, were being scheduled directly from production lines to dockside, and did not equal the theater's anticipated rates of expenditure. Stocks already in the theater plus quantities en route thus represented total resources. The theater, the War Department cautioned, must plan its expenditures in the light of these facts, and with the full understanding that any expenditures in excess of the department's announced resupply rate must be supported from theater reserves and could not be replaced from the United States. The War Department saw little possibility of improvement within the next ninety days, for it estimated that any increase in production would be matched by the theater's rising troop strength.[65]

The War Department's response to the recent appeal had obviously been a stopgap measure which had required scraping the barrel of U.S. ammunition stocks. It had not altered long-range prospects. In view of the War Department's previous warnings the latest forecast therefore should hardly have occasioned surprise. But the argument in October over the reasonableness of ETOUSA's requests had obviously obscured, for the theater at least, the fact that ammunition simply was not available in the desired quantities. Theater officials were incredulous over the latest forecast, and in expressing their disappointment repeated their previous argument that past expenditures should not form the basis for computing the theater's needs, ignoring the hard fact of production shortfalls.[66]

A few days later the theater, seemingly unaware of the recent communications from the War Department, submitted its requirements for December loading on the basis of rates which it had recommended more than a month before. General Somervell made no attempt to conceal his annoyance with the theater's action which, he said, "had been taken without any regard whatsoever to the information which had been supplied you." He suspected that either the theater was refusing to face reality or that its left hand did not know what its right was doing, and he asked that it recompute its needs on the basis of minimum needs rather than optimum supply conditions.[67]

The theater in the meantime had passed the War Department's 6 November forecasts on to the army groups,

---

[64] Cbl WARX-54796, WD to ETO, 30 Oct 44, Cbls SS&P Planning Div 201.02, A46-371.

[65] Cbl WX-58388, Somervell to Eisenhower, 6 Nov 44, SHAEF G-3 O&E 471 Ammunition.

[66] Cbl E-62083, ETO to WD, 10 Nov 44, Cbls, SS&P Planning Div 201.02, A46-371.

[67] Cbl E-63957, ETO to WD, 15 Nov 44, and Cbl WARX-64914, WD to ETO, 18 Nov 44, Cbls, SS&P Planning Div 201.02, A46-371; Ltr, Somervell to Lee, 18 Nov 44, Hq ASF European Theater—1st Half 1944.

where they were received with understandable dismay.[68] The resupply potential announced by the War Department had reduced the day of supply rate by an average of 50 percent in calibers of 155-mm. and larger, and in the case of 105-mm. howitzer ammunition to 18 rounds per gun as compared with the ETOUSA rate of 40 and the MTOUSA rate of 50.[69]

The 6th Army Group, because it received its support through the Mediterranean theater and enjoyed a slightly higher day of supply rate in most items than units in the north, estimated that the new rates would provide only about one third of its actual needs and would therefore inevitably affect the scale of operations. The reduction was all the more serious, it noted, because air support was habitually voided by adverse weather.[70]

General Bradley, translating the predicted resupply rates into tactical capabilities, estimated that the ammunition on hand and in sight for the next month would permit the 12th Army Group to continue its current offensive until about 15 December. Reserves would be practically exhausted by that date, and the resupply rate for the two critical calibers—105-mm. and 155-mm. howitzer ammunition—would then force his armies to revert to static operations. It would be insufficient even for such operations, he thought, if his forces faced an enemy capable of offensive action. Crossing the Rhine with such supply was out of the question unless enemy resistance collapsed.[71] On 22 November General Eisenhower transmitted this estimate word for word to the Army Chief of Staff.[72]

The Supreme Commander, recognizing that the ammunition problem had now reached a crucial stage, had already decided to send two high-ranking officers to Washington to place before General Marshall the exact supply situation and indicate the effect of the shortages on projected operations. In preparation for that mission he instructed General Clay, who had only recently arrived in the theater and was then commanding Normandy Base Section, to make a thorough study of the theater's assets and estimated expenditures, and asked Maj. Gen. Harold R. Bull to analyze the tactical implications of the supply outlook.[73] The two officers immediately proceeded to the 6th and 12th Army Group headquarters to lay before them information avail-

[68] Cbl S–66466, SHAEF to 12 and 6 A Gp, 12 Nov 44, 12 A Gp 471 Ammunition.

[69] Gen Bd Rpt 58, pp. 31–32.

[70] Sixth Army Group History, November, p. 45.

[71] Memo, Bradley for Eisenhower, 21 Nov 44, 12 A Gp 471 Ammunition. A COMZ analysis estimated that the War Department's announced resupply potential would permit the following expenditures of 105-mm. howitzer ammunition through April 1945: 26.8 rounds per weapon per day if no reserves were maintained, 25 rounds with a five-day reserve, and 23.3 rounds with a ten-day reserve. This assumed an average of 2,341 weapons of this caliber for the period. The 12th Army Group's desired rate for the 105-mm. howitzer was 52.2 rounds. For the 155-mm. gun the corresponding figures were 19.1 rounds without a reserve, 18.3 rounds with a five-day reserve, and 17.5 rounds with a ten-day reserve. The total number of weapons of this caliber was assumed to average 345. The desired rate for the 155-mm. gun was 23.25. Ltr, Lord to Crawford, 17 Nov 44, sub: Ammo Availabilities through 1 May 1945, EUCOM 471 Ammunition Availabilities III, Special Files.

[72] Cbl S–67807, Eisenhower to Marshall, 22 Nov 44, SHAEF G–4 471 Ammunition.

[73] TWX EX–65520, ETO to 6 and 12 A Gps, 20 Nov 44, 6 A Gp 471–1 November 44.

TABLE 6—12TH ARMY GROUP ARTILLERY AMMUNITION EXPENDITURES, 6 JUNE–22 OCTOBER 1944, COMPARED WITH DAY OF SUPPLY RATES

Weapon	Rounds Fired by 12th A Gp			Day of Supply Rate (rounds per gun per day)	
	Total	Per day per gun in action	Per day per gun in A Gp	12 A Gp desired rates	ETO Day of Supply rate
105-mm. howitzer	3,012,568	27.0	21.9	60	40
4.5-in. gun	131,837	17.9	13.3	30	28
155-mm. howitzer	827,462	17.0	13.8	40	25
155-mm. gun SP	62,306	12.6	9.1	25	25
155-mm. gun M1	219,070	14.7	11.7	25	25
8-in. howitzer	8,067	7.9	5.9	25	20
8-in. gun	1,086	5.2	3.1	12	15
240-mm. howitzer	15,478	5.7	3.6	15	7

Source: Memo, Hinds for Moses, 19 Nov 44, Memos—Moses, SHAEF 12 A Gp G–4, Folder 86.

able to the theater and to obtain data on supply in the combat zone.

The reaction was a familiar one. Presenting detailed statistics and charts, the 6th Army Group showed how miserly it had had to be in its expenditures for weeks in order to build up the three units of fire necessary for the November offensive, and how, after firing these stocks in the initial attacks, it had again had to tailor its operation to the small trickle of supply. Continued resupply at the same rates, it maintained, would permit nothing better than an active defense.[74]

General Moses had gathered together the 12th Army Group expenditures record in the main types of artillery ammunition for the entire period from 6 June to 22 October to show that in no case had expenditures been possible at the theater day of supply rate and in most cases had been only one third to one half of the desired expenditure rates. *(Table 6)* General Bradley still hoped to be provided with sufficient ammunition to permit firing at the "desired rates" during the current offensive, or until mid-December. But the desired rates, on which the 12th Army Group had based requirements in the past, were obviously unattainable for long-range supply and actually had little relationship to the average expenditures over a long period. Considering the number of inactive divisions and the periods of regrouping and light fighting, General Bradley estimated that a long-term maintenance rate, if maintained for several months, would permit the accumulation of ammunition in sufficient quantities to cover the heavy requirements during periods of offensive operations. He immediately queried the armies on this proposal and then recommended the following long-range maintenance rates in

[74] Memo, 6 A Gp for Bull and Clay, 23 Nov 44, sub: Ammo Consumption Rqmt Analysis, EUCOM, SHAEF Plans and Operations, 23 Nov 44–26 Jan 45; 6 A Gp G–3 Sec Final Rpt, p. 95.

place of the old desired expenditure rates for the four most critical items:

105-mm. howitzer	40	rounds/gun/day
155-mm. howitzer	40	
155-mm. gun	23.5	
8-inch howitzer	25	

In every case the rates recommended by the 12th Army Group commander still exceeded the resupply potential announced by the War Department. Moreover, the army group also wanted fourteen days' reserves established in ADSEC depots. Considering the need for a theater reserve and working margins in the Communications Zone, General Moses concluded that the outlook was "pretty sad." The history of ammunition supply thus far indicated to him that the theater had been permitted to embark on Operation OVERLORD without any certainty of receiving sufficient ammunition to carry on operations against continued stiff resistance. He suspected, furthermore, that the War Department, despite the theater's repeated protestations, had fallen into the habit of editing the theater's requisitions on the basis of past expenditures. These, he maintained, had no bearing whatever on the problem of future supply, and he strongly opposed having General Clay use such data in his discussion with the War Department or having him attempt any estimate as to what the armies could or could not do with the ammunition made available at the new rates. In Moses' opinion Generals Clay and Bull should do no more than present the theater's stated requirements and assert that failure to meet them would seriously hamper or halt offensive operations.[75]

Generals Bull and Clay flew to Washington at the end of November, and within a few days General Bull reported generally satisfactory results for the mission. The theater's immediate crisis was to be resolved by three expedients: some ammunition found by the ASF which could be rapidly reconditioned was to be shipped promptly; by various shortcuts and special handling the delivery time for all ammunition was to be reduced; and an all-out effort was to be made to assemble all components on hand and thus increase the total production output for December and January. These measures promised to improve the supply potential in all calibers, and particularly in 105-mm. howitzer ammunition, in which the increase was expected to sustain the desired maintenance rate through April 1945. In addition, the War Department was to make an effort to bring new capacity then under construction into production at an earlier date than then scheduled. General Bull reported excellent co-operation from the staffs of General Somervell and the Chief of Ordnance, and returned to the theater satisfied that the War Department was making every effort within its power to meet ETOUSA's needs.[76]

The War Department's steps to boost the production of field artillery ammunition actually antedated the theater's most recent appeal by several months. In the

[75] Memo for Record, Moses, 21 Nov 44, sub: Ammo Rates, Memo, Moses for Bull, 21 Nov 44, Memo, Moses for Bradley, 22 Nov 44, and Memo for Record, Moses, 26 Nov 44, sub: Visit of Bull and Clay, all in Memos—Moses, SHAEF 12 A Gp G–4 Folder 86; Ltr, Moses to Bull, 26 Nov 44, OPD 471 ETO Sec. I, Cases 1–16.

[76] General Clay did not return to the theater. Cbl W–71199, Bull to Smith, 1 Dec 44, SHAEF G–4 471 Ammunition; Memo, Bull for Eisenhower, 3 Dec 44, sub: Critical Ammo Items, Gen Bd file 471/1 Ammunition Supply for Field Artillery.

fall of 1943 the War Department had ordered a cutback in production under the pressure of criticism from a Congressional committee because of excess accumulations of stocks, particularly in the North African theater. The excess in North Africa had resulted from the automatic shipment of ammunition on the basis of empirical day of supply data which failed to reflect the relatively inactive status of weapons over long periods of time. Early in 1944 the demands for ammunition rose precipitately as the result of the increased tempo of fighting on all fronts, and particularly as the result of unexpectedly high expenditures in Italy and ETOUSA's upward revisions of its requirements for the coming invasion. These developments led the Planning Division of the ASF, after a thorough survey of the ammunition situation, to predict a critical shortage in mortar and medium and heavy artillery ammunition by November.

On this forecast the War Department in April began allocating ammunition on the basis of the number of active weapons in each theater. Within another month, after additional studies and recommendations from the various ASF divisions, the War Department General Staff assigned the highest priority to the construction of additional production facilities for ammunition, and also for guns. The War Production Board immediately issued the required directives to make basic materials and machine tools available. Tooling up for ammunition production was a complicated precision job, however, the manufacture of the 155-mm. shell alone requiring about forty separate operations. Even experienced manufacturers ran into trouble on such jobs, as the lag in production of 8-inch ammunition had shown. Meanwhile the War Department pressed for the maximum output with existing facilities, making the necessary manpower deferments and even furloughing men from the service to work in munitions plants.[77]

These measures were only beginning to be reflected in production increases when the theater made its urgent appeals in November, and the actions which the War Department could take to effect an immediate acceleration in the flow of ammunition were limited to the expedients mentioned above. Earlier in November, in an attempt to provide additional incentives to production in existing facilities, and to publicize the urgency of the ammunition situation, the Secretary of War had suggested that the theater send back artillery crews as special emissaries for a tour of production centers. Late that month, during the Bull mission, the theater responded by flying one mortar crew and two artillery gun crews, comprising twenty-seven enlisted men, to the United States.[78]

(4) *Ammunition Supply in December and January*

The emergency measures taken as a result of the Bull mission were to have

[77] History of Planning Division ASF, App. 18–A, Ammunition Supply for the European and Mediterranean Theaters, prep by ASF, 15 Aug 45, pp. 25, 32, 242–45; Cbl 302158Z, AGWAR to SHAEF, 30 Nov 44, EUCOM 471 Ammunition, General, III; Rpt, Hiland G. Batcheller, Critical Programs: A Report to the WPB, 14 Nov 44, Hq ASF, President–White House–Exec Office 1944, Somervell file.

[78] Cbl WX–58187, AGWAR to SHAEF, 6 Nov 44, and Cbl EX–62078, COMZ to AGWAR, 24 Nov 44, both in SHAEF AG 471–5 Summary of Ammunition Position.

no effect on ammunition availability in Europe for several weeks. Meanwhile the armies met the enemy's December onslaught with stocks already on hand and did some of the heaviest firing to date. Credits went out the window in the face of emergency needs of hard-pressed units, and allocations to the subordinate corps and divisions were suspended. Instead, Third Army, for example, adopted the practice of issuing informal status reports of critical items which corps commanders then used as a guide to expenditures. For the most part firing was unrestricted. In the First Army, expenditures of 105-mm. howitzer ammunition rose to 69 rounds per gun per day as compared with 13 rounds in the period of the pursuit and a previous high of 44 in November. For the 155-mm. howitzer expenditures were at the rate of 44 rounds as against 8 and 29 respectively in the earlier periods.

In the most of the major categories the firing in December was the heaviest of any month thus far. The result was to increase the gap between authorized and actual levels in the theater to the widest it was to reach during the war. In the case of the high explosive shell for the 105-mm. howitzer M2, the December expenditures, totaling 2,579,400 rounds, reduced theater stocks to 2,524,000 rounds against an authorized level of 8,900,000. In terms of days of supply at War Department rates this stockage represented only twenty-one days as against the authorized seventy-five.[79]

In the 6th Army Group expenditures did not soar until early in January, when the enemy launched his counteroffensive in the Hardt Mountains area. Firing continued heavy with the launching of the operation to clear the Colmar Pocket. The sustained firing during the month placed a heavy drain on army group reserves. At the end of the operation on 8 February 6th Army Group concluded that it would have to conserve ammunition in the heavier calibers for a full thirty days before undertaking additional offensive operations. Allocations were accordingly cut to one half of the SHAEF day of supply rate for the next month. Unallocated ammunition was used to rebuild the army group reserve.[80]

The heavy firing in December and January was thus supported only by drawing heavily on reserves. The flow of ammunition was anything but plentiful, and the theater continued to resort to various expedients to supplement normal supply and augment the armies' fire power. Late in November the First Army, using personnel from the 32d Field Artillery Brigade, formed two provisional battalions in order to make use of forty-eight German 105-mm. gun-howitzers and 20,000 rounds of captured

[79] In weight of rounds expended, the high explosive shell for the basic field artillery piece—the 105-mm. howitzer M2—topped all other ammunition items. Expenditures of this shell eventually accounted for 34.5 percent of the total tonnage of artillery ammunition expended by U.S. forces during the war in Europe (498,200 tons of a total of 1,446,400). Shells for the 105-mm. howitzer M1 accounted for another 14.7 percent (213,300 tons) of the total tonnage. History of Planning Division ASF, App. 18–A, p. 93.

[80] TUSA AAR, II, Ord, 8; FUSA Rpt of Opns, 1 Aug 44–22 Feb 45, Bk. III, p. 26; Ammunition Supply Report Western Europe, 1–31 December, prep by Ammo Off, Ord Sec FUSA, 24 Jan 45, Gen Bd file 471/1 Ammunition Supply for Field Artillery; 6 A Gp G–3 Sec Final Rpt; TWX BX–23700, Devers to SHAEF G–3, 30 Jan 45, SHAEF SGS 472 Artillery and Ammunition, II.

ammunition turned over to it by the Communications Zone. The captured guns and ammunition were used to good effect in harassing and interdictory missions during the Ardennes fighting. First Army's 155-mm. gun battalions also made use of approximately 7,500 rounds of captured 15.5-cm. ammunition.[81] Meanwhile the Communications Zone arranged with 21 Army Group for the loan of one hundred 25-pounders along with sixty days' supply of ammunition to the 12th Army Group, which divided them between the three armies.[82]

While the additional shipments arranged for by General Bull helped shore up the theater's ammunition position, they were insufficient to permit firing at the 12th Army Group's desired long-range maintenance rates. In December the theater informed the field commands that the supply potential for 105-mm. howitzer ammunition would be 26 rounds per gun per day as compared with the desired 45, for the 155-mm. howitzer 19.5 as against the desired 33, for the 155-mm. gun 13 instead of 25, and for the 8-inch howitzer 5.5 rather than 25.[83] In practice the theater actually bettered this forecast somewhat in the three months beginning with January, the maintenance rates for the four critical calibers averaging 29.6, 23.4, 13.3, and 5.5 respectively.[84]

But the supply potential continued to fall short of desired maintenance rates, and the entire ammunition situation remained tight. Early in January General Somervell, after a personal survey of the situation in the theater, appeared fully convinced of the theater's needs and asserted in no uncertain terms that the resources of the United States must be applied to whatever extent was necessary to the urgent production of as much ammunition of critical calibers as could be produced in the shortest possible time. "There are not enough 'A's' in all the alphabets in the United States," he cabled Washington, "to point up the necessity for this too strongly." [85] The critical need for ammunition in the European theater had a direct influence on the War Department's decision, made only a few days later, to cancel plans for the mass production of pilotless aircraft (the JB-2), intended for use against industrial targets in Germany, because of the inroads such a program would have made on labor and materials then committed to the manufacture of field artillery ammunition.[86]

The adoption of the long-range maintenance rate as a means of forecasting

[81] FUSA Rpt of Opns, 1 Aug 44–22 Feb 45, Bk. III, p. 28.

[82] The loan was renewed for another sixty days at the end of February. Cbl EX-85176, COMZ to 21 A Gp, 11 Jan 45, Cbl SD-1159, 21 A Gp to COMZ, 4 Mar 45, Memo, Arty Sec 12 A Gp for G-4, 4 Jan 45, and Memo, Arty Sec to Ord and G-4, 26 Feb 45, all in 12 A Gp 472 Cannons and Field Pieces.

[83] Gen Bd Rpt 58, pp. 35–36; Ltr, Gen Leven C. Allen to CG ETO, 16 Dec 44, sub: Long-Range Maintenance Rate of Supply of Mortar and Arty Ammo, 12 A Gp 471 Ammunition, V.

[84] For the 8-inch howitzer the figure 5.5 is deceptively low, for the rate is based on a large number of weapons which were never active. Expenditures by the 12th Army Group in this caliber actually averaged 13 rounds per active gun per day. Gen Bd Rpt 58, pp. 37–38.

[85] Memo, Marshall for Byrnes, 13 Jan 45, and attchd Cbl, Somervell to Styer, 11 Jan 45, CofS 400, Sec. I, Cases 2–35.

[86] Memo for Record, Lt Col C. E. Hutchin, 24 Dec 44, Cbl 82397, Marshall to ETO, 24 Dec 44, and Cbl 25880, Marshall to Eisenhower, 23 Jan 45, all in OPD 471.6 TS Sec. II, Cases 33–37.

future supply was accompanied by further changes in the control of ammunition distribution within the theater. After discussions earlier in the month, SHAEF on 20 December announced that it was assuming control of all ammunition resources in the theater, and proposed to exercise its control by establishing a maintenance day of supply rate and a reserve for each weapon. The reserve, it was proposed, would be held by the theater and would be made available to SHAEF to meet unforeseen contingencies or to reinforce the operations of a specified army group. The maintenance day of supply was in reality the available day of supply potential for each weapon and was calculated by considering as available all stocks then in the Communications Zone and SOLOC and the anticipated future supply from the zone of interior. SHAEF at first attempted to establish its control retroactively to 1 December. This placed the 6th Army Group in the position of having already overdrawn its allocations for the month, and much of its ammunition had already been fired. General Devers protested that the control had been imposed too abruptly and had caused dislocations, and asked for special allocations. SHAEF therefore postponed inauguration of its control to January. The SHAEF reserve was actually established by the middle of that month.

The new system established a maintenance day of supply for each weapon in the theater, thereafter known as the SHAEF Maintenance Rate. In effect, the inauguration of the system recaptured all ammunition in Army supply points and redistributed all ammunition on a weapons basis. While the newly established maintenance rates were not as high as desired, they were actually somewhat better than the resupply predictions of the War Department because of the spreading out of existing stocks in the theater. Furthermore, the system provided a guaranteed rate of resupply to the army groups and thus provided a sound basis for expenditure planning.[87]

Meanwhile the 12th Army Group had decided to establish a reserve of its own. It will be recalled that under the credit system adopted in November it had been agreed that the armies should create their own reserves on the basis of the thirty-day forecasts issued every ten days, and that no ammunition should be held back by the army group. Early in December General Moses proposed that the army group assume responsibility for "assessing the hazards" of ammunition supply behind the armies and thus leave the army commanders freer to consider purely operational problems. Under Moses' proposal the army group, after reviewing the expected rates of supply announced by the War Department, would establish a minimum reserve of its own of seven days in ADSEC depots, and credit the armies with all other available ammunition up to the total of the expected supply rate in accordance with operational activity factors. When available, additional ammunition would be credited to the armies to enable them to establish a seven-day reserve of their own. Under this procedure the armies

[87] Memo, Twitchell for Bull, 7 Dec 44, sub: Availability of Ammo, Memo, Bull for Crawford, 7 Dec 44, Memo, Twitchell for Bull, 20 Dec 44, sub: Conf on Availability of Ammo, and TWX S-75735, SHAEF to A Gps, 19 Jan 45, all in SHAEF G-3 O&E 471 Ammunition; 6 A Gp G-3 Sec Final Rpt, pp. 97-99.

would be told that they could expect to learn about ten days in advance the amounts to be credited to them; the practice of advising them of credit expectancy for a thirty-day period was to stop. Moses believed that this system would afford greater safety and flexibility to the army group commander in the use of critical items in planned operations, and would also relieve the armies of worry about supply in the rear of the forward depots. It would also eliminate the troubles arising from predictions of future supply which often could not be fulfilled.[88]

General Bradley approved the system outlined by his G–4. On 16 December the army group submitted to theater headquarters its desired long-range maintenance rates, which henceforth took the place of the old desired expenditure rate in calculating requirements.[89]

The army group reserve was not actually established until mid-January.[90] At that time the ammunition control system worked roughly as follows: On the basis of resupply forecasts furnished by the War Department and weapons lists provided by the Communications Zone, SHAEF computed the current maintenance rates. The quantities actually to be made available to each army were determined by the number of each type of weapon with the armies and by the activity factors established by the army group. The Communications Zone wrote credits on ADSEC depots for each army, and the armies then made withdrawals as needed against these credits.[91]

In January the supply of ammunition for the 4.2-inch chemical mortar was placed on an allocation and credit basis similar to that set up for artillery ammunition. Ammunition for this weapon had never been plentiful, but the shortage was seriously aggravated during the winter when large quantities in both the theater and the zone of interior were found to be defective and had to be impounded. The defect was found to be in the fuze, causing barrel explosions. Pending the receipt of reconditioned stocks from the United States, impounded shells in the theater were released only when absolutely necessary, and for a time the weapon was fired only by use of the lanyard.[92]

One of the most troublesome problems which plagued ammunition supply through the entire period of operations was the problem of the segregation of ammunition by lot number. Under a system of mass production in many plants there is no guarantee that all ammunition of a single type will have the same ballistic characteristics. Ammunition must therefore be segregated or grouped according to performance characteristics, particularly with regard to range. It is desirable of course, to keep the number of lots delivered to a single battalion as small as possible and, conversely, the number of rounds per lot as large as possible.

Lot segregation was not a new prob-

[88] Memo, Moses for Bradley, 11 Dec 44, sub: Notes on Ammo Supply, Memos—Moses, SHAEF 12 A Gp G–4 Folder 86.

[89] Memo for Record, Moses, 13 Dec 44, sub: Ammo Supply, Memos—Moses, SHAEF A Gp G–4 Folder 86.

[90] Diary of 12th A Gp G–4, 12 Jan 45.

[91] Gen Bd Rpt 100, p. 27.

[92] Memo, for G–4, 23 Jan 45, SHAEF G–4 471 Ammunition 1945; 12 A Gp G–4 AARs for Jan and Feb 45.

lem, and attempts had been made to cope with it before the invasion. Tests carried out in the United Kingdom had shown that variations between lots and even within lots were too great for safety in the close support of infantry, and some nonstandard lots were therefore rejected. Some 800,000 rounds of 105-mm. howitzer ammunition, the type used in greatest quantity for close support, were classified before the invasion. But this quantity was quite inadequate, and units eventually had to be provided classified, unclassified, and even previously rejected ammunition.[93] On the Continent the extended discharge over beaches, the continued receipt of many small mixed lots from the United States, the lack of transportation needed for the rehandling of ammunition once it was on the ground, and the October speed-up in unloading, accompanied by the forwarding of ammunition in bulk, all militated against the maintenance of lot integrity.

The 1st Army Group had originally set as a goal the delivery of ammunition in lots of at least 500 rounds to individual battalions. But this was rarely if ever achieved. In October Third Army found that one supply point with 7,445 rounds of 105-mm. howitzer ammunition contained 308 lot numbers; a tabulation of receipts in a single depot in a period of three days revealed that 545 separate lots had been received. In another three-day period late in September one field artillery battalion reported drawing 131 lots of ammunition averaging only eighteen rounds each.

Just before the November offensive First Army undertook to segregate and record by lot number all ammunition under its control, a task which involved an expenditure of 25,000 man-hours of labor. It attempted to segregate ammunition as far as possible into multiples of 150 rounds. But the stock of 105-mm. howitzer ammunition alone contained more than 1,200 lot numbers. Ordnance officials finally concluded that the medium battalions would simply have to accept a proportion of unsegregated ammunition with each issue of segregated shells.[94]

This highly unsatisfactory situation naturally brought complaints from the field. Artillery commanders reported that it was impossible to determine the behavior of ammunition by registration. The resulting inaccuracy of firing consequently necessitated the adoption of a safety factor certain to prevent short rounds from falling on friendly troops. But such measures also voided the benefits of close supporting fires.

In one respect the problem of ammunition handling resembled the problem of handling bulk rations. Care had to be exercised at every stage along the lines of communications to maintain the integrity of original loads or blocks of supply. As in the case of balanced rations, the Communications Zone had first tried to eliminate a major cause of trouble at the loading end where ammunition was often stowed without reference to lots. This fault was largely eliminated by the fall of 1944, and ships arriving in the theater carried sizable quantities of individual lots, which were block stowed. In November, with the

[93] Gen Bd Rpt 58, pp. 11–12, 76.

[94] *Ibid.*, p. 76; Gen Bd Rpt 100, p. 22; FUSA Rpt of Opns, 1 Aug 44–22 Feb 45, Bk. III, p. 27.

improvement of port discharge and transportation, some improvement was also evident in the handling of this ammunition within the theater. By the end of that month ammunition was arriving at three major ports—Cherbourg, Morlaix, and Le Havre. Ammunition detachments were stationed at each of these ports to prevent the breaking up of lots during unloading, and further attempts were then made to ensure the shipment of solid blocks—an entire shipload where feasible—to a single depot. At the forward depots the Advance Section in turn attempted to reconsign rail cars containing solid loads of one lot forward to the armies. The Advance Section reported savings of between forty and sixty trucks per day at each of two depots in December by following this practice.[95] This campaign gradually brought improvement. But complete lot integrity was never achieved, and the problem plagued ammunition supply operations until V-E Day.

Lot segregation was but one aspect of an essentially complicated supply problem. Estimating ammunition requirements suffered from the handicap of most military logistics in that requirements fluctuated with the course of tactical operations, which were largely unpredictable. Neither the field commands nor the Communications Zone foresaw the developments of August and September 1944, which had the effect first of reducing expenditures and then of creating a precipitate rise in requirements for the attack on the West Wall defenses.

But the problem actually went deeper. Unlike the problem of estimating the need for POL and most equipment, in the case of Class V there was much disagreement as to what constituted an adequate day of supply or average rate of fire. Field commanders could always justify a much higher expenditure rate than the War Department, faced with multiple production problems, was either willing or able to support. In brief, there rarely, if ever, is enough ammunition to satisfy what field commanders consider their legitimate needs.

Unfortunately the problem was needlessly aggravated by the lack of mutually understood ground rules regarding accountability of stocks in the pipeline and in the theater. In any event, the persistent uncertainty over supply prospects had its inevitable effect on operational planning, hobbling all action in October, and casting a shadow over most planning for several months.

[95] Ltr, Brig Gen R. M. Howell, 9th Arty Comdr, to CG 9th Div, 24 Oct 44, sub: Variation in Lot Numbers of Arty Ammo, with Inds, Incl COMZ reply, 26 Dec 44, EUCOM 471 Ammunition General, III; Ltr, COMZ to CGs COMZ Secs, 15 Nov 44, sub: Ammo Shipping Practices, and Ltr, ADSEC to CG COMZ, 19 Dec 44, sub: Ammo Arrival Condition Rpt, both in ADSEC 471 Ammunition; Ltr, Hodges to Bradley, 10 Dec 44, sub: Segregation of Fld Arty Ammo by Lot Number, with 1st Ind, 12 A Gp to COMZ, 20 Dec 44, 12 A Gp 471 Ammunition.

CHAPTER X

The Troop Build-up, August 1944–March 1945

(1) The Flow of Divisions

Logistic difficulties in the fall of 1944 logically should have had an adverse effect on the scheduled build-up of U.S. forces in the European theater, at least on the Continent. On the contrary, the flow of American units was actually accelerated in the face of the bad port and transportation situation in the fall of 1944, and the build-up of the original troop basis was completed substantially ahead of schedule. In the end, a much larger number of troops went to the European theater than had been planned.

Theater officials had considered the possibility of an acceleration in the build-up as early as June. At that date, however, there was little in the way of experiential data to indicate the advisability of such a course. A decision still had not been reached at the time of the breakout at the end of July, although General Eisenhower by that time had asked the War Department to add one division to the scheduled September shipments in order to maintain the level in the United Kingdom at four.

At the end of July there were 22 divisions in the theater, 18 of which were on the Continent, and plans at the time called for the shipment of about 4 divisions each month from the United States to complete the build-up of 47 by the end of January 1945.[1] By the first week of September (D plus 90) there were 26 divisions in the theater (exclusive of the 3 in southern France), 20 of which were then on the Continent as compared with the scheduled 21.

Meanwhile, early in August, when the tactical situation suddenly took an unexpected turn, the War Department made the first of several proposals to speed the flow of divisions from the United States. Anticipating the early capture of the Brittany and Loire ports and the likelihood that ETOUSA might want to expedite the flow of troops from the United States directly to the Continent, it offered to advance the shipment of two infantry divisions (the 26th and 104th) by about two weeks, from early September to late August, if ETOUSA could make an immediate decision.

General Eisenhower, who was anxious

[1] Memo, Col C. F. Thompson for Chief Current Opns Br, 26 Jul 44, sub: Availability of U.S. Divs for ETO, Interim Rpt 1, SHAEF G–4 320.2 Strength 44, I; See also *Logistical Support I,* 128–29, 456.

to exploit the favorable situation then developing in France, promptly accepted the offer. He recognized that there might be administrative difficulties, for there were doubts about the capacity of both the British ports and continental beaches and ports to handle additional divisions before the end of September, but it was agreed that the difficulties could be overcome somehow.[2]

The advancement of two divisions by two weeks hardly constituted a radical alteration in plans. But the War Department quickly followed up with a proposal much more far-reaching in its implications. On 11 August, two days after the theater had agreed to the first speed-up, the War Department proposed to ship in September not only the three divisions which remained on the schedule for that month, but all five divisions on the October schedule as well, thus advancing the entire build-up by a month and five divisions. In offering the speed-up it indicated that in most cases the only limiting factor was the time needed to pack equipment and move it to the ports for loading.[3]

General Eisenhower was elated with the prospects for a speedier build-up of additional combat strength. But supply officials received the offer with strong misgivings, claiming that it threatened serious complications in the cargo shipping situation. The Communications Zone earlier had announced a minimum requirement for 285 shiploads of cargo plus the equivalent of another 27 loads on tanker decks and in War Shipping Administration ships for September arrival. The War Department had already indicated in July that it could provide only 200 of that number because of the shortage of shipping. Protests from the theater were of no avail. In fact, early in August, coincident with the first acceleration in divisional movements, the War Department announced a further cut of 21 ships, reducing the allocation to 179 cargoes. Within a few days, to the astonishment of the theater, it proposed to cancel another 20, which would have reduced the September arrivals to 159.

The Communications Zone vigorously protested these proposals, arguing that the intended cancellations would result in withholding supplies and equipment of the highest operational priority from the theater. It now asked the War Department to restore as many ships as possible up to 250, which it regarded as an irreducible minimum. General Eisenhower supported this request, at least to the extent of restoring the two most recent cuts, totaling 41 ships. He was anxious to have the additional speed-up in the flow of divisions, however, and indicated that the theater was prepared to accept the difficulties it would entail.[4]

Transportation officials in Washington did not share ETOUSA's anxieties over cargo shipping, pointing out that current sailing schedules would more

[2] Cbl WAR–77333, Marshall to Eisenhower, 8 Aug 44, Cbl S–67204, Eisenhower to Marshall, 9 Aug 44, both in OPD 370.5 Sec. XIII, Cases 448–86; Cbl S–57189, Eisenhower to Marshall, 9 Aug 44, Eyes Only Cbls, Smith Papers; Ltr, Bull to Smith, 9 Aug 44, sub: U.S. Build-up, SHAEF G–4 320.2 Strength 44, I.

[3] Cbl 79344, Marshall to Eisenhower, 11 Aug 44, OPD 370.5 Sec. XIII, Cases 448-86.

[4] Cbl EX–43086, Lee to Lutes, 13 Aug 44, and Cbl S-57530, Eisenhower to Marshall, 13 Aug 44, Cbls, SS&P Planning Div 201.02, A46–371.

than match the theater's most optimistic forecasts of its discharge capabilities. They assured the theater that they could easily step up sailings at any time if its discharge rate indicated that the port backlogs were being reduced. Nevertheless, the War Department relented and met a large part of the theater's request, raising the September allocation to 220 ships plus an equivalent of 35 loads on tanker decks.[5]

At this time—mid-August—the War Department also forecast the availability of divisions for later shipments—that is, beyond September—and asked the theater whether it wanted the gains extended by advancing the sailing dates of divisions originally scheduled for later shipment. In fact, it actually offered a further acceleration in the flow, for it proposed to ship in October all 3 divisions originally scheduled for November and, in addition, 3 of those scheduled for December. This would advance the build-up by almost two months and by a total of 6 divisions, and would bring the U.S. build-up in Europe to 42 divisions sometime in November. Again the War Department asked for a prompt decision because of the need to start the reconditioning, packing, moving, and loading of equipment.[6]

As could be expected, the latest proposal aroused new fears among those responsible for the logistic support of U.S. forces. COMZ staff officers were particularly apprehensive over the effect which the speed-up would have on the theater's balance of service and combat troops, for the War Department had given no assurance that the divisions would be accompanied by adequate service troops. In fact, it had repeatedly told the theater that it could not increase the flow of engineer, medical, and signal units in line with the shipments of combat units. General Eyster, the ETOUSA G–3, estimated that the proposed shipments would lower the division slice to about 38,000 men by the end of October and would cause a serious thinning out of logistic support. Colonel Whipple, the SHAEF logistic plans chief, supported this view, asserting that such a reduction would be acceptable only if the new divisions were to relieve combat-weary units, which would therefore not have to be supported at full combat scales. General Stratton, the G–4, predicted in addition that critical shortages of supplies, notably in ordnance and signal Class II items, and in ammunition, would develop.[7]

The speed-up also had serious implications with regard to the continental ports. Receiving, staging, and equipping the sixteen divisions which would arrive in the nine weeks between 9 September and 12 November would impose a heavy administrative burden in the port areas, involving troops which were badly needed to support the forces already in the field. Shipments arriving directly from the United States would have to be accepted either at the beaches or at

[5] Cbl WAR–80524, TC Planning Div WD to Lee, 14 Aug 44, Cbl WAR-81787, Gross to Somervell (in ETO), 16 Aug 44, and Cbl WAR–81853, Lutes to Lee, 16 Aug 44, all in Cbls, SS&P Planning Div 201.02, A46–371.

[6] Cbl W–82481, AGWAR to SHAEF, 17 Aug 44, SHAEF G–4 Troop Flow 121/1 GDP–1, Folder 64.

[7] Ltr, Eyster to Theater Comdr, 19 Aug 44, sub: Increased Divisional Flow for Sep and Oct 44, and Memo, Whipple for SHAEF G–4, 19 Aug 44, sub: U.S. Troop Flow, SHAEF G–4 Troop Flow 121/1 GPD–1, Folder 64.

Cherbourg. But the Normandy beaches were expected to deteriorate rapidly in September. Cherbourg, with extremely limited capacity, was the only port thus far restored possessing facilities suitable for handling heavy cargo, and its use for troop units would involve the diversion of important cargo.[8] On this basis alone Colonel Whipple advised limiting the shipment of divisions to one per convoy in October instead of the two proposed. At this time U.S. forces were already crossing the Seine, and Whipple accurately foresaw that logistic difficulties would shortly become the major factor limiting operations. He recommended that the provision of additional service troops be given a higher priority than the movement of combat units, and, in anticipation of the inevitable delays in the development of rail transportation, that the introduction of motor transport be given greater emphasis.[9]

Despite these warnings the theater decided to accept the War Department's proposed schedule of shipments, its reply of 20 August stating only that it "understood" that the flow of service troops would not be prejudiced.[10]

With the acceptance of the War Department's plan of shipments through October only five divisions of the original forty-seven-division troop basis remained to be scheduled for shipment. By mid-August it was taken for granted, however, that the European theater would get the nine divisions in the undeployed reserve which the War Department had tentatively earmarked for ETOUSA a month before. Fourteen divisions therefore remained to be shipped to Europe after October, and the question now arose as to how these should be scheduled.

Thus far the War Department had anticipated no difficulties in carrying out the speed-up, for all the divisions involved had been maneuver-trained at one time or another, and could be adequately outfitted by withholding equipment from units remaining in the United States. Thereafter, however, the pinch would begin to be felt in both training and equipment. Late in August the War Department informed the theater of the training status of each of the remaining divisions, and also listed the equipment shortages which would affect the combat readiness of units remaining to be shipped. It offered alternative shipping schedules, the more accelerated of the two involving some sacrifice in combined maneuver training, and left it to the theater to choose between the two. It offered to make some revisions in training and production schedules based on the theater's estimate of its needs.[11]

Theater officials hesitated to ask for the speedier build-up at the expense of combined maneuver training. But in

[8] Min, Mtg at SHAEF on Shipment of U.S. Div to the Continent, Napier presiding, 26 Aug 44, SHAEF G–4 337 Conference 1944. In addition, accepting large troopships there was considered hazardous. British officials agreed that operational needs justified the risk, however, and gave their approval. Memo, COS Brief of Action Rpt for Eisenhower, 2 Sep 44, sub: Acceptance of Large Troopships at Cherbourg, SHAEF SGS 800 Harbors, Opening, Use, Construction.

[9] Memo, Whipple for Crawford, 19 Aug 44, sub: U.S. Troop Flow, SHAEF G–4 Troop Flow 121/1 GDP–1, Folder 64.

[10] Cbl S–58041, Eisenhower to Marshall, 20 Aug 44, OPD 370.5 Sec. XIII, Cases 448–86.

[11] Cbl WAR–86126, Marshall to Eisenhower, 24 Aug 44, and Memo, Handy for CofS, 24 Aug 44, OPD 370.5 Sec. XIII, Cases 448–86.

the end it requested the best possible flow schedule, subject only to the condition that units be fully equipped.[12] On this basis the War Department offered revised build-up schedules indicating the readiness dates for divisions with varying levels of training. It revealed that the major equipment shortages were in combat ordnance items and would affect mainly the readiness of armored divisions.[13] With this information the theater decided on a somewhat accelerated flow, which called for 1 division in November, 5 in December, 3 in January, 3 in February, and 1 in March.[14] This schedule made it possible for all divisions shipped after October to have the equivalent of one month's maneuver training at their home stations.[15]

Apparently anticipating some of the internal difficulties which might attend the accelerated flow of divisions, the theater commander meanwhile considered the possibility of diverting two or three divisions from northern France and routing them through Marseille, where port capacity was expected to be more adequate. The War Department, when queried on 5 September as to the shipping implications, replied on the following day that the switch could be made, and presented alternate schemes for carrying out the suggestion. Much depended on whether the theater desired general-purpose vehicles to accompany the units, for the vehicles of the October divisions had already been released for preshipment and were consigned to northern ports. The War Department suggested that ETOUSA arrange a loan of vehicles from the North African theater, which it could repay later. But General Devers proposed to solve the problem by providing vehicles from stocks already available in southern France and by borrowing from other units.

Meanwhile the War Department had emphasized that it must have a decision within four days (10 September) in order to implement the plan. On 7 September SHAEF cabled its acceptance of the plan to ship the 11th Armored Division and the 99th and 103d Infantry Divisions through Marseille. Transmission of the message was delayed an entire week, partially as the result of the disruption in communications occasioned by the movement of both COMZ and SHAEF headquarters to the Continent at this time. On 13 September, having received no word from ETOUSA, Washington notified the theater that its proposal to divert the above-named divisions was no longer possible. The War Department offered to divert four other divisions, however, if notification was received by the 21st. ETOUSA asked that three of the four divisions be shipped to Marseille as proposed, and the War Department accordingly made arrangements to divert the 14th Ar-

[12] Cbl FWD-13393, SHAEF to Marshall, 29 Aug 44, SHAEF AG 370.5-6 General 1944.

[13] Cbl WAR-24823, 2 Sep 44, OPD Cbl Files.

[14] One of the 9 divisions in the undeployed reserve, the 10th Light Mountain, was deleted from the ETOUSA troop basis at this time at the theater's request. This accounts for the scheduling of 13 instead of 14 divisions. Another division was submitted later.

[15] Cbl FWD-14659, SHAEF to AGWAR, 12 Sep 44, SHAEF AG 307.5-6 General 1944; Cbl 28646, Marshall to Eisenhower, 11 Sep 44, OPD 370.5 Sec. XIV, Cases 487-510; Cbl, Marshall to Eisenhower, 13 Sep 44, and Note for Record, OPD, 1 Sep 44, sub: Divisional Flow and Theater Troop Basis, both in OPD 370.5 Sec. XIII, Cases 448-86.

TABLE 7—OVERLORD DIVISIONAL BUILD-UP
D PLUS 90 TO D PLUS 210 [a]

Date	Total Theater Build-up		Continental Build-up			
	Original Plan	Actual	Original Plan (Plan X)	Plan Y [b]	Plan Z [b]	Actual
D plus 90 (4 September 1944)	24	26	21			[c] 20
D plus 120 (4 October 1944)	29	33	25	27		30
D plus 150 (3 November 1944)	34	41	30	34	35	34
D plus 180 (3 December 1944)		44			41	39
D plus 210 (2 January 1945)		49				46

[a] Exclusive of the three divisions in Operation DRAGOON.
[b] See *Logistical Support I*, 454–55.
[c] Exclusive of the 82d and 101st Airborne Divisions, which had returned to the United Kingdom.

mored and the 100th and 103d Infantry Divisions, scheduled for early October shipment, to southern France.[16] The three divisions arrived at Marseille between 20 and 29 October.

The stepped-up flow of divisions in the north soon brought its problems, as supply officials had predicted. Eight divisions arrived in France in September. Three of them crossed over from the United Kingdom; 5 came directly from the United States, the first troop convoy arriving at Cherbourg on 7 September with elements of the 26th and 104th Infantry Divisions. These arrivals raised the theater's strength to 33 divisions on D plus 120 (4 October), 4 divisions above the build-up originally planned, and brought the continental strength up to 30 for a gain of 5 over the build-up scheduled for that date. (*Table 7*)

Increasing difficulties attended the introduction of this extra combat strength onto the Continent in the face of the acute transportation shortage and the failure to take Brest. The wisdom of the policy was soon questioned in view of the inability of the theater to support additional units in combat. Four of the divisions (the 104th, 44th, and 102d Infantry and the 10th Armored) did not see action for at least five weeks, and one (the 9th Armored) remained uncommitted as a unit until mid-December. Lack of equipment, some of which had to be discharged in the United Kingdom, was a major contributing cause.

Divisions continued to cross over to the Continent from England in the next months. Except for the three diversions to Marseille, however, not a single di-

[16] Cbl FWD–13887, SHAEF to WD, 5 Sep 44, Cbl WARX–26901, Handy to SHAEF, 6 Sep 44, Cbl WAR–29983, Handy to SHAEF, 13 Sep 44, Cbl FWD–14996, SHAEF to WD, 16 Sep 44, Cbl WAR–32604, WD to ETO, 18 Sep 44, Cbl FWD–15279, ETO to WD, 20 Sep 44, all in Cbls, SS&P Planning Div 201.02 United Kingdom, A47–6; TWX BX–16089; Devers to SHAEF, 10 Sep 44, SHAEF AG 370.5–6 General 1944; Cbl 27747, Marshall to Eisenhower and Devers, 9 Sep 44, Note for Record, 6 Sep 44, sub: Diversion of Divs Through Marseille, Note for Record, 13 Sep 44, sub: Diversion of Div to Southern France, Cbl 32284, Marshall to Eisenhower, 17 Sep 44, Cbl 34375, Marshall to Eisenhower, 21 Sep 44, and Note for Record, 17 Sep 44, sub: Diversion of Div Through Marseille, all in OPD 370.5 Sec. XIV, Cases 487–510; Mil Shipments Priority Mtg, 9 Sep 44, SHAEF AG 337–18.

TABLE 8—DIVISIONAL BUILD-UP IN THE EUROPEAN THEATER, 1942–1945

Unit	Arrival in Theater[f]	Arrival on Continent[f]	Entry into Combat	Days of Combat
29th Infantry	11 October 42	6 June 44	7 June 44	242
5th Infantry	9 August 43	9 July 44	16 July 44	270
101st Airborne	15 September 43	6 June 44	6 June 44	214
3d Armored	15 September 43	24 June 44	9 July 44	231
28th Infantry	18 October 43	24 July 44	27 July 44	196
2d Infantry	19 October 43	8 June 44	8 June 44	303
1st Infantry[a]	5 November 43	6 June 44	6 June 44	292
2d Armored [b]	24 November 43	9 June 44	2 July 44	223
9th Infantry[b]	27 November 43	10 June 44	14 June 44	264
82d Airborne	9 December 43	6 June 44	6 June 44	194
8th Infantry	15 December 43	4 July 44	8 July 44	266
4th Armored	8 January 44	13 July 44	28 July 44	230
4th Infantry	28 January 44	6 June 44	6 June 44	299
30th Infantry	22 February 44	14 June 44	15 June 44	282
5th Armored	23 February 44	25 July 44	2 August 44	161
6th Armored	25 February 44	18 July 44	28 July 44	226
90th Infantry	5 April 44	8 June 44	10 June 44	308
79th Infantry	17 April 44	14 June 44	19 June 44	248
83d Infantry	19 April 44	19 June 44	27 June 44	244
35th Infantry	26 May 44	8 July 44	11 July 44	264
7th Armored	13 June 44	10 August 44	14 August 44	172
80th Infantry	7 July 44	3 August 44	8 August 44	239
94th Infantry	11 August 44	5 September 44	17 September 44	183
3d Infantry[c]	15 August 44	15 August 44	15 August 44	233
36th Infantry[c]	15 August 44	15 August 44	15 August 44	227
45th Infantry[c]	15 August 44	15 August 44	15 August 44	230
95th Infantry	17 August 44	19 September 44	20 October 44	151
17th Airborne	25 August 44	24 December 44	25 December 44	45
9th Armored	27 August 44	25 September 44	16 December 44	91
26th Infantry[d]	7 September 44	19 September 44	12 October 44	199
104th Infantry[d]	7 September 44	7 September 44	24 October 44	178
44th Infantry[d]	15 September 44	15 September 44	24 October 44	230
102d Infantry[d]	22 September 44	23 September 44	26 October 44	178
10th Armored[d]	23 September 44	23 September 44	2 November 44	124
84th Infantry[d]	1 October 44	1 November 44	18 November 44	152
12th Armored	2 October 44	9 November 44	7 December 44	102
11th Armored	5 October 44	17 December 44	23 December 44	96
99th Infantry	10 October 44	6 November 44	9 November 44	151
75th Infantry	20 October 44	13 December 44	25 December 44	94
103d Infantry[e]	20 October 44	20 October 44	11 November 44	147
100th Infantry[e]	20 October 44	20 October 44	9 November 44	163
78th Infantry	25 October 44	22 November 44	13 December 44	125
14th Armored[e]	29 October 44	29 October 44	20 November 44	133
106th Infantry	1 November 44	26 November 44	10 December 44	63
87th Infantry	13 November 44	3 December 44	13 December 44	134
8th Armored	21 November 44	6 January 45	23 February 45	63

Unit	Arrival in Theater[f]	Arrival on Continent[f]	Entry into Combat	Days of Combat
66th Infantry	26 November 44	26 December 44	1 January 45	91
69th Infantry	13 December 44	26 January 45	11 February 45	65
76th Infantry	21 December 44	17 January 45	19 January 45	95
63d Infantry[e]	14 January 45	14 January 45	6 February 45	119
42d Infantry[e]	18 January 45	18 January 45	17 February 45	106
70th Infantry[e]	18 January 45	18 January 45	3 February 45	83
65th Infantry[d]	21 January 45	21 January 45	9 March 45	5
89th Infantry[d]	22 January 45	22 January 45	12 March 45	57
13th Armored[d]	29 January 45	29 January 45	10 April 45	16
71st Infantry[d]	6 February 45	6 February 45	12 March 45	49
13th Airborne	6 February 45	10 February 45	No combat	None
16th Armored[d]	11 February 45	11 February 45	5 May 45	3
20th Armored	17 February 45	21 February 45	24 April 45	8
86th Infantry[d]	3 March 45	3 March 45	29 March 45	34
97th Infantry[d]	3 March 45	5 March 45	1 April 45	31

[a] First arrived in European Theater of Operations 7 August 1942 and participated in North African and Sicilian operations. The 1st Armored and 34th Infantry Divisions also came to the European theater in 1942 and, like the 1st Division, went to North Africa in November of that year. Neither of them ever saw action in the European Theater of Operations, however, and they are excluded from this list.

[b] Saw action earlier in North African theater.

[c] Landed in southern France from North African theater as part of DRAGOON invasion force.

[d] Entered northern France directly from the United States.

[e] Entered southern France directly from the United States. Infantry regiments of 63d, 42d, and 70th Divisions arrived early in December 1944 and saw action in advance of the divisions proper.

[f] All arrival dates—for both the theater and Continent—are the dates on which the main echelon of the division headquarters closed. The arrival of a division usually extended over several days and in some cases several weeks. There are minor discrepancies between the arrival dates given here and those given on page 457 of *Logistical Support I* for the months of June and July 1944 because different sources were used.

Source: Order of Battle: Divisions, European Theater, prep by Hist Sec ETOUSA, 1945.

vision was again accepted directly from the United States in continental ports until January 1945. (*Table 8*)

In October the War Department proposed a further modification in the build-up plan. Inspired in part by plans which the Combined Chiefs of Staff were considering for an all-out effort to end the war before the winter, and in part by the desire to relieve combat-weary infantrymen with fresh troops, General Marshall suggested an immediate speed-up in the shipment of the infantry regiments of divisions scheduled for later shipment. The idea was that the units be rotated with regiments of divisions already in the field. Under this plan the organic supporting and service troop units of the divisions would arrive later and eventually marry up with the infantry. Marshall proposed advancing the shipment of the regiments of all twelve of the infantry divisions remaining to be shipped.

The theater immediately approved the scheme for the infantry elements of three divisions (the 87th, 75th, and 106th), portions of which were already loading. But it was not enthusiastic about applying the idea to all the remaining divisions. General Bradley, in particular, raised objections in the conviction that it was bad practice to break up units, and that it would create both training and equipment problems for the rumps of the divisions in the United States. Moreover, he feared that advancing the infantry elements of additional divisions would affect striking power at the very time when the logistic situation would support a general offensive.[17]

General Eisenhower forwarded the 12th Army Group commander's views to the Army Chief of Staff, pointing out, in addition, the logistic troubles of the theater. ETOUSA was already incapable of supporting the divisions available to it, and would be unable to do so until Antwerp was opened; it had accepted the earlier accelerations, he said, only so that it could rotate units and thus relieve tired troops. Eisenhower therefore advised against a speed-up of infantry regiments beyond the three already under way.[18]

General Marshall believed it extremely important to relieve combat-weary infantry with fresh units and asked the theater commander to reconsider. At the same time he apparently wished to allay any fears about a delay in the shipment of the remaining portions of the divisions. He indicated that the three divisions whose regiments were being advanced could sail by mid-November, and that the remaining elements of the other nine divisions could follow their regiments by mid-December.[19]

[17] Cbl WARX-46072, Somervell to NYPOE, 13 Oct 44, Cbls, SS&P Planning Div 201.02, United Kingdom A46-371; Memo for Record, Bradley, 13 Oct 44, SHAEF 12 A Gp G-4 Memos—Moses, Folder 86; COMZ G-4 Plant and Communications Diary/Jnl, 15 Oct 44, ETO Adm 145C.

[18] Cbl S-63358, Eisenhower to Marshall, 21 Oct 44, SHAEF G-4 381 Troop Unit Basis 1944, I.

[19] Cbl 50676, Marshall to Eisenhower, 22 Oct 44, OPD 370.5 Sec. XV, Case 511.

This proposed schedule completely ignored the theater's logistic dilemma, whose seriousness the War Department obviously did not yet comprehend. In reply, the theater pointed out that it could not possibly accept the heavy elements of the divisions at the proposed rates, either on the Continent or in the United Kingdom, for the equipment involved alone amounted to nearly 500,000 tons of cargo.[20] As matters stood, nine regiments of three infantry divisions then in the United Kingdom were already available as relief regiments for nine of the eighteen divisions in the 12th Army Group. The nine regiments of the 87th, 75th, and 106th Divisions were due to arrive in the United Kingdom at the beginning of November and would provide similar reinforcements for all the remaining divisions in the 12th Army Group.

General Eisenhower proposed, therefore, that the regiments of only three of the remaining nine divisions earmarked for ETOUSA be shipped in advance of their divisions in November in order to ensure a reserve of three complete infantry divisions for the theater. In view of the port situation on the Continent and the lack of accommodations in the United Kingdom he asked that no additional units be shipped except for the "residues" of the 87th, 75th, and 106th Divisions, already scheduled for mid-November. To provide similar reinforcement of the 6th Army Group he asked that the regiments of three additional divisions be shipped directly to Marseille late in November. This schedule, he pointed out, would provide a

[20] Cbl S-63616, Eisenhower to Marshall, 23 Oct 44, Eyes Only Cbls, Smith Papers.

relief regiment for each of the 25 infantry divisions in France plus 2 extra, and would also provide a strategic reserve of at least 3 infantry divisions in the United Kingdom in addition to whatever armored units were not committed.[21] This schedule General Eisenhower considered the maximum allowed by the theater's tight logistic situation. Any plan calling for a faster flow, he asserted, would necessitate receiving units on the Continent, where they would have to be held near the ports.[22]

The theater commander asked the War Department for a week's deferment on a decision regarding the three divisions remaining in the United States and on the shipment of the heavy equipment of those whose infantry elements were being accelerated. On 1 November he made known his wishes regarding those units. Because of continuing administrative difficulties, particularly the delay in the opening of Antwerp, he asked that the residues of only three divisions (the 66th, 69th, and 76th) be shipped to the United Kingdom in December. For January he requested the shipment directly to southern France of the residues of the three divisions (42d, 63d, and 70th) whose infantry elements were already being preshipped to Marseille, and the shipment of three complete divisions (the 13th Armored and the 65th and 89th Infantry) to northern France on the assumption that Antwerp would then be open. This would complete the shipment of the residues of all nine divisions whose infantry had been preshipped. ETOUSA asked that all February shipments, including one airborne, one infantry, and two armored divisions, be sent directly to northern France. Three of the twelve divisions whose infantry elements the War Department had urged be advanced were to be shipped as whole divisions according to this plan, although the theater agreed to accept the infantry elements of these three divisions as well if arrangements for their preshipment were already too far advanced. But the War Department agreed to follow the suggested schedule.[23]

As of this date—1 November—there were 41 divisions in the theater, 34 of them (exclusive of the original 3 in the DRAGOON force) on the Continent and 7 in the United Kingdom. The effect of the earlier accelerations was strikingly evident, for the over-all build-up was now ahead of its original schedule by 7 divisions. Logistic difficulties on the Continent had forced several of the divisions to stop over in the United Kingdom temporarily, however, and the 34 divisions on the Continent thus far actually represented a gain of only 4 over earlier build-up plans. (*See Table 7.*)

The shipping schedule outlined by the theater early in November was fol-

[21] The plan, as accepted by the War Department, called for the following schedule of shipments: the infantry regiments of the 87th, 75th, and 106th Divisions in October and the residues of these units to the United Kingdom in November; the infantry regiments of the 69th, 66th, and 76th Divisions to northern France and the infantry regiments of the 63d, 42d, and 70th Divisions to southern France in November.

[22] Cbl S–63876, Eisenhower to Marshall, 24 Oct 44, OPD Cbl Files.

[23] Cbl S–64961, SHAEF to AGWAR, 1 Nov 44, OPD Cbl Files; Cbl 57703, Marshall to Eisenhower, 2 Nov 44, and Note for Record, 2 Nov 44, sub: Shipment of Divs to ETO, OPD 370.5 Sec. XV, Case 511.

lowed substantially as agreed in the next two months. But not all of the preshipped infantry regiments were committed in a relief role as originally conceived. Theater officials, none too enthusiastic over the idea from the start, did not consider the experiment successful. Removing infantry regiments from their parent organizations and placing them under strange headquarters, even temporarily, created problems of both command control and supply. In the end the theater reconstituted at least two of the nine divisions as complete units in England before transferring them to the Continent.

Two final changes were made in the theater build-up, one involving an additional acceleration and one a further augmentation of the troop basis. Early in January 1945 the War Department offered to advance the sailing dates of the four divisions on the ETOUSA troop list still in the United States. The theater accepted the proposal, and the 71st Infantry, 13th Airborne, and the 16th and 20th Armored Divisions, previously scheduled for shipment in February, each eventually gained about a week in its departure. This brought the number of divisions shipped to the European theater to fifty-six, plus the three of the DRAGOON force.

Meanwhile General Eisenhower, impressed by the offensive spirit the enemy was displaying in the Ardennes and alarmed over the possibility of enemy reinforcement from the east, made an urgent appeal for additional combat strength, even suggesting the consideration of diversions from other theaters. Only two divisions not already allocated to ETOUSA still remained in the United States, both of them earmarked for the Pacific. The Joint Chiefs nevertheless approved the allocation of the two divisions to the ETOUSA troop basis. Both divisions (the 86th and 97th Infantry) arrived early in March, completing the build-up and raising the theater's final strength to sixty-one divisions.[24] Of this total, one—the 13th Airborne—was never committed, and two—the 16th and 20th Armored Divisions—saw only a few days of combat.[25]

The speed-up in the shipment of U.S. forces to the European theater had its repercussions in the United Kingdom as well as on the Continent. ETOUSA had expected to close out U.S. installations in the United Kingdom fairly rapidly after the launching of OVERLORD in accordance with the Reverse BOLERO or RHUMBA plan. The transfer of supplies to the Continent had lagged from the start, however, and had made it necessary for U.S. forces to retain depot facilities in England much longer than planned.[26] The stepped-up flow of divisions from the United States created more serious complications, for many of

[24] Cbl S–74039, Eisenhower to CCS, 7 Jan 45, SHAEF SGS 381 Post–OVERLORD Planning; Cbl WARX-88482, Marshall to Eisenhower, 8 Jan 45, OPD 381 1943–45; Cbl WAR–88705, Hull to Eisenhower, 8 Jan 45, OPD Cbl Files; Memo, Leahy for FDR, early Jan 45, OPD Exec Office File 9; Cbl S–74327, Eisenhower to Marshall, 9 Jan 45, OPD Cbl Files; and Memo, J. F. M. Whiteley, Deputy Asst CofS G–3, for CofS, 9 Jan 45, sub: Allocs of Additional Inf Divs to the ETO, SHAEF G–3 370.01 Troop Build-up, II.

[25] Forty-two U.S. divisions were shipped to France in World War I, and an additional one was organized in France. Of the total, thirty-one saw combat.

[26] See Chapter XIV, Section 4, below, for an account of the progress in clearing the U.K. supply depots.

the units had to be diverted to the United Kingdom, where handling of their equipment added to the burden of English ports and inland transport, and where many troop accommodations had already been turned back to the British.

On the assumption that divisions would proceed from the United States directly to the Continent beginning in September, the RHUMBA plan had provided that U.S. field forces should be completely cleared out of the United Kingdom and that the total U.S. strength there should be reduced to about 650,000 (including an expected 137,000 hospital patients) by the end of October 1944.[27] Shipments to France proceeded more or less according to schedule throughout the summer, and at the end of September only 34,000 field force troops, including only one division, remained in England. Total U.S. strength there was down to about 688,000 men. (*Table 9*) The acceptance of eight divisions in France that month completely saturated Continental port, beach, and transportation facilities, however, and when two more divisions (the 84th Infantry and 12th Armored) arrived in European waters at the end of the month there was no choice but to divert them to the United Kingdom.

Both U.S. and British officials had foreseen the possibility that a few divisions might have to be accepted in the United Kingdom in view of the new schedule of shipments, and they had made arrangements to accommodate up to three at a time. But they hardly expected the flood which was to follow. In the next three months twelve divisions had to be accepted in the United Kingdom—in fact, all the divisions shipped from the United States in that period except for those diverted to Marseille. Early in October British officials, recognizing the operational necessity therefor, reluctantly agreed to make additional facilities available so that seven divisions could be accommodated. By the end of the month they had been prevailed upon to increase the number to nine. In the meantime, however, they asked the Combined Chiefs to defer sailings in order to limit the number accommodated in the United Kingdom at any one time to six. The diversions had already caused port discharge as well as accommodations problems, for they involved the handling of vehicles and other equipment as well as personnel.[28]

It was partly because of these considerations that SHAEF urged the War Department not to carry out the acceleration of infantry regiments which it proposed in October. In submitting its counterplan on 1 November SHAEF noted that to keep more than six divisions in the United Kingdom would necessitate the withdrawal to the United Kingdom of service troops badly needed on the Continent.[29] The number of divisions in the United Kingdom actually rose to seven at one time in October.

Transfers to the Continent improved somewhat in November. But there still were five divisions in the United Kingdom at the end of the month, and U.S.

[27] RHUMBA Plan, Hq ETO, 3 May 44, EUCOM 381 Rhumba Special File.

[28] Cbl 85685, Troopers to SHAEF, 20 Oct 44, SHAEF AG 370.5–6 General 1944; Mil Shipments Priority Mtgs of Sep and Oct 44, SHAEF AG 337–18.

[29] Cbl S–64691, SHAEF to AGWAR, 1 Nov 44, SHAEF G–4 381 Troop Unit Basis 1944, I.

Table 9—Theater Strength by Major Component
May 1944–April 1945 [a]

Date		Total	Field forces	Air forces	Communications zone	Non-operating[b]	GFRS[c]	Divisions[d]
31 May 44	Theater	[e] 1,526,965	640,635	426,819	366,310	93,201	([f])	20
30 June 44	Theater	[g] 1,625,000	670,709	430,660	378,296	54,017	101,528	22
	Continent	([h])	([h])	([h])	([h])	([h])	([h])	13
	U.K.	([h])	([h])	([h])	([h])	([h])	([h])	9
31 Jul 44	Theater	1,770,845	725,259	447,818	413,056	66,426	118,286	22
	Continent	860,649	563,638	88,251	181,548	27,212	([e])	18
	U.K.	910,196	161,621	339,567	244,603	144,405	([e])	4[i]
31 Aug 44	Theater	1,905,261	755,603	449,688	479,359	85,275	135,336	26
	Continent	1,075,681	610,780	100,028	310,747	2,787	54,250	20
	U.K.	829,580	144,823	347,660	180,537	82,488	81,086	6
30 Sep 44	Theater	2,041,023	854,148	442,711	500,804	92,040	151,320	31
	Continent	1,353,079	820,407	132,726	329,288	19,104	51,554	30
	U.K.	687,944	33,741	309,985	171,516	72,936	99,766	1
31 Oct 44	Theater	2,196,785	1,006,190	435,384	538,636	90,604	125,967	40
	Continent	1,566,224	908,522	154,496	391,592	19,368	92,246	33
	U.K.	630,561	97,668	280,888	147,044	71,240	33,721	7
30 Nov 44	Theater	2,588,983	1,259,295	450,370	623,048	141,269	115,001	47
	Continent	1,906,441	1,114,455	176,533	475,495	58,182	81,776	42
	U.K.	682,542	144,840	273,837	147,553	83,087	33,225	5
31 Dec 44	Theater	2,699,467	1,315,201	443,634	657,129	172,078	111,375	52
	Continent	2,022,749	1,196,999	178,151	513,349	52,331	81,919	49
	U.K.	676,718	118,202	265,483	143,830	119,747	29,456	3
31 Jan 45	Theater	2,829,039	1,402,060	440,517	654,525	195,337	136,600	55
	Continent	2,179,026	1,333,132	179,285	519,653	62,671	84,285	55
	U.K.	650,013	68,928	261,232	134,872	132,666	52,315	0
28 Feb 45	Theater	2,935,000	1,506,000	440,000	666,000	164,000	159,000	59
	Continent	2,329,000	1,448,000	196,000	540,000	49,000	96,000	59
	U.K.	606,000	58,000	224,000	126,000	115,000	63,000	0
31 Mar 45	Theater	3,051,000	1,633,000	458,000	633,000	158,000	169,000	61
	Continent	2,553,000	1,617,000	214,000	529,000	65,000	128,000	61
	U.K.	498,000	16,000	244,000	104,000	93,000	41,000	0
30 Apr 45	Theater	3,059,942	1,613,004	482,785	652,779	128,305	183,096	61
	Continent	2,628,082	1,612,734	259,223	544,005	70,194	141,926	61
	U.K.	431,860	270	223,535	108,774	58,111	41,170	0

[a] *Source:* Progress Reports, Hq ETOUSA Jun 44–May 45, ETO Adm. U.S. Forces in Southern France, including the three divisions in Operation Dragoon, are included for the first time in the 30 November figures. These data are preliminary, unaudited figures prepared for command purposes and, while differing slightly from the audited DA AG strengths, have been used throughout this volume because of the subdivision into air, ground, and service troops. This breakdown is unavailable in DA AG reports.

[b] Mainly hospital patients.

[c] Ground Force Replacement System.

[d] Discrepancies between these figures and Table 8 as to the number of divisions on the Continent and in the United Kingdom are accounted for by the fact that divisions were often split as of the end of the month. Dates of arrival in Table 8 are for headquarters, regardless of the location of the bulk of the division.

[e] Does not include 14,430 men in Iceland on 31 May 1944.

[f] Included in nonoperating total for this date.

[g] Does not include 13,444 men in Iceland on 30 June 1944.

[h] Breakdown unavailable for June.

[i] Includes the two airborne divisions which had returned to the United Kingdom.

troop strength there at one time during that month exceeded 700,000, of which 165,000 comprised field forces.[30] The situation did not improve noticeably with the opening of Antwerp. In December the Communications Zone, strained to meet urgent operational commitments attending the sudden change in the tactical situation when the Germans broke through in the Ardennes, asked the British to accept additional troop convoys. The request did not entail the release of additional accommodations, but it did mean postponing relief for the congested British ports and railways. British officials agreed to accept an additional convoy as a matter of paramount military necessity. No additional divisions actually were routed through England. The last division to be accepted in the United Kingdom (the 76th Infantry) debarked on 21 December, and by mid-January 1945 all U.S. divisions had been transferred to the Continent.[31]

The difficulties attending the acceleration of the divisional build-up demonstrated rather pointedly the futility of attempting to commit more combat strength than was logistically supportable. For at least three months in the fall of 1944 U.S. administrative resources were simply unequal to the task of supporting all the available divisions in combat. Logistic planners had in fact before D Day predicted with remarkable accuracy that port discharge and transportation deficiencies rather than the availability of divisions would be the factors limiting the strength that could be committed against the enemy beginning at about D plus 120 (early October). Several divisions consequently remained idle after arriving in the theater, some of them in the United Kingdom, and some on the Continent in the vicinity of the ports.

(2) Service and Supporting Troops

Part of the problem of accelerating the build-up lay in the inadequacy of physical facilities on the Continent—mainly ports and transportation. Almost as important was the shortage of service troops. Plans had called for a division slice of about 40,000 men, 15,000 of which were to consist of corps and army troops (both service and supporting units) and 10,000 of communications zone troops. On this basis the War Department in February 1944 had authorized a theater troop strength of 2,390,000 men, including air forces.

The War Department had insisted that requests for service troops be kept at the absolute minimum when the troop basis was prepared in 1943 and the technical service chiefs considered their final allotments inadequate in many categories. Moreover, there was no certainty that they had requested the various types of troop units in proper proportion, for requirements could not be accurately foreseen. The troop basis therefore had little stability, and literally hundreds of changes were eventually made to meet the shifting needs.

As could be expected, many deficien-

---

[30] Mil Shipments Priority Mtg, 4 Nov 44; Conf of 24 Nov 44, Chorpening for Vaughan, SHAEF G–4 337 Comd and Stf Conf 1944, I.

[31] Mil Shipment Priority Mtgs, 23 and 30 Dec 44, SHAEF AG 337–18; Min, U.K. Base Stf Mtg, 26 Dec 44, First Army 337; Office of the Theater Historian, Order of Battle of the United States Army, World War II, European Theater of Operations, OCMH.

cies and some surpluses developed. Every service chief had to resort to the expedient of using units in work for which they had not been organized or trained, and in some cases had to deactivate units in order to get manpower for more urgently needed types. The experience of the engineers was fairly typical. Lack of sufficient depot troops forced the Corps of Engineers to assign an engineer general service regiment to depot operations even before D Day. Failure to close out the depots in the United Kingdom after D Day aggravated the shortage and led to the assignment of additional construction units to depot and maintenance operations. Twelve base equipment companies likewise were employed entirely in depot and maintenance operations and never performed the functions for which they were trained. Shortages of petroleum distribution companies, resulting from the rapid extension of the pipelines, were also met in part by the assignment of general service regiments to assist in the operation of the pipelines, and finally by the conversion of engineer combat battalions. Similarly, to meet a severe shortage of forestry companies, the engineers relied increasingly on general service regiments and combat engineer battalions to help in logging operations, and on civilian labor and prisoners of war to augment available units.[32]

The entire problem had been greatly aggravated in the summer and fall of 1944, at first by the sudden and rapid extension of the lines of communication, and then by the speed-up in the shipment of divisional formations without adequate supporting tails. COMZ officials had voiced their fears that the division slice would suffer a reduction when the transfer of divisions from the United Kingdom was suddenly accelerated in July. Their fears were hardly justified at the time, for the continental slice at the end of that month came to well over 43,000, and the theater slice to more than 50,000. The proportion of service troop support in France actually rose somewhat in August, when only two divisions were added to the continental strength. This did not necessarily mean that the slice was in proper balance as to types of units, and acute shortages were in fact already developing, particularly in transportation, depot, and maintenance units.

Added misgivings over the adequacy of service troop support arose as the result of the War Department's augmentation of the theater's troop basis by nine divisions, and as a result of the decision made in August to speed the shipment of divisions from the zone of interior without assurance that they would have their required complement of service and supporting troops. The Communications Zone at first estimated that the theater would need about 80,000 additional service and supporting troops for the increased divisional strength.[33] In mid-August it submitted a longer list to SHAEF, admitting that it had included certain increases not justified by the additional divisions because they were needed to remedy existing shortages.

[32] *Final Report of the Chief Engineer, ETO,* I, 131–33, 316–17.

[33] [Capt. John E. Henderson] The Procurement and Use of Manpower in the ETO, Pt. IX of The Administrative and Logistical History of the ETO, Hist Div USFET, 1946, p. 187.

General Bull, anticipating the objections these requests would raise in Washington, asked the G–4 and the medical, signal, and engineer officers at SHAEF to go over them carefully in light of operational experience thus far and operations planned for the future. The G–3 was especially conscious of the several warnings which the War Department had sounded on manpower limitations, and felt that every effort should be made to limit requests for additional personnel by exploring the possibilities of converting units in which there were surpluses and in utilizing liberated manpower for service activities. Any augmentation of the existing troop basis, he felt, would have to be met by the conversion of units already in the theater or by the elimination of a like number of personnel from the troop basis.[34]

Only a few days later the War Department in fact pointed out where ETOUSA could make an important saving in manpower. The troop basis agreed to earlier had authorized the European theater 198 antiaircraft artillery battalions, 111 of which were either already in the theater or en route. The War Department felt that in view of the air superiority which it enjoyed ETOUSA could well afford to cancel some of the flow scheduled for the next few months so that the surplus could be converted into units for which there was greater need.[35] The theater agreed, and shortly thereafter reduced its requirement for antiaircraft artillery to 146 battalions. It also informed the War Department that it was considering other reductions and that it intended to convert certain units already in the theater, such as smoke generator units, to other use.[36]

The cancellation of antiaircraft artillery units represented a saving of about 38,000 men, which ETOUSA expected to apply as a credit for units which it needed to meet deficiencies in the troop basis. It particularly desired additional truck drivers and counterintelligence troops, for which it had already submitted a request. But the War Department promptly disabused the theater of any such idea, and informed ETOUSA that it was simply deleting the 38,000 men from the troop basis. By War Department calculations, the European theater as of 1 August already had an overstrength of nearly 145,000 men, which constituted nearly half of the entire overstrength of the Army. ETOUSA, the War Department pointed out, still enjoyed a high priority for men and supplies, even at the cost of withholding badly needed units from other theaters. But in view of the fact that U.S. resources were insufficient to meet all

[34] Ltr, COMZ to SAC, 20 Aug 44, sub: Additional Sv and Supporting Troops Made Necessary by Additional Divs, SHAEF AG 370.5–6 General 1944; Ltr, Bull to G–4 *et al.*, 22 Aug 44, sub: Troop Basis for ETO, SHAEF G–3 O&E 370.5 Troop Basis 1944, I.

[35] Cbl W–88018, AGWAR to SHAEF, 28 Aug 44, SHAEF G–3 O&E 370.5 Troop Basis 1944, I.

[36] Cbl FWD–14526, SHAEF to Marshall, 11 Sep 44, SHAEF AG 370.5–6 General 1944. Early in October the theater was permitted to restore six battalions of antiaircraft artillery to the troop basis from surplus declared by the North African theater on the plea that needs for the defense of Antwerp had reduced theater resources to a dangerously low level. Cbl S–62891, SHAEF to AGWAR, 10 Oct 44, and Cbl WX–50722, AGWAR to SHAEF, 23 Oct 44, SHAEF AG 370.5–6 General 1944.

requirements, the theater would hereafter have to submit detailed justification for any requests for augmentation of its existing troop basis.[37]

Theater officials were plainly chagrined by the War Department's refusal. They had understood that the theater would get additional units by presenting justification for them and by offering cuts in existing allotments, conditions which had been met. Now the War Department had offered nothing in return for its sacrifices. Why, it asked, were further deletions necessary, and precisely what was the War Department's policy? [38]

General Bull supported the Communications Zone in its stand. The need for additional service and supporting troops, he argued, had been clearly established by the experience of the past few months, in particular by the unexpected extension of the lines of communication. The pursuit across northern France, for example, had necessitated the overstrengthening of all truck and troop transport companies. This, the theater felt, was certainly sufficient justification for the request for additional drivers.[39]

For the moment, at least, the War Department was unmoved by these explanations. Once again it called attention to the generous way in which the European theater had been treated with respect to the world-wide allocation of manpower. The simple fact was that savings had to be made. Where units originally authorized were no longer required there was no choice but to recover such resources for the satisfaction of other needs. In accord with this necessity the War Department announced that it was placing the recently canceled antiaircraft artillery units in the undeployed reserve to help meet the need for service and supporting troops for the additional nine divisions recently allotted the European theater. It asserted, moreover, that there was nothing inviolate about troop bases; they could be reduced as well as augmented. And it made clear that it had no intention of relinquishing its authority to control activation, inactivation, augmentation, and reorganization in such a way as to ensure the maximum exploitation of the nation's manpower resources. The implication was clear that the theater still did not appreciate the seriousness of the nation's manpower difficulties.[40]

Meanwhile the theater had revised its requirement for service and supporting troops needed for the nine additional divisions, the amended request totaling about 100,000 men. ETOUSA claimed that it no longer possessed an overstrength from which the needed units might be formed, for it had absorbed all surplus manpower in the process of

[37] Cbl WAR–38881, AGWAR to ETO, 29 Sep 44, OPD 370.5 Sec. XIV, Cases 487–510; Memo, Brig Gen O. L. Nelson, Asst Deputy CofS to OPD, 2 Sep 44, sub: Strength of the ETO 31 July 44, OPD 320.2 Sec. XX, Cases 365–90; Ltr, SHAEF to COMZ, 13 Sep 44, 1st Ind to COMZ Ltr of 20 Aug 44, Chief of Staff 400, Sec. I, Cases 2–35.

[38] Cbl EO52674, ETO to AGWAR, 6 Oct 44, and Cbl WAR–41104, AGWAR to ETO, 4 Oct 44, SHAEF G–3 O&E 370.5 Troop Basis 1944, I.

[39] Ltr, Bull to Smith, 8 Oct 44, sub: Theater Troop Basis, and Cbl E–51516, ETO to AGWAR, 1 Oct 44, SHAEF G–3 O&E 370.5 Troop Basis 1944, I.

[40] Cbl WAR–46185, AGWAR to ETO, 13 Oct 44, SHAEF G–3 O&E 370.5 Troop Basis 1944, I; Memo, Handy for Hull, 18 Oct 44, OPD 370.5 Sec. XIV, Cases 487–510.

augmenting truck companies so as to permit twenty-four-hour operation, and in the organization of provisional MP units to handle the large bag of prisoners of war and guard supply shipments against pilfering. The activation of sixteen MP battalions and of sixteen prisoner of war overhead detachments had absorbed about 13,500 men. As a matter of bookkeeping the War Department now added the MP battalions to the theater's troop basis and deleted a like number from units carried in the undeployed reserve. In the case of the prisoner of war detachments it merely legalized an increase in the theater's overhead allotment.[41]

The War Department eventually authorized 123,000 men to provide the service and supporting troop complement for the nine divisions. But it continued to balk at the theater's request for augmentations to the existing troop basis, particularly the request for 16,000 additional drivers, pointing out that ETOUSA had offered no compensating deletions. It finally approved the request for driver augmentation teams, but only on condition that the increase be accompanied by the inactivation of ten field artillery battalions which had been earmarked for ETOUSA in the undeployed reserve. The War Department wanted it clearly understood that the theater would be unable to get the field artillery units later if it elected to accept the driver augmentation.[42]

Any proposal to delete combat units from the troop basis naturally concerned the field commands. Theater headquarters therefore took the matter up with 12th Army Group, asking that it review its needs and recommend cancellations to compensate for the desired augmentations. Twelfth Army Group's first reaction was to oppose any cut in the field force troop basis, at the same time arguing the absolute necessity for the requested augmentations. General Bradley apparently felt, however, that the need for additional service and supporting units outweighed that for combat units, and, after reconsidering, concurred in the surrender of the field artillery battalions in order to cover the driver augmentations. In fact, the field forces were willing to accept the inactivation of additional combat units in case the other augmentations they desired—involving about 10,000 men for army, corps, and division headquarters, corps and divisional signal and MP units, and military intelligence personnel requested by 6th Army Group—could be obtained in no other way. Twelfth Army Group nominated the 20th Armored Division, due to arrive in February, for inactivation, should that be necessary.[43]

General Bull was not ready to accede to these proposals without a careful con-

[41] Ltr, ETO to WD, 2 Oct 44, sub: Additional Sv and Supporting Troops for Nine Added Divs, ETO 320.2 Strength and Troop Basis, I; Note for Record, OPD, 23 Oct 44, sub: Constitution and Activation of Certain MP Units, OPD 320.2 Sec. XXIII, Cases 456–76; Cbl WARX–50661, Handy to Smith, 20 Oct 44, SHAEF Cbl Log (in), Smith Papers.

[42] Memo, Brig Gen J. DeF. Barker, Deputy Chief Theater Gp OPD, information extracted from Memo for Lord, dated 30 Nov 44, and Cbl WAR–71217, AGWAR to ETO, 1 Dec 44, SHAEF G–3 O&E 370.4 Troop Basis 1944, I.

[43] TWX E–72663, COMZ to 12 A Gp, 10 Dec 44, and TWXs QX–30668 and QX–30669, 12 A Gp to ETO, 13 Dec 44, all in SHAEF G–3 O&E 370.5 Troop Basis 1944, I.

sideration of the consequences. He immediately asked General Crawford for a thorough review of the service support needs in the light of future operational plans in hopes of forestalling the inactivation of some or all of the field artillery battalions. The G-3 particularly questioned the need for 16,000 additional drivers. He admitted that driver augmentations had been necessary at the time of the Red Ball Express, but suspected that they would constitute a luxury in view of the increasingly serious manpower shortages.[44]

Bull's request led to the issuance of a directive to all the technical services asking them to review their service troop requirements. Once again the continuing competition over the slicing of the manpower pie was clearly evident. As might have been expected, G-4 officials and the technical service chiefs were unanimous in their opposition to any sacrifice of service units, and they were not lacking in arguments. The issue, as Crawford put it, really resolved itself into the necessity to decide between the G-3 and the G-4 estimate as to the course of future operations—that is, between the G-3's estimate that continued heavy resistance well into the summer of 1945 required that the theater have additional field artillery at its disposal, and the G-4's anticipation of a possible break-through into Germany, which would require that truck transportation be brought up to its maximum efficiency and potential so that it could support a sustained drive over extended lines of communication. Reinforcing the G-4 viewpoint was the urgent requirement to clear the mounting tonnages being discharged in the ports. Crawford noted, moreover, that there was little prospect that ammunition supply would improve sufficiently by the summer of 1945 to permit the 8-inch howitzer battalions already in the theater to expend ammunition at the 12th Army Group's desired rates. This prospect, plus the lateness of their availability, seemed to depreciate the value of the ten battalions.

The service chiefs were equally appalled at the thought of losing manpower to compensate for other activations. Most of them could go beyond defending their current troop bases and show that they were already deficient in manpower and strained to provide the theater's minimum service needs. The chief quartermaster, for example, argued that his requirements would rise, if anything, because it would be impossible to rely on prisoners of war as extensively for labor after entering Germany. In the view of the service chiefs, the need for every unit in the troop basis had long since been thoroughly justified. Service troop needs, they argued, had been kept to the lowest possible figures in the preparation of the theater troop basis. The War Department, admitting that it had not provided sufficient service troops in the world-wide troop basis, had made additional cuts. Finally, ETOUSA's troop basis had been drawn up on the assumption that it would have to support only four armies. But it had had to help make up deficiencies in the DRAGOON force, and had also borne the brunt of deletions to provide unforeseen non-T/O overheads for SHAEF and

[44] Memo, Bull for Crawford, 15 Dec 44, sub: Additional Drivers for QM Truck Companies, SHAEF G-3 O&E 370.5 Troop Basis 1944, I.

other headquarters. In short, the technical services considered any deletions of service troop units as inadmissable.[45]

The carefully documented case presented by the G–4 and the technical service chiefs apparently failed to convince the G–3. General Bull continued to oppose the inactivation of both the field artillery battalions and the 20th Armored Division, and thought the matter should be examined further. Late in December he asked the theater G–3, General Eyster, for additional information on which to base a decision.[46]

The Communications Zone, hoping at least to settle the matter of the field artillery inactivations, pressed the two army groups to come to an agreement on their desires regarding the various augmentations and reorganizations they had requested for army, corps, and division headquarters. More specifically, it wanted to know whether the army groups would carry out the proposed changes at the expense of combat units or through the cancellation of the truck drivers, and, if so, what units they would be willing to delete to compensate for the changes.[47]

The two army groups, apparently taking courage from General Bull's hesitation, now replied that they could not sacrifice any combat strength after all. Both commands listed the augmentations they desired for the planned reorganizations without offering any sacrifices of combat strength. The 12th Army Group asserted, moreover, that the provision of additional personnel to permit truck companies to operate sixteen to twenty hours per day was absolutely mandatory. Withdrawal of the driver augmentation teams, it said, would merely result in a requirement for additional truck companies. The army group's solution was simple: it recommended that ETOUSA ask the War Department to reconsider the presently established ceiling for the theater and authorize the needed troops.[48]

The army group replies hardly answered the theater's question. They indicated, moreover, that the field commands were totally ignorant of the appeals which the theater had already made to the War Department, and of the conditions which the latter had laid down for any additions to the theater's troop strength.[49] Nevertheless, the 12th Army Group refused to compromise and again tossed the problem back into the lap of theater headquarters. All the requested augmentations, it asserted, were essential and of equal priority; the de-

[45] Memo, Crawford for Bull, 21 Dec 44, Memo, G–3 (Opns) for G–4 *et al.*, 18 Dec 44, sub: Deletions from U.S. Troop Basis to Compensate for Essential Allotments, Study, Col William Whipple, 20 Dec 44, sub: Additional Drivers for QM Truck Companies, Memo OCQM for G–4, 21 Dec 44, sub: Deletion of QM Units From Theater Troop Basis, Memo, OCOO for G–4, 21 Dec 44, sub: Depletion U.S. Troop Basis to Compensate for Essential Allotments, Memo, Chief of SvC Troop Br G–4 COMZ for G–4 SHAEF, 21 Dec 44, sub: Deletions From U.S. Troop Basis to Compensate for Essential Allotments, Ltr, G–4 SHAEF to G–3, 26 Dec 44, sub: Deletions from U.S. Troop Basis to Compensate for Essential Allotments, all in SHAEF G–3 O&E 370.5 Troop Basis.

[46] Memo, Bull for Eyster, 29 Dec 44, sub: Fld Force Troop Basis of the ETO, SHAEF G–3 O&E 370.5 Troop Basis.

[47] TWX EX–80877, COMZ to 12 A Gp, 1 Jan 45, SHAEF G–3 O&E 370.5 Troop Basis.

[48] TWX BX–22462, 6 A Gp to 12 A Gp, 7 Jan 45, and TWX, 12 A Gp to COMZ, 21 Jan 45, both in SHAEF G–3 O&E 370.5 Troop Basis.

[49] TWX EX–89366, COMZ to 12 A Gp, 23 Jan 45, SHAEF G–3 O&E 370.5 Troop Basis.

letion of as many as 10,000 combat troops (involved in the deactivation of an armored division) would seriously unbalance the field force troop basis. Army group saw no solution except an increase in the troop basis. Once again it urged the theater to make an effort to convince the War Department that this would be necessary.[50]

SHAEF G–3 officials had come to similar conclusions from an independent study. The original service and supporting slice of 27,680 men, they argued, was the minimum and proper requirement for the support of each division, based on the past eight months' operations. On 31 December the theater had 1,392,100 service and supporting troops. This number provided a slice of 26,800, which was barely adequate for the support of the fifty-two divisions then present. The theater troop basis currently authorized a strength of 1,535,600 service and support troops, which would support not more than fifty-five divisions on the basis of a 27,680-man slice. The support of sixty-one divisions, which recently had been authorized the theater, consequently required the addition of another 153,100 men to the troop basis, raising the total ground and service force basis to 2,404,800 men. G–3 officials calculated that the increase should be apportioned in the ratio of 25.8 percent for combat and supporting troops, 68 percent for service troops, and 6.2 percent for miscellaneous overhead and administrative services.[51]

The outlook for obtaining such an augmentation was dark indeed. By late January, in fact, matters seemed to have reached an impasse, the theater at that time having requested augmentations totaling nearly 29,000 men, against which the War Department had offered about 3,600 men from the undeployed reserve without requiring compensating deletions. In the meantime the signal service had requested an additional 18,000 men; but it was obvious that that request would not get favorable consideration.[52]

At the end of January, ETOUSA, despairing of getting any concessions from the field forces, informed SHAEF that it had about decided to ask the War Department for an increase in the troop basis.[53] Meanwhile, however, it had also determined on certain downward revisions of its earlier requests. Based on a restudy of its requirements it now decided to pare its total request by about 6,000 men, from 28,700 to 22,600, the bulk of the saving to be effected through a reorganization of some of the theater's truck companies. Against this requirement the War Department was ready to make available to the theater 3,656 men without compensating deletions, 5,600 as the result of inactivations of artillery units in the undeployed reserve, 11,056 in miscellaneous units in the United States not specifically earmarked for the European theater, and 2,771 men resulting from a transfer of certain artillery units to the Mediter-

[50] TWX QX–31078, 12 A Gp to ETO, 26 Jan 45, SHAEF G–3 O&E 370.5 Troop Basis.

[51] Unsigned Memo on Troop Basis, G–3 SHAEF, 26 Jan 45, SHAEF G–3 O&E 370.5 Troop Basis.

[52] Memo, Twitchell for G–3, 24 Jan 45, sub: Rpt on Status of Fld Force Troop Basis, and Ltr, Twitchell to Bull, 23 Feb 45, sub: Additions to ETO Troop Basis, both in SHAEF G–3 O&E 370.5 Troop Basis.

[53] Cbl EX 92817, COMZ to SHAEF, 31 Jan 45, SHAEF G–3 O&E 370.5 Troop Basis.

ranean theater, for a total of 23,083. Minor adjustments led to a slightly different allocation, but agreement was finally reached with War Department officials early in February.[54] The sacrifice of the field artillery battalions proved the proper decision. German resistance was completely broken by mid-March, and the theater's most urgent need thereafter was for service support, particularly in transportation.

Little hope remained that the theater would get the 153,100 men estimated as needed to fill out the troop basis for all sixty-one divisions. General Bull concluded that on the basis of past experience the theater could expect no additional help from the zone of interior, and guessed that it would have to meet the deficit from its own resources.[55] General Crawford believed the shortages might be offset by more extensive use of civilians and prisoners of war, and by additional conversions of the less critical military units into the types needed. The theater had, in fact, already taken steps to utilize prisoners on a bigger scale.[56]

The theater's service troop problem had been aggravated by basic weaknesses in the composition of French forces in the 6th Army Group, and also by the problem of distribution within the theater caused by the transfer of combat strength from one area to another. The problem of properly apportioning service troops between the north and the south arose as soon as the DRAGOON forces came under SHAEF's control, in connection with both the transfer to the 6th Army Group of three divisions already in the theater and the proposed diversion to Marseille of three divisions scheduled for arrival in October.

A misunderstanding immediately arose as to the intent of the diversion to Marseille. The 6th Army Group assumed that the three divisions were to be added to the forces in southern France, and its G–4, Brig. Gen. Clarence L. Adcock, went to Paris in mid-September to submit a list of the service troops required by the 6th Army Group to support the units in combat. The 12th Army Group was under the impression that the three divisions were being routed through Marseille only because of congestion on the northern lines of communication, and naturally opposed the loss of combat units originally scheduled for assignment to its control, to say nothing of the prospect of sacrificing service and supporting units as well. In view of its own logistic difficulties, however, it could hardly deny the logic of committing those divisions in an area where there was greater likelihood of providing adequate support, although the diversion entailed a delay in the build-up of the newly arrived Ninth Army. As COMZ and SHAEF supply

[54] *Ibid.;* Ltr, Whiteley to Smith, 4 Feb 45, sub: Augmentations to Units on Theater Troop Basis and Atchd Study of Rqmts and Credits for Augmentations to U.S. Forces in ETO, 4 Feb 45, TWX E–95086, COMZ to 12 A Gp, 6 Feb 45, Cbl S–79719, SHAEF to COMZ, 11 Feb 45, Ltr, Twitchell to Bull, 12 Feb 45, sub: Augmentations to Theater Troop Basis, Cbl EX–87333, COMZ to AGWAR, 12 Feb 45, all in SHAEF G–3 O&E 370.5 Troop Basis.

[55] Ltr, Bull to Crawford, 14 Feb 45, sub: Troop Basis for ETO, SHAEF G–3 O&E 370.5 Troop Basis.

[56] Memo, Crawford for Bull, 21 Feb 45, sub: Troop Basis for ETO, and Ltr, Crawford to Eyster, 24 Feb 45, sub: Troop Basis for ETO, both in SHAEF G–3 O&E 370.5 Troop Basis.

officials pointed out, the divisions could well be spared, since they were in excess of the number that could be supported in the north.[57]

Of much greater concern to supply staffs was the proposed transfer of service and support units. Such transfers obviously could have no other result than to reduce the scale of support in the north, where supply was already strained, and would therefore be contrary to SHAEF's declared policy directive giving operations in the north the highest priority. With these considerations in mind Colonel Whipple, the SHAEF chief of logistical plans, recommended that the transfers of service troops southward for the support of more than the ten divisions originally allotted to operations in southern France be approved only insofar as they could be spared without impairing the efforts of the 21 and 12th Army Groups.[58]

SHAEF meanwhile confirmed the 6th Army Group's claim that the three disputed divisions were intended for use in the south. Both the Communications Zone and 12th Army Group had already drawn up tentative lists of units which they were willing to have transferred in the event SHAEF should so decide, and on 22-23 September representatives of SHAEF, the Communications Zone, and the two army groups met at Lyon to agree on the exact number of units of each type that were to be transferred to southern France.[59]

In the course of the conference General Devers' chief of staff, Maj. Gen. David G. Barr, admitted that the requests originally made by the 6th Army Group included units required to meet earlier shortages in the DRAGOON troop basis. These were quickly discounted, for the conference had no authority to remedy basic deficiencies or compensate for original shortages. Moreover, 12th Army Group certainly would have objected to such adjustments, or would at least have countered with claims of its own. Nor was there, in view of the overall shortages in the theater, any thought of providing the diverted units with anything like "normal" administrative tails specified in logistical planning factors. Instead, the principle was generally followed of considering the total number of units of particular types available in the theater and making an equitable redistribution so as to provide approximately equal support to all divisions in the theater. On this basis SHAEF on 29 September ordered the transfer of a specific number of service and supporting troop units, totaling about 29,000 men, to 6th Army Group for the support of the six divisions scheduled for transfer or diversion to southern France—that is, the XV Corps, consisting of the 2d French Armored and the 79th U.S. Infantry Divisions, plus one division to be designated later,[60] and the three di-

[57] Memo, Col B. A. Holtzworth, Chief Orgn Br G-3 12 A Gp, 14 Sep 44, sub: DRAGOON SvC Troop Rqmt; Memo, Osmanski for Whipple, 16 Sep 44, sub: Mtgs in Paris on Diversion of Troops to Marseille, SHAEF G-4 Diversion of Service Troops DRAGOON, Box 1, Folder 51.

[58] Memo, Whipple for G-4, 19 Sep 44, sub: Diversion of Log Resources for Support of DRAGOON Forces, SHAEF G-4 381 ANVIL.

[59] Memo, Osmanski for Whipple, 24 Sep 44 sub: Discussion With Hq 6 A Gp—Diversion of Supporting Sv Troops, SHAEF G-4 Diversion of Service Troops DRAGOON, Box 1, Folder 51.

[60] The 12th Armored Division, which did not arrive until November.

visions which were to arrive at Marseille the next month (the 100th and 103d Infantry and 14th Armored Divisions).[61]

Hopes were high at this time that the southern line of communications might develop surplus capacity which could help sustain forces in the north. General Adcock, in arguing for additional service units for the southern army group, had asserted, in fact, that the ports and line of communications of southern France had a potential capability of supporting the entire Third Army in addition to the recently augmented DRAGOON force of sixteen divisions.[62] With this prospect in mind, COMZ officials of both the European and Mediterranean theaters met at Dijon on 11–12 October to consider additional transfers of service troops for the purpose of developing the maximum capacity of the Rhône line of communications. SOLOC planned to develop rail facilities in southern France sufficiently to handle 20,000 tons per day, 4,000 of which would be in excess of its own needs. For this purpose it badly needed engineer and signal troops, truck companies, and port battalions, its total requirement coming to about 38,000 men.[63]

SOLOC's bid again raised the question as to the relative adequacy of logistic support in the two areas. Although it considered the request reasonable, the Communications Zone at first opposed any additional transfers on the ground that they could not be spared without seriously affecting the support of the 12th Army Group. Operations in the north had higher priority than those in the south, and the Communications Zone was making every effort in October to rebuild the supply structure supporting the 12th Army Group so that the offensive could be resumed early the next month.

Supply officers at SHAEF reacted similarly. Colonel Whipple, after a restudy of the respective troop bases of the northern and southern forces, concluded that any additional diversions would provide the southern forces with more nearly adequate logistic support than those in the north. Completion of the transfers and diversions then in progress, he calculated, would result in division slices of about 39,000 and 34,500 men respectively for the OVERLORD and DRAGOON forces. These were admittedly short of the 40,000 and 42,500 respectively originally planned for the two areas.[64]

Two developments had occurred, however, which were completely disrupting plans for the adequate support of U.S. forces in the north. The recent accelerations in the flow of divisions from the United States were so unbalancing the troop basis that forces in northern France would be short five average administrative tails by December. Furthermore, the rapid extension of lines of communications in August and September had so thinned out the

[61] Ltr, SHAEF to COMZ, 29 Sep 44, sub: Transfer of Supporting and Sv Units, SHAEF G–4 Diversion of Service Troops, DRAGOON, Box 1, Folder 51; Notes on Conf at Hq SOLOC, 17 Jan 45, ETO 334 Confs at SOLOC.

[62] Memo, Osmanski for Whipple, 24 Sep 44.

[63] Notes on Conf, 11–12 Oct 44, SHAEF G–4 337 Conference 1944, I.

[64] The planned 40,000-man slice was made up as follows: 14,953 men in the division, 6,666 supporting combat troops in the army and corps, 8,028 service troops in the army and corps, and 10,263 service troops in the communications zone, not including those in the United Kingdom.

available logistical resources that the planned slice, designed to operate a road line of communications of no more than 150 miles, could no longer adequately support a division at the German border. Colonel Whipple estimated that thirty divisional tails were adequate to support only about twenty divisions in the existing circumstances. On the other hand, twelve tails in southern France, he argued, could support an equal number of divisions in the Belfort gap, as planned. Summing up, he asserted that the 12th Army Group must get first consideration as long as its operations had higher priority.[65]

The Communications Zone, probably because it better appreciated the logistic straits of the theater and the value of any additional line-of-communications capacity that might be developed, was willing to meet at least part of the 6th Army Group request. It suggested the transfer of about 15,000 men, some of them to come from its own resources, some from the 12th Army Group. But protests from the latter resulted in the major portion of the allotment being made from the Communications Zone. The SHAEF G-4 finally determined exactly what units were to be transferred on the basis of operational priorities, and on 21 October ordered the transfers.[66]

Although the 12th Army Group had avoided a major raid on its own resources, it realized that transfers from the Communications Zone would ultimately have their effect on the combat zone. It was particularly concerned about the loss of ordnance maintenance units, and stated that the effect would soon be evident unless replacements were provided. Moreover, it suspected, with some justification, that the 6th Army Group bid had again included troop units which were not required solely for the planned development of the Rhône line of communications, but to correct original deficiencies in its troop basis. If such was the case, the 12th Army Group requested that the theater ask the War Department for additional service troops to compensate for reductions in the planned proportion of service to combat troops. But ETOUSA disapproved this request, having only recently been turned down on the matter of the credit for the deleted antiaircraft artillery battalions.

The recent transfers continued to worry the 12th Army Group, which feared future raids. In its view, the theater was not facing up to the basic deficiencies in the troop basis, and it was dissatisfied with what it considered a haphazard manner of meeting crises by shuffling resources between the various areas in the theater. Uncertainty as to future action on service troops obviously troubled the field commands, just as uncertainty over supply was also affecting operational planning at this time.

---

[65] Memo, Whipple for Crawford, 12 Oct 44, sub: Diversions of Sv Troops to DRAGOON, and Draft Memo, Crawford for CofS (apparently drafted by Whipple), n.d., sub: Sv Troops for DRAGOON Force, both in SHAEF G-4 381 ANVIL 1944, I.

[66] Ltr, Lord to CG COMZ NATO, 14 Oct 44, sub: Additional COMZ Troops Required in Southern France, Ltr, Lord to SHAEF, 15 Oct 44, same sub, Memo, Whipple for G-4, 17 Oct 44, sub: Diversion of Sv Troops to DRAGOON, Ltr, Whipple to G-4, 18 Oct 44, sub: Sv Troops to DRAGOON, and Ltr, SHAEF to COMZs ETO and NATO, 21 Oct 44, sub: Sv Troops for DRAGOON, all in SHAEF G-4 Diversion of Service Troops DRAGOON, Box 1, Folder 51.

The army group wanted a definite allocation of units available or due from the United States so that it would know what it could count on having. SHAEF recognized the reasonableness of the request and late in November instructed the Communications Zone to make a firm allocation of service units between the field forces and the Communications Zone. Such an allocation was finally made early in February 1945.[67]

Early in November 6th Army Group again had requested additional service troops, mainly because of the situation in the First French Army. The French had been willing enough to form combat divisions, but had been notoriously remiss about activating and training the required complements of service and supporting troops. U.S. forces consequently had been forced to provide a larger and larger share of First French Army's logistic support in southern France.[68] SHAEF considered the new requests, admitted their reasonableness and desirability, but finally disallowed them. The strained logistic situation, it noted, plus the fact that operational priorities continued to favor forces in the north, simply prohibited additional transfers from the north.[69]

Despairing of obtaining additional troops, General Larkin, the SOLOC commander, now urged both Delta Base Section and CONAD to organize a much fuller utilization of German prisoners of war. SOLOC was already making extensive use of Italian service units.[70] Attempts were also made to employ civilian labor, but the demand far exceeded the supply. The Seventh Army had found, for example, that the Germans had removed most of the able-bodied men from Alsace. What manpower remained was either needed for the civilian economy or was being recruited by the French military services.[71]

Determining precisely what the scale of service support on the northern and southern lines of communications was at any one time was a highly controversial matter, as was always the case where statistics were involved. On 1 December, for example, the division slice within the 6th Army Group exceeded 30,000 men while that of the 12th was in the neighborhood of only 27,000. Compensating for this seeming imbalance, however, was the fact that each division on the northern line of communications was supported by about 11,300 service troops in the communications zone, while in the south the corresponding figure, according to ETOUSA, was only 8,600. The total theater slice on the two lines of communications therefore appeared to be roughly in

[67] Ltr, Maj Gen Leven C. Allen, CofS 12 A Gp, to CG ETO, 25 Oct 44, sub: Sv Troops for 6A Gp, with Inds 3, 16, and 24 Nov 44, and 5 Feb 45, ETO 381/320 Troops—OVERLORD 1944.

[68] Marcel Vigneras, *Rearming the French,* UNITED STATES ARMY IN WORLD WAR II (Washington, 1957), a volume in the Special Studies subseries, treats the subject of French forces in detail. See especially Chapter VII.

[69] Stf Study, G–3 Opns SHAEF, sub: Availability and Distribution of Supporting Units, 14 Nov 44, and Ltr, Smith to Devers, 23 Nov 44, sub: Additional Combat Support for 6A Gp, SHAEF G–3 War Diary.

[70] Ltr, Larkin to CGs CONAD and Delta Base Sec, 30 Nov 44, SOLOC Gen Larkin Out File.

[71] See Ch. XVIII, Sec. 2, for a summary of the attempts at local procurement of labor. Ltr, Larkin to CGs CONAD and Delta Base Sec, 30 Nov 44, SOLOC Gen Larkin Out File; Seventh Army G–4 Periodic Rpt 2, 10 Dec 44, SUSA 319.1 G–4 Weekly Periodic Rpts.

balance, totaling 38,000 in the north and 39,000 in the south.

But there were other factors to consider. The ETOUSA G–3 pointed out in December that the SOLOC totals did not include about 20,000 men in the Italian service units, which were employed in support of the forces in southern France. Their inclusion, on the basis of the nine U.S. divisions then operational in the 6th Army Group, raised the slice of the southern France forces from 8,600 to 10,800.[72] General Larkin was quick to challenge the basis on which this comparison had been made. For one thing, it had taken no account of the eight divisions of the First French Army, the support of which was fully as much a responsibility of SOLOC as was the support of the Seventh U.S. Army, but for which the French had thus far provided only 18,300 service troops. Moreover, he noted that Italian service units could not be counted as having the same effectiveness as trained U.S. military units. More properly, Larkin maintained, SOLOC, with a strength of 106,464 service troops (88,-164 U.S. and 18,300 French) was supporting 17 divisions (9 U.S. and 8 French), the COMZ slice thus averaging only 6,262 men.[73]

In the meantime two further augmentations of the 6th Army Group's combat strength, one of them temporary, had added to SOLOC's service troop difficulties. Early in December the nine infantry regiments of the 42d, 63d, and 70th Divisions arrived at Marseille minus their normal divisional service support. Later in the month the 6th Army Group acquired another division (the 87th Infantry) as the result of the shift in the army group boundary arising out of the situation in the Ardennes. Sixth Army Group's strength thus rose to twenty-one divisions. The latter acquisition turned out to be temporary; but the nine infantry regiments, although initially intended for employment on a rotational basis, were the advance elements of three full divisions. SOLOC and 6th Army Group therefore took the occasion to request additional service troops.

As of mid-January nothing had yet come of this request, and General Devers again appealed for additional service troops, noting that the separate infantry regiments had been employed constantly since their arrival without adequate service or combat support, and that the arrival of the remaining elements of the divisions within the next few days would shortly necessitate the support of the complete units. Members of the SOLOC staff repeated this plea at a conference with Generals Lee and Somervell a few days later, and noted that additional French divisions were also scheduled for activation and commitment. SOLOC officials claimed that the COMZ service troop slice in southern France had dropped to 4,425 men, and argued that southern France was critically short of service troops despite the maximum use of prisoners, Italians, and civilians.[74]

SHAEF at this time made an additional allocation of service units based

[72] Ltr, Eyster to Larkin, 18 Dec 44, SOLOC Gen Larkin Out File.

[73] Ltr, Larkin to Eyster, 27 Dec 44, SOLOC Gen Larkin Out File.

[74] Cbl B–22667, 6 A Gp to SHAEF, 13 Jan 45, ETO Cbls, ETO Adm; Notes on Conf at Hq SOLOC, 17 Jan 45, ETO 334 Confs at SOLOC.

on the assignment of 12 U.S. divisions to the 6th Army Group. Almost simultaneously it transferred the equivalent of another corps to the 6th Army Group for the Colmar operation, which gave the 6th Army Group a strength of 16 U.S. and 9 French divisions. SHAEF made a supplementary loan of service units for the support of this additional combat strength, ordering the 12th Army Group to release about 12,000 troops for this purpose with the assurance that they would be returned.[75]

Upon the completion of the Colmar operation in February 6th Army Group asked that it be permitted to retain certain of the units, especially combat engineers, which had been attached when it had taken over a portion of the Third Army front in December. But SHAEF refused, and by the end of February the divisions temporarily attached to 6th Army Group, along with the accompanying service and support troops, were returned northward.[76]

In the meantime ETOUSA completed the theater-wide reallocation which had been undertaken some months before at the urging of both army groups. By the end of February, therefore, a satisfactory redistribution was under way on the basis of a strength of twelve U.S. divisions in the 6th Army Group and forty-five in the 12th.

The service troop allocation was again under revision when hostilities came to an end. At the end of April the 6th Army Group had a combat zone slice of 30,500 as against a slice of 26,500 in the 12th Army Group. The over-all COMZ slice at that time came to 8,919 on the Continent, and to 10,700 in the theater as a whole.[77]

Effecting an equitable distribution of service troops within the theater was difficult at best. Varying local conditions and circumstances on the two principal lines of communication made it impossible to assign units purely on a mathematical basis, and the ratio of combat strength between the two army groups was constantly being upset. Both army groups, and particularly the 6th, were understandably impatient with the interminable delays in adjustments which were called for with the shifts in combat units, and argued for a procedure which would provide a more automatic shift in logistic support with the alterations in combat strength. The difficulties attested, moreover, to the fact that the southern forces had not been completely integrated into the European theater structure. In logistic matters SOLOC had long maintained ties with the Mediterranean theater. In some respects complete integration with the European theater was never accomplished.

---

[75] Cbl S–74958, SHAEF to 6 A Gp, 14 Jan 45, SHAEF G–3 War Diary; 6 A Gp G–4 AAR for Jan 45, p. 5; Cbl S–77528, SHAEF to A Gps, 3 Feb 45, ETO Cbls, ETO Adm.

[76] Cbl EX–24137, 6 A Gp to SHAEF, 10 Feb 45, and Cbl S–78852, SHAEF to 6 A Gp, 13 Feb 45, ETO Cbls, ETO Adm 407–08; 6 A Gp G–4 AAR for Feb 45, p. 9.

[77] [Henderson] Procurement and Use of Manpower in the ETO, pp. 249, 252–55.

CHAPTER XI

The Manpower Problem, August 1944–February 1945

(1) Rumblings of a Replacement Problem

Of all the logistic problems that plagued ETOUSA in the fall of 1944 the shortages of ammunition and replacements undoubtedly caused the greatest anxiety. In their development and chronology the two problems were closely parallel. In both there was speculation as to possible shortages even before D Day; in both a crisis developed in the fall of 1944, necessitating emergency measures and longer-range plans to ensure adequate support for the last months of the war.

The theater's first difficulties with replacements, in July, had resulted partly from the fact that losses in infantry, especially infantry riflemen, had been considerably higher than forecast for the first two months of operations, and partly from the fact that the War Department had not shipped replacements in the various branches in the proportions agreed to before D Day. The theater did not actually lack replacements at the end of July. But its stockage at that time was completely out of balance as the result of the disproportionately heavy losses in infantry riflemen. Converting men of other branches to infantry obviously could not solve the immediate difficulty. The theater therefore turned to the War Department for emergency shipments, at the same time asking that it increase the proportion of infantry riflemen in all future replacement training.[1]

The July experience served to focus attention on a larger manpower problem. The Army had already exceeded its authorized strength of 7.7 million men, and a serious shortage was developing in the Army as a whole. This development had in fact been the subject of repeated warnings from the War Department beginning as early as September 1943, when the Chief of Staff called attention to the manpower ceiling under which the Army thereafter had to operate, and warned that there would have to be greater economy in the use of men. General Marshall at that time suggested that it might be advisable to establish an investigating agency on the model of the War Department Manpower Board, which was then conducting extensive investigations of the entire Army estab-

[1] See *Logistical Support I*, 460–63.

lishment in the zone of interior with a view to releasing unneeded personnel for more urgent assignments. Visitors from the War Department to overseas theaters, he said, reported the impression that there was an unnecessary extravagance in the use of manpower in service installations, and he deemed it essential that there be a continuing review of the theater's needs relative to changing missions so that manpower could be transferred and utilized more efficiently, or recovered and transferred to more urgent tasks.[2]

In January 1944 Marshall had again called attention to the critical manpower situation developing in the United States, suggesting additional measures the theater could take to help solve the problem. Marshall observed that the manpower shortage was being aggravated by the mishandling of two groups of men: physically imperfect men who could still render useful service were being discharged, and men physically qualified for general assignment were being used in limited assignment positions. The Army, he said, would simply have to make better use of the manpower it already had. Basically, this meant conserving and properly using the important resource which it possessed in limited assignment personnel.[3]

The necessity for action along these lines was again emphasized in February 1944 as the result of a survey which Col. George R. Evans, chief of the Classification and Replacement Branch of The Adjutant General's Office, made of the entire replacement situation in the European theater. Evans urged the Communications Zone to direct all its units and installations to survey their personnel with the aim of identifying individuals physically qualified for field duty (other than those occupying key or highly technical positions) who could be replaced by men physically disqualified for full field service. The Communications Zone was to earmark such men for assignment to field force units as physically handicapped individuals were made available for reassignment to the Communications Zone.[4]

In April 1944 at a G–1 conference in Washington attended by representatives from both the European and North African theaters,[5] War Department officials tried to impress even more strongly upon the theaters the necessity for action along these lines. General McNarney, the Deputy Chief of Staff, rightly suspecting that the theaters still did not appreciate the seriousness of the manpower shortages, again made it clear that the Army had reached its authorized strength of 7.7 million men and that the acquisition of new troops henceforth would be restricted to the numbers required to maintain that strength. This meant that new demands for men not already provided for in the troop basis would have to be met by reduction elsewhere. The day had passed when personnel could be obtained for the asking.

[2] Cbl R–3412, Marshall to CG ETO, 23 Sep 43, ETO 381 Troop Basis 1943.

[3] Ltr, Marshall to Eisenhower, 6 Jan 44, ETO GFRC 300.5 Circulars, Hq Replacement System 1944.

[4] Memo, Evans for G–1 ETOUSA, sub: Observations and Recommendations, FFRS, ETOUSA, 31 Jan 44, ETO GFRC Planning file.

[5] The ETOUSA representatives were Brig. Gen. Oscar B. Abbott, the G–1, Brig. Gen. Ralph B. Lovett, the adjutant general, and Maj. Gen. Paul R. Hawley, the chief surgeon.

War Department officials were particularly critical of the North African theater, which apparently had been extravagant in its use of manpower for rear area services and which had failed to take effective measures to transfer able-bodied men from the supply services and retain them for combat. They were determined that the experience in that theater should not be repeated in Europe, and insisted that the theater not only adopt the War Department's policies on the conservation of manpower, but that it organize its replacement system along lines prescribed by the War Department so that those policies could be carried out effectively. Later in April NcNarney went to the United Kingdom and repeated these warnings at a theater command and staff conference.[6]

Despite these admonitions, plus strong criticism of General Lee for his opposition to the War Department's recommendations, the theater did not take effective action. ETOUSA had already adopted the policy of retraining limited assignment men who were physically able to serve usefully in some other military capacity. But it shrank from taking the necessary measures to remove general assignment men from service units and retrain them, and it resisted pressure to establish the kind of agency which General Marshall had originally recommended to scrutinize the use of manpower in the theater.

ETOUSA found one reason after another to postpone such distasteful work. General Devers, who was still theater commander in the fall of 1943, had originally opposed the idea on the ground that operational plans had not crystallized sufficiently to permit a thoroughgoing survey of troop requirements.[7] In June 1944 Maj. Gen. Ray W. Barker, the SHAEF G–1, offered a plan for a comprehensive survey of manpower problems, covering not only the matter of more effective utilization of limited assignment men, the release of general assignment personnel, and the use of prisoners of war and liberated manpower, but involving a thorough examination of the theater's organization with a view to uncovering and eliminating duplication of function and responsibilities. As part of the plan he proposed the establishment of a theater manpower board which would operate directly under the theater commander with wide powers to investigate all the ramifications of the manpower problem and make specific recommendations as to where savings should be carried out.[8]

The G–1's proposal appears to have been made in the true spirit of the War Department's directives, and was the first attempt to come to grips with the problem realistically. General Bull, the G–3, concurred in the plan. But General Crawford, the G–4, expressed strong disapproval of the idea, arguing that such a job was a command function, and that no "committee" could carry out such a survey which would not use up more

[6] Remarks by McNarney and Transcript of G–1 Conf on Repls and Pers Control, 3 Apr 44, and Ltr, Abbott to Lee, 15 Apr 44, sub: Rpt on Conf Held in Washington, D. C., both in SHAEF AG 200.3–1 Replacements 1944; Notes on Comd and Stf Conf ETO, 18 Apr 44, EUCOM 337/3 Conferences, Staff Weekly 1944, I.

[7] Cbl W–5527, Devers to Marshall, 8 Oct 43, ETO 381 Troop Basis 1943.

[8] Memo, Barker for Smith, 28 Jun 44, sub: Theater Manpower Bd, SHAEF G–1 320/2 Manpower, Establishment, Equipping.

manpower than it might save.[9] An amended proposal, which Barker submitted in answer a few weeks later, also encountered objections from the G–4, and for the moment, at least, the matter was dropped.[10]

Personnel officers at General Lee's headquarters also opposed the creation of a manpower board, mainly on the ground of uncertainty as to the future COMZ organization on the Continent. Late in July they found additional support for this argument as the result of the addition of eight divisions to the ETOUSA troop basis, which was expected to involve the activation of additional service units within the theater. In any case, the Communications Zone preferred to leave to the section commanders the responsibility for combing out general assignment men and replacing them with limited assignment personnel.[11]

It was a misreading of human nature, to say the least, to expect commanders to carry out measures which would obviously be to their own disadvantage, and it was a policy which in the end proved totally inadequate, as might have been expected. For the time being, the Communications Zone preferred to postpone the difficult business of screening general assignment men out of the service forces, and confined itself to issuing general pronouncements that the "wastage or improper use of manpower will not be tolerated in this theater," and toothless injunctions that men would be "assigned to positions in which they can render the maximum service." Such directives, while outwardly conforming with the War Department's prodding on the subject, were hardly specific enough to be enforced, and were in fact easily circumvented.

The crisis of July provided the theater with a dramatic reminder of its manpower problem, and the War Department took the opportunity to express its impatience with the theater for what it regarded as poor planning as well as poor administration of manpower resources. The War Department's main criticism at that time focused on the succession of revised requisitions which had followed the discovery of shortages in infantry. The War Department regarded this as evidence of poor planning,[12] and McNarney at the time expressed doubts as to the competence of Lee's G–1.[13]

The theater's actions were partially defensible, at least so far as the shortages in July were concerned. Lee could point to two extenuating circumstances: the prolonged hedgerow fighting had taken a toll of infantrymen which the field commands had not foreseen; and shipments from the zone of interior had

[9] Memo, Crawford for G–1, 30 Jun 44 sub: Theater Manpower Bd, SHAEF G–1 320/2 Manpower.

[10] Memo, Barker for G–3 and G–4, 11 Jul 44, sub: Conservation of Manpower, and Crawford's reply, 14 Jul 44, SHAEF G–1 320/2 Manpower.

[11] Memo, Franey for Lee, 13 Jul 44, ETO GFRC file on Replacements. The view that the screening out of able-bodied men should be left to the section commanders was also held by the Replacement Section of the Adjutant General's Office, ETO. Memo, AG Repl for G–1, 7 May 44, in reply to proposal by ETO G–1 (29 April 1944, subject: Disposition of Limited Service Men), that a board be established to conduct continuous surveys of the SOS for this purpose, to be directly responsible to the deputy theater commander, EUCOM 322 Replacement Units, II (a).

[12] See *Logistical Support I*, 461–63.

[13] Ltr, McNarney to Lee, 9 Aug 44, CofS 322 Sec. I, Cases 1–30.

been unbalanced as to type, even by accepted War Department planning factors—the infantry shipments of May, June, and July containing only 35, 58, and 50 percent respectively of riflemen as against a previously accepted factor of 64.3. Lee maintained, in addition, that the troop build-up had been more rapid than planned, although this was a tenuous argument insofar as the months of June and July were concerned.[14]

McNarney admitted that the European theater had been shortchanged on infantry riflemen, and explained the unbalanced make-up of the May–July shipments by the necessity to meet the North African theater's expected requirements for the southern France operation, and by the fact that the War Department had been forced by popular demand to place certain restrictions on the age at which combat replacements would be shipped to overseas theaters.[15]

Washington's concern over the unreliability of the European theater's estimates of future requirements was understandable. It was on the basis of these that the output of the training centers had to be planned, normally five to six months in advance of actual need. Again and again, according to McNarney, the theater's forecasts of needs had proved inaccurate, with the result that the War Department was forced to resort to painful improvisation in order to meet the theater's needs. There was, of course, no foolproof formula for estimating replacements needs. The Deputy Chief of Staff acknowledged this, asking only that the theater adjust the estimates of its needs as promptly as possible to actual experience.[16]

Behind the frustration over the unreliability of planning estimates lay the suspicion that the theater was not making the best use of its men. The War Department therefore continued to prod ETOUSA on the subject of using its available manpower to better advantage. Late in August it notified all theaters that it would be able to meet replacement requirements as currently estimated through December 1944, since replacements scheduled for shipment in that period were already in training. But it gave unequivocal warning that beginning in January 1945 it would be able to provide only a portion of the theater's estimated needs. It reminded the theaters, moreover, that War Department policy required that they provide a training and assignment system for men no longer physically capable of performing their previous duty assignments, for men physically capable of performing combat duty who were withdrawn from COMZ units, and for the conversion of surpluses in particular arms and services.[17]

Basically, the theater's problem boiled down to one of finding enough physically qualified men to meet its combat losses, and it had three possible sources which it could exploit to meet this need: (1) overages in types other than infantry which it might retrain; (2) theater overstrength; and (3) general assignment men

---

[14] Ltrs, Lee to McNarney, 25 Aug and 23 Sep 44, CofS 322 Sec. I. Cases 1–30.

[15] Ltr, McNarney to Lee, 29 Sep 44, CofS 322 Sec. I, Cases 1–30. See Robert R. Palmer, Bell I. Wiley, and William R. Keast, *The Procurement and Training of Ground Combat Troops*, UNITED STATES ARMY WORLD WAR II (Washington, 1948), pp. 205–09.

[16] Ltr, McNarney to Lee, 10 Oct 44, CofS 322.

[17] Ltr, WD to Theaters, 23 Aug 44, sub: Overseas Repls, SHAEF G–3 370.092 Reinforcements 1944.

in the supply services of the Communications Zone and Air Forces who could be withdrawn for conversion to a combat arm and replaced by men no longer physically qualified for such assignment.

The theater had already had some experience in the retraining of men, albeit a very limited one. In April 1944 agreement had been reached with the War Department to raise the proportion of infantry in the replacement pool from 64.3 percent to 70.3. Since it was already too late at that time to make adjustments in the May shipments, the theater took steps to retrain as infantrymen approximately 2,500 men, representing overages in other branches and replacements being improperly used, in an effort to establish what it regarded as a safe level of infantry replacements by D Day.[18]

Meanwhile the theater had also laid down the first outline of a policy on the utilization of limited assignment men. Shortly before the invasion, on the suggestion of the theater G–1, arrangements were made to establish machinery within the replacement system to receive, classify, and redistribute all personnel returning from hospitals and rehabilitation centers, to retrain limited assignment men, and to distribute to appropriate branch replacement depots all recovered general assignment men who were to be retrained for combat assignments. The theater commander approved setting aside certain facilities at the American School Center at Shrivenham, England, to be used by the replacement system for these purposes. It was realized from the start, however, that the number of men who would become available for limited duty by return from hospitals would far exceed the Communications Zone's normal losses, and that limited assignment personnel could be absorbed only through the release of able-bodied men for combat. Exactly how this was to be accomplished was a matter of considerable disagreement.[19]

Policy on limited assignment men was further clarified and developed during the first months on the Continent. The armies agreed, for example, to absorb limited assignment men up to 5 percent of their strength, some of whom would of course have to be retrained for new duties. Limited assignment troops from the Communications Zone were to return from hospitals and rehabilitation centers directly to their former units without requisition and be carried as temporary overstrength until absorbed by normal attrition. Limited assignment men from the combat zone were to be retrained for new assignments and absorbed by the Communications Zone.[20]

Everyone thus apparently appreciated the necessity to utilize limited assignment personnel. But no truly effective measures were as yet being taken to withdraw general assignment men from the Communications Zone to make room for

[18] History of the Ground Force Replacement System, ETO, Pt. I: The Replacement System From the Date of Activation to D Day, prep by the GRFC, ETO Adm, Ch. III, Sec. III, pp. 10–11, ETO Adm 334.

[19] Memo, G–1 for AG Repls, 29 Apr 44, sub: Disposition of Limited Sv Men; AG Repl reply, 7 May 44, EUCOM 322 Replacement Units, II (a); Memo, G–1 for G–3 ETO, 23 May 44, sub: Reassignment Center, Replacement System, Ltr, Hq ETO to SAC, 1 Jun 44, sub: Facilities at American School Center for Repl System and 1st Ind from SHAEF, 9 Jun 44, all in EUCOM 322 Replacements, II (b).

[20] Memo, Lee for Lord, 9 Jul 44, sub: Manpower Board, GFRC File on Replacements.

them. This was the crux of the entire manpower problem, for the Communications Zone and the air forces constituted the largest sources of able-bodied men in the theater.

Up to the beginning of July there apparently was no serious concern within the theater over a possible replacement shortage in the near future. In fact, the theater actually reduced its September replacement requisition by 15,000 men at that time, and also canceled its requisition for August in all branches except infantry. Its efforts to recover personnel for use as replacements was limited to initiating a survey of the Communications Zone to determine whether any excess of personnel existed, and to issuing a directive to section commanders to release such overstrength or excesses.[21] Section commanders were understandably reluctant to release men at this time in view of the uncertainty as to requirements in connection with the organization of the Communications Zone on the Continent. Consequently the release of men to the Replacement System both in number (about 4,800) and quality was disappointing. Brig. Gen. Walter G. Layman, chief of the Replacement System, complained that many of the men were not fit for training as riflemen. Section commanders obviously were not releasing their best men for conversion.[22]

The July crisis suddenly made the replacement situation a matter of much greater urgency. The theater not only made frantic appeals to the War Department for emergency shipments and for a much higher percentage of infantry-trained replacements, but also took additional steps to produce replacements from its own resources. Since the theater's need was urgent, the quickest dividends obviously promised to come from the conversion of men from combat arms other than infantry, of which there was an excess of more than 20,000, rather than from service personnel. The first step, therefore, was to take approximately 4,000 replacements representing overages in the branches of field artillery, tank destroyer, and antiaircraft and convert them as quickly as possible to infantry.[23]

Shortly thereafter in accordance with earlier War Department directives to reduce the number of basic privates in T/O units, the theater ordered that men so released, regardless of arm or service, also be made available for retraining as infantry rifle replacements. Up to that time the services and major combat commands had been allowed to activate new units utilizing the personnel made available through such reductions.[24] In addition, the theater notified the Replacement System[25] that men originally

[21] Memo, Franey for Lee, 13 Jul 44, GRFC File on Replacements.

[22] Ltr, GFRS to Deputy Theater Comdr, 17 Jul 44, sub: Shortage of Inf Repls, as cited in [Capt. Robert J. Greenwald] Human Logistics, a study of the reinforcement system, prep in Hist Sec ETO, Jan 45, p. 82.

[23] History of GFRS, Part II: D Day to V-E Day, Ch. IV, pp. 1–2.

[24] Memo, Hq ETO to Chiefs of Gen and Special Stf Secs, 4 Sep 44, sub: Use of Repls for Local Activations of Units and Installations, EUCOM 322 Replacements Units, II (a).

[25] For convenience, the term Replacement System is used throughout in reference to the command variously designated as the Field Force Replacement System, the Ground Force Replacement System, the Replacement System, and the Ground Force Reinforcement Command.

trained as infantry replacements in categories other than rifleman would be made available in certain numbers for retraining as riflemen by the Replacement System.[26]

These various measures bore their first fruit in August, when the Replacement System retrained about 5,500 men as infantry riflemen. In September 4,500 men completed conversion training. Most of the retraining up to this time was done in the United Kingdom, although some retraining had started at Le Mans. About 3,300 limited assignment men were being trained in new skills at Shrivenham, where the entire facilities of the American School Center were now being used for that purpose.[27]

These efforts undoubtedly represented progress in the desired direction, but they constituted only a beginning toward meeting the theater's needs for infantrymen, toward training and absorbing the mounting numbers of limited assignment personnel, toward reducing the excessive stocks in certain branches, and toward reducing the theater's overstrength. At the end of September the theater still reported overages in every category except infantry riflemen, the excesses in the combat arms alone totaling nearly 34,000. The shortage in riflemen totaled 7,000, although the replacement pool as a whole held 119,000 men and was substantially above its authorized strength.[28]

The theater's overstrength, not only in replacements but in its T/O units and overhead, made ETOUSA especially vulnerable to criticism by the War Department. Washington had called attention to the theater's excessive overstrength before, claiming that it exceeded 130,000 men at the end of July. More than half of it consisted of overages in replacements, resulting mainly from the fact that losses in many categories had been lower than estimated. Overstrengths in overheads and T/O units could be attributed to several things, among them the fact that all infantry basics had not been withdrawn from ground and service troops, that some units had not reorganized under the latest T/O's and that accelerated needs for continental installations had caused overheads to be exceeded at least until the U.K. installations could be closed out.

The War Department had been willing to overlook some overstrength, but by September it concluded that the theater was not doing enough to eliminate it despite categoric instructions to do so. It feared that large overstrengths were becoming a permanent feature of the manpower picture in all the theaters. The War Department was particularly disturbed over the possibility of a permanent accumulation of excess replacements in some categories, which it re-

[26] Ltr, CG COMZ to CG GFRS, 27 Jul 44, sub: Retraining of Basic Privates, History of GFRS, Ch. V, p. 37; Ltr, Hq ETO to CG GFRS, 11 Aug 44, sub: Retraining of Pers as Inf Repls SSN 745, ETO GFRC 353 Training 1945.

[27] Memo, CG GFRS for G–1 ETO, 1 Oct 44, ETO GFRC 370.092 Reinforcements, Oct 44 to Jun 45; [Henderson] The Procurement and Use of Manpower in the ETO, Pt. IX of the Administrative and Logistical History of the ETO, p. 140; Ltr, GFRS to Comdt American School Center, 23 Aug 44, sub: Training of Limited Assignment Pers, ETO GFRC 001.1 Schedules and Memos GFRC 1945.

[28] Memo, CG GFRS for G–1, 1 Oct 44, ETO GFRC 370.092 Reinforcements.

garded as the worst type of wastage.[29]

Late in September General McNarney sent the Army Inspector General, Maj. Gen. Virgil L. Peterson, to the theater to survey the manpower situation personally. General Peterson reported that the theater's replacement pool had a strength of about 119,000, nearly 49,000 men in excess of the 70,000 authorized. An additional 20,000 men who formerly had been replacements had been assigned as overstrength to various units, including 10,500 with truck companies, 6,700 in airborne divisions, and 2,250 with the engineer special brigades, which continued to operate the Normandy beaches.[30]

As was so typical wherever statistics were involved, War Department figures were widely at variance with those of the theater. Its total overstrength, the theater claimed, actually stood at 68,000 on 30 September, as compared with the War Department's figure of 131,000 for 31 July. General Eisenhower admitted that the difference did not result from any reduction in strength over the two-month period, but rather from "lack of a common basis of calculation." Once again, as demonstrated in the case of ammunition, it was clear that the theater and the War Department were not following uniform accounting practices. In the matter of replacements, moreover, the theater had been operating on the basis of an allowed replacement pool of 170,000, which, it insisted, had been authorized before D Day, in contrast with the War Department's figure of 70,000.

The theater actually reported a replacement strength of nearly 200,000 men, although this included replacements on requisition for October and November, men in transit, and replacements for the air force. On the basis of an authorized ceiling of 170,000 it admitted to an overstrength of slightly less than 30,000 as compared with the War Department's claim of 49,000, or even 69,000 counting the former replacements now listed as overstrength in various units. In any case, the theater felt that the War Department should take cognizance of the fact that a large portion of its personnel classed as replacements at any given time consisted of "dead stock" in that it was not actually available for use as replacements. Included in this category were air force troops, referred to as "happy warriors," who were either awaiting shipment to the zone of interior or were en route to or from the United States; men being retrained; and men earmarked for activation of new units. At the end of September, according to theater figures, men in these categories accounted for 52,000 of the 125,000 ground force replacements physically present in the theater. Only about 73,000 men were actually available for replacement purposes. Because of this, General Eisenhower asked for a "certain tolerance" between the authorized ceiling and actual stockages.[31]

But the War Department was not convinced that the theater was making the best use of its manpower resources. That

[29] Remarks by McNarney at Gen Council Mtg, Office of the DCofS, 4 Sep 44, and Ltr, McNarney to Eisenhower, 5 Sep 44, both in SHAEF G–4 320.2 Strength 1944, I; Memo, OPD DCofS, 8 Sep 44, sub: Strength of ETO, and Memo, Gen Henry, G–1, for CofS, 9 Sep 44, sub: Reduction of Army Overstrength, both in OPD 320.2 Sec. XX, Cases 365–90.

[30] Memo, Peterson for DCofS, 5 Oct 44, WD CofS 322 Sec. II, Cases 1–30.

[31] Ltr, Eisenhower to Handy, 17 Nov 44, SHAEF G–3 O&E 320.2 Theater Overstrength, I.

the "word" on manpower conservation had not reached everyone concerned, or at best was not thoroughly understood in the theater, was indicated by the fact that commanders continued to request authorization to activate new units using manpower available to them in the form of overstrengths despite the theater's measures designed to recapture such men. Early in September the theater again expressly forbade the use of personnel for such purposes, and emphasized that all troops in the theater in excess of T/O's were to be considered as replacements regardless of whether they were in the Replacement System, attached to units and installations, or assigned as overstrengths. It again prohibited the use of such personnel for any purposes other than as loss replacements, and served notice that it would not approve requests for local activations involving use of such men.[32] General Peterson particularly questioned the theater's authority to legitimize the overstrengthening of units with replacements, as it had done in the case of truck drivers, and its taking advantage of deletions from its troop basis to activate other units from manpower available to it within the theater, as it was trying to do in the case of the forty-nine antiaircraft battalions.[33]

With the report of such practices in mind, General McNarney in mid-October again pressed the theater to retrain as infantry all surplus replacements in other branches and to force general assignment troops out of jobs that could be performed equally well by men no longer physically qualified for combat. Anticipating the opposition which the latter would undoubtedly evoke, he told top commanders that they would simply have to break down the natural resistance of subordinates to the withdrawal of personnel.[34]

During October the theater continued to direct its efforts toward rebuilding its depleted infantry pool and toward correcting the maldistribution produced by the casualty experience of June and July. Early in the month it learned from the War Department that shipments from the United States would total less than 19,000 in November, representing a reduction of about 10,000 in ETOUSA's requisition for that month. This allotment was to include a high percentage of infantry, however—15,000 in regular infantry plus 1,400 infantry paratroops and 400 nisei infantrymen for the 442d Infantry Regiment. The War Department justified the reduction on the assumption that ETOUSA would as previously planned fall heir to about 10,000 replacements which the Seventh Army was to turn over when ETOUSA assumed responsibility for the logistic support of forces in southern France on 1 November.[35]

ETOUSA first concluded that the reduction would not necessarily be critical, although it would lower the infantry pool level which the theater considered essential. Much depended on whether

[32] Memo, Hq ETO to Gen and Special Stf Secs, 4 Sep 44 sub: Use of Repls for Local Activations of Units and Installations, EUCOM 322 Replacement Units, II (a).

[33] See Ch. X, Sec. 2, above, on the matter of the antiaircraft artillery units.

[34] Ltr, McNarney to Lee, 10 Oct 44, ETO GFRC 370.092 Reinforcements.

[35] Cbl WARX–41410, AGWAR to ETO, 5 Oct 44, SHAEF G–3 370.092 Reinforcements 1944, I.

NATOUSA made available sufficient replacements for the support of Seventh Army.[36]

It became evident in the next few days, however, that Seventh Army was already having manpower difficulties, to say nothing of bringing with it a dowry of 10,000 men. A critical manpower situation had developed in the Fifth Army in Italy as the result of recent heavy casualties and the War Department's refusal of a large part of its requisition for October and November, and NATOUSA warned General Devers on 7 October that it might not be able to provide replacement support for the Seventh Army through October.[37] In fact, a few days later General Clark asked Devers, who was still serving as deputy commander of NATOUSA, that shipments of NATOUSA replacements set up for Seventh Army be diverted to Fifth Army and, if it was already too late to stop their shipment, that ETOUSA be asked to ship 3,000 men by the fastest means available.[38] Devers replied that Seventh Army faced an equally serious shortage. He relayed Clark's appeal to ETOUSA, however, and offered to release all scheduled support for the southern France forces from NATOUSA, except for men returned from hospitals, if ETOUSA could assume responsibility for the support of Seventh Army about two weeks earlier than scheduled—that is, on 15 October instead of 1 November.[39]

The shipment of officer replacements from NATOUSA had already ceased, and Devers at this time asked ETOUSA for 400 infantry officer replacements, which ETOUSA agreed to furnish.[40] When ETOUSA learned of the situation in the North African theater, it informed the War Department that the situation in Seventh Army definitely made it essential that the entire requisition for November be met. In fact, it warned that it might have to request an increase.[41]

ETOUSA appeared resigned to assuming responsibility for the support of Seventh Army earlier than planned, and now also went to the aid of its neighbor in the Mediterranean. Convinced by both Clark and General Alexander that the operations of the Fifth Army had a direct bearing on the forces likely to be committed on the western front, ETOUSA indicated its willingness to meet NATOUSA's request for 3,000 men, if the War Department would make good the loss in its November shipments.[42] Lt. Gen. Thomas T. Handy first insisted that there was no need for such a transfer, arguing that the War Department was meeting NATOUSA's needs and had already shipped about

[36] Cbl EX-53721, COMZ to AGWAR, 11 Oct 44, SHAEF G-3 370.092 Reinforcements 1944, I; Memo, ETO G-1 to GFRS, 13 Oct 44, History of GFRS, Ch. VI, pp. 279-80; Brig Gen Henry J. Matchett, CG GFRS, for G-1 COMZ, 8 Oct 44, ETO GFRC 370.5 Troop Movement File B 1944-45.

[37] Cbl 35544, White of NATO to Devers, 7 Oct 44, 6 A Gp 322.96-1.

[38] Cbl, Clark to Devers, 9 Oct 44, 6 A Gp 322.96-1.

[39] Cbl BX-17480, Devers to Clark, 10 Oct 44, and Cbl B-17491, Devers to Eisenhower, 10 Oct 44, 6 A Gp 322.96-1.

[40] Cbl, 6 A Gp to ETO, 10 Oct 44, and CBL EX-54104, ETO to 6 A Gp, 13 Oct 44, 6 Gp 322.96-1.

[41] Cbl EX-54575, COMZ to AGWAR, 15 Oct 44, SHAEF G-3 370.092 Reinforcements 1944, I.

[42] Cbl S-62499, SHAEF to AGWAR, 16 Oct 44, and Cbl S-62508, Smith to Alexander and Clark, 16 Oct 44, SHAEF AG 200.3-1 Replacements 1944.

2,200.[43] The Fifth Army's need was urgent, however, and it appeared that shipments coming from the United States would not reach Italy before the end of the month.[44] On 19 October General Smith conferred with the G–1's of both NATOUSA (General White) and 6th Army Group (Brig. Gen. Ben M. Sawbridge), and decided that ETOUSA should and could help the Fifth Army through the critical period. He recommended to the War Department that ETOUSA be allowed to go ahead with the shipment.[45] Handy then gave his approval to the proposal, and within the next few days approximately 3,000 replacements were air transported to Italy. Handy notified ETOUSA that the War Department was adding 5,000 infantry replacements to its November requisition, which would more than compensate for the proposed diversion and would also compensate in part for the fact that ETOUSA was assuming responsibility for the support of Seventh Army earlier than planned. At the same time he warned that the War Department would be hard pressed to meet anticipated December requests, and again enjoined the theater to practice the utmost economy and to accelerate its retraining program.[46]

By the end of October the theater's replacement situation appeared appreciably brighter, thanks in part to the measures which the theater had taken to retrain men as riflemen, but also to the somewhat larger shipments from the zone of interior and to the substantially smaller losses of infantrymen which attended the highly mobile warfare of August and September and the lull in operations in October.

As requested by the theater, riflemen comprised a higher percentage of the total infantry replacements in the next few months, the percentage rising to 68 in August and over 80 in September.[47] Meanwhile, battle casualties, after totaling 51,400 in July, dropped to 42,500 in August, to 42,000 in September, and to 31,600 in October, despite the increasing size of the forces committed.[48] Operations during the pursuit brought a heavy demand for armored force replacements, particularly tank commanders, and for vehicle drivers rather than infantrymen.[49]

Throughout August and September the branch distribution of replacements in the theater's pool had continued to be badly out of balance, and there were substantial surpluses in branches other than infantry. In mid-August, for example, of a total stockage of 67,000 replace-

[43] Cbl WAR–47563, Handy to Eisenhower, 16 Oct 44, OPD Cbl Files.

[44] Cbl MA–1732, Alexander to SHAEF, 17 Oct 44, SHAEF Out Cable Log, Smith Papers.

[45] Cbl S–63030, Smith to Handy, 19 Oct 44, 6 A Gp 322. 96–1.

[46] Cbl WX–49381, Handy to SHAEF, 20 Oct 44, SHAEF AG 200.3–1 Replacements 1944; Cbl WX–49282, AGWAR to SHAEF, 20 Oct 44, SHAEF G–3 370.092 Reinforcements 1944, I; History of GFRS, Ch. VI, 283.

[47] Replacement Study, August 1944–April 1945, prep by G–1 ETO, Set 45, EUCOM 322 Replacement Studies 1944–45.

[48] Nonbattle casualties totaled 17,000 in August, 21,000 in September, and 28,400 in October. Army Battle Casualties and Nonbattle Deaths in World War II, Final Report prep by Statistics and Accounting Br AG, Dept of Army, OCMH, p. 32; [Henderson] Procurement and Use of Manpower in the ETO, p. 45.

[49] Aide Memoire, unsigned, 11 Aug 44, sub: Repl Situation, Incl to Memo, LST for Franey, 18 Aug 44, ETO GREC File on Replacements.

ments available to the theater, only 20,000 (30 percent) were infantry-trained, and of these only 3,250 were riflemen. At the time about 9,000 infantrymen were in the process of conversion to riflemen.[50]

In October the Replacement System continued its efforts to correct this maldistribution. The theater's aim was to establish and maintain at all times a pool of 70,000 replacements. On the basis of the casualty experience up to September it had decided that 78.3 percent (54,800) of this pool should consist of infantrymen, as compared with the earlier 64.3, and that 70 percent of the infantrymen (or about 38,000 men) should be rifle-trained. On 1 September the branch distribution was badly out of balance, although some progress was made in rebuilding the pool of rifle-trained replacements. On that date the theater's stockage of infantrymen had risen to about 42,000 as against its announced requirement of 55,000. Of these only about 15,000 had the much needed MOS 745 classification, the occupational specialty number of a rifleman.[51]

The retraining of an additional 14,400 men in September and October, combined with the smaller losses of those months, did much to bring the branch distribution of the theater's pool into better balance. By 1 November the theater had built up its stockage of infantry riflemen to 30,000. Despite the fact that this did not represent the announced target, and despite War Department injunctions and warnings, the theater authorized a substantial cutback in the retraining program for November.[52]

The balance that had been reached was actually a very precarious one. It had been achieved largely by the retraining of other replacements—that is, surpluses in infantry other than riflemen and in other combat branches. The theater had as yet made no real effort to tap its principal remaining source of general assignment men—that is, the air force and the Communications Zone. Viewed in the light of the measures which eventually had to be taken, those of the summer of 1944 hardly constituted more than stopgap measures and failed to go to the heart of the manpower problem.

(2) The Storm Breaks, November–December 1944

If there was any complacency over the manpower situation at the end of October it vanished quickly in the next few weeks. The launching of major Allied offensives in November under conditions of cold, wet weather had a dual impact on casualty figures: battle casualties, which had come to only 31,600 in the preceding month of relative inactivity, rose to 62,400 in November; meanwhile, nonbattle casualties, which had totaled 28,400 in October, suddenly reflected a high incidence of trench foot and rose to 56,300. Total casualties for the month thus exceeded 118,000 men. (*Table 10*)

Before this trend had become evident the theater had received discouraging

[50] Memo, Statistics Br SGS ETO, 14 Aug 44, ETO Adm 235 Ordnance Ammo.

[51] Memo, G–1 COMZ for CG COMZ, n.d. [Sep], sub: Repls, USFET 322 Replacements; Replacement Study, August 1944–April 1945, prep by G–1 ETO, n.d., EUCOM 322 Replacement Studies 1944–45.

[52] Stf and Comd Conf, COMZ, 5 Jan 45, EUCOM 337/3 Conferences, Staff Weekly 1944, I.

TABLE 10—BATTLE AND NONBATTLE CASUALTIES, JUNE 1944–MAY 1945

Year and month	Total	Battle	Nonbattle
1944			
June	---------	39,367	(*a*)
July	63,424	51,424	12,000
August	59,503	42,535	16,968
September	63,179	42,183	20,996
October	59,981	31,617	28,364
November	118,698	62,437	56,261
December	134,421	77,726	56,695
1945			
January	136,747	69,119	67,628
February	91,545	39,414	52,131
March	101,156	53,209	47,947
April	87,209	41,058	46,151
May (1–8)	14,178	2,028	12,150

[a] No data.

Source: For battle casualties, Army Battle Casualties and Nonbattle Deaths in World War II, Final Report, p. 32; for nonbattle, [Henderson] The Procurement and Use of Manpower in the ETO, p. 45.

and at the same time conflicting information regarding the future availability of replacements from the United States. Early in November the War Department gave ETOUSA a long-range forecast indicating that shipments in December would total 43,350, of which 35,000 would consist of infantrymen, and that shipments in the four succeeding months would average 44,650, of which 36,000 would be in the infantry branch. It cautioned, however, that these figures represented the maximum capability, that unforeseen requirements in other theaters might force downward revisions, and that theater plans should be sufficiently flexible to meet such an eventuality. It again urged, therefore, that ETOUSA make the maximum contribution to its own replacement capabilities by a vigorous retraining program.[53]

Theater replacement officials, comparing the War Department's forecast with the theater's requisitions, noted that ETOUSA might suffer a cumulative shortage of more than 53,000 infantrymen by the end of February. The theater at the time possessed a pool of approximately 61,000 infantrymen, of which about 38,000 were riflemen. That pool, it calculated, might easily be eliminated by the end of December if casualties were higher than then estimated, which indeed they were.

Surveying the potentialities within the theater, the G–1 of the Replacement System, Col. Walter C. Cole, concluded that there were three sources from which the estimated requirements for infantry riflemen might be met. These were: (1) general assignment men received from the Communications Zone in exchange for limited assignment men; (2) casuals and replacements other than infantry still available in the Replacement System in excess of actual needs; and (3) infantrymen in the three line-of-communications regiments. The theater had already taken steps to recover some of the infantry-trained general assignment men in its three line-of-communications regiments, having recently ordered 60 percent of their strength released to the Replacement System and replaced by limited assignment men.[54]

[53] Ltr, WD to ETO, 8 Nov 44, sub: Overseas Repls, WD AG 320.2 (2 Mar 43) (36) Sec. 1A Overseas Replacement System—Estimates.

[54] The three regiments were the 29th, 118th, and 156th. Ltr, Hq ETO to Hq Comdt SHAEF, 29 Oct 44, sub: Inf Enlisted Repls, SHAEF AG 200.3–1 Replacements 1944; and Memo, Lt Col H. A. Twitchell for Bull, 25 Nov 44, sub: Repls, SHAEF G–3 370.092 Reinforcements 1944, I.

The prospect of men in any of these categories becoming available in the immediate future was poor. The conversion of general assignment men withdrawn from service units was obviously the most remote. The quickest return could be realized through the withdrawal of the infantrymen from the three line-of-communications regiments, which was expected to yield about 5,000 men. Even these required a three-week refresher course and would not be available until December. The Replacement System recommended, in addition, that an initial increment of 4,000 men from the armored force, field artillery, chemical warfare, coast artillery, and ordnance branches be converted to riflemen. In order to select 4,000 men and still ensure a sufficient stockage to meet the theater's pool requirements, the Replacement System now proposed that all coast artillery, field artillery, ordnance, and quartermaster casuals be declared free replacements if no requisitions for them were received after their arrival in their appropriate assignment depots. Until this time the Replacement System had followed the policy of holding casuals so that they could be returned to their former units. This had often tied up sizable numbers of men when requisitions did not arrive from their former units.[55]

The theater G–1, Colonel Franey, approved these recommendations immediately, and in fact urged the Replacement System to increase the retraining program to a minimum of 20,000 within the next three weeks, even at the risk of a later inability to meet a portion of the requirements in branches other than infantry and armor. He also suggested that more prompt relief could be provided by the retraining of an appropriate number of infantry specialists, for reports had indicated that the ratio of infantry riflemen to infantrymen was still badly out of balance. It was estimated that in the latter category there still was a surplus of more than 22,000.[56]

Meanwhile the theater had received a disturbing communication from the War Department containing vague references to the possibility of canceling the remainder of requisitions not yet shipped.[57] By the theater's interpretation, it involved a reduction of nearly 36,000 infantry replacements previously scheduled for November and December arrival. Such a cut, it asserted, would almost certainly jeopardize the success of operations.[58] Moreover, the War Department had stated that the theater henceforth would have no choice as to the number of replacements in each branch, but would have to accept the War Department's distribution with regard to classification by arm or service.

On the matter of cancellations, as it turned out, there had been a misunderstanding, and the War Department quickly assured ETOUSA that it was not canceling any previously approved

[55] Memo, Cole for CG GFRS, 19 Nov 44, sub: Study, and Memo, Gillespie, CofS GFRS, for CG ETO, 19 Nov 44, sub: Requirements for Inf Riflemen, both in GFRC 370.092 Reinforcements May 44 to Apr 45.

[56] Memo, Franey for GFRS, 22 Nov 44, GFRC 370.092 Reinforcements.

[57] Cbl EX–64610, AGWAR to COMZ, 17 Nov 44, SHAEF G–3 370.092 Reinforcements 1944, I.

[58] Cbl E–66027, COMZ to AGWAR, 23 Nov 44, and Memo, Barker for Bull, 22 Nov 44, sub: Repls, both in SHAEF G–3 370.092 Reinforcements 1944, I.

requisition.[59] On the matter of branch distribution, however, the War Department stood firm, at least for the moment. Theater officials had noted that only 49.2 percent of all infantrymen shipped from the zone of interior consisted of riflemen as against requests that 56.5 percent be so trained, and asserted that the War Department was continuing to train men in the branches of coast artillery, tank destroyer, field artillery, chemical warfare, ordnance, quartermaster, and transportation in excess of needs. The War Department maintained that it was training replacements in accord with the arm and service breakdown of the best estimates of total replacement needs in all theaters, and that it was impracticable to change training programs to meet the frequent changes in theater estimates.[60]

The developments of November showed that the War Department and the theater had not yet overcome the language barrier on the subject of replacements and that there still was need for arriving at a common basis of understanding. Late in the month the theater therefore sent a group of officers to Washington to discuss the replacement problem and to make certain that the War Department clearly understood ETOUSA's situation. The Bull Mission, which also discussed the ammunition shortage while in Washington,[61] went over the entire replacement problem with War Department officials in the first days of December.

In some ways the conference underscored past misunderstandings and differences. General Henry, the War Department G–1, sensed that the theater and the War Department even at this late date were employing different personnel accounting methods. War Department figures pictured the theater as being well off in the matter of replacements. They showed, for example, that the theater's replacement pool on 31 October contained 160,000 men, of which 62,000 were infantrymen. Even allowing for 50,000 battle casualties in November, it showed that the theater would have 125,000 replacements at the end of the month, of which 55,000 would be infantrymen.

These figures were based on a definition of theater replacement resources which the theater had never accepted—that is, that the true replacement resources of the theater were the sum total of all replacements in the Replacement System plus overheads and existing overstrengths in theater units.[62] At the request of the theater the War Department had only recently authorized ETOUSA a replacement pool of 80,500 men, specifying that this number must include all replacements, whether in depots, in transit, in training, or as part of unit overstrengths, whereas the the-

[59] Cbl W–67832, AGWAR to ETOUSA, 24 Nov 44, SHAEF G–3 370.092 Reinforcements; Memo for Record, unsigned, 24 Nov 44, sub: Repl and Requisitions of ETO for Nov and Dec, OPD 370.5 ETO Sec. XIV, Cases 471–95.

[60] Ltr, Matchett to AG ETO, 20 Nov 44 sub: Comparative MOS Rpt of GFRS of ETO, with atchd Comparative MOS Rpt, Cumulative Arrivals and Requisitions, 1 Jan–30 Sep 44, EUCOM 322 Replacement Units 1944, 3B; Cbls E–64802, COMZ to AGWAR, 18 Nov 44, and W–67325, Witsell to G–1 ETO, 23 Nov 44, SHAEF G–3 370.092 Reinforcements 1944, I.

[61] See Chap. IX, Sec 3, above.

[62] Ltr, ETO to AG WD, 15 Jan 45, sub: Theater Rate Tables, with atchd Monthly Report of GFRS, ETO, EUCOM 320.2 Theater Rate Tables; Memo, Barker for Smith, 3 Dec 44, sub: Replacements, SHAEF G–3 370.092 Reinforcements 1944, I.

ater had desired a pool of that size over and above the "dead stock" which normally accounted for a large percentage of the total replacement population of the theater. The War Department refused to authorize such a pool.[63]

War Department and theater officials finally ironed out their differences over accounting procedures. But the War Department took the opportunity to express its dissatisfaction with the theater's retraining program, making special note of the continuing large surplus in branches other than infantry. Again it made clear to ETOUSA representatives that the theater's requirements simply could not be met from the zone of interior. The War Department could show that it had definitely made greater efforts to recapture personnel for replacement purposes than had the theater. All newly inducted men qualified for overseas service were already being trained as replacements; every man in the Army Ground Forces and Army Service Forces whom it was practicable to withdraw from other jobs was undergoing conversion to infantry; and 40,000 men were at this time being withdrawn from the Army Air Forces for the same purpose. Even with these efforts the War Department estimated that it could barely meet overseas losses in the next few months.[64] "This personnel business," as General Handy put it, "is one of the worst headaches we have."

Washington officials were also highly critical of the gross discrepancies between the theater's long-range estimates of requirements and the requisitions it finally submitted. The requisition for December, for example, represented an increase of 10,000 over the estimates submitted in August, and for January the theater had finally asked for 67,000 men as compared with its earlier estimates that it would need about 40,000. Taking all the theaters together, these discrepancies had been as high as 100 percent for ground forces and 140 percent in infantry alone. Such increases posed an obvious dilemma for the War Department, which had to plan the training of replacements several months in advance.

In the end the War Department agreed to add about 18,000 men to ETOUSA's January shipments, which raised the allocation for that month to 54,000. The theater on its part acknowledged the need to expand its retraining program in order to convert the excessive surpluses in branches other than infantry, and also agreed to meet more of its infantry officer requirements through appointments from the ranks and by retraining from other arms and services. The War Department agreed to consider shipping limited assignment men to the theater should this be necessary to meet the withdrawals of general assignment men from service units, and also to provide training cadres for the theater's conversion program and for an officer candidate school.[65]

---

[63] Cbl EX-61989, COMZ to AGWAR, 10 Nov 44, and Cbl WX-64610, AGWAR to COMZ, 17 Nov 44, OPD 370.5 ETO Sec. XIV, Cases 471-95.

[64] For the efforts made in the zone of interior, see Palmer, Wiley, and Keast, *op. cit.*, pp. 200-225.

[65] Ltr, Eisenhower to Marshall, 27 Nov 44, OPD Exec Office files; Memo, Henry for Bull, 2 Dec 44, sub: Repls SHAEF G-3 370.092 Reinforcements 1944, I; Memo, Henry for CofS, 3 Dec 44, sub: Repls, CofS 320.2 Sec. IV, Cases 126-75; Summary of Conclusions Reached and Actions Taken With Respect to the ETO Replacement Problem [ca. 5 Dec 44], CofS 322 Sec. II, Cases 31-45; Ltr, Handy to Eisenhower, 5 Dec 44, ETO Special File prep by G-1 Adm 17 Nov 44-1 Jan 45.

Shortly after General Bull returned to the theater General Eisenhower instructed the major commands of the theater to comb out men who could be replaced by limited assignment troops or from units which could operate at less than T/O strength.[66] The Communications Zone initially planned to meet its commitment under these orders by releasing about 15,000 men from service units, in some cases at least making them temporarily understrength. Section commanders were enjoined to comply fully with the order, particularly as to reporting positions which could be filled with limited assignment men.[67]

At this time the ETOUSA Replacement System had three major retraining courses under way designed to produce additional infantry rifle replacements from the theater's own resources: (1) a twelve-week basic infantry course organized to convert men from arms and services other than infantry to infantry riflemen; (2) a three-week refresher course for general assignment men withdrawn from the three line-of-communications regiments; and (3) a six to eight weeks' basic infantry course for the retraining of infantrymen other than riflemen. Two courses were then being conducted for officers: (1) a three-week basic infantry refresher course for infantry officers withdrawn from noncombat assignments, and (2) a twelve-week basic infantry course for the conversion of officers in other arms and services to infantry.[68] Approximately 16,000 enlisted men and 500 officers were in training under this program in mid-December, although the output for the month actually came to less than 6,000.[69]

This program was now due for a substantial expansion. The main impetus came from an unexpected direction. In mid-December the German counteroffensive in the Ardennes suddenly shocked the theater into action which it had repeatedly postponed. Losses in the first two weeks were to raise battle casualties to 77,700 for the month of December. Nonbattle casualties totaling 56,700 were to bring the total losses to 134,400 men. (*See Table 10.*) The replacement situation consequently became more critical than ever.

An accelerated conversion program offered no solution to the immediate problem, and the theater therefore was forced to take drastic emergency measures to meet the heavy losses which attended the battle raging in the Ardennes. Field commanders immediately voiced their concern over the heavy losses, which rose to an estimated 50,000 men in the 12th Army Group within the first week, more than 40,000 of which were infantrymen.[70]

---

[66] Cbl S–70984, Eisenhower to Handy, 15 Dec 44, OPD Cbl Files.

[67] Memo, Lee for G–1, 15 Dec 44, sub: Reinforcement Strength, and Memo, Franey for Lord, 16 Dec 44, EUCOM 322 Replacement Units 1944–45, 3B; Stf Mtg, 19 Dec 44, EUCOM 337 U.K. Base Stf Mtgs.

[68] Ltr, GFRS to G–3 COMZ, 6 Dec 44, History of GFRS, Ch. V, p. 61.

[69] Memo, GFRS for G–1 ETOUSA, 15 Dec 44, History of GFRS, Ch. V, p. 64; [Henderson] Procurement and Use of Manpower in the ETO, p. 45.

[70] Memo, Bull for Smith, 25 Dec 44, sub: Repls, SHAEF G–3 370.092 Reinforcements 1944, I.

The theater met the crisis mainly by two expedients: (1) speeding up the delivery of replacements already in the theater pipeline, and, more importantly, (2) stripping units not yet committed. As soon as the seriousness of the situation was apparent, theater headquarters took steps to have all available replacements moved to the armies as quickly as possible. As a result, approximately 2,100 replacements arriving at Marseille were flown to the Third Army on Christmas Day, and another 2,500 were flown to the Third and Ninth Armies the following day.

Meanwhile, in an unprecedented drastic action SHAEF on 20 December had ordered the basics of all nine regiments of the 42d, 63d, and 70th Infantry Divisions, then in the 6th Army Group area, withdrawn and released to the Third Army for use as infantry replacements.[71] At 219 men per regiment, this produced nearly 2,000 men. A few days later Supreme Headquarters decided to strip the 69th Division as well, and ordered 25 percent of the T/O enlisted strength of each of the three regiments, totaling about 2,200 men, released and shipped to the Continent by air for the First and Ninth Armies.[72] Additional measures taken within the first week included the withdrawal for immediate shipment to front-line units of 5,000 men then being retrained by the Replacement System.[73] By 25 December the theater through these various measures had arranged to provide about 30,000 men to the three armies of the 12th Army Group.[74]

Allocation of replacements between the 12th and 6th Army Groups gave overwhelming priority to the former, which was heavily engaged in the counteroffensive. Within the 12th Army Group deliveries at first favored the Third Army somewhat in order to bring General Patton's forces up to the fullest possible strength for the counterattacks they launched a few days before Christmas.[75]

Theater headquarters also considered other possible actions, such as the withdrawal of combat trained men from engineer combat battalions and general service regiments, and of basics, or even infantry, from divisions which had not yet been committed.[76] But these measures were not taken. Instead, on 26 December the theater made an appeal for volunteers for retraining to infantry. Among those who responded to the appeal were 2,250 Negroes, some of whom accepted a reduction in grade to qualify, since only privates and PFC's were declared eligible. By mid-March, when the training of volunteers was suspended,

[71] Ltr, Bull to Smith, 19 Dec 44, sub: Urgent Repls, SHAEF G–3 370.092 Reinforcements 1944, I; Cbl S–71778, SHAEF to COMZ and A Gps, 20 Dec 44, SHAEF AG 200.3–1 Replacements 1944.

[72] Cbl 8–72407, SHAEF to 69th Div COMZ and CGRS, 25 Dec 44, and Cbl S–76596, SHAEF to 6 A Gp, 26 Jan 45, SHAEF AG 200.3–1 Replacements 1944.

[73] Cbl EX–79059, ETO to Barker and Franey in Washington, 26 Dec 44, SHAEF G–3 370.092 Reinforcements.

[74] Cbl S–72409, SHAEF to 12 A Gp, 25 Dec 44, SHAEF AG 200.3–1 Replacements.

[75] Ltr, Bull to Smith, 14 Dec 44, sub: Alloc of U.S. Repls, SHAEF G–3 370.092 Reinforcements; Cbl EX–76107, COMZ to A Gps, 19 Dec 44, SHAEF AG 200.3–1 Replacements; Memo, Col Richard L. Gillespie, DCofS GFRC, for CofS GFRC, 21 Dec 44, in History of GFRS, Ch. VI, 297.

[76] Memo, Eyster for Twitchell, 24 Dec 44 sub: Possible Inf Rifle Repl Sources, SHAEF G–3 370.092 Reinforcements.

nearly fifty platoons of Negro infantrymen had been formed.[77]

Meanwhile two of the armies had instituted emergency retraining programs to produce replacements from their own resources. The Seventh Army, informed that it stood little chance of getting more favorable treatment in the matter of allocations for possibly a month or more, late in December withdrew nearly 4,000 men from its service units for conversion to infantry riflemen, replacing them with limited assignment men provided by the Replacement System. Using its own training personnel and the facilities of the Replacement System's second depot at Thaon, France, the Army retrained nearly 4,000 men in the next month.[78] Third Army had inaugurated a similar program early in December, transferring about 6,500 general assignment men to its own replacement battalion at Metz for retraining. But the pressing need for replacements after the middle of the month led to the commitment of both training units and trainees before the conversion could be properly completed.[79]

The enemy counteroffensive had of course seriously aggravated an already bad manpower situation, and theater officials concluded within a few days of the launching of the offensive in the Ardennes that the emergency measures then being considered could not see the theater through its immediate difficulties. On 19 December, therefore, General Eisenhower decided to send a second mission to Washington. Its goal, as he put it, was to effect a "better understanding and insure that we are speaking the same language" in the matter of replacements, and to show how critical the manpower situation had become despite the "drastic actions" which the theater had recently taken.[80]

The War Department welcomed the proposed visit, but it forewarned ETOUSA that there was little hope of improving on the shipments already scheduled for the immediate future.[81] A few days later, on 22 December, General Barker, Brig. Gen. Joseph J. O'Hare, and Col. James M. Franey, respectively the SHAEF, 12th Army Group, and COMZ G–1's, and Col. Lyle T. Shannon, chief of the Reinforcement Section of the G–1 Division, ETOUSA, flew to Washington to present the theater's case for additional men. Discussions with War Department officials, among them Maj. Gen. Stephen G. Henry, the G–1, General Handy, the Deputy Chief of

[77] Memo, G–1 for CofS, 27 Dec 44, sub: Retraining of Colored Pers as Inf, Ltr, Lee to Comdrs of Colored Troops COMZ, 26 Dec 44, sub: Volunteers for Training and Assignment as Reinforcements, Ltr, ETO to Sec Comdrs, 28 Dec 44, sub: Retraining of Colored Pers as Inf Riflemen, Ltr, Hq ETO to CG COMZ, 9 Feb 44, Memo, Lee for Lear, 1 Feb 45, Memo, Lear for Shannon, 10 Feb 45, and Ltr, ETO to A Gps and GFRC, 10 Mar 45, sub: Employment of Colored Reinforcements, all in EUCOM 322 Replacement Units 1944–45, 3B; [Henderson] Procurement and Use of Manpower in the ETO, pp. 103–04, 143–44, 146.

[78] TWX, 6 A Gp to SUSA, 22 Dec 44, TWX AX–15635, SUSA to 6 A Gp, 25 Dec 44, TWX BX–21949, 6 A Gp to SUSA, 27 Dec 44 and TWX CPX–16078, SUSA to 6 A Gp, 31 Dec 44, all in ETO 319.1 Transportation (Telephone Conversations) 1 1944–46; [Greenwald] Human Logistics, pp. 44–45, 86.

[79] [Henderson] Procurement and Use of Manpower in the ETO, pp. 70–71, 147, 150–51.

[80] Cbl S–71574, SHAEF to AGWAR, 19 Dec 44, SHAEF Out Cbl Log, Smith Papers.

[81] Cbl W–80589, Handy to Eisenhower, 21 Dec 44, SHAEF AG 200.3–1 Replacements.

Staff, and General Porter, the G–3, got under way at the Pentagon on 23 December.

Essentially, the conference covered no ground that had not been covered in the meeting held earlier in the month. Despite the desperate plight of the theater resulting from the heavy fighting in the Ardennes, ETOUSA officials actually found themselves more on the defensive than ever. The ETOUSA representatives argued with some validity that the Ardennes battle was seriously aggravating its manpower shortage. General Barker showed, for example, that the theater would be short about 17,000 riflemen by the end of the month and that divisions would be down to about 78 percent of their rifle strength by that time. Beyond this, however, ETOUSA's case was weak and vulnerable. The claim, for example, that ETOUSA's replacement problem had been aggravated by the necessity to go to Fifth Army's aid in Italy and to assume responsibility for the support of the Seventh Army sooner than planned had no validity, for, as Pentagon officials were quick to point out, the War Department had more than made good these losses. The claim that ETOUSA had had a retraining program in operation for some time and that it was drawing on COMZ units for general assignment men and even reducing the strengths of service units could hardly be supported. ETOUSA's conversion program actually had gotten under way in earnest only within the past month. The ETOUSA representatives admitted, in fact, that the theater had not appreciated the seriousness of the manpower situation until November.

The theater based its main plea for special consideration on the claim that Germany's defeat had the highest priority, and that victory in the European theater within the next four or five months hinged on the support which the War Department would give ETOUSA. Barker pleaded for special consideration at least through March, when the accelerated retraining program would be in full swing and producing results. General Handy pointed out that the War Department had given the European theater favored treatment all along. The War Department's allocations showed, moreover, that the European theater was scheduled to continue to get the lion's share of the available manpower for the next four months. In infantry alone ETOUSA's share would average nearly 75 percent of the world-wide allocation.

It was precisely by virtue of this favored treatment, Washington officials argued, that ETOUSA now had the biggest and best remaining pool of manpower on which to draw. The bulk of potential combat replacement material was now overseas. There no longer was any reserve of combat troops in the United States on which the War Department might depend in an emergency, for all major formations were now committed for deployment overseas. Moreover, the War Department had already withdrawn all potential replacement material from the air force and service forces in the zone of interior and was in the process of doing the same in the outlying defense commands.

Practically the only source remaining in the United States was the manpower becoming available through induction. This was actually a dwindling asset so far as providing able-bodied men was concerned, for the pool of men in the 18–26 age group which had not been deferred for industrial or farm purposes was now depleted, and only a limited number of men was entering that age group every month. The War Department had in fact agreed with Selective Service to re-examine about 30,000 men previously rejected as borderline cases, and was contemplating relaxing physical standards. The quality of replacements, in other words, would not be as high as in the past.

Earlier in the month the War Department had agreed to increase shipments to the theater by about 18,000 men, but it now explained that it had been able to do so only by cutting the training cycle from seventeen to fifteen weeks and by shortening the furloughs customarily given men scheduled to go overseas. In increasing the January commitment, therefore, the War Department warned that it was "borrowing from the future," for it was merely speeding up the delivery of men in the replacement pipeline and not increasing the total number available.[82]

War Department officials could not refrain from contrasting the relatively drastic steps taken in the United States to recover manpower with the rather unimpressive performance of the theater. General Handy noted that in terms of the ratio of combat units to their support, ETOUSA's divisions were the most expensive of any theater. And why, he asked, was it necessary for any unit in the European theater to be understrength when the theater's T/O units and overhead were nearly 20,000 men overstrength?

It was plain that War Department officials considered the theater wasteful in its use of manpower, and they were more pointed than ever before in their remarks concerning its failure to take more timely action on the retraining of its overstrengths and the general assignment men in its noncombat units. The warnings had been clear as to the War Department's declining capability to furnish replacements. How, they asked, could the theater wait until late in November to accelerate its retraining program? In General Handy's view, the solution was obvious: "You just have to comb them out."

In the end the conference produced only one important offer of assistance to the theater: the War Department offered to provide the nucleus of a training cadre for an officer candidate school in France. Beyond this it said only that the European theater would continue to get the largest share of the 80,000 replacements which the War Department hoped to provide to meet world-wide requirements each month. It had made unequivocally clear—this time by refusing to make any further concessions—that the theater henceforth must look to its own resources and ingenuity. In effect, therefore, the problem was thrown back into the theater's lap.[83]

[82] It had done the same with field artillery ammunition in response to the theater's pleas late in November. See Ch. IX, above.

[83] Min, Mtg to Discuss ETO Repl Situation, 23 and 28 Dec 44, ETO Special File prep by G–1 Adm, 17 Nov–1 Jan; Memo, Gilmer, Chief European Sec Theater Gp OPD, for Hull, 27 Dec 44, sub: Repl Status of ETO, OPD 320.3 Sec. XXVII, Cases 516–53; Cbl 77780, Lee to WD, 23 Dec 44, and Cbl 82668, Marshall to Eisenhower, 27 Dec 44, OPD 370.5 ETO Sec. XV, Cases 496–513.

(3) The Theater Acts, January–February 1945

The War Department's refusal to promise additional replacements had one very salutary effect, which probably could not have been produced in any other way. ETOUSA, finally convinced that the War Department meant what it had repeatedly asserted about the shortage of manpower, now took steps to accelerate the retraining program which it had belatedly undertaken late in November.

On 8 January General Eisenhower announced the creation of a U.S. Theater Manpower Section, responsible directly to himself, to supervise and control the entire conversion program. The Manpower Section, initially drawn entirely from the SHAEF G–1 Division, was charged with the final determination of both the numbers and categories of personnel to be withdrawn from each component command for transfer to the Replacement System, and was empowered to issue all the necessary instructions to effect such withdrawals and to control the allocation of both limited assignment and general assignment men to the various commands. Emphasizing the seriousness of the manpower situation, General Eisenhower instructed that all necessary records of the various headquarters be made available to the section, and ordered that nothing be allowed to interfere with the success of the program.[84]

In general, it was expected that the Theater Manpower Section would implement the recommendations of the group of officers from the War Department Manpower Board which was about to carry out a thorough survey of the theater's manpower. General Eisenhower at this time authorized the group, headed by Maj. Gen. Lorenzo D. Gasser, to proceed anywhere in the theater to carry out this mission.[85]

General Eisenhower at first specified that the Theater Manpower Section was to operate under the direct supervision of the U.S. element of the SHAEF G–1 Division. Shortly thereafter, however, he decided to combine all manpower affairs, including procurement, training,

[84] Ltr, Eisenhower to Major Comds, 8 Jan 45, sub: Manpower, SHAEF AG 200.3–4 Reinforcements (Reconversion Program–Policy) 1945.

[85] General Gasser and his aides began their survey with detailed job analyses of units in the U.K. Base, where they uncovered substantial duplication of services and surpluses of men. Several officers of the Gasser board, under the direction of Brig. Gen. Robert C. Rodgers, later undertook a similar survey of continental installations and headquarters of the Communications Zone and pointed out where substantial savings in manpower could be made. Memo, Eisenhower for CGs USAF and COMZ, 12 Jan 45, sub: Authority for Gasser, Memo, Gasser for OPD WD, 27 Jan 45, sub: Plan for Surveying U.K. and COMZ Activities on the Continent, Memo, Gasser for Smith, 12 Jan 45, sub: Opns of WD Manpower Bd in ETO, and Ltr, Lear to CGs COMZ *et al.*, 8 Mar 45, sub: Authority for ETO Manpower Bd, all in ETO 322.01/3 Manpower Board and Manpower Section; Memo, Gasser for Lear, 2 Mar 45, sub: Rpt of Progress, ETO Sec WD Manpower Bd, ETO AG 319.1/1 Daily Progress Reports, I, Ltr, Rodgers to Gasser, 17 Mar 45, sub: Changes in AF Orgn, Recommendations to Effect Further Pers Savings, and Memo, Brig Gen J. C. Drain, Chief Misc Facilities Br WD Manpower Bd, for Gasser, 17 Mar 45, sub: Recommended Additional Pers Reductions in the U.K. After 15 Apr 45, both in ETO 200.3 Manpower; Ltr, ETO to CG USSTAF, 6 Apr 45, sub: Survey of Pers Requirements USSTAF in U.K. by WD Manpower Bd, and Ltr, ETO to Lee, 6 Apr 45, sub: Pers Requirements U.K. Base, both in EUCOM 320.2 Strength Reports 1944–45; Ltr, ETO Manpower Bd to Lear 23 Apr 45, sub: Progress Rpt, and Ltr, Rodgers to Lear, 5 May 45, sub: Pers Survey of Normandy Base Sec, both in EUCOM 322 Manpower.

and morale, under one office, and created the new position of Deputy Theater Commander for that purpose, naming Lt. Gen. Ben Lear to the position. Lear had commanded the Second Army in the United States from 1941 to 1944 and had succeeded Lt. Gen. Lesley J. McNair as commanding general of the Army Ground Forces in July 1944 when McNair was killed in Normandy. General Marshall had put Lear at General Eisenhower's disposal to carry out his invigorated replacement training program. The theater commander specifically delegated to Lear the authority to "coordinate, control, and direct" the activities of the Theater Manpower Section, which in effect meant authority over all matters dealing with the economic use and proper handling of men.[86]

The creation of the Theater Manpower Section and the appointment of General Lear to supervise all manpower activities immediately raised questions regarding the role and authority of the theater G–1, which already was handling much of the staff work on manpower. General Lear promptly announced that he intended to establish the Theater Manpower Section as a completely separate entity operating directly under his jurisdiction, and that he intended to avail himself of the assistance of the various agencies of the theater staff already concerned with manpower, in line with this interpretation of the theater commander's desires. Lear announced his intention of naming Colonel Shannon, the head of the Reinforcement Section of G–1 ETOUSA, as chief of the Theater Manpower Section, and of reinforcing the section as necessary with other officers from both the SHAEF and ETOUSA G–1 Divisions.[87]

The Communications Zone questioned this interpretation, objecting particularly to the loss of control over the Reinforcement Section of its G–1 Division, which did much of the staff work on manpower—gathering data on requirements and availability and determining the number of men that had to be withdrawn for retraining. General Lee naturally desired to retain control over this important part of the entire manpower machinery. The matter was finally brought to General Eisenhower's attention, and the Supreme Commander referred the entire problem to his G–1, General Barker.[88]

General Barker believed strongly in the need for an agency at the highest level that would impartially and vigorously prosecute the manpower program now contemplated, and for an agency that should not be subject to control by any interested party. He also considered it necessary that all theater-level staff work on the manpower problem be brought under the control of one agency. The Replacement System had already been placed under the supervision of General Lear. General Barker considered it logical that the Reinforcement Section of the ETOUSA G–1 Division, which was closely tied up with the func-

[86] Ltr, CG ETO to Major Comds, 23 Jan 45, sub: Duties and Responsibilities of the Deputy Theater Commander, EUCOM 322 Manpower. See also Ch. XIII, Sec. 3, for Lear's position and authority.

[87] Memo, Lear for Lee, 29 Jan 45, sub: U.S. Theater Manpower Sec, EUCOM 322 Manpower.

[88] Memo, Lord for Franey 30 Jan 45, and Memo, Lee for Lord, 1 Feb 45, EUCOM 322 Manpower.

tions of the Replacement System and the new Theater Manpower Section, should also pass to Lear's control. In his opinion, leaving it with the Communications Zone would defeat the intent of the whole program, for it would deny General Lear control over a vital element of the entire structure and result in divided responsibility.

General Eisenhower approved his G–1's recommendations. On 5 February, therefore, General Barker notified Lee that Lear was to have control of all activities concerned with the direction of the Theater Manpower Section and the Replacement System. In line with this decision the theater-wide function of the Reinforcement Section of G–1 ETOUSA, with the necessary personnel, were now to be transferred to the Theater Manpower Section. To be included in the transfer was the section's current head, Colonel Shannon, who was to become chief of the Theater Manpower Section. A suitable cadre was to be left with ETOUSA G–1 to carry on the work on the Communications Zone's own manpower problems. On 24 February General Lear officially announced the reconstitution of the Theater Manpower Section as a separate staff section responsible directly to himself as Deputy Theater Commander, and outlined its duties and responsibilities.[89]

Creating the Theater Manpower Section and granting it undisputed authority in the field of manpower represented a triumph for General Barker's ideas, for the SHAEF G–1 had argued with little success for such an agency as early as June 1944. General Barker had clearly seen the danger of having the Communications Zone, by virtue of its alternate role as theater headquarters, possess a major voice in a matter in which it had an important vested interest and which now required vigorous and impartial treatment. The entire argument once again pointed up the anomalous position of one headquarters attempting to play a dual role, and indicated the need for a disinterested agency which would exercise real theater-wide surveillance and control.

In the meantime the theater had also taken action to get the expanded retraining program into operation, laying down policy on withdrawals and eligibility, specifying the objectives of the retraining program and the kind of training to be given, and issuing the first directives on the release of general assignment men to the replacement system. With one exception, the training courses to be given remained the same. But the Replacement System now ordered a substantial shortening of the courses in an effort to speed up the delivery of replacements. The three-week refresher course previously specified for men taken from the three line-of-communications regiments, for example, was now shortened to one week; similarly, the six- to eight-week course for the conversion of infantry other than riflemen to MOS 745 now became a two-week course; the course for conversion from other arms and serv-

[89] Ltr, Osborne to Lear, 2 Feb 45, sub: G–1 SHAEF Conf on Manpower, and Ltr, Barker for Eisenhower, 2 Feb 45, sub: Theater Manpower Sec, ETO 322.01/3 Manpower Board and Manpower Section; Memo, Barker for Lee, 5 Feb 45, sub: Status of Theater Manpower Sec, and Ltr, Lear to Major Comds, 24 Feb 45, sub: Manpower, EUCOM 322 Manpower.

ices was reduced from twelve to four weeks; and the course for the conversion of officers from other arms and services from twelve to six.[90]

The scope of the new training program was revealed in mid-January, when the theater commander ordered the Replacement System to increase the enlisted infantry retraining facilities to a capacity of 40,000 men, and to establish an officer training school capable of producing 1,900 infantry officer replacements per month. Of the 1,900, 400 were to consist of officers retrained from other branches in a course of six to eight weeks' duration, and 1,500 were to be provided through an officer candidate school course of twelve weeks. The theater ordered the expanded program to be put into effect by 15 February.[91]

The theater had also issued the first calls for the release of general assignment men for retraining. On 1 January General Lee assigned specific quotas to the various COMZ sections, the total COMZ commitment for the first five weeks coming to 21,000 men. For the initial increment the theater directed USSTAF to transfer 10,000 men beginning late in January. The army groups and the U.S. element of SHAEF were to make smaller contributions. The first quotas were established fairly arbitrarily in order to get retraining under way without delay. The Theater Manpower Section later established a more systematic method of determining quotas, based on the maximum understrengths

at which units could operate, the number of limited assignment men available to maintain them at minimum operational strength, and so on. The separate retraining being conducted by the armies was now ordered stopped in order to ensure control of replacement requirements by branch, which had been thrown out of balance as a result of the several conversion programs.[92]

At this time theater headquarters also laid down the policy to govern withdrawals. In general, all physically qualified white enlisted men under the age of 31 assigned to noncombat units were declared eligible for transfer to the Replacement System for retraining to infantry. Key specialists who possessed

[90] Ltr, GFRC to 16th Repl Depot, 1 Jan 45, History of GFRS, Ch. V, p. 67.

[91] Ltr, Hq ETO to CG GFRC, 13 Jan 45, sub: Retraining Program, ETO 353 Training.

[92] Ltr, Hq ETO to CG USSTAF, 16 Jan 45, sub: Inf Reinforcement Training, SHAEF AG 200.3-4 Reinforcements (Reconversion Program—Policy) 1945; Cbl E-81246, Lee to SOLOC *et al.*, 1 Jan 45, and Cbl EX-87604, ETO to A Gps, 17 Jan 45, ETO Adm 398.

highly specialized skills, who were not in excess of minimum requirements, and who could not be suitably replaced, as well as medical enlisted men in infantry and armored regiments and battalions, were exempted. Men who had passed their 31st birthday and were serving in combat units were not to be removed from combat units, nor were physically qualified men to be reassigned to noncombat units after hospitalization solely for reasons of age.[93]

In each case the theater planned to replace the men withdrawn with limited assignment men. It hoped to ease the transition at least partially by authorizing the assignment of limited assignment men for on-the-job training for a week to a month before the withdrawal of men for conversion training. It realized from the start, however, that this objective would probably not be attained, and warned the various commands to be prepared to operate understrength pending the receipt of replacements.[94] Meanwhile, it notified the War Department that it would need about 25,000 limited assignment men in addition to those available in the theater in order to expedite the release of general assignment personnel for retraining. The War Department could not promise to meet the requirement in full, but it took immediate action to recover personnel for this purpose in the United States, assigning quotas to the Army Service Forces, the Army Ground Forces, and the Army Air Forces.[95]

ETOUSA also asked the Mediterranean theater if it possessed any surplus limited assignment men it could release. General McNarney offered to make 3,000 men in this category available, and scheduled the first shipment early in February. In fact, MTOUSA, after a War Department inquiry regarding its overstrength in replacement, also offered to provide ETOUSA with an additional 3,000 general assignment replacements, thus reciprocating ETOUSA's favor of the preceding October. Needless to say, ETOUSA accepted with alacrity, and the shipment was scheduled for 15 February.[96]

The officer shortage constituted a special problem, and the Officer Candidate School was the major addition to the retraining program which ETOUSA already had placed in operation. The infantry officer replacement problem had grown progressively worse during the war. As in the case of enlisted replacements, the War Department had warned the theater in October that it could not

[93] Ltr, Hq ETO to Major Comds, 17 Jan 45, sub: Manpower, ADSEC 320.2 Strength.

[94] *Ibid.;* Ltr, Hq ETO to CG USSTAF, 16 Jan 45, sub: Reinforcement Training, SHAEF AG 200.3-4 Reinforcements (Reconversion Program—Policy) 1945.

[95] Cbl EX-84377, 9 Jan 45, and Cbl W-21658, AGWAR *to* ETO, 16 Jan 45, both in SHAEF AG 200.3-1 Reinforcements (Requisitions, Availabilities, Training); Note for Record, OPD, 14 Jan 45, sub: 25,000 Limited Assignment Repls Desired by ETO, and Memo, Henry for CGs Major Comds, 22 Jan 45, sub: Limited Assignment Requirements for CZ ETO, both in OPD 370.5 ETO Sec. XVIA, Case 521.

[96] Cbl S-76296, SHAEF to MTOUSA, 23 Jan 45, Cbl FX-91140, MTOUSA to SHAEF, 25 Jan 45, Cbl EX-93002, COMZ to MTOUSA, 31 Jan 45, and Cbl FX-19843, MTOUSA to COMZ, 2 Feb 45, Cbl FX-22498, MTOUSA to COMZ, 7 Feb 45, Cbl EX-95983, ETO to MTOUSA, 8 Feb 45, Cbl FX-26639, MTO to ETO, 14 Feb 45, all in SHAEF AG 200.3-1 Reinforcements (Requisitions, Availabilities, Training); Cbl WX-26191, AGWAR to McNarney, 24 Jan 45, SHAEF G-3 370.092 Reinforcements 1944.

continue to fill ETOUSA requisitions, for the training of officer candidates was actually being curtailed for lack of qualified candidates. Consequently it had warned the theater that it must meet a greater and greater portion of its needs through conversions and direct appointments.[97]

The theater actually had one of the best possible sources of officer material in its combat-tested and experienced noncommissioned officers. But it had not fully exploited this resource, in part because of the policy which did not assure commanders making battlefield appointments that they could retain such men in their own commands. Casualties rose sharply in November, and the theater estimated that officer losses would total 2,500 per month if operations continued at the current pace. Approximately 40 percent of casualties could be expected to return to duty, leaving a net loss of about 1,500. With the War Department promising only 600–700 replacements per month, this meant that the theater would have to furnish 800–900 from its own resources. The 12th Army Group G–1 painted an even more pessimistic picture, forecasting officer casualties of nearly 3,700 for the month of December. Appointments had averaged fewer than 400 per month for the entire theater.[98]

This trend led the Replacement System to recommend the establishment of an Officer Candidate School in the theater. In January, as part of the greatly expanded retraining program, the theater commander ordered the establishment of an Officer Candidate School capable of training 1,900 infantry officers per month. With assistance from the War Department in the form of a badly needed training cadre dispatched from the Infantry School at Fort Benning, Georgia, the theater Replacement System announced the establishment of an officer training center at Fontainebleau, France, on 21 January, naming Col. Harold E. Potter as commandant. The training schedule provided for three classes, each of 240 men, to start every week.[99]

Field commanders were urged to take advantage of their authority to make more direct appointments. The heavy losses of December, plus a change in policy which allowed units to retain the officers they commissioned, overcame some of the earlier reluctance.[100] But the number of appointments continued to be disappointingly low, and replacement officials forecast sizable shortages in the infantry officer category in view of the fact that the officer candidate program could not begin to graduate officers until May. Early in March, at the urging of his G–1, General Eisenhower again urged both army group commanders to appoint more officers from the ranks, pointing out that this was the only method of meeting the shortages in the

[97] Cbl WX–49282, AGWAR to SHAEF, 20 Oct 44, SHAEF G–3 370.092 Reinforcements 1944.

[98] Ltr, Hq ETO to A Gp and Army Comdrs, 30 Oct 44, sub: Inf Off Repls, 6 A Gp 322.96–1; Memo, Barker for Bull, 22 Nov 44, sub: Repls, Memo, O'Hare for Bull, 22 Nov 44, and Memo, Col F. J. de Rohn, Chief Training and Experiments Sec G–3 SHAEF, for G–3, 22 Feb 45, sub: Battlefield Appointments, all in SHAEF G–3 370.092 Reinforcements 1944.

[99] History of GFRC, Ch. II, p. 136; History of the Theater Manpower Section, pp. 7–8, ETO Adm 560.

[100] Memo, Twitchell for G–3, 2 Mar 45, sub: Off Reinforcements, SHAEF G–3 370.092 Reinforcements 1944.

OFFICER CANDIDATE CLASS, FONTAINEBLEAU, *performing detailed stripping of the M1 rifle.*

next few months.[101] The Replacement System had earlier instituted a three-week indoctrination course at the Fontainebleau training center for officers who had been commissioned directly.[102]

By mid-January the goals of the conversion program had been clearly outlined. But the program did not achieve real momentum until the next month. At the end of January 13,600 men were in training; a month later the number had risen to 33,400. The actual conversions in these two months totaled 7,685 and 8,193 respectively.[103] The fruits of the program were still some way off, and the replacement situation therefore remained tight.

Meanwhile, replacement requirements continued high. Battle casualties, which

[101] Ltr, Barker to Smith, 7 Mar 45, sub: Off Repls, and Ltr, Eisenhower to Bradley and Devers, 11 Mar 45, both in SHAEF G-3 370.092 Reinforcements 1944.

[102] Ltr, Hq ETO to CG GFRC, 7 Feb 45, sub: Retraining Program, ETO 353 Training; Gen Bd Study 3, Reinforcement System and Reinforcement Procedures in ETO, p. 26.

[103] [Henderson] Procurement and Use of Manpower in the ETO, p. 140.

had risen to 77,700 in December, dropped to 69,100 in January, but nonbattle casualties rose from 56,000 in December to 67,600 in January. Total casualties, coming to 136,700 men consequently exceeded the losses of the preceding month.[104]

Late in January the theater estimated that the shortage of infantrymen within the armies alone totaled 82,000, of which nearly 50,000 were in riflemen.[105] The outlook for the future was hardly encouraging despite the special efforts now being made. Earlier in the month the War Department had again relented somewhat, revising its capabilities upward to 44,000 men for May and 46,000 in June. To achieve these figures, however, it noted that it would have to call upon the Army Air Forces for an additional 15,000 men. These would consist largely of students in training for air crews and of highly trained technicians, and the Army Air Forces warned that their transfer would seriously affect air operations in the European theater. The War Department preferred not to make these withdrawals. ETOUSA manpower officials insisted on the additional replacements, however, pointing out that the theater's conversion capabilities would begin to diminish after June.[106]

Early in February the Replacement System estimated that its needs in the three months beginning with March would average 90,000 men. Seemingly unaware of the recent announcements from Washington, it recommended that the War Department be asked to furnish 91,300 in March, 88,375 in April, and 51,300 in May.[107]

The theater had continued to allocate replacements between the two army groups on the basis of their relative divisional strengths and operational missions. Early in January, in the midst of the Ardennes battle, SHAEF ordered the available infantry riflemen replacements allocated to the 12th and 6th Army Groups in the ratio of 8 to 1. Armored replacements, in which the shortage was also serious at this time, were allocated between the two commands in the ratio of 10 to 1.[108]

The 6th Army Group considered this division inequitable, particularly in view of the intensified fighting that had attended the enemy's offensive in the south beginning on New Year's Day. General Devers immediately asked for reconsideration, therefore, claiming shortages of 15,000 enlisted men and 500 officers in the Seventh Army. SHAEF defended its original allocation on the basis of operational priorities, although it notified the 6th Army Group that it would get 3,000 men which the Mediterranean theater had offered to ship; this would in effect change the ratio from 8 to 1 to 5.7 to 1. Hospital returnees did not figure

[104] *Ibid.*, p. 45, and Army Battle Casualties and Nonbattle Deaths in World War II, p. 32.

[105] Memo, Bull, 26 Jan 45, sub: Repls, EUCOM SHAEF Plans and Operations 23 Nov 44–26 Jan 45.

[106] Cbl W–30564, Handy to Eisenhower, 1 Feb 45, and Cbl S–77790, Lear to AGWAR, 4 Feb 45, both in SHAEF AG 200.3–1 Reinforcements (Requisitions, Availabilities, Training); Memo, Osborne for Lear, 5 Feb 45, sub: Manpower Reinforcements From U.S., ETO AG 319.1/1 Daily Progress Reports I.

[107] Memo, GFRC for G–1 ETO, 8 Feb 45, ETO GFRC 200.3 Personnel Requisitions File B.

[108] Whiteley, Deputy G–3 SHAEF, Stf Study, 3 Jan 45, sub: Allocs of U.S. Reinforcements, with approval by Smith, 7 Jan 45, SHAEF G–3 370.092 Reinforcements 1944.

in the allocations, for they automatically went back to their original units.[109]

As had happened so frequently before, the calculations of the two commands were based on conflicting claims. SHAEF had referred to shortages of only 5,200 infantry riflemen in the 6th Army Group, while the latter claimed a deficiency of 13,300. Moreover, 6th Army Group entered an additional bid for approximately 5,000 officers and men, representing losses which the 42d, 63d, and 70th Divisions had suffered as the result of emergency releases in December.[110] SHAEF refused to alter the allocations for January at so late a date, and promised instead to make adjustments in the February allocations if they were warranted.

Late in January SHAEF and 6th Army Group reconciled their conflicting claims. SHAEF acknowledged that its previous allocations had been inequitable. In compensation it now assigned the February allocation of approximately 50,000 infantry replacements to the 12th and 6th Army Groups on a 3 to 1 ratio. Armored replacements were to be divided equally between the two commands. The 6th Army Group's situation improved greatly in the next few weeks, owing mainly to smaller losses. In March, therefore, SHAEF, taking into consideration the new operational priorities and missions, again threw proportionately greater support to the 12th Army Group, allocating infantry replacements, totaling 60,000 men, at a 5 to 1 ratio and armored replacements, totaling 2,100 at a 4 to 1 ratio.[111]

(4) The Replacement System in Operation

For the most part U.S. forces in the European theater attempted to keep units at their full T/O&E strength through the provision of individual replacements, delivered to units either while they were in action or when they were out of the line and refitting. The relief of battle-weary regiments by units shipped in advance of other divisional components had only a limited application and did not constitute a significant exception to this policy.

The provision of replacements was the mission of the Replacement System, a separate command operated by the SOS/COMZ under the staff supervision of the theater G–1. Planning for such an organization got under way in the spring of 1943, and ETOUSA directed the SOS to establish a replacement system the following fall. On 24 November the SOS announced the establishment of the Field Force Replacement System, naming Col. Walter G. Layman as its chief. One replacement depot, with five battalions, located at Lichfield, was already in operation in England, but the

[109] TWX BX–22756, 6 A Gp to SHAEF, 12 Jan 45, and Memo, Nevins for Smith, 13 Jan 45, sub: Reinforcements for 6th A Gp, both in SHAEF G–3 370.092 Reinforcements; TWX S–74930, SHAEF to 6 A Gp, 14 Jan 45, ETO Cbls ETO Adm 404.

[110] Cbl BX–23339, 6 A Gp to SHAEF, 23 Jan 45, and Cbl BX–23377, 6 A Gp to SHAEF, 25 Jan 45, both in ETO Cbls ETO Adm 405.

[111] Cbl S–76474, SHAEF to 6 A Gp, 25 Jan 45, ETO Cbls ETO Adm 405; Memo, Whiteley for G–1 SHAEF, 30 Jan 45, sub: Alloc of Reinforcements, Twitchell, Stf Study, 7 Feb 45, sub: Alloc of U.S. Reinforcements, TWXs EX–98008 and EX–97974, COMZ to A Gps, 13 Feb 45, and TWX FWD–17358, SHAEF to ETOUSA, 26 Feb 45, all in SHAEF G–3 370.092 Reinforcements.

newly created Field Force Replacement System, as such, did not actually come into operation until January 1944. The command was subsequently redesignated Ground Force Replacement System, then the Replacement System, and in December 1944 its name was finally changed to Ground Force Reinforcement Command.[112]

The War Department showed an early interest in the newly established organization. Increasingly concerned over the manpower shortage developing in the United States, it was determined that the Replacement System in the European theater should be established along lines that would insure the best possible manpower management. As noted above, in January 1944 it sent Colonel Evans, chief of the Classification and Replacement Branch of The Adjutant General's Office, to survey the theater's progress in establishing adequate replacement handling machinery.

Evans found several faults with the policies which the theater had laid down for the embryo replacement system and made several recommendations, based on experience in the Mediterranean, for the operation of the system on the Continent. In line with the warnings which the War Department was beginning to issue at this time regarding the developing manpower shortage, he also recommended that all units and installations of the SOS be directed to survey their personnel to identify men occupying other than key or highly technical jobs who were physically qualified for field duty, and that these men be earmarked for assignment to field force units as limited assignment men were made available to replace them. As shown earlier, this recommendation met with little enthusiasm in a theater feverishly preparing for the Normandy invasion.

Evans' main criticism dealt with the matter of the Replacement System's place in the command and organizational structure. The order establishing the system had provided for a "chief" of the Field Force Replacement System and had left its position in the command structure somewhat nebulous. Certain supervisory powers over the depots, for example, had remained with the various base section commanders. The latter, according to Evans, were not confining their supervision to the functions orally agreed to. Moreover, there appeared to be no central staff agency, such as The Adjutant General's Office in the War Department, through which all directives, requests, and information would be routed. In short, Evans thought the command and staff channels were vaguely defined, and he argued strongly for giving the chief of the Replacement System command status so that he would have complete and sole authority over all ground force replacement troops. As an example of the inflexibility of the system as it was then being operated, he cited the practice of earmarking all replacements arriving in the theater for either field force or SOS assignments, with no provision for interchange of men with specialties common to two or more arms or services, a policy which did not permit the maximum utiliza-

[112] SOS GO 122, 24 Nov 43, EUCOM 322 Replacement Units II (a). For the history of the planning of the replacement system see file ETO SOS AG 322.96 Replacement Troops Jun 42 to Aug 43.

tion of skills when and where they were needed.[113]

ETOUSA officers who attended the War Department conference on manpower early in April 1944 took exception to the requirement that the theater accord command status to the Replacement System, and asked that the theater be allowed to establish the Replacement System as it thought best. Brig. Gen. Oscar B. Abbott, then the ETOUSA G–1, explained that the theater commander, through the chief of the Field Force Replacement System, exercised effective control over the location, training, and assignment of units and personnel of the system and over the flow and distribution of replacements, and denied that the base section commanders had any control over such manpower. He saw no need to give the chief of the system any additional powers, such as court-martial jurisdiction or supply, and asked that no changes be made at that late date.

War Department representatives did not press the issue for the moment, appearing satisfied with assurances that the head of the Replacement System, whatever his title, possessed sufficient powers to exercise a centralized control over the handling of replacements and to produce the desired results.[114] General Marshall was not satisfied with this arrangement, however, and a few days after the conclusion of the conference stated his desires in the matter in a lengthy cable to General Eisenhower. The North African experience had convinced him of the absolute need for a single commander with sole responsibility for the operation of the theater's Replacement System in conformity with both theater and War Department policies. It was imperative, he maintained, that the commander of the Replacement System have control of all casual personnel; that he direct and co-ordinate training programs for the recovery and proper utilization of men coming out of hospitals and for the retraining of able-bodied men in the Communications Zone to make them available for duty with combat units; and that he take aggressive action to prevent the accumulation and stagnation of men in depots. Furthermore, he must exercise control of loss replacements sent to the theater to prevent their diversion for other purposes.

These lessons, according to the Chief of Staff, had been learned at high cost in North Africa and were based on a mass of cumulative evidence. Referring to the opposition which the theater had so recently expressed to the War Department's views, he noted that he would not tolerate a "stiff-necked attitude" in opposition to an essential change dictated by experience with which ETOUSA officials were not acquainted.[115]

General Eisenhower assured the Chief of Staff that his wishes would be carried out, and noted that Colonel Layman, the chief of the Replacement System, exercised complete control over the system, and for all practical purposes actually

[113] Memo, Evans for AG ETOUSA, 13 Feb 44, sub: Observations and Recommendations, FFRS, SHAEF AG 200.3–1 Replacements (Replacement Requirements) 1944; Memo, Evans for Lovett, 12 Feb 44, sub: Final Rpt, GFRC Planning File.

[114] Notes on Conf, 8 Apr 44, SHAEF AG 200.3–1 Replacements 1944.

[115] Cbl W–22651, Marshall to Eisenhower, 13 Apr 44, ETO Eyes Only Cbl File.

was a commander, for the base sections retained supervision over the depots only in matters of housekeeping and supply.[116] Early in May General Marshall's views on the whole matter were embodied in a formal directive to the theater, which specifically laid down War Department policy on the subject. On 18 May the theater headquarters in turn announced these policies as effective within the theater and gave the Replacement System command status.[117]

By D Day there were thirteen depots in the Replacement System, comprising fifty-three battalions and 168 companies. The OVERLORD build-up plan called for the transfer of five of these depots to the Continent by D plus 90. In general, the scheme provided for a depot in support of each army, one battalion in direct support of each corps, and additional depots to operate reception centers or to serve as stockage or training installations.

Deployment to the Continent took place substantially as planned. In line with the general scheme of command for the early phases of the invasion, however, control of the Replacement System in France was initially decentralized. For the assault stage one battalion was attached to each corps. In mid-June First Army took control of the replacement units which had arrived in Normandy. First Army tried, in fact, to retain indefinite control over the replacement units which supported it. In accordance with the plan, however, control of all replacement units reverted to the Replacement System after the drawing of an army rear boundary early in August, so that centralized control was once more established over the entire system.

Four depots were then in operation on the Continent. By mid-September the Replacement System in France had grown to six depots, with twenty-six battalions. By mid-November the transfer of all but two of the depots remaining in England, and the incorporation of the depot which had arrived via southern France in support of the Seventh Army, raised the strength of the Replacement System on the Continent to ten depots. No further expansion of the system was contemplated. But the inauguration of the big retraining program in January 1945 led to the activation of two additional depots on a provisional basis, which raised the Replacement System's strength on the Continent to twelve depots. Two depots remained in England throughout the period of operations.[118]

There were five types of depots. One type provided direct support to the armies and was located well forward. At the end of January 1945 there were four of these: the 18th Replacement Depot, located at Tongres, Belgium, serving the Ninth Army; the 3d Depot, at Verviers, Belgium, serving the First Army; the 17th Depot at Angervillers, France, serving the Third Army; and the 2d Depot, at Thaon, France, serving the Seventh; a separate battalion, the 51st, at Charle-

[116] Cbl E–24656, Eisenhower to Marshall, 23 Apr 44, ETO Eyes Only Cbl File.

[117] Ltr, WD to ETO, 4 May 44, sub: Opn of Theater Repl System, and Ltr, Hq ETO to Major Comds, 18 May 44, sub: Opn of Repl System ETO, both in SHAEF AG 200.3–1 Replacements.

[118] History of GFRS, Pt. II: D Day to V-E Day, Ch. IV, pp. 1–2; [Henderson] Procurement and Use of Manpower in the ETO, pp. 17–26.

ville, France, supported the Fifteenth Army. Two depots acted primarily as receiving stations for hospital returnees, or casuals. These were the 19th, at Etampes, France, and the 10th, at Lichfield, England. Three depots served as intermediate or "stockage" pools. These were the 14th, at Neufchâteau, France, generally in support of the Third and Seventh Armies; the 11th, at Givet, Belgium, in support of the Ninth and First Armies; and the 6900th Provisional Depot, established in January at Verviers, Belgium.

Two depots and a separate battalion served solely as reception agencies. These were the 12th Depot in the United Kingdom, the 15th Depot at Le Havre, and the 54th Battalion at Marseille. The 12th, located at Tidworth, England, eventually became a retraining center. Three other depots came to serve purely as training centers by January 1945. These were the 9th Depot at Fontainebleau, for officers and officer candidates, and the 16th Depot, at Compiègne, and the 6960th Provisional Depot at Coetquidon, both for enlisted men. In addition, Training Center No. 1, at Shrivenham, England, trained limited assignment men for new duties.[119]

The rapid build-up of the Replacement System on the Continent was accompanied by a shift in the "processing" of replacements from England to the Continent. Most replacements continued to debark in the United Kingdom until the end of the year because of inadequate port facilities on the Continent. Beginning as early as August, however, replacements arriving at northern U.K. ports moved directly to Southampton by rail and re-embarked for the Continent, most of them going ashore at the Normandy beaches. The opening of Le Havre eventually led to the establishment of the main reception facilities at that port.

Whether he arrived directly from the United States or came via the United Kingdom, the average combat replacement made four stops along the continental pipeline before he finally was assigned to a unit. If he debarked at Le Havre, for example, as most replacements did beginning in November 1944, he went directly to the 15th Depot located just outside the port. The 15th acted purely as a reception center or transit area, providing barely more than a roof for the men while they awaited transportation forward. Normally the stay at the reception depot lasted only overnight, sometimes only a few hours. The replacement's next stop was the intermediate or stockage depot, where the first processing was begun. It was at this stage that the infantry replacement was issued a rifle, and an attempt was made to meet other individual equipment shortages and to bring service records up to date. Here the first attempt was also made to orient him on the status of current operations, and lectures were given on medical hygiene, including advice on the prevention of trench foot. The stockage depot also provided the replacement the first opportunity to collect back pay since his departure from the United States. The stay in the stockage depot varied. Men were sometimes on their way within forty-eight hours, as was common during

[119] [Henderson] Procurement and Use of Manpower, pp. 26–29.

ORIENTATION LECTURE FOR ENLISTED MEN, *19th Reinforcement Depot, February 1945. Visiting the class are Lt. Gen. Ben Lear (carrying swagger stick, left background) and General Eisenhower.*

the December crisis. But they often stayed much longer, and the delays at this stage were the source of much of the dissatisfaction of the individual replacement.

From the stockage depot the replacement went next to an army depot. If he came from the 14th Depot at Neufchâteau, for example, he next found himself at either the 17th Depot, which supported the Third Army, or the 2d, which supported the Seventh. In general, the processing at this stage of the journey was simply a continuation of that already begun, and included a check of a man's medical record and the filling of remaining equipment shortages. When directed by the army, the replacement moved on to the forward battalion supporting a particular corps, and finally to a specific unit.[120]

The operation of this human pipeline was bedeviled by many difficulties and became the target of endless reproach. In general, complaints fell into two broad categories. Those originating with the field commands generally dealt with such subjects as the unsatisfactory quality of replacements, endless equipment shortages, and theater policy on such matters as requisitioning and returning casuals. Those originating with the individual replacement dealt mainly with the subject of the physical discomforts and mental distress attending the interminable delays involved in moving from

[120] Ibid., pp. 28–40; [Greenwald] Human Logistics, pp. 7–48.

reception depot to the unit of final assignment.

The lot of the individual replacement was not an enviable one at any time, and it was particularly hard in the first several months of operations on the Continent. Housing facilities at the depots offered little or nothing in the way of creature comforts, and transportation for a long time took the form of either open trucks or slow-moving trains made up of unheated and crowded "40 and 8's," neither of which permitted adequate sanitation or messing facilities. Most replacements accepted the physical discomforts of the forward journey without complaint in anticipation of quickly reaching the unit of their ultimate assignment. Thousands of replacements, particularly infantry riflemen, did in fact pass through the system with relative speed and quickly found "homes" with units. Contrary to theory, however, thousands of others found themselves detained for unconscionably long periods at some point along the way.

For the latter, life in a "repple depot" was a constant battle against boredom, frustration, and worry. The Replacement System was ill-prepared, particularly at first, to keep men occupied or comfortable. Time therefore hung heavily on their hands. Left with nothing to do, and without knowledge as to his future assignment, the replacement pictured a black future for himself and expanded every rumor. Association with casuals usually did not help matters, for casuals, unfortunately, often delighted in feeding the new man's imaginary fears with tall tales about the enemy's cunning and the small chance of survival. Realizing the bad effect which this had on the morale of new men, the Replacement System eventually adopted the policy of separating casuals from new replacements during their stay in the system.[121]

Most of the depots forward of the ports eventually offered training to replacements in transit. But such training was often makeshift and pointless, and not always of a high caliber, for its conduct depended largely on officers who themselves were replacements or casuals and who gave only grudging co-operation. Meanwhile much of the processing at each stage along the replacement's forward journey, involving endless paper work, interviews, and short-arm inspections, seemed meaningless and unnecessary.[122]

The speed with which a replacement might finally be assigned and delivered to a unit and thrown into combat often contrasted sharply with the protracted delays in the depots. Common sense suggested that a man should be integrated into his unit while the unit was at rest or in reserve. Field service regulations had in fact once advocated such a policy. But it was common practice for replacements to be absorbed into a unit without knowing much more than its name. In fact, replacements often joined their units at night, without even seeing the faces of the men with whom they were to fight, and in some cases without learning the names of their squad or platoon

---

[121] [Henderson] Procurement and Use of Manpower in the ETO, pp. 40, 91–92; [Greenwald] Human Logistics, pp. 34–37; Ltr, Hq, GFRS to COs Repl Depots, 27 Nov 44, sub: Segregation of Casuals and Repls, in History of GFRS, Ch. VI, pp. 290–91.

[122] [Henderson] Procurement and Use of Manpower in the ETO, pp. 40–43, 170–72; [Greenwald] Human Logistics, p. 18.

INFANTRY REPLACEMENTS CHECKING EQUIPMENT *at a forward battalion before leaving to join a specific unit.*

leaders. Such practices not only multiplied the confusion and anxiety for the individual replacement, but created an additional hazard for the unit as a whole. Proper guidance and orientation of the replacement at this stage of his journey was undoubtedly more important than at any other, for it was at this stage that he became a member of an organized group and had to meet the first test of combat. Many units appreciated this and made special efforts to facilitate the integration of new men and help them make the necessary psychological adjustment. Some divisions established permanent processing agencies designed to introduce the replacement to his unit, acclimate him to his new surroundings, orient him on the habits of the unit, and check his equipment.[123]

The problem of the morale of the individual replacement received increasing attention during the fall of 1944 and was one of the factors which eventually led to the creation of a separate office for the control of all manpower affairs under General Lear. The Replacement System itself was aware of many of the deficiencies in the handling of replacements. In November Brig. Gen. Henry J. Matchett, who had succeeded General Layman as commander of the Replacement System upon the latter's death in England on 24 September, invited constructive criticism and

[123] [Greenwald] Human Logistics, pp. 52–56.

suggestions for improvement in the operation of the system. Late in December Maj. Gen. Charles H. Bonesteel, chief of the Inspectorate Section under Lear, with the aid of other officers, conducted a systematic investigation of the many complaints over the handling of replacements.[124]

Some of the deficiencies were outside the power of the Replacement System to remedy. In the case of others, such as the inadequacy of training at the depots, the Replacement System was undoubtedly severely handicapped by the lack of training units and equipment, and by its necessity to rely on casuals and replacements for actual instruction. Nevertheless it made a conscientious effort late in 1944 to alleviate the hardships attending the replacement's progress along the replacement route, including the organization of more meaningful training programs, provision of better housing and recreational facilities, and the provision of somewhat better facilities during the trip forward. Included in the latter were kitchen cars capable of heating rations, and permanent train commanders and medical noncommissioned officers to take the place of those previously drafted from the ranks of casuals and replacements. These measures, while alleviating some of the physical hardships attending the journey through the replacement system, actually did little to ease the replacement's mental anxiety. The handling of thousands of men was unavoidably an impersonal matter, and most replacements did not overcome the feeling that they were orphans until they finally found a home with a unit.[125]

The very term "replacement," in the view of some, had a bad psychological effect on those to whom it was applied, for it had a connotation of expendability. In an attempt to overcome this, theater headquarters late in December ordered the use of the term discontinued and the term "reinforcements" substituted on the ground that replacement personnel should be considered as a combat reserve. The War Department had no objection to the change, but it rejected the suggestion that it adopt the new term for use throughout the Army, and reminded the theater that the treatment of replacements before their entry into combat was more important to morale than calling a "rose by another name." It was at this time that the theater also redesignated the Ground Force Replacement System the Ground Force Reinforcement Command.[126]

At the suggestion of the War Department, ETOUSA in March took a more

[124] Early in October 1944 General Matchett authorized Maj. Arthur Goodfriend, Editor in Chief of *The Stars and Stripes,* to survey the workings of the Replacement System from within. Posing as "Private Arthur Goodwin," and furnished with orders as an infantry replacement, Goodfriend entered the system via the 2d Transit Area at OMAHA Beach on 8 October and went the entire route of a replacement, ending with assignment to the 26th Infantry Regiment of the 1st Division. His observations are contained in, Rpt to the CG of the Replacement System, 27 Oct 44, sub: Experiences in Army Ground Force Replacement System, EUCOM 322 Replacement Units 1944–45, IIIB. The later reports of General Bonesteel and other officers can be found in EUCOM GFRC 333.3 Inspections by General Inspectorate Section ETOUSA. See also Ltr, Matchett to Armies, Corps, and Divisions, 9 Nov 44, History of GFRC, Ch. IV, pp. 286–89, and [Greenwald] Human Logistics, pp. 99–102.

[125] [Greenwald] Human Logistics, pp. 16–21.

[126] Memo, Franey for Hughes, 1 Dec 44, EUCOM 322 Replacement Units, II (a); ETOUSA GO 131, 28 Dec 44, SHAEF AG 200.3–1 Replacements 1944; [Henderson] Procurement and Use of Manpower in

positive, if belated, measure to improve the morale of replacements. Early that month General Joseph W. Stilwell, commanding general of the Army Ground Forces, proposed that the War Department ship infantry replacements in squad- or platoon-size units rather than as individuals, and that it earmark such units for specific divisions before their departure from training centers in the United States. One obvious advantage to such a scheme was that it facilitated control, discipline, and training during movement through the Replacement System. More important, groups of men who had learned to know each other and had trained as a team could be assigned intact to units.

ETOUSA accepted the basic idea of the proposal, although it objected to the idea of earmarking units in advance for specific assignment. On 10 March it announced that henceforth all replacements would be organized into four-man groups, three of such groups forming a squad, four squads a platoon, and four platoons a company. The Replacement System announced that insofar as practicable it would organize casuals, limited assignment men, and men in the training depots, as well as shipments from the zone of interior, in this manner. While there was little chance that entire companies, or even platoons, could be maintained as units and so assigned, and while the organization was not made binding on the units which eventually received them, it was intended that every effort be made to maintain the integrity of at least the smallest grouping—that is, of four men—for its entire time in the Replacement System. Since the plan did not go into effect until barely a month before the end of hostilities. there was little to indicate how successful it might have been.[127]

Hospital returnees, or casuals, constituted an important source of manpower, comprising nearly 40 percent of all personnel passing through the Replacement System.[128] But they also constituted a special problem. As a veteran, the casual, when thrown into the Replacement System, had but one objective, which was to return to his former unit. Any delay in setting him on his way he considered intolerable. It irked him to have to go through the same routine as new replacements, and in many cases he communicated his embitterment to the green and impressionable newcomer. For the most part, commanders were equally anxious to have their old men back.

The most desirable procedure would naturally have been to return the casual

the ETO, pp. 3–4; Memos, Maj Gen I. H. Edwards, G–3, for DCofS, 17 and 23 Mar 45, sub: Redesignation of Certain Repl Units, Chief of Staff, 322, Sec. II, Cases 31–45.

[127] Cbl WAR–46854, Stilwell to C. G. Christiansen at SHAEF, and Cbl S–81733, SHAEF to Stilwell, 13 Mar 45, both in ETO Adm 410; Stf Study, Col H. W. Riley, Actg G–3 ETO, 24 Mar 45, sub: Cir 25, GFRC, 10 Mar 45, in ETO 319.1/1 Daily Progress Reports, II; Ltr, ETO to A Gps, 8 Apr 45, sub: Orgn, Training, and Flow of Reinforcements, EUCOM 322 Replacement Units, II (a); History of GFRC, Ch. VI, pp. 306–10.

[128] The casualty experience of the first three months showed that out of every 100 casualties, 70 were hospitalized, the remaining 30 representing killed, captured, or missing. Of the 70, 45 could be counted on to return to general assignment duty, 11 to limited assignment duty. Time of hospitalization varied, but an average of 46 of the 70 returned to duty within 120 days. [Henderson] Procurement and Use of Manpower in the ETO, pp. 84–85.

automatically to his old unit. Unfortunately, such a practice would have conflicted with one of the basic tenets of the theater's manpower policy—namely, the proscription against overstrength in units. Theater replacement policy, as laid down at the time of the invasion, had in effect admitted the desirability of such a procedure, but had made only a limited concession to the idea in practice. It had decreed that men discharged from hospitals who were still fully qualified to perform the duties of their MOS would be returned to their former units "whenever practicable," which meant only if requisitions were on hand from those units to fill vacancies. The basic reason for disallowing the automatic return of casuals was that it would result in overstrengths in some units at the same time that others were short of men.[129]

In view of the prospective manpower shortages, this policy undoubtedly had much validity. But, as events showed, it failed to reckon the results in terms of morale. The prolonged argument over the return of casuals illustrated the difficulty of reconciling the demand for economy in the management and use of manpower with the desire to accommodate the field commands and the individual replacement.

The theater had had occasion to confirm its adopted policy on returning hospital casuals within the first month of the invasion. Late in June, and again early in July, the commander of the Replacement System reported his concern over a growing accumulation of casuals against which no requisitions had been submitted by their former units, and asked what the disposition of these men should be. The theater G–1 pointed out that the policy on this subject was clear: unless vacancies existed in their former units, the men in question must be considered available for use in filling requisitions from other units. The G–1 conceded, however, that such personnel should not be used to fill requisitions from other units as long as other replacements were available for that purpose.[130] An exception had already been made in the case of all field grade officers and enlisted personnel of the first three grades in the case of nondivisional units.

The month of August brought an unexpected complication. Casualties were relatively light that month, with the result that the number of hospitalized men returning to the Replacement System temporarily exceeded the rate at which vacancies were occurring. This further reduced the chances that a man would return to his former unit. In an attempt to overcome some of the injustices which might result from this situation the 12th Army Group succeeded in getting priorities established by which preference for return to a unit was given to those men who had served with the particular unit the longest time. The Replacement System also adopted the policy of automatically returning general assignment casuals to the depot sup-

[129] Ltr, Hq GFRC to Repl Depots, 13 Jun 44, sub: Disposition of Offs and EM Returning From Hospitals, as cited in History of GFRS, Ch. VI, pp. 376–77; [Henderson] Procurement and Use of Manpower, pp. 85–86.

[130] Ltr, Hq GFRS to COs Repl Depots, 13 Jun 44, sub: Disposition of Offs and EM Returning From Hospitals, and Memos, GFRS for G–1 ETO, 3 Jul 44, and G–1 for GFRS, 6 Jun 44, all in GFRS Hospital Personnel Returning to Duty, May 44–Jun 45; Memo, GFRS for G–1, 25 Jun 44, as cited in History of GFRS, Ch. VI, pp. 377–78.

porting the army from which they had come, which increased to some degree the chances that men would return to their former units.[131]

The theater's policy on casuals was closely tied up with theater requisitioning policy. Under the terms of requisitioning regulations a division was prohibited from requesting replacements for an understrength regiment if the division as a whole was overstrength, although former members of the regiment needing the replacements might be languishing in the Replacement System after their discharge from hospitals.[132] Transferring men from one unit to another within the division was obviously no solution.

Field force commanders appealed again and again for the abandonment of this policy and asked that casuals be returned to their old units automatically. To the casuals themselves, detainment in the replacement depots was incomprehensible. In desperation many a casual took matters into his own hands and returned to his unit at the risk of being charged AWOL rather than sweat out official orders. Nevertheless the theater in November reaffirmed the standing policy on casuals: they were to be held in forward depots only ten days. If no requisition justified by actual vacancies was forthcoming from the men's former units within that period, they were to be considered free replacements and used to fill any requisition received by the depot.[133]

In actual practice some replacement battalions tried to return all casuals from divisions to their former units, even though this often entailed holding men beyond the authorized time limit and thus violated theater policy. Many felt that it was better to do this than to assign men to new units and risk having them go AWOL. Failure of the armies to enforce theater policy encouraged men to do exactly this.[134]

In any case, the manpower crisis brought on by the enemy onslaught in December finally changed all this. Early in January 1945, on General Devers' suggestion, SHAEF authorized the automatic return of all casuals to their former units in the 6th Army Group. Later in the month it suspended the policy requiring requisition for such casuals for the 12th Army Group as well.[135]

The suspension of the policy requiring requisitions for casuals applied only to enlisted men. Field grade officers had been excepted from the rule earlier. One group of casuals therefore still remained subject to the theater's original policy—that is, company grade officers. Later in February the replacement system, recognizing the bad effect which prolonged

[131] Ltr, 12 A Gp to Army Comdrs, 8 Sep 44, sub: Priorities for Returning Hospital Returnees to Their Former Units, as cited in History of GFRS, Ch. VI, pp. 379–80.

[132] Memo, Col Waine Archer, Chief Combat Lessons Br G–3 ETOUSA for CG GFRS, 1 Dec 44, GFRC 322 Combat Observer's Report 1944–45. See also Ltr, Bull to Chief Training Sec G–3 SHAEF, 13 Nov 44, and AGF Observers' Rpts in OCT, both in SHAEF G–3 Reinforcements.

[133] Ltr, Hq GFRS to COs of Depots, 25 Nov 44, sub: Return of Casual Pers, GFRC Hospital Pers Returning to Duty; Memo, Franey for CofS GFRS, 22 Nov 44, GFRC 370.092 Reinforcements May 44–Apr 45.

[134] [Greenwald] Human Logistics, pp. 49–50.

[135] Cbl EX–22330, 6 A Gp to SHAEF, 4 Jan 45, and Cbl S–73767, SHAEF to 6 A Gp, 5 Jan 45, ETO Cbls, ETO Adm 403; Cbl EX–88999, SHAEF to 12 A Gp, 21 Jan 45, in History of GFRS, Ch. VI, pp. 383–86.

retention in depots was having on the morale of this group, recommended that all general assignment company grade officers from the armies be returned to them without requisition. The theater commander approved the proposal on 13 March.[136] Two weeks later the theater finally rescinded the original regulation entirely, thus abandoning a policy for which there had long since been little support and which, in fact, had been widely disregarded.[137]

One exception remained. In February the theater had decreed that the automatic return-to-duty policy would not be applied to men who had sustained three or more wounds, each of which had required ten days' hospitalization. Thereafter such individuals were assigned to noncombat duty unless they specifically requested that they be returned to their old units.[138]

Up to the time of the manpower crisis late in 1944 the theater policy on casuals applied equally to service force and field force men. Beginning in January 1945, however, service force casuals, if hospitalized more than sixty days, were automatically dropped from the rolls of their organization and upon release from the hospital were considered free replacements. If still classified general assignment, they were subject to conversion training.[139]

[136] Ltr, CG GFRS to Lear, 26 Feb 45, sub: Return of Company Grade Gen Assignment Offs to Armies, and Ltr, Hq ETOUSA to GFRC, 13 Mar 45, same sub, as cited in History of GFRS, Ch. VI, pp. 387–88.

[137] Cir 35, Hq ETO, 29 Mar 45, GFRC Hospital Pers Returning to Duty.

[138] [Henderson] Procurement and Use of Manpower in the ETO, pp. 88–89.

[139] Memo, CG GFRS for COMZ G–1, 16 Jan 45, as cited in History of GFRS, Ch. VI, pp. 384–85.

Meanwhile field commanders protested the theater's policy on requisitioning and overstrength for the reason that it often operated to keep units considerably below their T/O strength. They pointed out that the proscription on requisitioning against anticipated casualties often resulted in the accumulation of losses for several days because of the lag in filling requisitions. One division noted that during periods of intensive combat, when its losses averaged 165 men per day, the shortages might add up to a thousand or more men by the time the replacements for the first day's losses arrived because of the normal lag of six or seven days. The inevitable result was a loss in combat effectiveness, for the heaviest casualties were mainly in infantry riflemen.

Finally, combat commanders also protested against the regulation that they requisition on the basis of assigned strength rather than effective strength. The number of men in a unit who were either AWOL, confined awaiting trial, awaiting reclassification proceedings, sick but not evacuated, or in the hospital, was always sizable. Under existing regulations a man who was AWOL, for example, had to be carried as part of the assigned strength of a unit for one year, and a man being hospitalized normally was carried on the rolls of a unit for sixty days. Units could not requisition replacements to make up for shortages resulting from such absences.[140]

The field commands had an opportunity to air their grievances on personnel policy at a conference called by the thea-

[140] Ltr, Gen Wyche to CG ETO, 14 Nov 44, sub: Combat Replacements, 6 A Gp 322.96–1.

ter G–1 early in November. The main concern, as always, was the problem of keeping units at authorized strength in view of the theater's existing regulations. Some commanders frankly admitted that they had circumvented theater regulations, requisitioning men in advance of actual need in order to overcome the lag in delivery. General Eisenhower's own representative, Maj. Gen. Everett S. Hughes, agreed that existing regulations only invited subterfuge. In fact, he favored the kind of dubious bookkeeping which some units, like First Army, had already resorted to in order to accomplish the desired end.[141]

Shortly thereafter the theater legitimized the practice which some units obviously were already following by authorizing units engaged in combat to requisition replacements forty-eight hours in advance of expected losses. But it refused to permit units to compute replacement needs on the basis of effective strength.[142]

The theater's regulations on casuals and requisitioning were probably a necessary part of any over-all manpower control and accounting system designed to conserve and economize. In actual practice, however, it became clear that policy on stocking, requisitioning, and use could not be applied as arbitrarily to men as to supplies and equipment. Moreover, it is questionable whether the savings effected by these regulations—savings which had implications for both the efficiency of units and the morale of individuals—were significant in comparison with the savings which might have been effected through an earlier implementation of War Department injunctions with respect to the recovery and conversion of able-bodied men in the Communications Zone and air force, and in various theater overstrengths and surpluses.

Providing replacements suffered from the handicap of most military logistics in that requirements are not easily calculated and "production" is not readily adjusted to changing demands. But the handling of the manpower problem was handicapped, in addition, by a basic flaw in the theater's command and organizational structure. Giving the Replacement System command status failed to ensure the kind of personnel management the War Department had in mind, for the Replacement System had no authority to recover general assignment men from the major subcommands for conversion training and to replace them with limited assignment men. Manpower management was a problem which touched every command in the theater. It needed to be handled at the highest command level, not by a command with an important vested interest in the situation. The fact that no effective measures were taken until the Supreme Commander personally intervened indicated that no true theater headquarters existed which could enforce War Department policy. Unfortunately the manpower problem, like the ammunition problem, was needlessly aggravated by the lack of mutually understood rules on accountability.

---

[141] Min, G–1 Conf at Hq ETO, 3–4 Nov 44, EUCOM 337 G–1 Conf, 3–4 Nov 44, I.

[142] TWX EX–62284, Eisenhower to Major Comds, 10 Nov 44, GFRC Requisitions Nov 43–May 45; Ltr (1st Ind), ETO to CG SUSAG, 25 Nov 44, sub: Combat Repl, 6 A Gp 322.96–1.

CHAPTER XII

The Logistic Structure Under Scrutiny

(1) The Communications Zone and the Field Commands

Not until February 1945 did the European theater fully recover from the logistic depression that had started in September. The deficiency in transportation, which had originally ushered in this period of hard times, and the port problem appeared well on the way toward elimination by late November with the improvement in rail operations and the opening of Antwerp. But other difficulties took their place. The displacement of the main logistic base northeastward, involving among other things a shift of service troops and rolling stock, brought new dislocations in the form of saturation of forward depots and congestion on the rail lines. Equally important, serious supply shortages gripped the theater during the fall. Taken together these difficulties made December the most crisis-ridden month of the war.

Late in November the accumulation of actual and prospective difficulties led General Eisenhower to ask Lt. Gen. Brehon B. Somervell to make Lt. Gen. LeRoy Lutes, Director of Operations, ASF, available for an on-the-ground survey of the theater's supply situation. Somervell agreed, and on 5 December General Lutes, accompanied by five other officers, flew to Paris to study the theater's difficulties. The party remained in the theater until 12 January. At that time General Somervell himself went to the theater for a briefer visit, returning to the War Department later in the month. Between them, Lutes and Somervell and their aides performed much the same type of mission they had carried out in the spring of 1944. By the time of their departure they had made a searching analysis of the theater's administrative structure and its operating methods.

Lutes' initial impression was favorable. After the first talks at theater headquarters he reported that the COMZ staff had matured noticeably since his last visit to the theater and that he found it a much more professional organization. General Smith, the SHAEF chief of staff, tended to confirm this impression, asserting that "no one could say that supply here has failed." Both he and General Crawford, the G–4, told Lutes that they believed the supply situation was now in hand and that the armies could be supported in the offensives planned for mid-December, although both were worried about the current congestion on the rail lines east of Paris. Lutes was to alter his impressions within the next few weeks, and both he and Somervell eventually found much to criticize in the theater's logistic structure.[1]

The Lutes and Somervell observations fell into two broad categories: (1) Those concerning the general lack of confidence in the Communications Zone which was widely prevalent in the field commands, and (2) those concerning the shortcomings in logistic management, or what might be termed the field of business administration in supply.

Lack of confidence in the Communications Zone was hardly a new phenomenon. Suspicions engendered by the differences over organization and planning in the U.K. period had never subsided,[2] and relations between the Communications Zone and the field commands were never completely cordial. Part of the mistrust undoubtedly stemmed from the traditional and probably unpreventable feeling that rear area troops were better supplied than those at the front, particularly in such items as clothing and food. The sudden famine in the more popular brands of cigarettes in the combat zone in November, brought about by distributional difficulties, was but one of several examples of COMZ indifference and inefficiency, in the view of the front-line soldier, which confirmed this attitude. However unwarranted such feelings may have been, the field commands never were completely satisfied that the Communications Zone appreciated the urgency of their needs or showed the proper zeal in meeting them. They made little attempt to conceal their suspicions, and on occasion voiced their dissatisfaction openly to the COMZ staff, condemning what they referred to as the "lethargy and smugness" of some of its members.[3] It is noteworthy that dissatisfaction with the service elements never reached such proportions on the southern line of communications.

The attitudes of the field commands varied, undoubtedly reflecting in some degree their training in the understanding of supply problems. First Army had always been the most outspoken in its criticism of the Communications Zone. General Hodges himself was intolerant of any supply deficiency and, according to Lutes, the least disposed to make any attempt to understand logistic problems. General Patton frankly stated that he trusted no one to the rear of the Advance Section and usually planned his operations in partial defiance of supply difficulties, making the most of available resources. Lutes believed that of the army commanders Simpson had the best understanding of supply. Perhaps because of this the Ninth Army commander did not expect miracles and trusted in the Communications Zone to support him. General Bradley was credited with being logistically minded, although conservative and disposed to make doubly sure of his supply by asking for large reserves, as he did in the case of ammunition. The field commanders in the 12th Army Group were unanimous in expressing confidence in the ADSEC commander, General Plank,

[1] Ltr, Somervell to Lee, 1 Dec 44, Ltr, Lutes to Somervell, 7 Dec 44, and Memo, Lutes for Somervell, 8 Dec 44, all in Hq ASF European Theater—Last Half 1944.

[2] See *Logistical Support I*, 264–68.

[3] Memo, Col James MacCormack, Chief Mov Br G–4 12 A Gp, for Moses, 1 Dec 44, 12 A Gp G–4 Letters and Memos Assorted, 1944.

who had always maintained the closest liaison with the combat commands and who apparently had to their satisfaction instilled in his entire command an urgency about meeting the needs of the combat forces. General Somervell felt that this "spirit of urgency" was lacking at other echelons of the Communications Zone and suspected that the lack of it was sensed by the armies and was one of the major sources of the distrust expressed by them. He emphasized this strongly at a staff conference in January, pointing out that the Communications Zone existed for one purpose alone—serving the combat forces—and that it must convince the field commands that it appreciated this. "A lack of confidence or antagonism," as Somervell put it, "has no place in a situation of that kind. I do not care who is right or who is wrong, the point is that we have to satisfy our customers and do so in a way which pleases them." [4]

As in the past, some of the difficulties could be attributed to a clash of personalities. This had been most evident in the relations between the Communications Zone and Supreme Headquarters, for the relations between Generals Lee and Crawford had never been completely free of strain. That the dealings between the two headquarters had improved some over the months was attributed largely to the diplomacy of Lee's chief of staff, General Lord, who personally conducted much of the business with SHAEF and also handled many of the complaints from the field.[5]

More important than any personality differences, and the source of at least some of the distrust between the field and service forces, was the continued dissatisfaction with the command structure as it affected supply. Although technically General Lee no longer held the title of deputy theater commander, his position had not really changed. While he himself no longer legally wore two hats, his staff did, for it doubled as the COMZ and ETOUSA staff. It had a dual role, therefore, and the army group commanders could not help feeling that it could not be counted on to give the desired priority to the demands of the combat forces where such demands conflicted with those of rear area troops. The thought that the Communications Zone, a command co-equal with the army groups, might be passing on the validity of their requests was particularly distasteful to them. They strongly suspected, in fact, that the COMZ-ETOUSA staff had been responsible for the "arbitrary" paring down of their recommended replacement and expenditure factors and was therefore responsible for the current shortages in ammunition and such items as radios and tanks.

The basic difficulty with the arrangement was likewise illustrated in the problem of allocating supplies between the two army groups. It appeared logical that theater headquarters should make such allocations. But the field commands considered the COMZ-ETOUSA staff a doubtful authority for that purpose and

[4] Memo, Lutes for Somervell, 17 Dec 44, sub: Preliminary Rpt, and Ltr, Somervell to Lutes, 27 Dec 44, in Hq ASF European Theater—Last Half 1944; COMZ Stf and Comd Conf, 19 Jan 45, EUCOM 337/3 Confs Staff Weekly I.

[5] Memo, Lutes for Somervell, 8 Dec 44, sub: Rpt, and Ltr, Lutes to Somervell, 7 Dec 44, both in Hq ASF European Theater—Last Half 1944.

preferred to look to the U.S. element at SHAEF for such decisions. Most of them would have preferred a setup which had been proposed several times before—that is, a strictly American GHQ, separate from SHAEF and primarily tactical in nature, but equipped to give over-all direction to the logistic effort as well.[6]

(2) Expediting Supply Deliveries

At the time of General Lutes' arrival early in December the theater's two most pressing problems were (1) the serious supply shortages, notably in ammunition, tanks, tires, general purpose vehicles, and field wire, and (2) its inability to handle, particularly in the forward areas, the large tonnages which the ports were now able to discharge and the railways to move forward. General Lutes was able to give immediate and material assistance on the former. As a top-ranking official of the ASF assessing the theater's situation at first hand he was in a position to add great weight to the theater's requests. While confirmation of the theater's needs should not have been necessary, General Lutes nevertheless was invaluable in obtaining additional releases, particularly of such items as tanks and trucks.

But additional allocations and releases could solve only part of the problem, for they promised to meet the theater's demands in only a limited number of items. It was estimated that production in the United States was falling short of demands in about 600 major items procured by the ASF, and there was little prospect of raising the output of most of these within the next six months. The only means by which supply to the combat zone could be increased in the near future, therefore, was by accelerating the flow of supplies already available or becoming available—that is, by shortening the delivery time of supplies already in the pipeline or coming off production lines. A time analysis of recent ammunition shipments had revealed that an average of 46 days elapsed between the date on which ammunition arrived in the theater and the date on which it was finally laid down in forward depots—23 days awaiting discharge in U.K. and continental waters, 15 days in actual unloading, and 8 days in movement from shipside to forward depots. Movement from U.S. depots to the theater required approximately 50 days. Total pipeline time from zone of interior depots to combat zone consequently averaged nearly 100 days.

The time consumed in waiting in European waters was already being cut, and theater officials estimated that both unloading and intransit time could also be reduced by special handling. The result would be a saving of as much as thirty days, or 65 percent of the pipeline time previously required within the theater. General Lutes believed that time consumed in shipment from U.S. depots to the theater could also be materially reduced by both special handling and speeded shipments. He pointed out that for each ten days that the pipeline could be shortened and kept shortened for critical items, an additional ten days of

[6] Memo, Lutes for Somervell, 17 Dec 44, Memo, Lutes for Somervell, 24 Jan 45, sub: Final Rpt on Visit to the ETO, and Memo, Lutes for Somervell, 8 Dec 44, sub: Rpt, all in Hq ASF European Theater—Last Half 1944; Memo, Maj J. R. Greenwood, OPD Observer for Hull, 6 Dec 44, OPD 319.1 Sec. VIII, Case 2.

production could be made available in the combat zone. This he believed would be of substantial assistance to the theater, at least in the immediate future. General Lutes proposed that a major saving in time could be made in this portion of the pipeline by the inauguration of an express shipping service, using either fast unescorted vessels or small fast convoys. The War Department had already undertaken to expedite the shipment of certain critical types of ammunition via fast freighter. Lutes now proposed that this service be expanded to include other highly critical items of supply in order to meet the theater's most urgent needs in the next few months. He also suggested that the delivery of other important supplies could be improved by a more careful selectivity in loading and by special handling. Greater selectivity in the loading of ships would prevent high priority items from being buried in slow-moving cargo, as had often occurred in the past.

General Somervell approved the idea of a rapid shipping service on 1 January and laid down the ground rules which were to govern its use. A few days later General Stratton, the COMZ G–4, in turn announced to the theater supply services the establishment of the rapid express service, known as REX, explaining its purpose and prescribing the conditions and procedures which were to govern its use.[7] To complement this service and thus ensure the most expeditious handling of high priority supplies throughout the supply pipeline General Lutes meanwhile had proposed to theater officials that a fast rail service be established between the ports and forward depots to handle small tonnages of urgently needed items. Such a service, known as the "Toot Sweet Express," was also established in January.[8]

(3) Supply Planning Procedures

The critical supply shortages of December had again called attention to two or three other aspects of theater logistics which had long been of direct concern to the War Department. From the point of view of the ASF the theater's tardy planning reflected a lack of appreciation of the time factor involved in procurement. Both the Communications Zone and the 12th Army Group were found to be remiss in this regard. General Lutes saw no evidence of planning by the army group beyond April 1945. Plans which extended barely four months forward really constituted nothing but current plans so far as supply was concerned and corresponded roughly to the time required to fill requisitions and deliver supplies from existing stocks in the United States. They made no allowance for any lead time involved in special procurement. Forward planning had to be projected a minimum of nine to twelve months in advance, Lutes pointed out, for items the need for which the ASF could not foresee.[9]

[7] See Chapter XIV, Section 3, below, for the later history of REX. Memo, Lutes for Somervell, 23 Dec 44, sub: Supply of Critical Items to Combat Theater, Cbl WARX–84997, Somervell to Lutes and Lee, 1 Jan 45, and Memo, Stratton for Chiefs of Svs COMZ, 5 Jan 45, sub: Rapid Express Shipping Sv, all in Hq ASF Official Rpt, Mission to ETO, 4 Dec 44–13 Jan 45; Memo, Lutes for Somervell, 17 Dec 44; COMZ G–4 History, I, 131–36.

[8] See below, Chapter XV, Sec. 1.

[9] Memo, Lutes for Somervell, 17 Dec 44, sub: Preliminary Rpt, Hq ASF European Theater—Last Half 1944; Memo, Lutes for Somervell, 24 Jan 45,

Closely related was the matter of the theater's requisitioning practices. War Department circulars had long since prescribed the ground rules governing requisitioning by the theaters, and had specified a form designed to provide an accurate monthly picture of the theater's stock position and to facilitate the editing of the theater's requests. The theater's failure to follow the prescribed procedures had long complicated the editing job at the New York Port, accounting for at least some of the delays of which the theater so frequently complained. It undoubtedly explained why the War Department and the theater were so often at odds over the quantities of supplies ETOUSA was entitled to. Early in 1945 ASF officers assigned to the COMZ staff made a thorough examination of the requisitioning practices of the theater's supply services and reported an utter lack of an adequate or uniform system of computing requirements. As a result, their analysis showed, the theater was requisitioning improper quantities of supplies.

The principal shortcoming of the theater's practices, and one which all the services shared, was the failure to provide the New York Port with the data it needed to edit requisitions intelligently as to quantity—that is, stock status data as to quantities of supplies "on hand," "due in," and "due out," and the additional information needed to determine the theater's "requisitioning objective." [10] The services should have included as due in, for example, quantities requisitioned on the New York Port although not yet delivered, quantities expected from other theaters, expected deliveries from local procurement, and also salvaged or reclaimed items returned to depot stocks. But the services followed no standard procedure in such reporting.

The same lack of uniformity was apparent in the troop bases which the services used to calculate requirements, and in the order and shipping time they used. Manpower figures frequently varied, depending in part on the troop basis used, and in part on conflicting totals of Allied military personnel, displaced persons, civilians, and prisoners of war. Order and shipping times ranged from 60 to 124 days, and a study of actual shipments revealed that the time between the submission of a requisition and the first deliveries of supplies in COMZ depots ranged from a minimum of 47 days in the case of medical supplies to a maximum of 145 days in the case of quartermaster supplies. The corresponding figures for the delivery of 75 percent of the supplies against a particular requisition were 60 and 234 days respectively.

Although the War Department had frequently called the theater's attention to irregularities in requisitioning methods, the theater apparently had never made a serious effort to enforce the rules. In February, inspired by the find-

sub: Final Rpt on Visit to the ETO, and Memo, Lutes for Eisenhower, 25 Dec 44, sub: Rpt on Supply Situation—Northern France, both in Hq ASF Official Report, Mission to ETO; Somervell's remarks at Comd and Stf Conf, COMZ, 25 Jan 45, Hq ASF ETO 1945, Somervell files.

[10] The requisitioning objective is the total quantity of an item authorized to be on hand and on order in the theater. Quantitatively it represents the sum of the maximum level of supplies and the order and shipping time.

ings of ASF officials, ETOUSA finally moved to standardize its entire requisitioning procedure in line with long-standing War Department wishes, defining specifically what supplies were to be included as on hand, due in, and due out, and eventually settling on a single order and shipping time of 120 days. The new procedure became effective on 1 April. Like several other measures, therefore, its adoption was too late to have any effect on logistical support before the end of hostilities.[11]

Differences over the theater's requisitioning habits inevitably involved the problem of replacement and consumption factors. Accurate maintenance factors were obviously important to the theater, since they were intended to ensure that supply was adequate to meet the ups and downs of combat requirements over a long period; they were equally important to the War Department, since they played a large part in determining future production. It was desirable that in the long run they coincide with actual needs, no more and no less.

The theater had complained repeatedly about the difficulties of getting prompt consideration of its recommendations for revisions of replacement factors, and chafed at the endless War Department requests for "justification." In despair at the interminable delays, it sometimes proceeded to requisition supplies on the basis of its recommended factors without waiting for approval. The War Department, its procurement program under frequent scrutiny by Congressional committees, was understandably cautious in authorizing upward revisions. Moreover, it suspected that the purpose of replacement factors was not universally understood in the theater, and that recommended revisions frequently reflected only fragmentary and short-term experiential data rather than long-term trends. Again and again the War Department held the theater's requests to be unsupported by experience. As an example, Somervell pointed out as late as March 1945 that the theater had asked for a replacement factor of 25 percent for the 4.2-inch mortar, whereas average monthly losses over a period of five months had not exceeded 10 percent.

ASF officials considered their misgivings confirmed when they found that, as in the case of requisitioning, the ETOUSA supply service had no uniform policy in determining revisions of replacement factors. During his visit to the theater early in 1945, therefore, General Somervell asked for a comprehensive review of maintenance factors. The COMZ G–4 thereupon assigned an officer with the necessary technical qualifications to study the problem and maintain liaison with the supply services. The latter in turn assigned men whose main duty it was to review and revise replacement factors. The review of replacement factors was a slow process, however, and the War Department continued to hold up approval of theater recommendations which were not backed by adequate justification based on consumption experience. Not until late in April did the ASF indicate satisfaction

[11] Ltr, Robinson to Somervell, 19 Mar 45, ASF 400.312 SAFE DCofS, SC, and Control Div, Drawer 2103; COMZ G–4 History, IV, 178–201.

with the reports being submitted by the theater.[12]

(4) The Depot System and Records Keeping

Meanwhile General Lutes had also begun to examine problems of distribution within the theater. The congestion on the railways and in the forward depots which had developed at the very time of his arrival early in December tended to highlight certain basic deficiencies in the theater's logistic structure, particularly with regard to its depot system, movement planning and control, and requisitioning and stock control practices.

Both Lutes and Somervell regarded the lack of adequate depots properly echeloned in depth as the greatest single weakness in the theater supply structure. The absence of intermediate depots meant that disproportionately large portions of its supplies continued to be stocked in base and forward areas. Although the Communications Zone had made some progress in moving stocks forward, approximately half of all the supplies in the Communications Zone still lay in Normandy and Brittany in mid-December.[13]

LT. GEN. BREHON B. SOMERVELL *(right), arriving at an airfield with Lt. Gen. John C. H. Lee, is met by Lt. Gen. Jacob L. Devers (left), 16 January 1945.*

The lack of intermediate depots would not have been so serious had there been a system of true base depots. But a large portion of the supplies in the port areas really lay in huge dumps, in which no facilities for segregation or classification existed, and in which the exercise of any degree of selectivity in forward shipments was therefore extremely difficult. The attempt to move some of these large stocks forward as

[12] Ltr, Robinson to Somervell, 19 Mar 45, Memo, Brig Gen T. M. Osborne, Dir of Rqmts and Stock Control Div, for Lutes, 27 Mar 45, Hq ASF 400.312 SAFE DCofS, SC, and Control Div; Ltr, Hq ETO to CG ASF, 9 Apr 45, sub: Necessity for Prompt Rev of WD Repl Factors, with 1st Ind, 21 Apr 45, EUCOM 400 Supplies, Services, and Equipment (Policy), VII.

[13] Memo, Lutes for Eisenhower, 25 Dec 44, sub: Rpt on Supply Situation—Northern France; for weekly figures on distribution of supplies see also Summaries of Br Chiefs' Mtgs, COMZ G–4 Plant and Communications Diary/Jnl, ETO Adm 145C.

transportation improved during the fall, and to clear some of the rising port backlogs, led to huge bulk shipments in December which the forward depots were unprepared to handle.

Theater officials had actually planned a depot structure along more orthodox lines. The existing structure's lack of resemblance to original plans they attributed in large part to the effects of operations in August and September. The speed of the advance and the resulting absorption of transport in the movement of the barest essentials had precluded the planned build-up of stocks at Rennes, Le Mans, and Chartres. In fact, supply installations at those places never developed into true intermediate depots, and were quickly relegated to the role of local issue depots, in which capacity they served for the remainder of the war.[14]

The shape which the depot system ultimately took reflected in large measure the tactical thinking and the prevailing optimism of the late summer. In the expectation that Allied forces would continue the drive to and beyond the Rhine, SHAEF instructed the Communications Zone not to build large depots in the Paris area. The Communications Zone therefore decided to forego the establishment of sizable intermediate depots and to build up large installations in the Advance Section instead. The principle of depth consequently gave way to the desire to be in the best possible position to exploit an early break-through to the Rhine.[15]

With the bogging down of operations late in September the Communications Zone made plans to establish intermediate depots in the Reims, Soissons, and Paris areas. For many weeks, however, shipments to those areas were on a negligible scale. Throughout the month of October all available transport was absorbed in the build-up of army and ADSEC stocks in preparation for the November offensive. No true base depots were established either. When such installations were proposed for Le Havre and Rouen in October, the chiefs of the supply services protested that they possessed insufficient personnel to man them. They also feared that the establishment of such depots would result in a relaxation in the efforts to keep the ports cleared.[16]

Early December therefore found the theater lacking adequate base and intermediate depots and its supply stocks poorly distributed. This could be attributed in part to decisions made in a period of greater optimism, in part to the fact that the day-to-day needs of the theater, including the necessary build-up for planned offensives, had absorbed all the energies of the Communications Zone throughout October and November and had precluded any major adjustment designed to effect a proper echelonment in depth of the theater's supplies. The poor distribution of supplies was well illustrated early in December when Third Army requisitioned 7,000 rounds of 155-mm. smoke shells

[14] COMZ G–4 History, II, 3.

[15] Memo, Lutes for Eisenhower, 25 Dec 44; COMZ G–4 History, IV, 90, and Memo, Lord for Eisenhower, 2 Jan 45, sub: Comments on Rpt on Supply Situation—Northern France, EUCOM 400 Supplies, Services and Equipment (Policy), VI; FUSA Rpt of Opns, 1 Aug 44–22 Feb 45, Bk. I, p. 163.

[16] COMZ G–4 History, II, 4, 39–40.

which it required for a river crossing, only to find that half of them would have to be trucked from a dump at OMAHA Beach, involving a round trip of about 800 miles.[17] The inadequacies of the depot system—particularly the lack of intermediate installations—were even more pointedly demonstrated when the enemy's December counteroffensive endangered the heavy concentrations of supplies in the forward areas.

Meanwhile, the absence of true base depots with an efficient stock records system meant that no selectivity in shipments could be exercised. The ports themselves frequently had to serve as retailers, although they were never intended to segregate and classify supplies. In December and January, with emphasis being placed on port clearance, huge quantities were forwarded in bulk to whatever depots could receive them and with little regard for actual needs in the forward areas. Under these circumstances high priority items often were lost in the shuffle. It was partly because of this that Lutes had recommended a red ball railway service between the ports and advance depots.

The heavy movements of January removed all doubts about the Communications Zone's ability to forward large tonnages. Its main problem now, as both Lutes and Somervell emphasized, was in exercising the necessary selectivity in its shipments. The prerequisite for this was the organization of base depots capable of receiving, sorting, and classifying supplies in such a way that they could be located and identified for selective forwarding.[18]

An adequate depot system was but one link in the logistic chain, albeit a vitally important one. Inseparably a part of the structure and essential to its efficient functioning were such things as movements planning and control, stock records keeping, requisitioning procedures, and documentation. Lutes and Somervell pointed out shortcomings in all of these fields. As will be shown later in detail, movements planning did not ensure the most efficient use of all the means of transportation in the theater, and failed to co-ordinate transportation with the capacities of the depots. No detailed plans had been made at all, Lutes found, for the transfer of supplies from Normandy Base Section to the Oise Intermediate and Advance Sections in the fall when the improvement in transportation made possible a drawing down of the large base dumps. Such plans would have entailed a study of the local labor needed to unload freight cars, and of depot capacities in the forward areas, and should have prevented the current congestion on the railways east of Paris and the tie-up of loaded rail cars in forward depots. Somervell felt, moreover, that the control of movements should be more centralized and should be vested in the transportation service,

[17] Memo, Col Lee A. Denson for Lutes, 11 Dec 44, sub: Ammo Supply ETO, Hq ASF Official Report, Mission to ETO.

[18] Final Rpt of Chief Engr, ETO, I, 200–01, 215–19; COMZ G–4 History, II, 10–11; Stf and Comd Conf COMZ, 19 Jan 45, EUCOM 337/3 Confs Staff Weekly; Memo, Lutes for Somervell, 24 Jan 45, sub: Final Rpt on Visit to ETO, Hq ASF Official Report, Mission to ETO.

the agency responsible for actual transport operations.[19]

Sound stock records, documentation, and requisitioning procedures were basic essentials of the logistic structure if supply officials expected to know at all times what supplies were on hand and where they were located, and to ensure that they arrived at their proper destination. All these procedures began to give trouble from the very start of continental operations. Pressure to unload supplies and get them across the beaches in the early months often required the entire effort of service personnel and left little time for keeping proper records. Accurate records keeping was further complicated in the early stages, when First Army was in complete command on the Continent, by the habit of army units of drawing supplies from dumps without submitting proper requisitions. Many such outloadings never became a matter of record, and consequently were not reflected in matériel status reports, which determined the quantities the theater might requisition from the zone of interior.

After First Army relinquished control of the rear areas Communications Zone ordered the services to inventory all dumps and depots. But the practices of the various services were anything but uniform, and the accuracy of stock records was widely doubted by the armies, which frequently discovered supplies in rear area depots which the Communications Zone denied existed. The documentation of shipments was likewise deficient, in part because it did not allow requisitioning agencies to identify shipments with specific requisitions, and in part because it often resulted in the diversion of supplies and their delivery to the wrong destination. The chief signal officer of the Communications Zone, for example, found it advisable to place one of his officers aboard every signal supply train to ensure its arrival at the intended destination.

The theater's internal requisitioning SOP's had, like some of the other administrative procedures, been a casualty of the pursuit period, and their corruption had contributed immeasurably to the supply difficulties of the fall months. SOP 7, the basic supply operating guide, had provided that the armies should draw the bulk of their needs from the Advance Section on the assumption that balanced stocks of supplies would be maintained in ADSEC depots. Unfilled portions of requisitions were to be extracted to base section depots supporting the Advance Section. But the developments of August and September had gradually rendered this procedure unworkable. The inability to maintain balanced stocks in the Advance Section soon led to the bypassing of that organization and the submission of more and more requisitions directly to the headquarters of the Communications Zone. The scattered location of dumps and the shortage of transport encouraged this trend toward the centralized handling of supply requests.

This development, plus the ineffectual stock control system, inadequate signal communications, and poor documenta-

[19] Memo, Lutes for Eisenhower, 8 Dec 44, sub: Rpt, Hq ASF European Theater—Last Half 1944; Memo, Somervell for Lee, 24 Jan 45, SHAEF G–4 319.1 Rpt—Gen Somervell, 1945, I. See Ch. XIV, below.

tion of shipments created an exasperating uncertainty as to the fate of requisitions. In effect, many were pigeonholed through the inability of the Communications Zone to fill them, and the armies frequently were left in doubt as to when, if ever, they would be filled. Late in September the Communications Zone instituted a system of "back ordering" which was intended to keep the armies better informed as to when they could expect delivery of past requisitions. But questions as to the priority which such supplies should have for lift and poor communications continued to beset the system.

Within the armies, meanwhile, there was no uniformity of attitude toward back-ordered supplies. The First Army, for example, continued to requisition previously ordered supplies on the assumption that back orders would not be filled. The other armies made varying allowances in their requests for the eventual delivery of back-ordered items.[20] Occasional freaks in supply, such as the appearance of a carload of anvils in an army depot, only heightened the misgivings of the field commands. On the other hand, reports of bungling sometimes proved wildly exaggerated. A "carload of pianos," appearing at the front in mid-winter, for example, and reported to the Communications Zone with considerable indignation, turned out on investigation to be a long-delayed shipment of field organs, for which space had suddenly been found when more urgently needed supplies were unavailable for movement.[21]

The Communications Zone made repeated attempts to improve the back-ordering system, and these efforts eventually had some success. However, backlogs of orders had become so confused that the theater finally asked the armies to cancel all outstanding requisitions and to submit new consolidated requisitions for items they still needed. Early in December the Communications Zone abandoned the old daily requisition for Class II and IV supplies, which had allowed insufficient time for orderly posting and back-ordering and had needlessly multiplied bookkeeping, and instituted instead a ten-day requisitioning period. By February the build-up of stocks in ADSEC depots had improved sufficiently to permit the armies to requisition directly on the Advance Section, as originally intended.[22]

(5) *Expedients*

The shortcomings of the Communications Zone's management procedures were only too accurately reflected in the practices which the field commands resorted to in self-defense. Impatient with the long delays in getting requisitions filled and with the unauthorized di-

[20] TWX EX-62723, Lee to Secs, 6 Oct 44, Memo, Hass for Moses, 3 Nov 44, sub: Backlog Shipments, Ltr, 12 A Gp to COMZ, 28 Oct 44, sub: Requisitioning and Mov of Supplies, Memo, Hass for Moses, 12 Nov 44, sub: Policy for Handling Requisitions, Memo Hass for Moses, 19 Nov 44, sub: Backlog Shipments, Ltr, 12 A Gp to Armies, 26 Nov 44, sub: Rpts of Back-Ordered Items, with 1st Ind, FUSA, 3 Dec 44, and Memo, TUSA, for 12 A G-4, 6 Dec 44, all in 12 A Gp Supplies—Backlog and Requisition, No. 128, Gen Bd Rpt 110, App. 5.

[21] Intervs with Col Alvin Viney, Deputy Comdr ADSEC, 24 Feb 50, and Lord, 9 Aug 51.

[22] Memo, Lord for Crawford, 31 Jan 45, sub: Gen Lutes' Memorandum to Supreme Commander dated 25 Dec 44, EUCOM 400 Supplies, Services, and Equipment (Policy), VI; COMZ G-4 History, IV, 85, 103; Gen Bd Rpt 27, pp. 64-68.

versions along the lines of communications, the armies sought various ways, some of them extra-legal, to ensure themselves against the uncertainties of supply. Most of the field commands, for example, padded their requisitions and attempted to build larger reserves than were desirable from the point of view of mobility as a cushion against unpredictable deliveries. Some resorted to requisitioning practices which were clearly contrary to existing SOP's. In the period of tonnages allocations, for example, the Communications Zone had specified that requisitions be processed through command channels so that the G-4's could exercise a more effective control. But many army service chiefs circumvented this channel and requisitioned directly through the technical services.

In October the irregularities in requisitioning were finally aired in the course of an official investigation of the First Army engineer's practices, with the result that the theater commander ordered the First Army not to honor requisitions which were improperly processed and routed. Both First Army and the 12th Army Group protested the ruling. In any case, the abolition of tonnage allocations in December permitted a return to the procedures originally laid down in SOP's—that is, requisitioning through supply service channels, as the armies preferred.[23]

The uncertainties of supply manifested themselves in still other ways. By October, for example, the armies had made it a common practice to employ expediters to hand carry their requisitions to Paris and to follow shipments through to final delivery. Many a service chief in the field commands, despairing of getting a requisition filled via the prescribed channels, sent personal representatives to COMZ headquarters to learn first hand about shipments long overdue and presumed lost. This practice had reached its extremes in the First Army, whose ordnance officer alone employed about one hundred men as field agents to follow through on ordnance requisitions to ensure the delivery of supplies. In General Somervell's opinion, this was the severest indictment that could be made of the Communications Zone, for it indicated that its entire logistic management was faulty.

The Communications Zone strongly denied the armies' need for such expediters, particularly by ordnance, which had one of the better stock control records in the theater. General Somervell nevertheless felt that the use of such "bloodhounds" reflected seriously on the Communications Zone's performance.[24]

[23] Memo, Stratton for CofS COMZ, 4 Oct 44; sub: Investigation of the Inspector General of Diversion of Supplies, Ltr, Col James H. Day to CG ETO, 20 Oct 44, sub: Rpt of Investigation of Unauthorized Acts in Filling Requisitions, Ltr, CG ETO to CG FUSA, 8 Nov 44, sub: Rpt of Investigation of Unauthorized Acts in Filling Requisitions, with 2d Ind, FUSA, 28 Nov 44, 3d Ind, TUSAG, 2 Dec 44, and 44th Ind, ETO to 12 A Gp, 13 Dec 44, all in 12 A Gp 400.312 Requisition for Supplies, II, Gen Bd Rpt 110, p. 9.

[24] Memo, Lutes for Eisenhower, 25 Dec 44, sub: Rpt on Supply Situation—Northern France; Memo, Somervell for Lee, 24 Jan 45; Stf and Comd Conf, COMZ, 19 Jan 45; Memo, Lord for Eisenhower, 2 Jan 45, sub: Comments on 'Rpt of Supply Situation—Northern France'; COMZ G-4 History, I, 2–21, II, 21; History of Normandy Base Section, pp. 82–83, ETO Adm; TUSA AAR, II, Sig, 13; Rpt, Day to CG ETO, 20 Oct 44, sub: Rpt of Investigation 12 A Gp 400.312 Requisitions for Supplies, II; FUSA Rpt of Opns, 1 Aug 44–22 Feb 45, Bk. IV, p. 155.

Somervell's criticisms on this score focused attention on a corollary practice of the armies—the sending of their own trucks to pick up supplies in the rear. Many an army service chief had come to the same conclusion as the surgeon of the First Army, who, finding that delivery via rail required from three to five weeks, decided that the only way to get quick delivery of highly critical items was to dispatch trucks directly to Paris. Early in October the Communications Zone had objected to this "foraging to the rear" and had instructed the base sections not to make issues directly to army units. Its inability to meet army needs with its own transportation, however, led it to cancel this restriction early in November. Later in the month the Communications Zone modified the stand further, and asked the armies not to send its trucks into the rear areas without prior clearance with the Communications Zone.

The practice of sending both expediters and trucks nevertheless continued, and General Somervell's criticisms in January led the COMZ G–4 to reconsider the entire problem. Brig. Gen. Morris W. Gilland, the new G–4, took a strong stand against allowing any agents in the rear areas, insisting that they tended to confuse and upset orthodox and businesslike supply procedures rather than to facilitate deliveries. He recommended, however, that no restriction be placed on the dispatch of army transportation to the rear in emergencies, for transportation available to the Communications Zone was not adequate to satisfy all demands for high priority movements. In mid-March the Communications Zone, confident that its transportation and operating procedures had improved sufficiently to meet all future demands, asked that both practices be stopped. Neither personnel nor equipment was henceforth to be sent into the Communications Zone to pick up supplies except by mutual agreement.[25]

The Communications Zone was aware of many of the weaknesses which ASF officials noted upon their arrival in the theater and had made efforts to correct the deficiencies in some fields. Both Lutes and Somervell had recognized, moreover, that the theater's difficulties were at least in part attributable to circumstances beyond its control—mainly the forced growth of the Communications Zone caused by the sudden extension of the lines of communications in August and September. The Communications Zone simply had not yet overcome some of the disruptions which had attended the assumption of tasks beyond its capabilities.

On the other hand, both ASF officers felt that the Communications Zone's staff work had been below the desired standard and that it had failed to give

[25] Ltr, COMZ to Normandy Base Sec, 5 Oct 44, sub: Third U.S. Army Trucks in Rear Areas, Ltr, COMZ to Normandy Base Sec, 3 Nov 44, sub: Issue of Engr Supplies to Army Units, with 1st Ind, n.d., Ltr, COMZ to Base Sec Comdrs, 26 Nov 44, sub: Army Trucks in Rear Areas, Ltr, Gilland to CofS, 25 Feb 45, sub: Dispatch of Trucks and Pers From Armies to Rear Areas, Ltr, Lee to Sec Comdrs and Supply Sv Chiefs, 18 Mar 45, sub: Supply and Transportation to the Armies, Ltr, Lee to A Gps, 18 Mar 45, Ltr, 12 A Gp to Armies, 31 Mar 45, sub: Forward Mov of Supplies, and Ltr, 6 A Gp to Armies, 31 Mar 45, sub: Supply and Transportation to the Armies, all in EUCOM 400 Supplies, Services, and Equipment, VI–VII; COMZ G–4 History, IV, 220.

sufficient attention to the details of logistic management mentioned above. Too often, in their judgment, the Communications Zone had let things drift, with the result that it was constantly rushing to put out fires which it should have had the foresight to prevent. This was clearly demonstrated in the necessity to rush service troops from the base areas to forward depots when it was found that the latter could not handle the increased flow of supplies late in the fall.

General Lutes had undertaken his mission to ETOUSA with some trepidation, realizing that the COMZ staff, to whom General Lee was intensely loyal, was a "tight corporation," sensitive to criticism. But, as he explained to the COMZ commander on his arrival, he felt obliged to determine whether any of the supply deficiencies in Europe could be laid to failures in the United States, and to ensure that the field commands, whose complaints had reached the War Department, did not blame the ASF for deficiencies for which the theater was responsible.[26]

Lee himself never ceased to be a controversial figure. General Eisenhower, like others on the SHAEF staff, was aware of the attitude of the field commands toward the Communications Zone and continued to have misgivings about the supply organization of the theater. Despite this uneasiness, the Supreme Commander chose not to make a change in the COMZ command in view of the lack of sufficient evidence of specific failures. Lee was probably unaware of many of the procedural deficiencies, for he was not normally concerned with the details of logistic administration. In any event the COMZ commander had a reputation for attaching more importance to the outward appearance than to the substance of things. He tended to underplay the "several small deficiencies in supply methods" which Lutes and Somervell had discovered, and to emphasize the "basic soundness" of the Communications Zone's organization and operating procedures.[27] To the field commands this attitude, plus his personal unpopularity, unfortunately tended to magnify the inefficiencies of the Communications Zone.

Both Lutes and Somervell were determined that the "several small deficiencies" should be corrected. SHAEF supported them in this resolve and asked the Communications Zone to make periodic reports on its progress on the Lutes and Somervell recommendations. General Lord submitted the first report on 23 January and continued to make weekly reports until early in March. Meanwhile General Somervell also insisted that the Communications Zone re-establish a Control Division within its headquarters to regularize reporting procedures and to keep the commanding general informed on the progress made toward the various objectives. On Lee's own suggestion, Somervell designated Brig. Gen. Clinton F. Robinson, who had accompanied the ASF commander on his visit, to organize the division.

Robinson worked hard at his assignment, and eventually reported that the COMZ staff had achieved a much more professional standard in its work. Early

[26] Memos, Lutes for Somervell, 20 and 22 Dec 44, Hq ASF European Theater—Last Half 1944.

[27] Memo, Lutes for Somervell, 31 Dec 44, Hq ASF European Theater—Last Half 1944.

in February these efforts were aided when General Gilland came up from the disbanded Southern Line of Communications to become the new G–4. General Gilland brought with him several members of the SOLOC G–4 Division, including his deputy, Col. R. W. Colglazier, his chief of plans, Col. Carter Page, and his chief of operations, Col. Charles Cobb. Carrying out the Lutes and Somervell recommendations was not easy, for there were many limiting factors and conflicting demands. But the Communications Zone made a conscientious effort to overcome its earlier difficulties and made substantial progress in this effort in the final months of the war.[28]

[28] Ltrs, Robinson to Somervell, 7, 16, and 25 Mar 45, and Ltr, Somervell to Robinson, 12 Mar 45, both in Hq ASF 321 European Theater, A47–81; Memo, Robinson for Lee, 8 Mar 45, Program (sic) Made on Items in Gen Somervell's Memorandum of 24 Jan 45, EUCOM 400.192 Miscellaneous Reports, VII; Memo, Vissering for Current Opns Sec G–4 SHAEF for Current Opns Br, 25 Jan 45, sub: Gen Lutes' Memo to Supreme Comdr, and Ltrs, Crawford to Smith, 8, 15, and 25 Feb 45, sub: Weekly Rpt on Maj Gen Lutes' Memo Dated 25 Dec 44, all in SHAEF G–4 319.1 Reports on Gen Lutes, I; Ltr, Somervell to Styer, 17 Jan 45, Hq ASF 201 Styer, A47–81. See EUCOM 400 Supplies, Services, and Equipment (Policy), VI, for weekly reports by Lord to SAC, 23 January to 26 February 1945.

THE LAST OFFENSIVE
FEBRUARY–MAY 1945

CHAPTER XIII

Tactical, Logistical, and Organizational Aspects of the Last Offensive

(1) Tactical Developments, 8 February – 8 May 1945

The Ardennes operation not only proved to be the enemy's last major offensive effort; it had weakened his defensive capabilities beyond recovery. Allied might, on the other hand, continued to grow, and by the end of January 1945 was greater, both in numbers and in the logistic capability to support sustained operations, than it had ever been before. Early in February the Allies launched offensives which did not relax in intensity until victory was finally won.[1]

The basic plan for the spring offensive had been laid down some time before. It called for closing to the Rhine and then crossing that barrier to envelop the Ruhr, the main effort to be made in the north. The first phase was conceived of as a series of hammer blows, beginning in the north. (*Map 6*)

In accord with these plans Field Marshal Montgomery's 21 Army Group launched the first attack in the battle for the Rhineland on 8 February. The First Canadian Army, operating from the vicinity of Nijmegen over ground either waterlogged by thaws or completely flooded, drove southeastward between the Maas and the Rhine, fighting one of the most bitterly contested battles of the war (Operation VERITABLE). The Ninth U.S. Army, also under 21 Army Group control, was by plan to have launched a complementary attack (Operation GRENADE) two days later, driving northeastward across the Roer to meet Canadian forces and clear the area west of the Rhine between Duesseldorf and Wesel. These plans were temporarily frustrated on 9 February when the enemy blew the critically important Roer dams just as the First Army's V Corps captured the controlling ground, creating a formidable water barrier and forcing the postponement of the Ninth Army's attack for two weeks. Ninth Army, aided by an extensive air interdiction operation and by a shattering forty-five minute artillery preparation carried out by the guns of three armies, finally launched its assault of the Roer River line on 23 February, putting twenty-eight battalions of infantry across

[1] For the full account of tactical operations in the final months see Fred J. Meyer and William G. Bell, The Last Offensive, a volume in preparation for the series UNITED STATES ARMY IN WORLD WAR II.

D. Holmes, Jr.

MAP 6

LUDENDORFF RAILWAY BRIDGE *over the Rhine River at Remagen captured by men of the 9th Armored Division.*

on the first day. After several days of hard fighting General Simpson's forces broke out of the bridgehead and began making excellent progress. On 2 March elements of one corps reached the Rhine, and on the following day elements of another joined with British forces at Geldern. By 5 March the Ninth Army had closed to the Rhine in its entire sector.

Farther south First Army also crossed the Roer on 23 February and within another five days the Erft River as well. General Hodges' troops pushed rapidly eastward in the next few days and closed to the Rhine. By 5 March most of the west bank of the Rhine north of Cologne was in Allied hands, and on the 7th Cologne itself fell. The capture of this important prize was overshadowed by a more dramatic event on the same day farther south in the First Army zone when the 9th Armored Division seized a railway bridge intact over the Rhine at Remagen. First Army promptly established a bridgehead there and reinforced it as rapidly as possible.

By this time operations to close to the Rhine were also under way still farther south. Third Army, beginning with probing attacks in accordance with an injunction from higher authority to limit its action to an "aggressive defense" during 21 Army Group's operation, steadily chewed its way through the Siegfried Line fortifications in its sector. By 23 February, the date of the Roer assaults farther north, Third Army had battered through the enemy's main defenses in the western Eifel, the area of the December counteroffensive. In the next ten days it captured Trier and advanced beyond the Kyll River. On 6 March the XII Corps erupted from its Kyll bridgehead and in less than two days advanced fifty miles along the ridge roads of the Eifel to the Rhine. By 10 March another sizable stretch of the Rhine's west bank, from Koblenz northward, was in Allied hands.

Third Army's quick success north of the Moselle made possible a favorable development in operations farther south. After a few days of mopping-up operations north of the Moselle, General Patton's forces on 13 and 14 March launched attacks eastward from Trier and southeastward across the lower Moselle, advancing rapidly across the rugged Hunsruck far to the rear of the German forces. On 15 March the Seventh Army joined the attack from the southeast, driving into the Saar industrial area. While General Patch's forces slugged their way through the heavily fortified West Wall in the south, Patton now sent armored columns racing south and east from the lower Moselle, rendering the entire enemy defensive position in the Palatinate untenable. Within a week of the Moselle crossing Third Army forces had cleared the entire Rhine from Mannheim to Koblenz. On 21 March the enemy's defenses in the Saarbruecken area finally crumpled, and Seventh Army broke through to link up with the Third Army and complete the clearance of the area west of the Rhine. By this time the First Army held a bridgehead at Remagen eight miles deep and six divisions strong.

While isolated enemy pockets west of the Rhine were being eliminated and while preparations went forward with great deliberation for a Rhine crossing north of the Ruhr, Third Army had brought its bridging forward in the wake of its advancing infantry, and on the night of 22 March, unaided by either artillery or air support, launched a surprise crossing at Oppenheim, fifteen miles south of Mainz. Within two days three divisions had crossed the Rhine and had carved out a bridgehead ten miles wide and nine miles deep. Other Third Army units made additional crossings which were quickly joined. Patton's forces rapidly expanded these bridgeheads in the next few days. By 28 March most of the army was east of the Rhine, and armored units had already crossed the Main and thrust forty miles beyond.

The 21 Army Group operation, employing units of both the Second British and the Ninth U.S. Armies, got under way on the night of 23 March. The northern crossing was aided by a huge air interdiction program designed to isolate the Ruhr, and on the morning of the 24th the assault was further supported by the dropping of two airborne divisions beyond the river. Resistance was heavy, but there was never any question of success, and by the end of the first day the Ninth Army held a bridgehead about nine miles wide and from three to six miles deep.

Reinforcing these successes, the First Army on 25 March launched attacks from the Remagen bridgehead and within the next three days drove eastward as far as Marburg and Giessen, and southeastward to link up with Third Army units north of Wiesbaden. Finally, additional crossings of the Rhine were made on 26 March, when the Seventh U.S.

Army crossed both north and south of Worms, and on 31 March, when the First French Army crossed near Speyer. By the end of the month, therefore, five Allied armies were firmly established east of the Rhine, the enemy's last great natural bulwark. Two U.S. armies, the First and Third, were already advancing up the Frankfurt corridor into the heart of the Reich.

In the first days of April all efforts were focused on the encirclement of the Ruhr. Contrary to the scheme of maneuver originally envisaged, tactical developments now favored a major effort in the center, where both armies of the 12th Army Group had already made deep thrusts beyond the Rhine, rather than in the north, where the bridgehead was still small. Accordingly, the First and Third Armies were now ordered to advance shoulder to shoulder in a northeasterly direction, the Third on the axis Hersfeld–Kassel, and the First on the axis Marburg–Paderborn, to make contact with 21 Army Group. General Devers' 6th Army Group was to protect the southern flank of this drive.

These missions were largely accomplished within two days. On 1 April the Third Army reached Kassel, and then shifted its attacks eastward across the Werra River. In the meantime the First Army, while guarding with its left along the Sieg River and the rough Rothaar Gebirge, jabbed due north through the Hessian hills with its armored and motorized right to join up with armored elements of the Ninth Army at Lippstadt. On 1 April the Ruhr "pocket," an area of 4,000 square miles and containing the entire German Army Group B, was finally a reality.

First and Ninth Armies immediately began regrouping in preparation for the final elimination of the huge pocket, the Ninth Army once more reverting to 12th Army Group control, and on 4 April General Bradley issued instructions for attacks by the three U.S. corps which contained the large enemy force from the south, east, and north. A corps of the Fifteenth Army had already assumed responsibility for the area west of the Rhine. The encircled enemy forces counterattacked repeatedly in the next few days, and heavy fighting ensued at some points. But all attempts to break out of the pocket proved abortive, and the ring was steadily drawn tighter. On 14 April First and Ninth Army forces met at Hagen, splitting the pocket in two. Two days later the eastern half collapsed, and on 18 April all resistance came to an end. The Ruhr pocket had yielded nearly a third of a million prisoners.

The advance eastward had continued after only a brief pause for reorganization. On 2 April General Eisenhower had issued instructions for the drive into central Germany, specifying that the main effort was to be made in the center with the intent of splitting the Reich and then destroying enemy forces on either side. The 12th Army Group, with twenty-two divisions initially available for the eastward drive, surged forward on 6 April on a 150-mile front. Resistance thereafter diminished rapidly, and the enemy's central front soon lost all semblance of cohesion. Armored units now slashed forward at will, leaving pockets of resistance to be eliminated by the infantry. Only in the Thuringian Forest and the Harz Mountains did the enemy offer serious opposition. By 18

April, the day on which the Ruhr pocket was eliminated, all three armies had reached the restraining line drawn along the Elbe and Mulde Rivers and through the cities of Chemnitz and Bayreuth. British forces in the meantime had cleared a part of the coastal area in the north and were not far from Bremen and Hamburg. General Devers' forces, in addition to protecting Third Army's flank, had driven southeastward as far as Nuremberg. Well over a million prisoners had been taken since the Rhine crossings.

These swift advances brought the Allies into imminent contact with Russian forces, which were fast approaching the Elbe River from the east, and which were also driving up the Danube. With Germany virtually split in half, General Eisenhower now halted the advance in the center in order to reinforce and provide logistic support for drives to the southeast and the north. Assigning both the First and Ninth Armies a largely defensive role along the Elbe and Mulde Rivers, General Bradley now swung the Third Army southeastward and launched a powerful thrust down the Danube Valley. Simultaneously with this operation General Devers' two armies also drove southward to destroy the myth of a German redoubt in the Alps, and far to the north Field Marshal Montgomery's two armies continued their sweep of the north German plain.

Resistance melted away rapidly in the final weeks, and by the end of the month the once formidable German war machine lay broken and overpowered. In the first days of May the 21 Army Group completed clearing the north German plain westward to Holland and eastward to Luebeck, Third Army drove into Austria and western Czechoslovakia, and General Devers' two armies drove southward and made contact with the Fifth Army at the Brenner Pass. First Army in the meantime had overcome the remaining pockets of resistance in the center, capturing Leipzig and closing up to the Mulde in its sector, and had made contact with Russian forces at Torgau. On 7 May Generaloberst Alfred Jodl, representing the German high command, signed the Act of Surrender which finally brought an end to hostilities in Europe.

(2) Logistic Factors in Planning the Last Offensive

The problem of crossing the Rhine was inevitably the central feature of all planning for the spring offensive, and its logistic aspects had a prominent place in all such planning. Detailed planning for the Rhine crossings had begun early in the fall of 1944, and every operation envisaging a possible break-through to the Rhine gave fresh impetus to such preparations.

In all the planning for the Rhine crossings the SHAEF staff invariably considered three areas in which crossings were deemed feasible—the Emmerich–Wesel area, the Cologne–Koblenz area, and the Frankfurt–Mannheim area. Terrain considerations and the importance of the Ruhr as a major military objective had led to the early assumption that the major effort would be made in the Emmerich–Wesel sector, north of the Ruhr. One of the early appreciations spoke of committing twenty-seven divisions in the northern bridgehead, leaving a slightly smaller number for a subsidiary offensive, either in the Cologne–

Koblenz or the Frankfurt–Mannheim sector. A later study considered the establishment of bridgeheads in all three areas, building up the northern operation to a strength of thirty-one divisions and exploiting only one of the others depending on the prospects for success.[2]

Crossing the Rhine was from the start looked upon as more than just another river crossing operation. Because of the width of the river the problems of the operation in some respects resembled those of a short sea voyage. As in the cross-Channel attack, speed in building up Allied strength in the bridgeheads was all important. The main logistic problem, consequently, was one of transportation. Good road and rail networks on both sides of the Rhine were obviously necessary. The Rhine bridges, like the Normandy beaches, would be the bottlenecks through which Allied forces and supplies would have to be funneled and would largely determine the rate of build-up and the size of forces that could be supported.

Planning, which had been interrupted by the enemy counteroffensive in December, was renewed with fresh vigor early in the new year. On 8 January SHAEF issued a directive to the major subordinate commands summarizing the assumptions on which planning for the spring offensive should proceed. Although SHAEF planners continued to consider the possibilities of attacks in all three of the areas mentioned above, the directive of 8 January dealt only with an operation north of the Ruhr. It emphasized that a major offensive would probably be launched in that area, and concluded that a maximum force of 36 divisions (15 British and 21 U.S.) could eventually be supported in the northern bridgehead. In arriving at this total SHAEF planners believed they were planning the maximum potential exploitation of logistical resources, and the chief of staff emphasized that the support of these forces would require a "drastic curtailment of ordinary standards of administrative convenience." They had figured maintenance requirements at 600 tons per division slice per day in the combat zone, including sixty tons of common supply items per division required by tactical air units. An important feature of planning from this time on was the decision to lay pipelines across the Rhine to carry three fourths of the POL requirements east of the river. Laying pipelines beyond the German border had previously been considered inadvisable. But since POL constituted a third or more of all tonnage requirements, the use of pipelines obviously would mean a tremendous saving in traffic over the Rhine bridges.[3]

Both the 12th Army Group and the Communications Zone concluded that the operation could be supported under the conditions outlined by the SHAEF

[2] For early planning estimates and studies on the Rhine crossings see the following files: SHAEF G–3 370–47, Plans—Rhine River; SHAEF G–4 381 War Plans General, II; 12 A Gp 800.23 River Crossings; SHAEF G–4 Logistical Planning—Future Operations; SHAEF G–3 Plans 1 Future Operations—1945.

[3] Logistical Implications of Operations in the Northern Bridgehead—Spring Offensive 1945, G–4 SHAEF, 7 Jan 45, SHAEF G–4 381 War Plans General 1945, I; Ltr, Smith to CGs COMZ, 21 A Gp, and 12 A Gp, 8 Jan 45, sub: Adm Plans and Preparations for Future Opns, 12 A Gp 370 Operational Planning; Brig Gen C. Ravenhill, "Logistics in Northwest Europe 1944–1945," digest in *Military Review*, XXVII (November, 1947), 77.

directive, and differed on only one major point. Apprehensions growing out of past experience led the Army group to ask for 650 tons per division slice per day as insurance against what it termed "misdeliveries," admitting that the smaller figure more nearly represented actual consumption needs. SHAEF, the Communications Zone, and 12th Army Group all based their estimates on the assumption that there would be no rail service into the bridgehead until about D plus 63, and that the armies would be able to transport their own supplies east of the river until D plus 50. The 12th Army Group figured that it might require truck support by the Communications Zone before that date only under conditions of rapid advance—that is, if points of delivery were thirty miles beyond the Rhine and fifty-five miles beyond the service areas. POL pipelines over the Rhine were counted on to be in operation by D plus 21.[4]

The SHAEF planning staff had also studied the logistic implications of an offensive in the south. An offensive north of the Ruhr would employ less than half of the divisions expected to be available for the spring offensive; logic seemed to dictate that preparations should be made for an offensive in the south to achieve the maximum strategic flexibility. A planning appreciation prepared in mid-January considered two possible alternatives for a southern offensive—either a subsidiary to the attack north of the Ruhr, or a larger effort assuming the use of less than maximum forces in the north. A study of logistic implications indicated that as many as fifty divisions could be supported in a southern bridgehead if it was developed to maximum capacity. This assumed, as in the north, that railways would be pushed close to the Rhine and that motor transport would be reserved for use forward of rail and pipeheads.[5]

General Eisenhower's plans, which contemplated offensives by both First and Third Armies, raised fears among British officials that a northern offensive might be prejudiced, and led them to insist on a more unequivocal commitment on a northern operation and a promise that an assault north of the Ruhr would not await the closing to the Rhine along its entire length. After discussions between Generals Eisenhower and Marshall at Marseille, and between General Smith and the Combined Chiefs at Malta at the end of January, the Supreme Commander early in February issued a new directive for long-range planning designed to dispel any doubts as to his intentions. In it he announced his intention of carrying out the main offensive north of the Ruhr, and he also gave assurances that crossings would be made as soon as feasible and without waiting to close to the river along its entire length. He made clear, however, that the possibility could not be ignored that circumstances might compel a switch of large forces to the Mainz–Karlsruhe area, and he considered it prudent to

[4] Ltr, Lord to Smith, 24 Jan 45, sub: Adm Plans and Preparations for Future Opns, EUCOM 381/2 War Plans General, I; Ltr, 12 A Gp to SAC, 25 Jan 45, sub: Adm Support and Preparations for Future Opns, and Ltr, Osmanski to Whipple, 6 Feb 45, sub: Adm Plans and Preparations for Future Opns, both in SHAEF G–4 Logistical Implications Future Operations—1945.

[5] Logistics of Operations in the Southern Bridgehead—Spring Offensive, G–4 SHAEF, 27 Jan 45, SHAEF SGS 381 Post-OVERLORD Planning.

make logistical preparations for such an eventuality and for the support of an alternate offensive.

General Eisenhower directed the Communications Zone to prepare to support such an offensive, operating northeast out of a southern bridgehead beginning as soon after 15 April as tactical conditions permitted, on the assumption that the force employed would eventually build up to a strength of forty divisions. Logistical resources were not considered adequate to permit the support of both offensives simultaneously at maximum scale—that is, thirty-six and fifty divisions. But the planners believed that the maximum strategic flexibility would be achieved if logistical preparations were made for the maximum force on either of the two axes. The 5 February directive specified that in any question of priorities, preparations for the northern offensive were to have overriding priority.[6]

The possibility of crossings in the Cologne–Koblenz area was no longer given serious consideration. The central sector had always been the least favored for both tactical and logistical reasons. Road communications to the northeast led through highly defensible terrain, and most of the important rail routes ultimately ran northward into the built-up area of the Ruhr. The maintenance of large forces northeastward in that area was considered infeasible without rail and therefore extravagant of engineering effort. Moreover, the area west of the Rhine in the Cologne–Koblenz sector did not lend itself to the establishment of the large depot and maintenance installations needed for the support of a major operation. A build-up east of the Rhine in that area consequently would be at the expense of support in the other bridgeheads.[7]

The Communications Zone proceeded to plan on the basis of the directive of 5 February, and on 2 and 3 March issued outline plans for the logistic support of spring offensives in the north and in the south respectively. The two plans differed very little. Both called for pushing rail support well forward into the army areas up to the time of the crossings and constructing rail bridges over the Rhine, and they designated the general lines along which the railways would be rebuilt east of the river. Both also called for POL pipelines across the river. As usual, the two advance sections were to provide support for the armies, and their role east of the Rhine was conceived as being highly mobile. Both planned to turn over their depots to other sections after the crossings—the Advance Section to Channel Base Section and Oise Intermediate Section and the Continental Advance Section to Oise—and to operate only mobile depots east of the Rhine and some dumps west of the river. Neither was to have territorial responsibility within Germany. In the main, the northern bridgehead was planned to receive its support from the Liège–Charleroi–Lille depots and the

[6] Ltr, Smith to CG COMZ, 5 Feb 45, sub: Adm Preparations for Spring Offensive, and Ltr, Crawford to Smith, 30 Jan 45, sub: Adm Preparations for Spring Offensive, both in SHAEF G–4 381 War Plans General 1945, I; see Pogue's *Supreme Command,* Chapter XXI, for a full discussion of the controversy over closing to the Rhine and the priority to be given an offensive in the north.

[7] Development of Operations, 1945, Appreciation 1, Planning Staff SHAEF, 11 Feb 45, SHAEF G–3 Plans L Future Operations—1945.

southern forces from the Verdun–Nancy–Toul–Langres–Dijon installations. The Communications Zone had from the beginning had apprehensions about supporting U.S. forces in the Wesel area because of congested communications and conflict with British installations, and had asked for additional depot space in the north and for running rights over some of the British-controlled railways. While some pause for regrouping and gathering of supplies was recognized as possible before the actual Rhine assault, the Communications Zone wisely directed that for planning purposes administrative preparations should assume that bridgeheads would be established immediately upon closing to the Rhine.[8]

Ironically, fate decreed the capture of the first bridgehead in the area least favored in all the planning estimates, partially upsetting the carefully laid plans for the spring offensive. Nevertheless, commanders at all echelons acted swiftly to exploit the unexpected capture of the Remagen bridge, and immediately studied its logistic implications. Plans were promptly made to extend rail service to the site from both the northwest (via Dueren) and the southwest (via Trier). Preparations for the latter were almost immediately canceled as the result of the collapse of the captured rail bridge. The build-up went ahead, nevertheless, and the bridgehead was quickly exploited eastward along the axis of the Autobahn, which led directly to objectives in the Giessen area, and to an eventual link-up with forces in the southern bridgehead area, where support could more easily be given by rail from the Metz–Verdun depots. POL pipelines never extended to the Rhine in the Remagen area, but a short pipeline was eventually built at Mehlem to pump gasoline from a west bank terminal to the storage tanks near the Autobahn east of the river. Plans now called for a build-up of thirteen to eighteen divisions in the Remagen bridgehead. Sixteen divisions were eventually supported on the central axis.[9]

Barely two weeks after the First Army's crossing at Remagen the Third Army's surprise assault raised the question of the advisability of strengthening the southern bridgehead at the expense of the northern. No decision was immediately made, although it was decided that the operation north of the Ruhr should not fall below twenty-four divisions in strength. The build-up in the north eventually reached thirty-one divisions, and that in the south 35.[10]

The actual course of operations beyond the Rhine strikingly emphasized the conservatism and caution of the planners, despite SHAEF's injunction against "over-insurance" in the matter of logistic support. The Allied build-up in the Rhine bridgeheads eventually totaled

[8] Ltr, Hq ETO to SvC Chiefs, CONAD, and ADSEC, 8 Feb 45, sub: Plng Directives, EUCOM 381/2 War Plans General; Outline Plans for Administrative Support of Spring Offensive in the North and South, Hq COMZ, dated 2 and 3 Mar 45, respectively, 12 A Gp 370 Operational Planning.

[9] Memo, Vissering for G–4 Exec, 10 Mar 45, sub: Outline Plan for Adm Support of Spring Offensive in the South, Memo, Whipple for G–4, 12 Mar 45, sub: Remagen Bridgehead, Memo, Osmanski for Mov and Tn Sec G–4 SHAEF, 18 Mar 45, sub: Railroad Dev, all in SHAEF G–4 381 War Plans Gen 1945, II.

[10] Appreciation of Effect of UNDERTONE's Success on Future Operations, SHAEF G–3, 21 Mar 45, G–3 Plans L Saar Operations.

eighty-two divisions, of which sixty-six were U.S.-supported, and was effected at a much faster rate than was originally thought feasible. Moreover, within the time limits adopted, Allied forces were supported at much greater distances than originally believed possible.

Forecasts of tactical progress which SHAEF planners made even after the Rhine assaults, assuming "optimistic conditions, with organized but weak enemy resistance," showed a bridgehead with a depth of only fifty miles north of the Ruhr and sixty miles in the south on 15 May, and extending generally along the line Luebeck–Magdeburg–Regensburg–Munich in mid-July.[11] Logistic plans had not counted on Rhine rail bridges being in use until nine weeks after the crossings, an estimate which proved far too conservative. Moreover, they had calculated maintenance requirements at substantially higher rates than either past or current experience warranted.[12]

Logistic planners at the Communications Zone fortunately had made preparations to support a much more rapid advance than was assumed in SHAEF forecasts. Undoubtedly recalling the experience of the preceding summer, the Communications Zone as early as 19 February had issued a planning directive and outline plan keyed to rapid advance conditions which called on both advance sections to plan for an emergency marshaling of truck transport in the event of an accelerated advance.[13] Logistical planners realized that such a coralling of transportation for use in the forward areas would mean a curtailment of other COMZ activities, such as port clearance and static interdepot hauls, and would result in a heavy drain on forward depots. But it was felt that the logistic structure was strong enough to withstand such a sacrifice for as much as thirty days without adverse effects, even though it might eventually have repercussions on port discharge.

The speed of the advance in the last month actually exceeded even these plans for "rapid advance." As in the summer of 1944, the prospect of great tactical gains led to an even more extraordinary marshaling of resources and sacrifice of normal operations than the Communications Zone had considered safe. But its foresight paid good dividends, and there was no hesitation in stretching resources to the limit when opportunity beckoned the armies forward. Success in supporting the final drive can be attributed in large measure to the flexible plans which COMZ organizations had worked out for the expansion of transport in the forward areas and the rapidity with which rail bridges were installed. The decision to prepare simultaneously for alternate offensives with maximum strength in both the north and south likewise proved wise, for these preparations made possible the prompt exploitation at Remagen and adequate support of the crossings in the Frankfurt–Mannheim sector. Once again, however, the Allies, conscious of the logistic difficulties which had hampered

[11] Study, Logistical Factors of Future Operations, April–September 1945—Tonnages, SHAEF 215/4 GDP–2, Tonnage Requirements.

[12] See Chapter XV, Section 5, below, for a summary of consumption factors.

[13] Emergency Plan for Support of Rapid Advance, Hq COMZ, 19 Feb 45, 12 A Gp 370 Operational Planning.

operations since September, appear to have underestimated their capabilities.

(3) Command and Organization, February – August 1945

ETOUSA's command and organizational structure had been substantially completed by the end of 1944, and, except for the dissolution of SOLOC, no further major alterations were made in either the field or service force commands. In the ground forces the only changes of importance concerned the operational control of the First and Ninth Armies and the role of the Fifteenth Army. The First Army, which along with the Ninth, had been placed under the control of 21 Army Group during the Ardennes battle, was returned to the 12th Army Group on 18 January 1945, after its juncture with the Third Army in the Houffalize area. The Ninth Army remained under Field Marshal Montgomery's control for the Roer and Rhine crossing operations, and finally reverted to General Bradley's control after the encirclement of the Ruhr early in April. The Fifteenth Army, which had been activated late in December, was limited for several weeks to an almost purely administrative and planning role, handling the staging and equipping of newly arrived units in the theater and planning for the occupation. Late in March it took command of the forces containing the Brittany ports, and early in April it relieved other 12th Army Group units of occupation duties west of the Rhine. Fifteenth Army had only a minor operational role in the final offensive beyond the Rhine.[14]

In the Communications Zone the only changes of importance were those involving adjustments in section boundaries and in COMZ headquarters organization which were made in response to special needs and in the search for greater efficiency. Changes in the territorial organization of the Communications Zone arose largely from tactical developments and from the continued shift away from the Normandy and Brittany areas in logistical affairs. Brittany Base Section had absorbed Loire Section as a district in December 1944. On 1 February 1945 Normandy Base Section in turn incorporated Brittany as a district. Brittany Base's headquarters was given a new mission. Brig. Gen. Roy W. Grower now took control of a sizable portion of the Continental Advance Section known as Burgundy District to assist in the support of the 6th Army Group. Burgundy District occupied a rather special status in the COMZ command and organizational structure. Although General Grower's command was operationally under the control of the Continental Advance Section, he reported directly to General Lee, and transfers of personnel between CONAD and Burgundy District could be made only on the authority of COMZ headquarters. In some respects Burgundy District more nearly resembled a section, for it enjoyed considerably greater administrative powers than were normal for a district.

The dissolution of SOLOC, which was formally accomplished on 12 February, did not alter this arrangement, but it finally brought all COMZ sections in both the north and south directly under the control of General Lee's headquar-

[14] Organization and Command in the ETO, II, 258–64.

ters.[15] In mid-February the Communications Zone thus consisted of eight sections—two advance sections providing direct support to the two U.S. army groups, three base sections controlling the coastal areas and operating the ports, and, in addition, the U.K. Base, Seine, and Oise Sections.

Steps were also taken early in 1945 to form a separate French line of communications to support the First French Army in the south. On 19 February Base 901 was officially activated as a subcommand of the Communications Zone with the mission of assisting CONAD and Delta Base Section in co-ordinating and supervising the supply of French forces. French units attached to CONAD and Delta Base Section actually remained under the control of these commands for operations, but for administrative purposes they now came under the control of Base 901, which was commanded by Brig. Gen. Georges Granier. It was intended that Base 901 should eventually assume complete responsibility for the support of French forces, but that stage was never reached.

French authorities had long agitated for the establishment of a French zone of interior, and SHAEF had promised as early as August 1944 to turn over control of French territory to the French Committee of National Liberation as rapidly as possible for that purpose. It took the first step late in October 1944, turning over the control of eight departments in the north to the French national authorities. In accord with promises made to General Charles de Gaulle, the Supreme Commander at that time proposed a similar transfer in the south. But General Larkin, the SOLOC commander, opposed such a step, arguing that port, signal, rail, and highway operations in the south were too critical to risk the possible interference of civil authorities. General Lee supported the SOLOC commander, and the proposed transfer was postponed.[16]

The last changes in COMZ territorial organization were occasioned by the advance across the Rhine. Early in March plans for the spring offensive announced the principle that the two advance sections would move into Germany with the armies and continue under COMZ control, but that they would have no area responsibility there. Both Oise and Channel Base Sections were to extend their control in the north, Oise absorbing all French territory, part of Belgium, and the Duchy of Luxembourg, and Channel taking over the remainder of Belgium and whatever area was necessary in the Netherlands. In the south Burgundy District was to become part

[15] See Ch. II, Sec. 2, above.

[16] SHAEF finally decided to extend the French zone of interior early in April 1945, when it transferred eleven additional departments to French control, some in the north, some in the south, and expressed the intention of turning over all French territory to French authorities in the near future except for the departments actually lying within the combat zone. Allied military installations, however, were to continue in those departments as long as needed, and the Communications Zone insisted, on the basis of experience with all-French railway operations, that military control over the railways and communications facilities was also to continue. General Larkin protected the extension of the French line of communications even with these provisos, arguing that complete control of all logistic facilities was needed for both the conduct of the war and for the redeployment program. In any event, hostilities ended before the intended transfer was made. Organization and Command in the ETO, II, 221–22, 271–72, 283–86.

of Oise Section and take over the territory released by CONAD, and Oise itself was to become an "intermediate" section, stretching the entire length of the front.

Implementation of the plan began on 21 March, when Burgundy District was absorbed by Oise. On 2 April Oise was redesignated Oise Intermediate Section in line with the aim of establishing intermediate depots which were intended to hold the major portion of the theater's stocks of supplies, as urged by General Somervell. Oise subdivided its huge territory into three districts—Burgundy, Luxembourg, and Marne, Burgundy shortly thereafter being renamed Lorraine District.

Meanwhile the process of freeing the two advance sections of all territorial responsibility was also completed. On 1 April, with large forces operating east of the Rhine, the Advance Section relinquished its territory to the Channel Base and Oise Intermediate Sections, continuing to carry out its supply responsibility to the 12th Army Group in Germany without area control. A week later CONAD likewise turned over its territory to Oise, although certain stations remained exempted to CONAD's control. Both the Advance Section and Continental Advance Section now established headquarters in Germany, the former initially at Bonn and later at Fulda, and the latter at Kaiserslautern. (*Map 7*)

Such was the COMZ territorial organization until V-E Day. Except for boundary changes and the establishment of the principle that the Communications Zone should exercise no area control in Germany, no major alterations had taken place since February. The introduction of an intermediate section in line with the recommendations of Generals Lutes and Somervell hardly constituted more than a change in names, for the organization of a system of depots in depth was only partially realized.

Command of the various COMZ sections had remained relatively stable, except in the case of Normandy, and only two changes were made in the final months, one of them only a few days before the end of hostilities. In mid-March Brig. Gen. Egmont F. Koenig came up from North Africa to take command of the U.K. Base from General Vaughan, who was given a new assignment. And, in the only other change, General Aurand, the commander of Normandy Base Section, was succeeded on 4 May by Brig. Gen. Jesse A. Ladd, who had been his deputy.[17]

Internally, the COMZ sections on the Continent developed along much the same lines as had the former base sections in the United Kingdom. Not all of them developed a district organization, although such territorial subdivision was authorized. The Advance Section, operated two fairly distinct lines of communications and depot complexes, one, known as ADSEC North, in support of the First and Ninth Armies, and one, known as ADSEC South, in support of Third Army. CONAD's operations were fairly well channeled along one line of communications. Oise Intermediate Section's organization into three districts has already been mentioned.

The old conflict over the division of authority between the section commanders and the chiefs of the technical serv-

[17] Organization and Command in the ETO, II, 266–67, 274–81.

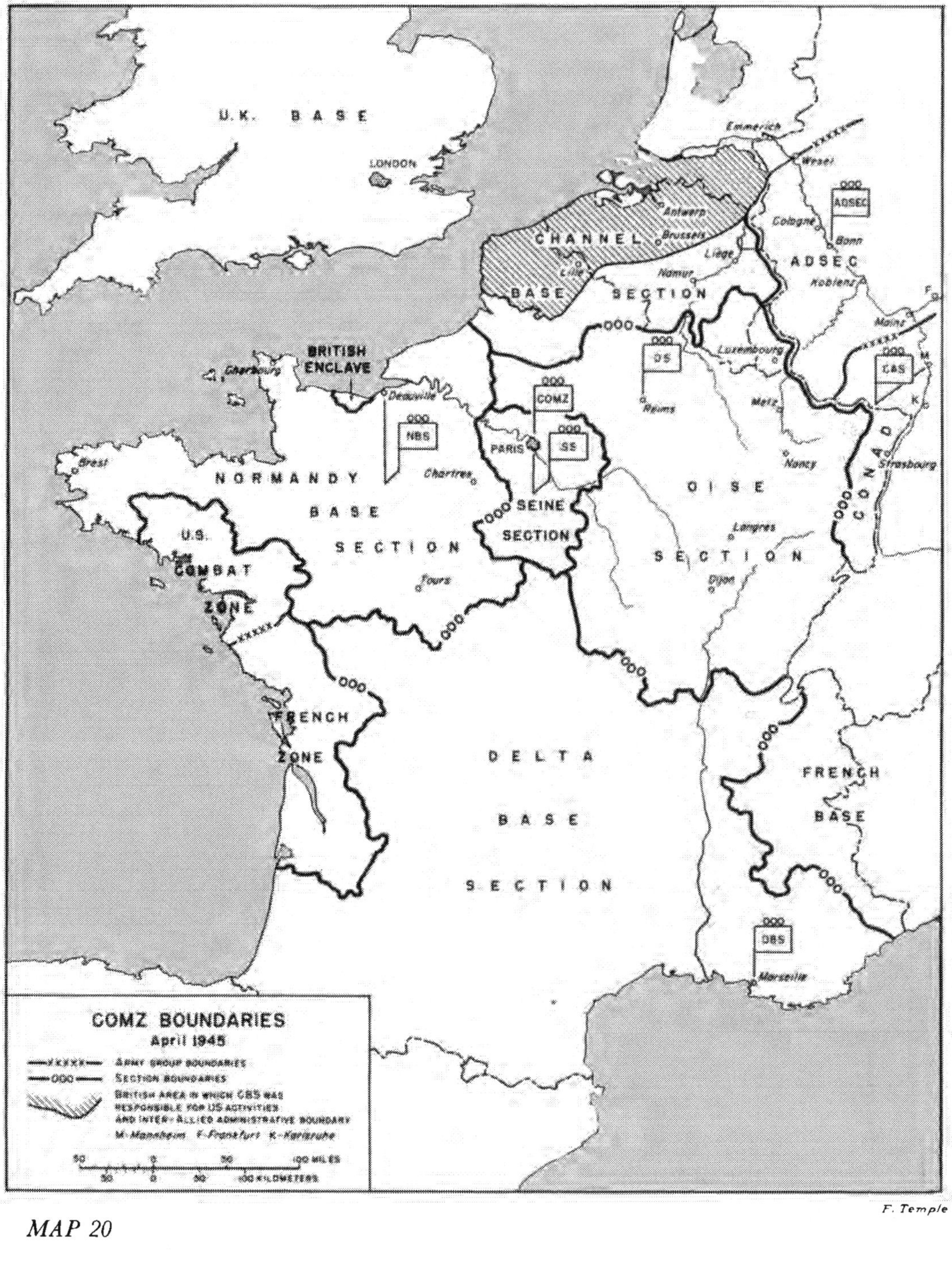

F. Temple

MAP 20

ices was never fully resolved and caused grief to the very end. General Lee had consistently favored the division of authority along territorial lines and preferred to delegate maximum responsibility to section commanders in the direct chain of command. As late as mid-April 1945, for example, a circular delineating COMZ organization and functioning attempted to fix more clearly the authority of section commanders over the "allocation, assignment, transfer, and command" of all COMZ personnel and units assigned to the sections. The control of depot operations, on the other hand, required a more nearly equal division of authority between section commanders and technical service chiefs. The operation of depots was the responsibility of the section commanders, and all matters relating to operating policies, procedures, and reports on operations were handled through command channels, while the technical control of stock and reporting procedures remained the responsibility of the chiefs of the supply services. In some matters, notably intersectional movements via rail and highway, and in the operation of the pipelines, the tendency was toward less control by the sections and for greater centralization of authority. In a system in which territorial and functional organization existed side by side there obviously was no single guiding principle which could be consistently applied. Many differences therefore had to be considered on individual merit and resolved on a pragmatic basis. Personal co-operation and teamwork obviously counted for much in a system which was administratively imperfect.[18]

Changes in the organization and functioning of the COMZ headquarters itself resulted in part from the Lutes-Somervell visit, in part from the necessity to meet certain *ad hoc* requirements, in part from the absorption of SOLOC, and in part from the anticipation of necessary posthostilities adjustments. The establishment of a control division followed directly from the criticisms and recommendations of General Somervell. Certain of the staff sections and services already possessed control divisions, but no agency had existed to enforce uniform practices and procedures for the entire command. Such an agency, established early in 1945, now undertook to keep a close check on progress in carrying out the Lutes-Somervell recommendations, and made frequent studies of the progress of the Communications Zone. On the recommendation of the ASF commander the Communications Zone also took steps to remove the G-4 office from the field of operations and restrict it to its more proper role as a staff section. Late in February the G-4's Transportation Section was transferred to the Office of the Chief of Transportation, which was thereafter given more complete control over all movements. The Office of the General Purchasing Agent was also removed from G-4 control and was again given separate general staff status. General Somervell's recommendation regarding the appointment of an assistant chief of staff for plans on the COMZ staff was not carried out as promptly. After some delay the theater finally created the position of deputy chief of staff for planning early in April, placing him directly under the chief of staff and charging him with broad co-

[18] *Ibid.*, II, 287-91.

ordination, supervision, and control of planning activities, while detailed planning itself remained the responsibility of the various staff sections and commands. Col. Ira K. Evans, who was requested from the War Department, was the first to be assigned to the post.[19]

The dissolution of SOLOC left certain problems of readjustment, both as to supply procedures and the disposition of personnel. Adjustments in supply procedure had long been under way, and the transition took place without too much difficulty. Responsibility for establishing shipping priorities for southern France finally passed to ETOUSA with the disbanding of SOLOC early in February. Many of SOLOC's staff officers were absorbed by ETOUSA-COMZ as deputies. In this way, for example, Brig. Gen. William H. Middleswart, chief quartermaster of SOLOC and the quartermaster officer of the original Special Observer Group in 1941, now returned to the theater after a long absence to become deputy chief quartermaster. In the first change in the ETOUSA-COMZ general staff since before the invasion General Gilland, who had been chief of staff of SOLOC, became the new G-4. General Stratton, after serving briefly as his deputy, went to a new post in the United States.

General Larkin, who had commanded SOLOC, went to Paris to become the deputy commander of the Communications Zone for operations. General Lord was also a deputy commander, as well as chief of staff, and it was not long before the question of their respective authority and responsibility arose. Late in February General Lee stepped in to clarify their relationship, specifying that General Lord's authority was to be in the administrative sphere and General Larkin's in the operational. General Larkin enjoyed seniority and was designated to assume command in the event of Lee's absence. A general order published on 26 February tried to make the distinction clear. Apparently some doubts remained, however, and on 12 March General Lord's title of deputy commander was dropped. General Lord shortly thereafter took command of the Assembly Area Command, charged with the planning and execution of the redeployment of U.S. forces at the end of hostilities. General Larkin thereupon took over the assignments which Lord had held since early in 1944—that is, as Chief of Staff, Communications Zone, and Deputy Chief of Staff, ETOUSA.[20]

A third category of changes in ETOUSA-COMZ headquarters organization had been occasioned by the arrival of General Lear in the theater, to which reference has already been made.[21] Lear was to hold a rather special position in the theater's organizational setup. Shortly after his arrival he was named deputy theater commander and authorized to act for the theater commander in "matters involving Theater administration." The appointment of a deputy theater commander with powers in the field of administration threatened to further obfuscate already poorly defined relationships and balances of power.

Theater officials were aware of the danger and took pains to define Lear's authority as precisely as possible. The letter

[19] *Ibid.*, II, 292-98, 302-04.

[20] *Ibid.*, II, 298-302.

[21] See above, Ch. XI, Sec. 3.

announcing his functions specifically limited General Lear's duties and responsibilities to the field of manpower and morale and made it clear that his functions did not extend to any authority over the theater chiefs of services. Most important of the various manpower and morale functions brought within Lear's cognizance was the problem of the reconversion training which was at its most acute stage at this time. Concern with this problem brought within Lear's purview the Ground Force Reinforcement Command, which underwent several changes in command late in 1944 and early in 1945. In addition, the assignment gave Lear direction and control of the General Inspectorate Section, the U.S. Theater Manpower Section, and the ETO Section of the War Department Manpower Board. Also included under his general supervision were the functions then being performed by Brig. Gen. Benjamin O. Davis in connection with Negro affairs, and by General Hughes, who was specially assigned to the deputy theater commander on matters of manpower, morale, and Negro troops.

The General Inspectorate Section, not to be confused with the Office of the Inspector General, had been created as a special staff section of ETOUSA late in 1944 as the result of complaints reaching the War Department of poor treatment of soldiers overseas. General Marshall had asked the theater commander in October 1944 to investigate the conditions reported. After considering two or three alternative methods of dealing with the problem General Eisenhower created a special staff section which reported directly to his chief of staff, General Smith. General Bonesteel, who had once commanded the Iceland Base Command, was appointed head of the General Inspectorate Section late in December 1944. Field teams appointed by the section made trips of inspection and had the authority to correct certain conditions on the spot. But for the most part the section confined itself to reporting conditions, at first to the SHAEF chief of staff, and then to the deputy theater commander, General Lear.

The U.S. Theater Manpower Section, which had been established early in January 1945 to cope with the serious infantry manpower shortage in the theater, was charged mainly with making allocations of critical personnel and with implementing the recommendations of the ETO Section of the War Department Manpower Board headed by General Gasser. The section had operated at first under the supervision of the U.S. element of the G–1 Section at SHAEF. Late in February it was reconstituted as a special staff section of ETOUSA directly under the deputy theater commander. Colonel Shannon, who had headed the section under General Barker at SHAEF, remained as chief of the section under Lear.

The control and discipline of Negro troops had always been a special problem in the theater, particularly in the Communications Zone, because a large percentage of Negro troops served with service units. General Devers had taken cognizance of the problem in 1943 and had instructed each subordinate command to appoint an officer to study Negro troop problems. The SOS at that time had appointed a special deputy provost marshal to co-ordinate all matters concerning such troops in the SOS. No co-ordination of policy at the theater level was attempted until January 1945, when the theater announced the appointment of General Davis, highest ranking Negro officer in the U.S. Army, as special adviser on matters relating to Negro troops and as head of a theater special staff section called the Negro Affairs Section. Like the Manpower Section, General Davis' office initially reported directly to the theater commander. Like the other agencies dealing with manpower and morale activities, this office now came under General Lear's jurisdiction.[22]

A fourth category of changes in ETOUSA-COMZ organization resulted from the necessity to plan for the post-hostilities period. In general, planning for the period after V-E Day was divided. ETOUSA-COMZ was responsible for redeployment planning and the U.S. component at SHAEF for planning the occupation of Germany. Redeployment planning got under way in November, 1944 when a Redeployment Planning Group was established as a special staff section of ETOUSA, operating under the direct supervision of the ETOUSA deputy chief of staff, General Lord. This section operated until March 1945, when a Redeployment Planning Committee, headed by Brig. Gen. George S. Eyster, the G–3, was created, on which each of the general staff sections of ETOUSA-COMZ was represented by a senior planning officer. Late in April a Redeployment Coordinating Group was added, operating directly under the deputy chief of staff, then General Larkin, for the purpose of keeping the theater commander informed on the status of redeployment. Earlier in the month the Communications Zone had established the Assembly Area Command with headquarters at Reims to handle the vast troop movements which were expected to take place after V-E Day. The new command had established camps and had already begun to process units when hostilities ended on 8 May.

The agencies which were to be concerned with the occupation of Germany functioned more directly under the U.S. component at SHAEF. The U.S. Group Control Council, established in August 1944, was the most important of these, for it was the nucleus for the organization which eventually participated with British and Russian representatives in forming a tripartite group for the control of Germany. Early in March 1945 the group ceased to be merely a planning agency and officially became a command, and under the direction of General Clay prepared to assume its role in the military government of Germany.[23] The Communications Zone entered into occupation planning in at least one important respect—the operation of the port of Bremen, for which a U.S. enclave in the Bremen–Bremerhaven area was marked out. On 11 April a separate COMZ command known as the Bremen Port Command was created to operate the port and administer the area. General Vaughan, who had once been in charge of the Bristol Channel ports and more recently had commanded the U.K. Base, was named to command the new

[22] Organization and Command in the ETO, II, 312–20.

[23] See Pogue's *Supreme Command* for details on the occupation organization.

organization and to plan for the time when the port would be turned over to U.S. forces. The Communications Zone was to assume other occupation responsibilities, but these were not added until after V-E Day.

One other change in the ETOUSA-COMZ staff organization took place as the end of hostilities approached. In anticipation of the need to keep troops occupied after V-E Day, the theater planned a great expansion in special service activities, one feature of which was an enlarged information and education program involving the use of European schools and universities as well as Army facilities. Early in April, on the recommendation of Maj. Gen. Frederick H. Osborn of the Information and Education Division of the War Department, the theater separated information and education activities from Special Services, creating a separate special staff section for the former. Information and education, or I&E, was to come under the supervision of the G–3 and was to be responsible for information, education, and orientation, including such organs as the American Forces Network, *Stars and Stripes,* and *Yank* magazine. General Osborn himself served as the first chief of I&E, but returned to the United States in mid-April and was succeeded by Brig. Gen. Paul W. Thompson. Special Services, as before under the supervision of the G–1, was to handle the more purely recreational activities such as motion pictures, music, the USO, theatrical entertainment, athletics, library services, and vocational arts and crafts. Brig. Gen. Oscar N. Solbert, the incumbent chief of Special Services, continued in charge of these activities.[24]

The theater's command and organizational structure presented something of a puzzle to the very end, particularly in regard to ETOUSA-COMZ's position and its relationship with other commands. Much of the difficulty stemmed from the fact that no genuine U.S. GHQ had ever really developed. At best, theater headquarters retained a kind of ectoplasmic character. Until the end of 1944 the functions of a GHQ had been divided between SHAEF and ETOUSA-COMZ. With the appointment of General Lear as deputy theater commander those functions were divided between three agencies.

Residing at SHAEF were the theater commander, his chief of staff, and several influential general staff officers, who determined over-all policy and rendered many decisions normally made by a theater staff. The official theater staff, however, resided with the commanding general of the Communications Zone at the ETOUSA-COMZ headquarters and served also as General Lee's staff. While the COMZ commander no longer held the title of deputy theater commander for supply and administration, his authority had not really changed, for his headquarters handled virtually all supply and administrative matters for the theater, including correspondence with the War Department, and it continued to issue theaterwide instructions on administrative matters in the form of ETOUSA orders and circulars. The au-

[24] Organization and Command in the ETO, II, 322–29.

thority of the ETOUSA-COMZ general staff was actually restricted somewhat by the actions of the U.S. component at SHAEF. But the special staff, and particularly the chiefs of services, exercised theaterwide control in their respective fields. Finally, a third agency possessing powers normally exercised by theater headquarters was the office of the deputy theater commander, General Lear, who had taken over certain functions regarding manpower and morale from the U.S. component at SHAEF, and authority over the Replacement System from the Communications Zone. As deputy theater commander General Lear could of course issue directives in General Eisenhower's name.

While Headquarters, ETOUSA, was theoretically the top U.S. echelon of command, the split in its authority left it in an ambiguous position. Important decisions were naturally usually made at SHAEF, where the theater commander tended to seek advice on most matters from the staff that was physically present. Consequently U.S. representatives at Supreme Headquarters, rather than the official theater staff residing in Paris, habitually served in the capacity of the senior U.S. staff. To the extent that they did so, and concerned themselves with U.S. as distinct from Allied matters, they added to ETOUSA's difficulties in establishing its primacy with the air force and army groups. The latter tended to regard ETOUSA as a coequal command and the chiefs of services as essentially COMZ officials, since they had been under the SOS-COMZ almost from the start and physically separate from the theater commander. "This view was intensified," as General Lord observed, "because of the fact that ETOUSA was largely the former SOS headquarters, for which combat elements traditionally have little respect." Instead, both the air force and the army groups tended to look to SHAEF as the next higher echelon of command, and as a consequence many matters were handled at the SHAEF level which were normally the province of the theater headquarters. This practice not only impaired ETOUSA's authority, but resulted in the expansion of the U.S. representation at SHAEF to much greater size than should have been necessary. Much worse, in the view of COMZ officials, decisions involving supply and administrative matters were made between SHAEF and the field commands, which were too far removed from the supply operation of the theater to be properly informed.[25]

The 12th Army Group likewise assumed administrative duties to an extent quite at variance with field service doctrine. General Bradley's staff had become accustomed to dealing with supply matters in detail even before the invasion, for it had been assigned responsibility for much of the administrative planning for the first ninety days. It had therefore fallen into the habit early of performing certain functions with regard to supply which were more properly a part of routine theater administration. It had continued this practice throughout the period on the Continent. Officers of General Bradley's staff admitted at the end of the war that the army group's deep involvement in supply matters had been improper. But it had been neces-

[25] *Ibid.*, II, 215–17, 336–37.

sary, they explained, because of lack of confidence in the Communications Zone and lack of direction from a true theater headquarters. The field commands could never regard the decisions emanating from ETOUSA-COMZ headquarters as anything but the decisions of General Lee's Communications Zone—decisions which in some instances, they felt, affected the field commands adversely. They took the view, moreover, that if the over-all co-ordination of administrative matters could not be performed by the theater commander or a U.S. commander who was clearly superior to the three major commands of the theater, it should logically be performed by the field force (army group) commander, since he was most directly responsible for the success of operations.[26]

A genuine theater headquarters eventually emerged in Europe, but only as the result of the adjustments incident to the end of hostilities. Beginning about March 1945 there was an increasing tendency to recognize that, technically at least, there was a difference between COMZ and ETOUSA headquarters, for the U.S. element of SHAEF assumed more and more the role of the real U.S. theater headquarters, and General Lee's headquarters more and more that of a strictly COMZ headquarters. The appointment of General Lear as deputy theater commander can be regarded as the first step in that direction, for it had removed some of the administrative setup of the theater from General Lear's control.

Early in March, in anticipation of the eventual elimination of the combined command at SHAEF, the Supreme Commander and his chief of staff formulated a plan which called for a clear-cut separation of COMZ and ETOUSA headquarters. It provided that the U.S. component at SHAEF was to become the theater general staff, and that the technical service chiefs would thereafter be resident at the new theater headquarters. The Communications Zone's mission, as before, was to provide administrative support for all U.S. forces on the Continent and in the United Kingdom and service functions in connection with redeployment; the theater would have the normal role of over-all direction.

In mid-April, three weeks before V-E Day, the theater commander actually issued orders designating the U.S. staff members at SHAEF as the acting general staff of ETOUSA, which included General Clay as deputy chief of staff, General Barker as G–1, General Betts as G–2, General Bull as G–3, General Crawford as G–4, and Brig. Gen. Frank J. McSherry as G–5. No change in missions or in the method of doing business actually occurred at this time. The previous incumbents of the above positions now became acting deputies on the theater general staff. They continued to serve in dual capacities, however, for they were almost immediately designated chiefs of the respective general staff positions in the Communications Zone. Most of the ETOUSA special staff offices remained for the time being as part of the Communications Zone.

[26] Memo, Moses, n.d. (c. Feb 45), SHAEF 12 A Gp G–4 Memos for Record, File 77; Memo, G.A.C. of G–4 Sec 12 A Gp for G–4, 3 Jun 45; Memo, Col McCormack, Chief Plans Br G–4 12 A Gp, for G–4, n.d., and Memo, Col R. C. Kyser, Chief M&I Br G–4 12 A Gp, for G–4, 21 May 45, sub: Functions of A Gp, all in 12 A Gp Functions of Army Groups, File 98A.

A modified dual staff setup continued, therefore, and did not come to an end until after V-E Day. On 12 May a theater general order announced a fairly complete separation of the theater and COMZ staffs, although certain of the special staff officers were still to serve in a dual role. The new arrangement did not go into effect until 1 July, however, when ETOUSA was officially redesignated U.S. Forces, European Theater (USFET). The Communications Zone came to an official end on 1 August, when it was redesignated the Theater Service Forces, European Theater (TSFET).

Headquarters, USFET, established in Frankfurt, Germany, became the true theater headquarters, maintaining a general control over all subordinate elements, including 12th Army Group, the Theater Service Forces, the U.S. Group Control Council, and U.S. Strategic and Tactical Air Forces (now redesignated U.S. Air Forces in Europe, or USAFE). TSFET headquarters, which remained for the time being in Paris, was made responsible primarily for administration outside Germany and Austria, for carrying out the redeployment plan, for the shipment of supplies and personnel into Germany, and for the liquidation of U.S. installations in the liberated countries.

The end of the war meanwhile had brought about the liquidation of the functions of the deputy theater commander, General Lear. Effective 15 May the Ground Force Reinforcement Command began to report directly to the theater commander, the Theater Manpower Section reverted to the supervision of the theater G-1, and the General Inspectorate Section formed the nucleus for the office of the theater Inspector General, absorbing also the functions of Generals Davis and Hughes.[27]

The closing months of the war had also brought a final change in the theater's boundaries. Late in January Lt. Gen. Joseph T. McNarney, commander of the Mediterranean theater, had recommended the transfer of Spain, Portugal, and the Azores to ETOUSA on the ground that the European theater was in a better position to control those areas in the event operations ever became necessary there. General Eisenhower concurred in the recommendation, and the War Department ordered the change effective 1 March. The Azores, which had been a separate command under the Eastern Defense Command and a month before had come under administrative control of the North Atlantic Division of the Air Transport Command, continued to be administered by the Air Transport Command.[28]

---

[27] Organization and Command in the ETO, II, 337-47.

[28] *Ibid.*, II, 347-49.

CHAPTER XIV

Movements and Distribution: Port Discharge and Clearance

(1) *Planning and Controlling Movements*

Movement problems rather than serious supply shortages dominated the logistic support of U.S. forces in the last few months of operations. The problem of movements went beyond the provision of adequate transportation. It involved the proper integration and co-ordination of several closely related factors such as port discharge, port clearance, and depot operations. The importance of movement control was never more evident than in early 1945, when the theater's supply organization faced the task of building up forward supply stocks for the resumption of the offensive, reducing the large supply backlog accumulating at the continental ports, and, in addition, handling the tonnages involved in the recently adopted program aimed at clearing the U.K. depots.

No effective co-ordination of movements was achieved in 1944. Some "programing" of movements was started in September, but it was largely confined to the daily allocation of tonnages designed to satisfy the minimum requirements of the armies. It did not provide an integrated plan for the efficient utilization of all types of transportation or for the matching of movements with combat zone requirements and with depot receiving and outloading capacities. Until the end of the year supply movements were either directed by the COMZ G–4 or were decentralized to the various COMZ sections, the "technical supervision" of the Transportation Corps being construed in its narrowest sense. Commanders of the sections had almost complete control over shipments originating in their own commands, and any instruction on freight movements coming from the Communications Zone had to be processed through time-consuming command channels. To the extent that the G–4 Section of the Communications Zone issued detailed instructions to the section commanders it had become an operating agency, a development which General Somervell criticized as contrary to sound command and staff doctrine.

The defects of this procedure became particularly noticeable late in the fall when the capacity of the transportation system began to exceed the barest maintenance needs of the armies and it became possible to build up forward stocks. Shipments were frequently made without regard for the capacity of depots or of particular rail lines, with resulting congestion at some points and inefficient utilization of facilities at others. "The essence of movement control," as one member of General Ross' staff remarked, "is to limit movements to the capacity of the 'bottlenecks' of the system and to press constantly for the elimination of such bottlenecks." Under existing practice the chief of transportation obviously could not effectively co-ordinate freight traffic in the theater because the Freight Branch of the Movements Division, Office, Chief of Transportation, was restricted largely to an advisory role. The Transportation Service was particularly handicapped so long as section commanders were free to use movement capacity left over after the G-4's priority allocations had been met, for this often resulted in shipments to depots already overburdened.[1]

The need for better integration of movements was eventually recognized, and in December 1944 the COMZ G-4 and the chief of transportation agreed on the first step in overhauling the existing system. All movements thereafter were to be "programed" a month in advance, taking into account the requirements and the capacities of both transportation and depots. Preparation of the monthly movement plan was carried out roughly as follows: The COMZ G-4 first consolidated all data on the supplies available for movement from the ports and allocated tonnages to the technical services. The latter in turn proposed a program for the movement of their allocated tonnages to the desired destinations. The G-4, after reviewing and consolidating these requests, forwarded them to the chief of transportation. He, after estimating transport capabilities, then recommended a specific movement plan for port clearance and shipments to depots, specifying the points of origin, daily tonnages by class of supply, destinations, and the types of transport—rail, water, or motor—to be used. The G-4, after final approval, then published the plan as the Monthly Movements Program on the 25th of the month preceding that in which it was to be effective. Although issued in the form of a directive, the program was intended to serve mainly as a guide and was subject to change. The first program, worked out under the direction of Col. Hugh A. Murrill, chief of the Control and Planning Division, Office, Chief of Transportation, went into effect on 1 January 1945.[2] (A typical Monthly Movements Program, that for March, is shown graphically on Map 8.)

In essence the movements plan attempted to accommodate the requirements of the services to movement capabilities and schedule all forward movements in such a way as to make the most economic use of transport facilities. The new procedure properly went beyond the integration of inland transport and depot reception facilities. The Transportation Corps also allocated shipping to the ports in such a way as to economize on land transportation and prevent wasteful back-hauling. Early in February

[1] Rpt to Ross by Movs Div OCofT, quoted at length in Hist Rpt of TC ETO, VII, Ch. II, 60–61.

[2] The southern line of communications was not included in movement plans until March.

the chief of transportation elaborated on this by instituting shipping allocation meetings for the purpose of preallocating supplies to their final depot destinations.

The adoption of the new system was undoubtedly a step in the right direction from the point of view of the Transportation Corps. In outlining the new procedure the COMZ commander emphasized that the success of the new procedure would depend on the channeling of all supply movement matters through the chief of transportation, who was to control the program through his Freight Branch.[3] Base section commanders no longer were free to ship wherever and in whatever volume they desired. Furthermore, planning movements a full month in advance permitted some flexibility in the use of transportation and in making necessary adjustments.

The Monthly Movements Program nevertheless proved an imperfect instrument of control and only partially fulfilled its purpose. Six weeks after the inauguration of the first program General Burpee, commander of the 2d Military Railway Service, complained that the services were ordering the shipment of supplies which the depots were unable to accept. Moreover, some depots had been established without consideration of rail trackage.[4] General Robinson who had accompanied General Somervell to the theater and had remained to help organize the ETOUSA Control Division, noted that the movement plans thus far had been based less on a determination of the armies' needs than on a desire to clear the ports, with the result that supplies were being shipped forward regardless of the need for them in the forward areas.[5]

The implication was that requirements in the forward areas were being subordinated to the convenience of movement facilities. Moreover, as originally established, the movements program provided for no control of interdepot movements, the curtailment of which became increasingly necessary as freight cars began to accumulate at forward depots in the final months of operations. The authority to handle such situations by embargoing shipments remained in the hands of the COMZ G–4. The chief of transportation could request such stoppages, but effective control over movements was partially nullified by the necessity to process requests through command channels, which precluded timely action. In March the G–4 finally agreed to a compromise solution which provided that all nonprogramed movements should be cleared through the Freight Branch of the Movements Division, Office, Chief of Transportation, which was to co-ordinate such movements with the G–4. In another concession the Office, Chief of Transportation, was granted authority to impose embargoes without awaiting confirmation from the G–4. The Transportation Corps thus moved another step nearer the goal

[3] Ltr, Lee to Gen and Special Stf Secs and Sec Comdrs, 23 Jan 45, sub: Procedure for the Control of Supply Movs on the Continent, EUCOM 400/1 Procedures for the Control of Supply Movements.

[4] Conf on Transportation in COMZ in Burpee's Office, 15 Feb 45, EUCOM 337 Confs Gen, III.

[5] Memo, Robinson and Col Charles R. Broshous, Chief Control Div COMZ, for Lee, 8 Mar 45, sub: Program [sic] Made On Items in Somervell's Memo of 24 Jan 45, EUCOM 400.192 Misc, Rpts, VII.

it had long sought—complete control over military freight traffic.

The monthly movements program undoubtedly represented an improvement over the relatively un-co-ordinated movements of 1944, for it ensured a more efficient use of all line-of-communications facilities and consequently a more orderly flow of supplies. But the Transportation Corps' control proved inadequate in the final months, for, as will be seen, it was unable even by use of the embargo to halt the accumulation of freight cars under load in the forward areas and serious congestion at other points.[6]

(2) *Discharge and Clearance*

While limiting factors of one kind or another plagued the logistic support of U.S. forces until the very end of hostilities, port discharge capacity had ceased to be one of them. Discharge targets were not consistently met in the last three or four months of operations, but this resulted mainly from restrictions forced by the inability to clear the ports and not by the lack of unloading capacity.

By the end of January 1945 all the major ports—the 4th at Cherbourg and Granville, the 6th at Marseille, the 5th and 13th at Antwerp, the 16th at Le Havre, and the 11th at Rouen—had become seasoned organizations and were demonstrating increasing efficiency in their operations. At Antwerp, for example, the speed of discharge in tons per ship per day rose from 456 in December 1944 to 1,100 in the final two months of hostilities; at Cherbourg the unloading rate increased from 384 in the earlier period to 716 in March and April; Marseille achieved a daily discharge rate of 910 tons per ship in the final months.[7]

The average daily discharge of 43,800 tons in January had reflected a slight drop owing to the embargoes on forward movements imposed during the Ardennes battle. In February daily unloadings at the continental ports, including those of southern France, rose to 53,100 tons, but still fell short of the target of 58,200. In March the ports more than met the target of 53,375 announced in the monthly movements program with a daily average of nearly 60,000 tons, and in April the ports continued to outdo themselves with average unloadings of 63,000 tons per day against a target of 54,000. (See Table 4 for monthly discharge figures.) They chalked up their peak performance on 13 April, when discharges exceeded 77,000 tons. Planned coal imports in the last few months ranged from 8,000 tons to 10,000 tons per day and bulk POL imports from 14,000 to 18,00 tons.[8]

All the ports continued to make improvements in their discharge facilities during the winter. But few major reconstruction projects were undertaken to enlarge unloading capacity. Only one

---

[6] Hist Rpt of TC ETO, VI, Ch. II, 58, 114, and VII, Ch. II, 64–67; COMZ G–4 History, II, 47.

[7] All discharge figures are exclusive of bulk POL, coal, and vehicles. Hist Rpt of TC ETO, V (October–December 1944), App., Chart 14, VI (January–March 1945), App., Chart 14, and VII (April–June 1945), App., Chart 14.

[8] Ltrs, Col Page, COMZ G–4 Plans Div, to CG COMZ, 17 Feb and 7 Mar 45, sub: Estimate of COMZ Supply Situation (Rpts 1 and 2), Tab A to Plans for Opn of COMZ Through Jun 45, no signature, EUCOM 381 Projected Operations of COMZ ETOUSA, 14 Mar 45; Supply Movement Program for March and April, 26 Feb and 25 Mar 45, respectively, EUCOM 400.22 Shipments Gen, 1944–45; SHAEF G–4 War Diary/Jnl.

new port—Ghent—was placed in operation.

Cherbourg, whose discharge had been pushed to an average of 14,300 tons per day in November, had a gradually diminishing role after the opening of Antwerp. Its discharge target was set at 7,000 tons in January, raised to 10,350 the next month, and then lowered to 7,600 tons in March and to less than 3,000 tons in April. These targets were usually met and often bettered. On 11 April a small enemy party from the Channel Islands landed on the west coast of the Cotentin with the intention of blowing a bridge on the main rail line and thus disrupting clearance of the port, but the attempt was defeated by an alert defense. By V-E Day Cherbourg had handled upwards of 2,600,000 tons of general cargo in addition to locomotives, rolling stock, vehicles, and bulk POL.[9]

Granville continued to operate as a subport of Cherbourg confining its operations almost exclusively to the unloading of coal. Its performance was erratic, and daily discharges averaged only about 1,500 tons in the last months. On 9 March the tranquillity and boredom of the port was suddenly broken by a surprise raid by an enemy force from the Channel Islands. Striking shortly after midnight, the well-armed party, estimated to have a strength of about 150 men, succeeded in completely surprising the port command, causing about eighty casualties and inflicting fairly heavy damage on port facilities and coasters. The Granville raid led to the general reorganization and tightening up of local defenses in all the rear areas of the Communications Zone. Granville was turned over to the French at the end of April.[10] Its discharge record is summarized in Table 3.

The port of Marseille averaged about 13,500 tons per day in February, March, and April, although unloadings exceeded 19,000 tons on three occasions, and on one of these—15 March—topped the 21,000-ton mark. The capacity of the port was actually estimated to be about 24,000 tons, but the limited rail net of the Rhône valley consistently prevented full utilization of Marseille's facilities. Late in February General Somervell expressed concern over this failure to take advantage of Marseille's surplus. But in the final months of operations there was ample port discharge capacity on shorter lines of communications, making it uneconomic to use Cherbourg or Marseille any more than necessary. Even vehicle ships, many of which had been assigned to Marseille, could now be provided berths and assembly facilities at the northern ports, nearer the ultimate destination of the equipment. Limitations of inland transport rather than discharge capacity determined the selection of ports.[11]

Neither Le Havre nor Rouen ranked with Cherbourg, Marseille, or Antwerp in capacity or performance. Nevertheless, they helped meet a desperate need for discharge capacity on shorter hauls, particularly in the fall of 1944, and Le

[9] Hist Rpt of TC ETO, VI, Chs. III, 12–20, and VII, Pt. I, 113–14.

[10] *Ibid.*, VI, Ch. III, 2–4, 22.

[11] SHAEF G–4 War Diary/Jnl; Ltrs, Somervell to Lee, 26 Feb 45, and Lee to Somervell, 9 Mar 45, EUCOM 800 Rivers, Harbors, and Waterways, II; Hist Rpt of TC ETO VI, Chs. III, 36–37, 45, and VII, Pt. I, 116.

Havre eventually handled large numbers of troops, first as a port of debarkation and later as a port of embarkation in the redeployment of ETOUSA forces after V-E Day. Le Havre averaged less than 6,000 tons per day in the last three months. Three fourths of its tonnage was brought ashore via lighters rather than by direct quayside unloading. At Rouen flood conditions on the Seine seriously hampered port operations. Barge movements, on which the port was heavily dependent for clearance, were completely halted for three weeks in February, and unloadings were kept down to 4,000 tons per day. In March and April the port averaged about 7,500 tons.[12]

The port of Ghent, which had been captured by British forces in September, was not brought into use until late in January. An inland port, Ghent had access to the sea by means of the twenty-mile Terneuzen Canal and the mouth of the Schelde. Like Antwerp, therefore, its use was denied until the enemy was cleared from Walcheren and the Beveland Peninsula in November. Ghent ranked as the second port of Belgium, and before the war had been counted among the ten busiest ports in western Europe. An extensive canal system connected it with Bruges, Ostend, Antwerp, and Liège. Damage to the port was not extensive. The main obstacles to rapid restoration were several destroyed bridges over the canal, damaged lock gates, sunken vessels, and the fact that some cranes had been removed. The harbor was also badly silted and had to be dredged to accommodate deep-draft vessels. British forces started repairing the port in December.

Ghent had not figured seriously in early COMZ port planning. In mid-January, however, U.S. and British officials agreed on a plan for its joint use, the main thought being that it would serve as a standby to Antwerp in case operations at the latter were interrupted. The arrangements were very similar to those made for Antwerp. A port executive committee was appointed to decide on allocations of space, the joint use of certain facilities, and so on. Initially the port's capacity was divided to allow a discharge of 5,000 tons per day for the British and 7,500 tons for the Americans. The 17th Port, then operating the Bristol Channel ports, was assigned the mission of working the U.S. sector.

Ghent had been used almost exclusively by barges, coasters, and small freighters, and there was doubt at first as to whether ocean-going ships could be accommodated. The first ship to enter on 23 January was a Liberty, however, whose 57-foot beam barely cleared the Terneuzen locks, and the port thereafter handled both Liberties and coasters regularly. Some ships had to be lightened by discharge to barges at Terneuzen, at the entrance to the canal, before they could proceed to the port. Ghent unloaded only about 2,500 tons per day in the first month of operations, but it more than doubled that record in March, and discharged an average of 9,300 tons in the final month before V-E Day.[13]

---

[12] Hist Rpt of TC ETO, VI, Chs. III, 49, 119–20, 132, and VII, Pt. I, 120, 221.

[13] Hist Rpt of TC ETO, VI, Chs. III, 136–42, 146–48, and VII, Pt. I, 263; Hist Rpt 11, Port Construction and Repair, OCE ETO, pp. 45–49; SHAEF G–4 War Diary/Jnl.

Antwerp, with tension relieved by the cessation of V-bomb attacks in March, gained in efficiency in the final months and by V-E Day was accounting for fully one third of all the general cargo discharged on the Continent. Operations there illustrated most pointedly how the problem of clearance rather than that of unloading capacity had come to dominate port operations on the Continent. Antwerp had felt the effect of the embargo on forward movements imposed during the Ardennes counteroffensive because of the shortage of intransit and depot storage facilities in the area, and was almost immediately forced to decelerate its unloading. Large quantities of supplies nevertheless began to accumulate there and in other port areas as well. At the beginning of February the backlogs at the various ports totaled 373,000 tons. An additional 312,000 tons lay in ships awaiting discharge.

Beginning in February monthly movements plans took cognizance of this trend, and deliberately set port-clearance targets higher than discharge targets. In February, for example, the chief of transportation established a clearance goal of 59,800 tons per day as against a discharge target of 58,200. Some progress in this program was made that month, when daily movements out of the port averaged 59,344 tons and unloadings averaged 57,000. At the end of February port backlogs totaled 294,000 tons, and supplies awaiting discharge totaled 225,000 tons. Antwerp alone accounted for 128,000 tons of cargo in port storage.

For the next two months plans called for a further reduction of port backlogs. The March plan set a clearance target of 59,800 tons per day as against a discharge rate of 53,375, and the April plan a clearance of 61,000 tons per day as against unloadings of 53,000. Cherbourg, for example, was given a discharge goal of less than 3,000 tons in the hope of reducing its port stocks. But little or no progress was made in March, for discharges generally exceeded the established targets, in part because of the program of accelerated shipments from the U.K. depots, and thus nullified whatever gains were realized in clearance. Toward the end of the month the program was completely disrupted when Allied forces, after crossing the Rhine, lunged forward in the final drive into central Germany. The support of combat elements over rapidly extending lines of communications necessitated the withdrawal of more and more transport from port clearance operations. The ABC Haul, operating out of Antwerp since late in November, was suspended on 26 March, for example, and its trucks were transferred for use in the forward areas. At the end of March port backlogs had risen to 300,000 tons.

Discharges meanwhile continued to rise, averaging 63,000 tons per day in April, with the inevitable result that supplies continued to pile up in the ports. By the end of April port backlogs exceeded 400,000 tons.[14] The accumulation at Antwerp alone accounted for 70 percent of the lag in clearance, supplies in intransit storage there rising from 96,000 tons at the end of March to 283,000 tons at the end of April. At that time, with Antwerp's discharges rising to 25,000 tons per day, SHAEF stepped

[14] G-4 Confs, SHAEF, beginning 24 Mar 45, 12 A Gp Supply Misc, No. 133.

STACKS OF WAR MATERIEL IN OPEN STORAGE *near Antwerp, March 1945.*

in and directed that 5,000 empty freight cars be transferred to the port from France and Belgium at the rate of 500 per day. The first of these cars arrived on 2 May. But the backlog continued to grow for a few more weeks. On 23 May Antwerp's quays and sheds held 340,000 tons of supplies. At that time the ABC Haul resumed operations, and clearance finally began to match imports.[15]

Failure to clear the ports could be attributed largely to the inadequacies of the depot system. The lack of depots, properly echeloned in depth, was listed by both Lutes and Somervell as one of the major deficiencies of ETOUSA's logistic structure. Like most limiting factors, its effect echoed in several directions. Inadequate reception capacity in the forward areas affected rail transportation by immobilizing freight cars, and this in turn threatened to hold port discharge down and to force embargoes on base depots.[16] Lack of depot capacity also tended to perpetuate the maldistribution of theater stocks, too large a portion of which was consistently held in rear areas.

By far the most serious difficulties were on the northern lines of communications. The focal point of the trouble was in the Antwerp area. As shown

[15] Ltrs, Page to CG COMZ, 17 Feb and 7 Mar 45, sub: Estimates of COMZ Supply Situation (Rpts 1 and 2); Supply Movement Programs for March and April, 26 Feb and 25 Mar 45, respectively, EUCOM 400.22 Shipments Gen, 1944–45; Hist of TC ETO, VI, Chs. VI, 8–9, 17–18, and VII, Pt. I, 18, 169, 263; Comd and Stf Conf, COMZ, 5 Mar 45, EUCOM 337/3 Confs, Staff Weekly, I; SHAEF G–4 War Diary/Jnl.

[16] Hist Rpt of TC ETO, VI, Ch. II, 142n.

earlier,[17] the Communications Zone had not at first appreciated the need for base depot facilities in the Antwerp area and had acquired only minimum storage facilities there. The lack of such base depot space meant that all cargo had to be dispatched whether it was needed in the forward areas or not, and a needless expenditure of transportation and tying up of rolling stock often resulted.[18]

The Ardennes counteroffensive forced the issue of providing additional space for U.S. forces in Belgium, and the Lutes and Somervell criticisms on the depot problem in general gave additional urgency to the problem in January and February. But progress was extremely slow. Early in March the COMZ Control Division, reporting on the progress on the Somervell recommendations, noted that only in the Dijon–Nancy area had intermediate depot sites been selected and a few depots placed in operation. At that time no base depots had yet been provided to serve the ports of Le Havre or Rouen.

Late in March the Communications Zone issued a plan for an echeloned depot structure, providing for issue, filler, base, and key depots as recommended by the ASF commander. The scheme called for issue depots (in the Advance Section) which were to move forward in the rear of the armies, filler depots (normally in intermediate section) which were to hold the main portions of theater stocks and replenish the issue depots, key depots for the storage and issue of certain items requiring centralized control, and base depots (normally in base sections) which were to receive and classify supplies cleared from the ports and to store excess theater stocks. The Communications Zone made some progress in carrying out this program, although the depots which it established often performed missions other than those intended. By mid-April it had established base depots for Le Havre and Rouen, and also for Ghent and Antwerp. At Le Havre most of the port storage was simply converted to base depot operations. Additional filler depots were established at Toul and Metz. Early in April the Advance Section turned over most of its installations to Oise Intermediate and Channel Base Sections. From that time the Advance Section had no real depots, since the Communications Zone exercised no area control east of the Rhine. The improvements of March and April undoubtedly eased the depot problem, but the attempt to establish an adequate structure was belated, and it was never completed as conceived.[19]

(3) Shipping

The elimination of the port discharge bottleneck meanwhile had been followed by marked progress in the solution of two closely related problems—the shipping tie-up, and the transfer of supply stocks from the United Kingdom to the Continent. The shipping backlog, which had mounted to 233 vessels in November, had already been substantially reduced by the end of the year. The War Department nevertheless continued to pare the theater's requests for sailings

[17] See above, Ch. VI, Sec. 2.

[18] Ltr, Larkin to Gen Graham, MGA 21 A Gp, 26 Mar 45, EUCOM 323.3 Depots.

[19] COMZ G–4 History, IV, 115, 145–47, 151–52; Mechanics of Supply in Fast Moving Situations, Gen Bd Rpt 27, pp. 38–41.

until early in February, when for the first time the number of ships being worked in the ports exceeded the number awaiting discharge. At the end of February, with less than one hundred idle ships in the theater, General Ross felt that ETOUSA was finally "out of the woods" on both discharges and forward deliveries. Within another month the bank of ships had fallen to fifty-seven.

General Somervell had noted the improvement with satisfaction and, with his confidence in the accuracy of the theater's estimates of shipping requirements restored, agreed to schedule future sailings as requested by the theater. The surplus port capacity promised to provide a much needed flexibility to the entire supply reception procedure, for it made possible the allocation of ships to the ports best adapted to the handling of particular types of supply, and nearest their ultimate destination, thus shortening both the line-of-communications hauls and the turnaround time for shipping.[20]

Late in March the requirements of U.S. forces beyond the Rhine began to draw transportation away from the port clearance operations, and this in turn threatened to limit discharge. In fact, tactical developments beyond the Rhine quickly caused reverberations along the lines of communications all the way back to the depots in the zone of interior. On 1 April General Gilland, the COMZ G-4, informed the New York Port that serious congestion had already developed in the base areas and notified him that additional withdrawals of trucks from port clearance might force a reduction in the sailings from the United States in order to prevent the formation of another backlog of shipping. The G-4 was giving advance warning that he might soon ask that convoys be phased back by five days, or possibly ten. Setting back the sailing of a convoy by five days would temporarily relieve the northern ports from handling about 60,000 long tons of cargo.

The New York Port was willing to postpone sailings as long as necessary, but asked for a fifteen-day notice so that the flow of cargo into the port could be adjusted to prevent the same kind of congestion which the theater G-4 was attempting to prevent in the continental ports. On 3 April the theater requested a five-day deferment of convoys scheduled for May arrival except for certain loads of vehicles, ammunition, bombs, rail cars, and locomotives.[21]

Apart from this expedient, which was designed as an accommodation to a temporary limiting factor, arrangements had long since been made to adjust the flow of supplies to the requirements of the posthostilities period. In October 1944 the War Department had directed the theater to mark all requisitions to indicate whether, upon the cessation of hostilities, they were to be filled and shipped or canceled. Requisitions marked STO (for stop) were to be automatically canceled with the arrival of V-E Day, and those marked SHP (for ship) were to be filled and dispatched.

[20] Comd and Stf Conf Notes, Hq COMZ, 23 Feb 45, EUCOM 337/3 Confs, Staff Weekly, I; Ltr, Lord for Lee to Somervell, 17 Feb 45, and Ltr, Somervell to Lee, 23 Feb 45, both in ASF 200.02 DCofS, S/C and Control Div, A47-6; Hist Rpt of TC ETO, VI, OCofT, 144.

[21] COMZ G-4 History, I, 145-49.

Many items, for which the need would continue after the end of the fighting, such as rations, packing and crating materials, medical supplies, and post exchange items, were to be arbitrarily marked SHP.

Early in April 1945 General Lord, anticipating an early end of hostilities, proposed an immediate review of scheduled shipments with the thought of cutting back the flow of supplies even before the end of hostilities. If the theater waited until V-E Day, he pointed out, large quantities of supplies would still be in the pipeline and would continue to arrive at continental ports. Prompt action would prevent the accumulation of unneeded supplies and also facilitate redeployment. He asked the G-4 to consider the advisability of asking the New York Port to halt immediately the loading of all supplies marked STO for thirty days. Supplies already in the pipeline, he figured, would easily maintain theater levels until 1 May, and if V-E Day was declared by 10 May the embargo would become permanent and no further action would be necessary. This course of action would have substantially reduced shipments of such items as tanks, ammunition, and artillery, and chemical, engineer, and Transportation Corps supplies, including much special project equipment. General Lord also directed that requisitions thereafter be based on post-V-E-Day consumption requirements and specified that zone of interior replacement factors be used in all computations.[22]

The G-4 was much less sanguine about the likelihood of an early V-E Day. Estimates of the SHAEF staff, he said, did not support such a view. SHAEF had opposed any cutbacks in ammunition requirements, even suggested the possibility that large-scale fighting might still be in progress "at this time next year." General Gilland therefore opposed an embargo. Meanwhile he listed the plans which the various supply services had made to reduce shipments from the zone of interior, although most of these entailed no cuts until V-E Day.[23]

The theater made a few cancellations in this period, but it followed a conservative course. Most of its requests were for deferments, with "holds" on most supplies to prevent their diversion elsewhere. The result was mounting congestion in the New York Port, which repeatedly called for more specific cancellations. In mid-April the theater agreed to the release of forty fast ammunition vessels which were needed elsewhere, but it did not cancel ammunition shipments. On 25 April, however, on the urging of the War Department that the theater again review its requisitions in the light of the current tactical situation, SHAEF finally approved a reduction in ammunition shipments from 6,300 tons per day to 4,500 tons in the period 15 May-15 June.

The War Department continued to prod the theater for cancellations in the final weeks, pointing out that its requisitions for ordnance, engineer, medical, and signal supplies far exceeded consumption rates. Finally, on 5 May, only

[22] Ltr, Lord to G-4, 6 Apr 45, sub: Reduction of Shipments to ETO, EUCOM 400.22 Shipments General 1944, II.

[23] Memo, G-4 for CofS, 8 Apr 45, sub: Reduction of Shipments to ETO, EUCOM 400.22 Shipments General 1944, II.

three days before the end, the ETOUSA G–3 asked that all shipments marked STO which were consigned to the European theater be canceled. These cancellations were estimated to involve about 1,280,000 tons of supplies. All STO supplies in zone of interior filler depots or at the port awaiting loading at the time of the cancellation were returned to base depots; all such cargo already stowed on vessels was allowed to go forward, but was eventually returned to the United States; STO items which had already been unloaded in the theater were kept segregated at ports and base depots and were ordered returned on the first available shipping. On 8 May the COMZ G–4 imposed an embargo on inland movement of such cargo, and ordered the supply services to furnish shipping instructions for its return to the United States. STO-marked supplies in the United Kingdom, whether in ports or depots, were likewise ordered shipped back to the United States. Shortly after V-E Day, the G–4 reported that seventy-five ships had either been returned to the United States intact or reloaded and returned under the STO procedure.[24]

The rapid express shipping service known as REX, inaugurated in January in accordance with General Lutes' proposal, had only a brief usefulness. ETOUSA made its first request for shipments under the REX procedure on 20 January. More than half of the first requisition, totaling about 35,000 long tons, consisted of ordnance equipment, the bulk of it comprising special purpose vehicles. ETOUSA and the New York Port immediately disagreed on the method of stowing REX supplies and on the distribution of such cargo between different types of ships. More important, it soon became apparent that the theater and War Department held conflicting concepts of the use to be made of the new shipping service. Included in the first requisition were such items as chaplains' flags, folding organs, hymnals, Purple Heart medals, and alarm clocks. Technically such items probably came within the category of supplies to which the Communications Zone itself had restricted the use of REX procedure—that is "items . . . for which there is an established and urgent requirement by the Armies"—for they had been consistently included in the armies' "critical items" list, "the shortage of which is likely to affect operations." To the War Department, however, the inclusion of such items was in clear violation of the intended purpose of REX, at least as conceived by General Lutes, the ASF, and the New York Port, and the War Department quickly enjoined the theater to exercise more care in its selection of items and to impose some restraint on its requisitioning agencies. General Lutes left no doubt in the minds of theater officials that the REX procedure was being abused, pointing out that supply chiefs had been allowed to request express shipments for more and more items which could not possibly be legitimately considered critical.

Early in February the New York Port spelled out in somewhat clearer language

[24] Cbl EX–35274, G–4 ETO to SHAEF, 20 Apr 45, and Cbl FWD–19863, SHAEF to ETO Ordnance, 25 Apr 45, ETO Adm 402 (ETO Cbls); Cbl WARX–71085, AGWAR to Lee, 21 Apr 45, SHAEF AG 400.22–1 Allocation of Tonnages—Policy, No. 2; COMZ G–4 History, I, 150–60.

the ground rules governing the use of REX, specifying that the procedure be limited to items immediately required for front-line combat operations, and, so far as practicable, to packaged supplies. Under these terms the first REX shipments left the United States late in March and arrived in the theater early in April. Approximately 120,000 long tons of REX cargo, 90 percent of it consisting of ordnance and signal supplies, were delivered to the theater before V-E Day, when the system was discontinued. A small amount of REX cargo, made up primarily of spare parts and medical supplies, went to Europe by air.[25]

(4) BOLERO in Reverse

The clearance of U.S. supplies from depots in the United Kingdom in accordance with the Reverse BOLERO or RHUMBA plan had been frustrated time and again. As early as July 1944 General Marshall, foreseeing a possible tendency on the part of theater officials to requisition supplies directly from the United States once they possessed ports on the Continent, enjoined the theater commander to "roll up the rear" as quickly as possible by first using up supplies available in the United Kingdom. This would save shipping and also simplify the liquidation of the American establishment in Britain at the end of the war. The War Department subsequently made clear, however, that it did not want an indiscriminate mass transfer of supply stocks to the Continent, for there was no point in moving stocks to the Continent which were not needed there. This might actually result in double handling in case they later had to be returned to the zone of interior.

General Eisenhower assured the Chief of Staff that all requisitions to the War Department would be critically examined and that all supplies usable on the Continent would be transferred from the United Kingdom as rapidly as possible.[26] Plans at the time provided that U.S. stocks in Britain should be reduced to 1,500,000 tons by 1 August 1944, and to 870,000 tons by 1 December.

The planned reduction lagged from the start. On 1 August the level of supplies in the United Kingdom stood at nearly 2,000,000 tons. On 1 December there still were 1,750,000 tons of supplies in U.S. depots in Britain, double the planned level. Failure to draw down the U.K. stocks were attributable in part to the fact that substantial tonnages—averaging about 275,000 tons per month—continued to arrive in the United Kingdom throughout the fall because of the limited capacity of the continental ports. Equally important, however, was the poor turnaround rate of coasters, which handled most of the cross-Channel traffic. An analysis of the performance of sixty-three coasters in the period 22 October–4 November, for example, revealed that the average round trip required 22.5 days against a planned 9.6. Selective discharge, the inability to clear the ports of Le Havre and Rouen, where

[25] Teletype Conf, Lutes (in Washington) with Heileman (in Paris), 27 Jan 45, ASF 319.1 ETO Transportation (Tel Convs), No. 1 1944–45; COMZ G–4 History, I, 131–43.

[26] Ltr, Marshall to Eisenhower, 1 Jul 44, SHAEF SGS 400.3/1 Supply Problems of Allied Advance; Memo, Lee for Chiefs of Supply Svs, 7 Jul 44, sub: OVERLORD Tonnage Allocs, SHAEF G–4 Supplies General; Ltr, Eisenhower to Marshall, 20 Jul 44, OPD Exec Office File 9.

many of the coasters were unloaded, bad weather, and extended lay-ups for repair contributed in varying degree to the poor record. There was some improvement during the winter, but performance was erratic, and at the end of February 1945 the turnaround time still averaged 17.5 days.[27]

Early in October, in an attempt to meet some of the more critical supply shortages on the Continent, the Communications Zone inaugurated a cross-Channel express service similar to the Red Ball procedure used in the early days of the invasion.[28] Three LST's were reserved for expedited shipments from the United Kingdom to Le Havre. But this service involved only small tonnages and made no dent on the large supply stocks in U.K. depots.

The clearance program finally got a major shot in the arm as a result of the January visit of General Somervell, who was highly critical of the failure to reduce the U.K. stocks. January itself proved to be one of the poorest months, for supply transfers that month came to only 154,000 tons as compared with 258,000 in November and 200,000 in December. General Lord at the time directed the chief of transportation to give his personal attention to the problem and to plan for the daily shipment of 7,500 tons to the Continent. A few days later, on 1 February, General Lee made known that he desired the substantial closing out of the U.K. depots by the end of May, and asked for the movement of 300,000 tons per month for four months to meet that goal. His U.K. Base Section commander, General Vaughan, doubted whether more than half that amount could be moved to the Continent and estimated that about 1,000,000 tons would either be left on the ground or later shipped to other places. Moreover, the shipment of 300,000 tons per month, he pointed out, would not result in a net reduction of stocks by that amount, for supplies were continuing to arrive in England.[29]

In any case General Lee's program, which required the shipment of 10,000 tons per day, necessitated the acquisition of additional shipping. Available at the time were twelve Hog Islanders [30] (which had been substituted earlier for Liberties for cross-Channel use), with a capacity of about 2,000 tons per day, and coasters with an effective lift of 3,200 tons per day. To make up the deficit the chief of transportation asked for thirty-six Liberty ships. His plan called for the bulk of the shipments (7,000 tons) to discharge at Ghent, and the remainder (3,000 tons) at Rouen. In support of the program the G–4 again directed the supply services not to requisition items from the United States which were known to exist in U.K. depots.[31]

February saw only a partial implementation of the plan, owing in part to the fact that the War Shipping Administration allocated only seventeen of the thirty-six Liberties requested, and in part to the shortage of British labor in the depots. The result was that only

[27] COMZ G–4 History, I, 119–22, 127.

[28] See *Logistical Support I*, 309–10.

[29] Comd and Stf Confs, COMZ, 9 and 23 Feb 45, EUCOM 337/3 Confs, Staff Weekly 1944, I.

[30] The World War I equivalent of the Liberty ship, built in large numbers at Hog Island, Philadelphia.

[31] COMZ G–4 History, I, 117–18.

180,000 tons were outloaded. Before the end of the month, however, additional coasters and Liberties were made available. Also, General Ross agreed to leave in the United Kingdom ten port battalions which had been scheduled for transfer to the Continent.[32] As a result, March dispatches rose to 370,000 tons. The high rate of shipments continued in April, when 392,000 tons were moved. Despite this improvement, U.K. stocks still exceeded 1,000,000 tons on V-E Day.

Late in April the Communications Zone had planned to step up the clearance program still further to 450,000 tons per month. But this goal was beyond the available lift, which could not be augmented. In fact, such a program hardly squared with the findings of the supply services in late April that only 550,000 tons of the supplies still in the United Kingdom could be used on the Continent. General Hughes, who had long served as a trouble shooter for the theater commander, got the impression from a personal survey that, contrary to the policy enunciated in the summer of 1944, supplies were being shipped that were not needed on the Continent.[33]

Shortly after V-E Day the COMZ chief of staff, General Larkin, placed an embargo on the shipment of all supplies except those for which there was an essential need on the Continent. This list initially included packing and crating materials and equipment needed for redeployment, rations, POL, coal, and vehicles, and later, civil affairs, American Red Cross, and special service supplies. The May and June shipments to the Continent totaled 270,000 and 150,000 tons, respectively. Earlier an embargo had been placed on shipments of certain supplies from the United States to the United Kingdom, the results of which became apparent in May, when receipts in the United Kingdom fell from an average 150,000 tons to 90,000. Discharges in June dropped to 35,000. At the end of September 1945 there still were slightly more than 500,000 tons of U.S. supplies in U.K. depots.[34]

---

[32] Hist Rpt of TC ETO, VI, Ch. II, 65T.

[33] Memo, Hughes for Larkin, 25 Apr 45, and Ltr, Larkin to Hughes, 9 May 45, EUCOM 400.22 Shipments General 1944.

[34] COMZ G-4 History, I, 118-19, 122, 128.

CHAPTER XV

Movements and Distribution: Transportation and Forward Deliveries

(1) The Railways

As in August and September 1944, transportation again became the major limiting factor in Allied operations in the last month of the war. But there was an important difference between the two periods. In March 1945 plans were deliberately made for a rapid and sustained drive into the heart of Germany. Moreover, U.S. forces now had the advantage of experience in the supply of highly mobile forces, and much greater resources for that type of warfare.

Of necessity, motor transport initially had to bear the main burden of supply support in a rapid advance. But the support of a sustained drive also required that the railways supplement truck transport as early as possible. Detailed plans were accordingly made for the extension of the railways into Germany, and especially for the enormous engineering efforts involved in the bridging of the Rhine.

By February rail operations west of the Rhine had become fairly routine, and deliveries were being made well forward into the army service areas. In January the Transportation Corps had inaugurated an improved express service to replace the Little Red Ball, the trucking service which had been delivering about 100 tons of urgently needed supplies daily from Cherbourg to Paris since mid-September. The "Toot Sweet Express," as the new service was called, was organized to handle only high priority freight, but was to make deliveries all the way from Cherbourg and Paris to the forward ADSEC depots. A train of twenty cars was to leave Cherbourg every day. At Paris, with a maximum of twenty additional cars, two trains would then be made up, one proceeding to Namur and one to Verdun. Space was allotted to the armies, the air force, and the Advance Section on the basis of bids screened by the COMZ G–4. Total running time was set at thirty-six hours. To maintain this schedule selected rolling stock was set aside for the express service and could not be reconsigned at the terminals. Unloading had to be carried out within six hours of arrival. The first Toot Sweet Express left Cherbourg on 21 January with 107 tons of freight, and in the first two months deliveries to the advance depots averaged about 385 tons per day.

Shortly after the express was inaugu-

TOOT SWEET EXPRESS READY TO LEAVE CHERBOURG. *Crewmen for the express receiving instructions before departure.*

rated, its northern terminus was changed to Liège, and later its southern terminus was moved to Bad Kreuznach. Although it handled relatively small tonnages, the Toot Sweet Express filled an important need and continued to operate for several weeks after V-E Day.[1]

The tactical situation prevented any important forward extension of the rail lines in February, although some rehabilitation was carried out northward from Liège toward Roermond and from Aachen both north and east. With the completion early in March of VERITABLE and GRENADE, the operations of the 21 Army Group in the north, work was immediately begun to push railheads even closer to the Rhine in preparation for the crossing of that obstacle. Construction proceeded generally in accord with the current tactical plans. It was planned to provide a double-track line and a single-track bridge over the Rhine for each of the armies in the 12th Army Group. On this basis the Advance Section completed engineer plans in January for the restoration of the following lines: for the Ninth Army the line Aachen–Muenchen-Gladbach–Geldern–Wesel; for the First Army the line Aachen–Dueren–Cologne; and for the Third Army the line Thionville–

[1] Another special delivery service dubbed the "Meat Ball Express" delivered perishables, chiefly meat, from Namur to the First and Ninth Armies on alternate days beginning early in March. COMZ G–4 History, II, 108–09; ADSEC Operations History, pp. 113–14; Hist Rpt of TC ETO, VI, Ch. IV, 10, and VI, Ch. VI, 85; Memo, Gilland for Crawford, 20 Apr 45, sub: Lutes' Memo for the Supreme Comdr Dated 25 Dec 44, SHAEF G–4 400.192 Supply Reports, I.

Trier–Koblenz. Rail bridges were initially planned at Wesel, Cologne, and Koblenz.[2]

ADSEC Engineer Groups A and C began rebuilding the rail bridges over the Roer near Baal and Dueren respectively as soon as the First and Ninth Armies had crossed that river, completing them on 11 March. Meanwhile the 1056th Port Construction and Repair Group, which had had so prominent a role in the restoration of Cherbourg and more recently had had its first experience in rail work with the construction of the bridge over the Meuse at Maastricht, began restoration of the railway northward from Baal toward Geldern and Wesel. Engineer Group C in the same period began extending the lines eastward from Aachen toward Cologne. This project was suspended when, with the capture of the Remagen bridge, priority shifted to the reconstruction of the line running southeastward from Dueren to the Remagen crossing. Plans were also made to open a single-track line to the bridgehead from Trier, but the collapse of the Remagen rail bridge on 17 March nullified these plans. The line from Dueren to Remagen was restored and then extended southward to Koblenz along the west bank of the Rhine.[3]

One feature of the rail net in the north had long been a cause of worry. All rail traffic in support of the First and Ninth Armies had to be funneled through the narrow bottleneck at Liège and was therefore extremely vulnerable to disruption by a few well-placed enemy bombs. Aerial bombing, strafing, and V-bomb attacks did in fact cause some damage and interrupted work on the Renory Viaduct, one of the three bridges in Liège, during reconstruction in December and January. SHAEF G–4 planners had advocated the opening of at least one additional rail route across the Meuse farther north, preferably at Maastricht. A bridge was rebuilt there, but a through route from the west via Hasselt was not restored.[4]

Farther south railroads were still some distance from the Rhine at the beginning of March, although rail support extended well forward into the Third and Seventh Army maintenance areas. The collapse of the enemy in the area east of the Moselle later that month finally opened the way for an extension of the railways to the Rhine, although rail service was not immediately avail-

---

[2] ADSEC Operations History, p. 134.

[3] In any case, SHAEF logistical planners considered an exploitation of the Remagen bridgehead northeastward toward the Ruhr as infeasible for both tactical and logistical reasons. Supporting operations northeastward from Remagen would have put an added burden on the northern line of communications and, in addition, had the disadvantage of poor communications and terrain. An advance southeastward from the bridgehead, on the other hand, would eventually lead to a link-up with the Mainz–Mannheim area where support could more easily be given by rail and pipeline from the Verdun–Metz depots. See note by G–3 Div SHAEF on Developments From the Remagen Bridgehead, First Draft, 9 Mar 45, and Memo, Osmanski for Whipple, 12 Mar 45, sub: Remagen Bridgehead, both in SHAEF G–4 Operations East of the Rhine and Remagen Bridgehead 162/12/7; Memo, Whipple for G–4, 12 Mar 45, sub: Remagen Bridgehead, and Memo, Osmanski for Mov and Tn Br G–4 SHAEF, 18 Mar 45, sub: Railroad Dev, SHAEF G–4 381 War Plans Gen 1945, II; FUSA Rpt of Opns, 23 Feb–8 Mar 45, I, 42; Hist Rpt of TC ETO, VII, Ch. IV, 23–24.

[4] Study, Mov and Tn Br SHAEF G–4, 14 Dec 45, sub: Rail Lines of Communication, SHAEF G–4 Rail Transportation 117/3 GDP–1, Folder 61; Railroad Reconstruction and Bridging, Hist Rpt 12, OCE ETO, pp. 81–82.

BRIDGE ACROSS THE RHINE AT WESEL *constructed by 1056th PC&R Group.*

able to the Rhine by the time of the crossing as it was in the north. Reconnaissance of the area then revealed that the line from Thionville eastward through Saarbruecken and Bad Kreuznach to either Bingen or Mainz could be restored with a much smaller engineering effort than the line northeastward to Koblenz. Engineer Group B began work on the line on 25 March and completed a single-track line to Mainz on 1 April.[5]

[5] ADSEC Operations History, p. 134; Memo, Osmanski for Mov and Tn Br G–4 SHAEF, 18 Mar 45, sub: Railroad Dev, SHAEF G–4 381 War Plans Gen 1945, II; Comd and Stf Conf Notes, Hq COMZ, 23 Mar 45, EUCOM 337/3 Confs, Staff Weekly, I.

Bridging the Rhine proved to be one of the major engineering tasks of the war in Europe, ranking with the engineer aspects of the Normandy assault and the reconstruction of the ports in magnitude and complexity. Planning had begun early in October 1944, when the chief engineer, Maj. Gen. Cecil R. Moore, held the first meeting with engineers of all the major headquarters to discuss Rhine bridging problems. General Moore's office immediately thereafter began to prepare the long lists of materials and equipment, and to disseminate intelligence and technical information. Because of the peculiarities

SINGLE-TRACK RAILROAD BRIDGE AT MAINZ *constructed by Engineer Group B.*

of the big river, even a flood prediction service was established. The assembly of construction materials at the proper places was in itself a formidable job, involving the inland movement of naval craft and the handling of steel beams up to ninety-two feet in length and more than a yard in depth, and pilings up to a hundred feet in length. Transportation of these "out-of-gauge" materials inland required the careful selection of routes and taking into consideration the turning radius of trailer loads and of bridge and underpass clearance.[6]

[6] *Final Report of the Chief Engineer, ETO,* I, 170, 224–25.

Advance planning and careful preparation had its reward in the dispatch with which the Rhine bridges were installed. Starting on 29 March and working round the clock, the 1056th PC&R Group, using two engineer general service regiments, a construction battalion, and several smaller units, completed the first rail bridge across the Rhine at Wesel in ten days. Ground reconnaissance immediately after the assault crossing had resulted in the selection of a site farther upstream than planned, and required the bridging of the Lippe River as well. But plans were flexible, and the change was made without difficulty. The main crossing over the Rhine was a twenty-three-span structure 1,753 feet long, the shorter bridge over the Lippe a six-span structure of 463 feet. Completion of the Wesel bridges immediately opened up a rail line of communications to Haltern, twenty-five miles east of the Rhine, and shortly thereafter to Muenster.

Engineer Group B in the meantime had reconnoitered three sites in the Third Army area, and on 4 April began construction of a railway bridge at Mainz. The 3,445-foot structure, 2,100 feet of which was of new construction, was also completed in ten days. In dedication ceremonies held on 14 April the bridge was named for President Roosevelt, who had died two days before. The bulk of the material used in its construction was brought forward about 150 miles from dumps in Luxembourg.

Farther south, in the area of the 6th Army Group, engineer units of the Seventh Army, working under the direction of the 1st Military Railway Service, began construction of a 937-foot rail bridge at Mannheim on 12 April and an 851-foot bridge at Karlsruhe on 17 April. The two structures were completed on 23 and 29 April, respectively.

Construction of a fifth Rhine railway bridge, at Duisburg in the Ruhr, was ordered late in April, but was not completed in time to play any part in the support of U.S. forces before the end of the fighting. After careful planning and assembly of materials, engineer units of Group A started construction on 2 May and completed the thirty-eight-span, 2,815-foot structure in a record six and one-half days.[7]

Logistic plans did not contemplate any substantial reliance on supply by rail beyond the Rhine until after mid-April. Army and ADSEC engineers nevertheless began the rehabilitation of lines east of the river immediately after the crossing, and made limited use of these lines to haul forward supplies transferred from trucks. In the north the line Muenster–Soest–Paderborn–Kassel was already in operation when the Wesel bridge was completed on 9 April, providing a continuous rail line of communications deep into Germany in support of the Ninth Army. First Army operated a line from Sarnou to Ingringhausen for five days beginning on 7 April, and then transferred operations to a new line from Kirchhain to Kassel to Ingringhausen. In the area of the Third Army the line Frankfurt–Friedberg–Giessen–Kassel was also open by 9 April. Rail extension almost kept pace with the armies in the next few weeks, and by V-E Day railheads were already in operation at Stendal and Magdeburg in the north, at Leipzig on the central line of communications, and at Regensburg and Stuttgart in the south.[8] (*Map 9*)

Rail traffic over the Rhine, which began on a small scale on 8 April, quickly overtook motor transport as the main long-distance carrier. Within ten days the railways were handling about

[7] ADSEC Operations History, pp. 147–50; *Final Report of the Chief Engineer, ETO,* I 284–85; Hist Rpt 12, OCE ETO, pp. 86–94; Seventh U.S. Army Rpt of Opns, III, 888.

[8] Annex K to G–4 Rpt, Cbl R–50547, ADSEC to TC ETO, 9 Apr 45, ETO Adm 413; FUSA Rpt of Opns, 23 Feb–8 May 45, II, 66, and I, 68; Ltrs, Whipple to G–4, 22 and 31 Mar 45, sub: Rail Dev in Support of the Southern Bridgehead, and Memo, Crawford for AG, 21 Apr 45, same sub, SHAEF Rail Transportation 116/3 G–4 Folder 50; Hist Rpt of TC ETO, VII, Ch. IV.

D. Holmes, Jr.

MAP 9

12,000 tons over the Rhine bridges, which approximately equaled the tonnage hauled by truck. By V-E Day, when they were handling 20,000 to 25,000 tons per day, they were accounting for fully three fourths of the total tonnage.[9] At that time twenty-six general service regiments, supplemented by PC&R groups, dump truck companies, and engineer combat battalions, were engaged in reconstruction work, fifteen of the regiments with the Advance Section and the 1st Military Railway Service, and the remainder in other COMZ sections. The 1st and 2d Military Railway Services at that time were organized into 7 grand divisions, with 24 railway operating battalions, 7 shop battalions, plus mobile workshops, MP battalions, and other ancillary units, and had 11,000 miles of track under their jurisdiction.[10]

The rapid extension of rail traffic through narrow bottlenecks and under conditions of extremely mobile tactical operations soon brought their operating difficulties. Both of the major Rhine bridges—at Wesel and Mainz—were single-track structures and soon became serious traffic bottlenecks, in part simply because they lacked the capacity to handle the volume of traffic demanded for support beyond the Rhine, in part because of poor traffic control. Both bridges were required to handle traffic for two armies. The Wesel bridge had to handle a part of the traffic for First Army as well as the Ninth; the Mainz bridge had to handle the remainder of the First Army traffic in addition to the Third's. Moving supplies for two armies over a single bridge inevitably raised problems of priority, and was further complicated by the multiplicity of agencies involved, including the Advance Section, its regulating stations, and the armies. Proper movement control was lacking at first, and a remedy was not found until control agencies were established—initially at Thionville and then at Mainz, on the southern route, and at Muenchen-Gladbach in the north—on which all interested agencies were represented.[11]

Other factors complicated operations on both lines and initially prevented the development of their full potential. In the south poor signal communications caused delays in calling trains forward and resulted in congestion beyond Saarbruecken. In addition, the bridge over the Main River at Hanau had only a limited capacity and caused many trains to be sidetracked in the Mainz area.

The bridge at Wesel had a potential capacity of 7,000–8,000 tons per day, and was initially reserved exclusively for American use. For nearly two weeks after its opening, however, it handled only about 4,500 tons per day, and to make matters worse British forces almost immediately appealed for an allocation of running rights. The 21 Army Group requests were denied at first, but after several appeals, SHAEF on 23 April allocated one train path per day for 500 tons. Within another week the allocation was increased. Traffic over the bridge improved late in the month, and in the

[9] TC Progress Rpts for April and May 1945, Charts 26 and 22, respectively, App. 7, of Hist Rpt of TC ETO, VII, Apps.

[10] Hist Rpt 12, OCE ETO, pp. 29–30; Hist Rpt of TC ETO, VII, Ch. IV, 25–26, 45–46; Ltr, Page, 13 May 45, sub: Estimate of COMZ Supply Situation (Rpt 4), ETO 381/400 Supplies (OVERLORD).

[11] ADSEC Operations History, pp. 151–52; Cbl R–51743, ADSEC to G–4 ETO, 28 Apr 45, ETO Adm 415.

week before V–E Day averaged about 10,000 tons per day. It was because of the inability of the Wesel bridge to meet both U.S. and British requirements that the construction of another at Duisburg was ordered on the 25th.[12]

Rail traffic suffered even more seriously from the failure to unload freight cars at the railheads in the forward areas and from the practice of selective forwarding. The congestion resulting from the latter was particularly serious in the south, where Third Army representatives, attempting to expedite the shipment of urgently needed supplies, accepted only portions of the army's requisitioned supplies for forward dispatch, sidetracking others to what amounted to dead storage. This practice eventually had its effect in the base and intermediate sections, where large numbers of loaded cars accumulated. The tendency to hold supplies on wheels in the forward areas was common in both army groups and produced an increasingly acute shortage of rolling stock. ETOUSA twice attempted to force the discharge of cars in the forward areas by temporarily restricting loadings in the Oise Section, and Supreme Headquarters also issued a warning about the possible effects of the critical shortage of cars on both military operations and the civil economy. But these measures had little effect, and the accumulation of cars under load continued. At the end of April 2,000 loaded cars were still on hand at former army railheads west of the Rhine alone, and the number of freight cars dispatched beyond the Rhine exceeded by more than 12,000 the number of empties returned.[13]

The difficulties at the Rhine bridges, the accumulation of loaded rail cars at other points on the lines of communications, and the attempt by the armies to institute partial acceptance or selective forwarding, all highlighted an old movement control problem—that is, the problem of controlling traffic between the Communications Zone and the combat zone, which were separate and coordinate commands. The point at which supplies passed from one command to the other—that is, the army rail and truckheads—was a critical point on the supply lines and a potential source of difficulty. Any consideration of the problem inevitably involved the regulating stations, whose basic mission was the control of traffic into the combat zone. A postwar review of their functioning revealed that there had been little uniformity in the concept which the various commands had as to their role or method of operating. In some cases the regulating officer became what amounted to an agent of the army with which he was serving and was

[12] Ltr, Crawford to Smith, 19 Apr 45, sub: Logistical Picture of Pending Opns, SHAEF G–4 381 War Plans Gen 1945, II; Note of Mtg at SHAEF Fwd, Gen Napier, 15 Apr 45, and TWX MGA–10, 21 A Gp to SHAEF G–4, 16 Apr 45, SHAEF SGS 400.3/1 Supply Problems of Allied Advance; Cbl R–51616, ADSEC to Larkin, 27 Apr 45, ETO Adm 415; ADSEC Operations History, p. 154. See also the SHAEF G–4 File 400.22, Shipments, Exchanges, and Transfers, I.

[13] Cbl FWD–19080, SHAEF to G–4 ETO, 1 Apr 45, Cbl E–32949, G–4 ETO to SHAEF, 15 Apr 45, Cbl EX–34381 G–4 ETO to COMZ Secs, 19 Apr 45, Cbl EX–38785, G–4 ETO to COMZ Secs, 29 Apr 45, all in ETO Adm 402 ETO Cbls; TWX FWD–19974, SHAEF to A Gps and COMZ, 26 Apr 45, SHAEF AG 617 Railway Case A, 1; COMZ G–4 History, II, 103–04; Comd and Stf Conf, COMZ, 20 Apr 45, EUCOM 337/3 Confs, Staff Weekly, I; Memo, Transportation Sec 12 A Gp to G–4 12 A Gp, 29 Apr 45, sub: Loaded Rail Cars, 12 A Gp Rolling Stock, No. 106.

utilized virtually as a transportation section of the army headquarters. In some cases the regulating stations exercised the greatest control over movements within the army service area rather than over movements into the combat zone itself. In any case, movements were often initiated without their authority or contrary to their instructions, and embargoes were imposed without their knowledge. However useful they were as expediters and in a liaison capacity between the armies and the Communications Zone, it is clear that they did not operate as contemplated in field service regulations. As agents of the Advance Section rather than the theater commander, as contemplated in regulations, the regulating stations lacked the necessary authority to control movements where conflict arose either between the armies (as it did at the Rhine bridges) or between the armies and the Communications Zone. As agents of the Advance Section, moreover, they were unable to control reserves on rail cars in the army areas and were limited to requesting command action to expedite the release of cars, requests which had to go through the Advance Section, the Communications Zone, and the army group, to the armies.[14]

The shortage of both locomotives and rolling stock had been a persistent limiting factor in rail operations. The most desperate shortage occurred in January, when as many as 800 trains were held up at one time for lack of motive power, despite the receipt by that time of about 1,200 locomotives from the United States. This situation saw substantial improvement in the succeeding months. The United States shipped more than 500 additional engines to Europe between January and the end of April. These shipments, plus the large number of rehabilitated French and Belgian engines, and a few hundred German units, brought the total number of locomotives in use at the end of April to about 11,500.[15]

The Allies had originally planned to ship 50,000 freight cars to the Continent. About 20,000 of these had been delivered by the end of 1944, and a total of 234,000 cars was then in use on the French and Belgian railways. At that time schedules called for the delivery during 1945 of 28,000 U.S. cars, which American and British forces planned to assemble at Marseille and Brussels. But while there were sufficient cars in the United States to meet the planned build-up, insufficient shipping was allocated to move them. Partly because of inadequate receipts, assembly of cars also fell far short of the targets.

On V–E Day there were approximately 250,000 cars in use on the Continent, of which 29,000 had been provided from the United States. These proved far from adequate under the conditions existing in April. The rapid extension of the lines of communication into Germany that month suddenly drained thousands of cars from France and Belgium and im-

[14] The postwar review of their employment revealed different opinions in the various commands. Continental Advance Section, for example, consistently opposed the establishment of regulating stations on the southern line of communications, insisting that they were superfluous organizations, and established one in March 1945, only after protesting an order from the Communications Zone that it do so. See Gen Bd Rpt 27, pp. 51–59.

[15] Of these, 1,736 were U.S.-built, and 7,717 were French.

mobilized additional thousands under load. Concern over the dire effect which this development threatened to have on both military operations and the civil economy led to General Eisenhower's personal intervention near the end of the month. Fortunately relief was in sight with the arrival of V-E Day shortly thereafter.[16]

In the last month of the war the 1st and 2d Military Railway Service had continued to operate the continental railways. Co-ordination of the two systems was provided through General Headquarters, Military Railway Service, which had been established under General Gray at the time SOLOC was integrated with the Communications Zone in February. As of that date, however, the railways were still operated in accordance with an SOP of July 1944, which had limited the authority of the Transportation Corps to "technical supervision." The Transportation Corps had long striven for undisputed theaterwide control over operations in view of the intersectional nature of rail operations. Early in April 1945 it finally realized this goal when a new SOP clarified the Transporation Corps' authority by unequivocally making it directly responsible for both operations and maintenance of way and equipment. Railway construction continued to be the responsibility of the Engineer Service. Under the new operating procedure only minor administrative authority was left to be exercised over the Military Railway Service and its attached units by the COMZ sections. This was confined to matters of supply of common items, general court-martial jurisdiction, hospitalization and evacuation, and financial transactions. In accord with the SOP, troop assignments were issued relieving Military Railway Service units from attachment to the sections and assigning them to Headquarters, Military Railway Service, which in turn assigned them to either the 1st or 2d Military Railway Service. In an attempt to bring the pilferage problem under better control the Transportation Corps also assumed responsibility for the security of supplies in transit. Military Police units previously assigned to the sections were accordingly assigned to Military Railway Service.[17]

(2) Motor Transport—XYZ

Motor transport operations had become fairly routine in February. The bulk of the COMZ truck units were then engaged in port clearance, rail transfer, and interdepot hauls. Of the various express services which had been organized from time to time by the Motor Transport Service, only the ABC Haul was still in operation.[18]

[16] Ltr, Crawford to Gale, 28 Dec 44, sub: Availability of Railroad Rolling Stock to Meet Pres and Long Term Requirements, Cbl S-76142, SHAEF to CCS, 22 Jan 45, Cbl 885 COS (W) 619, AMSSO to JSM, 4 Feb 45, Cbl W-49694, CCS to SHAEF, 8 Mar 45, Memo, Napier for CAO SHAEF, 20 Apr 45, and Ltr, Gale to Mayer, 1 May 45, all in SHAEF AG 617 Railway Case A, I; CAO Mtgs, 19 Jan, 2 Feb, 16 and 30 Mar 45, SHAEF AG 337-2 CAO Mtgs; Memo, Robinson, Dir Control Div ASF, and Cbl, Broshous, Chief Control Div COMZ ETO, for Lee, 8 Mar 45, sub: Program [sic] Made on Items in Somervell's Memo of 24 Jan, EUCOM 400.192 Misc Rpts, VII; COMZ Progress Report for April 1945 (Transportation Section), App. 2 of Hist Rpt of TC ETO, VII, Pt. III.

[17] ETO SOP 32, July 44, App. 10 to Ch. IV, Hist Rpt of TC ETO, V, and ETO SOP 32, 3 Apr 45, App. 4 to Ch. IV, pp. 19-20, Hist Rpt of TC ETO, VI; see also Hist Rpt of TC ETO, VI, Ch. VI, 24 and VII, Ch. IV, 4-5.

[18] See above, Ch. V, Sec. 1.

The month of March brought a gradual shift in emphasis. The advance to the Rhine resulted in a substantial acceleration in truck movements in the forward areas, and the crossings of that river toward the end of the month brought line-of-communications hauling by motor transport to an unprecedented tempo, which was maintained until the end of hostilities.

Motor transport not only supported the advance to the Rhine, but handled the initial build-up in preparation for the Rhine crossings in both the north and south pending the extension of the railways. This proved a much greater task in the area of the Third Army, for rail service could not be pushed forward to the Rhine in time for the crossings, and, even after it was made available, suffered from bottlenecks in the Saarbruecken area. Throughout April, therefore, a portion of Third Army's supplies had to move forward by truck over the narrow and winding roads of the Palatinate.[19] Preparations for the Rhine crossings also presented the Motor Transport Service with hauling tasks it had never faced before, including the movement of a small navy to the Rhine. Tank transporters were used to move several types of assault craft, including the unwieldy LCM, which weighed 46 tons and was 72 feet long, 14 feet wide, and 18 feet high. The movement from seaports to the Rhine required careful route reconnaissance because of the problems of overhead clearance and bridge loads.[20]

But these operations were completely overshadowed by the unprecedented scale on which trucking was organized to support the final offensive beyond the Rhine. The elaborate and thorough preparations for the use of motor transport in support of this drive contrasted sharply with the impromptu manner in which the Red Ball was brought into being in the summer of 1944. Planning for the operation got under way early in February, when the Communications Zone directed the two advance sections to determine what plans the two army groups had for future action in the event of a break-through and to determine what their supply requirements would be. At the same time it directed the Transportation Corps to survey motor transport resources and to make specific plans for the marshaling of all transport in support of a rapid advance.[21]

On 3 March the Motor Transport Service issued an administrative order outlining in full detail the plan for the organization, operation, and control of motor transport, which in its execution came to be known as the "XYZ Operation." The operation took its name from the three-phase scheme by which the plan would be implemented in accord with rising requirements. Assuming a two-day turnaround in each case, transport was to be allocated to move 8,000 tons of dry cargo per day under Plan X, 10,000 tons under Plan Y, and 12,000 tons under Plan Z. On a one-day turnaround deliveries could be double these amounts. In addition, a POL tanker fleet, with a daily capacity of 4,100 tons in all three phases on a one-day turnaround, was to move bulk POL forward from pipeheads

[19] ADSEC Operations History, pp. 140–41.

[20] Hist Rpt of TC ETO, VI, Ch. V, 23.

[21] COMZ G–4 History, III, 60.

TRUCK-TRACTOR AND 40-TON TANK TRANSPORTERS *carrying LCM's for Rhine River crossings, March 1945.*

and railheads, initially from Maastricht, Thionville, and Saaralbe.

The Transportation Corps was far better prepared to undertake a mission of such scope than it had been in August 1944 from the point of view of experience, organization, and available equipment. In March 1945 the Communications Zone had 226 truck companies at its disposal on the Continent. In contrast with the situation in 1944, a substantial number of these were either 10-ton tractor-semitrailer combinations or 10-ton diesels, heavy-duty types which were best suited for long-distance hauling. Since January the Communications Zone had received sixty-four new heavy companies, fourteen of them consisting of diesel companies transferred from the Persian Gulf Command. The larger capacity of the heavy duty companies—rated at double that of a 2½-ton unit—actually gave the Communications Zone the equivalent of 316 2½-ton companies.[22]

Plans provided for the use of 55 companies under Phase X of the coming operation (2,750 vehicles), 67 under Phase Y (3,350 vehicles), and 81 under Phase Z (4,050 vehicles). In each case the bulk of the motor transport was to consist of the larger 10-ton companies, whose capacities were equivalent to 101, 125, and 150 2½-ton companies respectively in the three phases. In order to employ this equipment to best advantage the Motor Transport Service planned to use the 10-ton tractor-trailer companies over the main long-distance hauls, and the 2½-ton and 10-ton diesel units primarily for branch line-of-communications work,

[22] The 10-ton tractor-semitrailer combination proved the most ideal for long-distance hauling. The diesel also performed well, but presented a fuel problem. Hist Rpt of TC ETO, VII, Ch. V, 40–41.

local operational missions, and in the physically less favorable areas. It aimed at having forty vehicles per company in service at all times. The bulk POL fleet was to consist of seventeen companies, some of them of the 2,000-gallon type, some of the 750-gallon type, and a few 10-ton companies with skid-mounted tanks, with four 750-gallon tanks to a trailer.[23]

The planned organization and control of this transport was a vital feature of the XYZ Operation, and reflected the experiences and lessons of the earlier express services. There was no longer any question as to the desirability of providing an effective centralized control over a service which promised to be intersectional in nature. All motor transport used in XYZ was assigned to the 6955th Headquarters and Headquarters Company of the Motor Transport Service (Col. Ross B. Warren), which remained the ultimate authority for control of the entire operation. To permit flexibility in operations over the several routes the Motor Transport Service delegated the actual control over field operations to subordinate agencies known as highway transport divisions (HTD), provisionally created for this purpose. Two highway transport divisions were initially activated, the 1st HTD to operate the two routes in support of the First and Ninth Armies, in the north, the 2d HTD to operate the two routes in support of the Third and Seventh Armies in the south. Basically, each highway transport division consisted of a quartermaster group (TD) headquarters, augmented with personnel from the Motor Transport Service. In the period before the haul became intersectional the divisions were to be under the direct control of the transportation officer of the advance section in which they were operating.

Operating and maintenance procedures also reflected the influence of earlier trials and errors, bearing a strong resemblance to the SOP's which had been worked out for the successful ABC Haul. At points of origin, for example, marshaling-yard type of operations was to be in effect, with all trailers handled in the same manner as were freight cars by the Military Railway Service. All movements would be made in serials consisting of tractors operating in platoon convoys. Quartermaster detachments attached to each HTD were to operate marshaling yards and dispatch vehicles at the point of origin, operate road patrols for over-the-road discipline, control, vehicle recovery, and route reconnaissance, and ensure the prompt turnaround of vehicles at points of destination. Only the HTD's were to issue movement instructions. The HTD's were thus intended to be the contral operating agencies in the field, co-ordinating loading and movements from point of origin to point of destination and issuing all movement instructions on the basis of over-all orders of the advance sections. Each truck company was made responsible for full preventive maintenance, and, reminiscent of old cavalry practice with respect to animals, drivers were ordered to carry out before resting specified maintenance procedures in bivouac, at servicing points, at halts,

[23] AdmO 1, 6955th Hq and Hq Co MTS (Provisional), 3 Mar 45, in Hist Rpt of TC ETO, VI, Ch. V, 10–13; see also, p. 3; for detailed plans on the organization and operation of XYZ see Annexes A–F to AdmO 1 in Hist Rpt of TC ETO, VI, App. 5, Pt. 1.

and at loading and unloading points under the supervision of serial detachments. A mechanic was to ride each serial, carrying a small supply of high mortality parts. Mobile ordnance maintenance units from the advance sections were to provide road patrols, make major repairs, and provide replacement vehicles.

The XYZ service was mobilized with unexpected speed. The operation got under way on 25 March, and the first deliveries were made to the Third Army over a line of communications averaging about 120 miles in length. Within a few days trucks of the Motor Transport Service were rolling forward with supplies for the other three armies. At the end of the first week the operation had already shifted into Phase Z and was making deliveries of approximately 12,000 tons per day. Motor transport had been marshaled with unprecedented speed, truck companies in the various COMZ sections having been earmarked early in March and alerted for movement on twenty-four-hour notice. Movement orders had been placed on file for every unit nominated for XYZ, and were simply fed to teletypists in the Office of the Chief of Transporation when the need for the units arose.[24]

The marshaling of transport did not stop with the implementation of the third phase of the plan, and the operation eventually far exceeded the planned scale of operations. Toward the end of April the number of truck companies on XYZ hauls rose to the equivalent of 244 2½-ton companies, and deliveries for a time averaged at least 15,000 tons per day.

Shortly after the operation got under way it was realized that, largely because of the lateral distance between routes, one HTD could not efficiently handle the transport for two armies. Early in April, therefore, a third HTD was organized to support the First Army, and in the south a reinforced quartermaster group (the 469th) was assigned to the Continental Advance Section to serve in similar fashion in support of Seventh Army. The latter truck service, operating under CONAD, was known as the Yellow Diamond Route.[25]

XYZ was in constant flux, with loading points repeatedly moving forward as railheads could be advanced, and unloading points moving forward as the armies advanced deeper into Germany. Starting points were initially at Liège for the Ninth Army, at Dueren for the First, at Luxembourg City for the Third, and at both Saarbruecken and Nancy for the Seventh. Hauls in the first week ranged in length from eighty miles in the case of Ninth Army to 160 miles in the case of the Third. By V-E Day loading points had been established as far forward as Bielefeld in support of the Ninth Army and Wuerzburg in support of the Third, although trucks of the 3d HTD (First Army) were still returning all the way to Dueren to pick up their loads. (*Map 10*) Trucks of the 1st HTD (Ninth Army) by that time were delivering their loads as far forward as Magdeburg, and those of the 2d HTD (Third Army) were carrying supplies beyond Regensburg.

---

[24] Hist Rpt of TC ETO, VI, Ch. II, 58.

[25] COMZ G-4 History, III, 62; Hist Rpt of TC ETO, VII, Ch. III, 13.

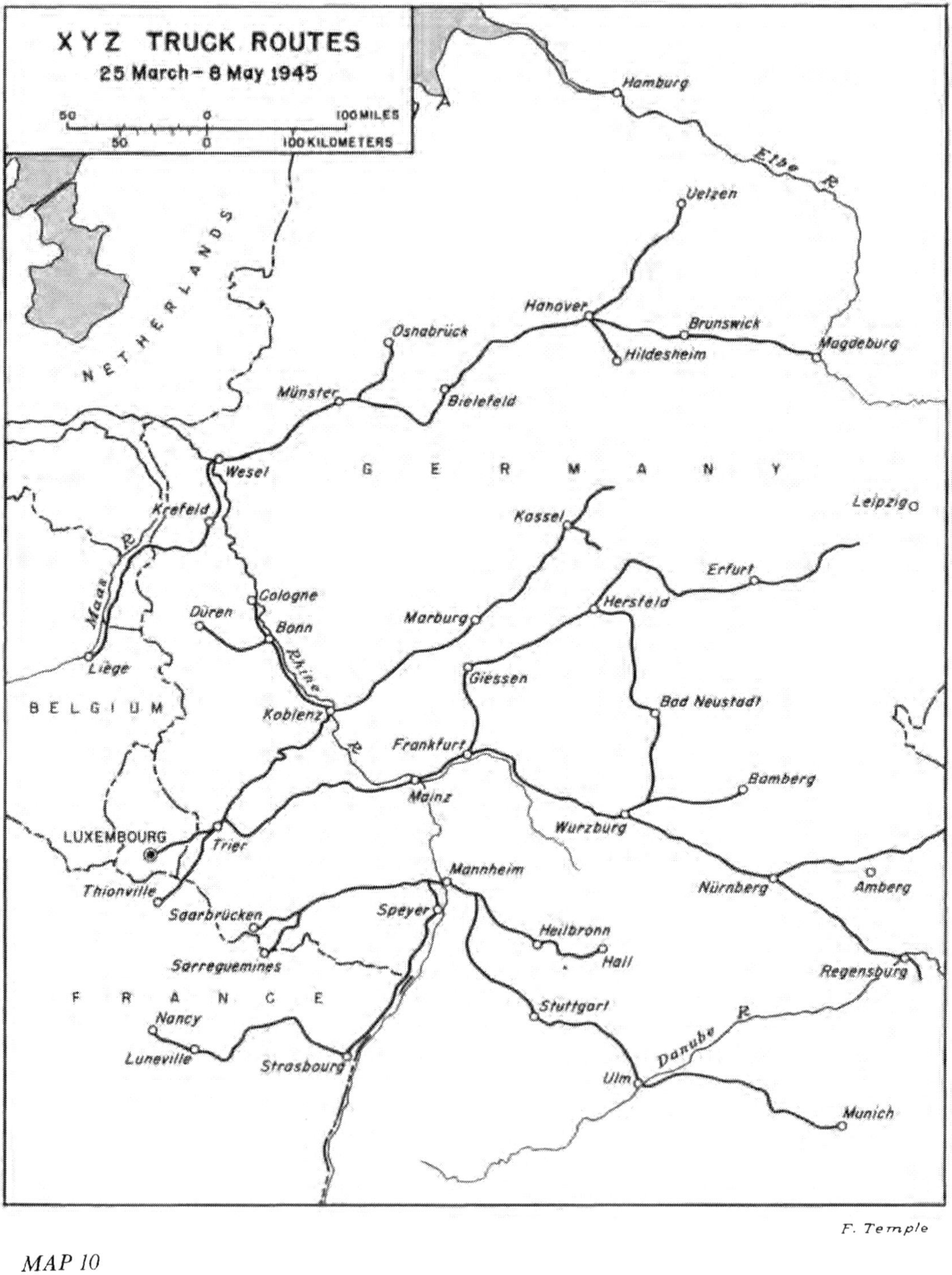

F. Temple

MAP 10

While hauls sometimes exceeded 200 miles (mainly in the First Army), the average was nearer 140. The distances involved on the XYZ runs were therefore generally much shorter than on the Red Ball Express. This resulted largely from the speed with which railheads were opened in the wake of the armies. By V-E Day the XYZ trucks had hauled 630,000 tons to the armies,[26] averaging 14,000 tons per day. The operation continued until the end of May. By that time it had handled a total of 872,000 tons for an average of about 13,000 tons per day.[27]

The XYZ Operation was a highly creditable performance, although it also had its hitches. Maintenance of vehicles was the biggest problem, as usual, despite the advance preparations and precautions. On the Third Army route, for example, a mechanic accompanied every convoy, four hours of maintenance were given to all vehicles before they were released for dispatch, and ordnance maintenance companies were assigned to service specific truck battalions. But spare parts were often lacking, with the result that the goal of a serviceability rate of forty vehicles per company was never quite achieved. On the Yellow Diamond Route the fortuitous capture of 1,000 German tires relieved one of the more serious shortages, and prisoners of war were extensively used to relieve drivers in preventive maintenance.[28] The 2,000-gallon tankers created one of the most difficult maintenance problems, for they developed many leaks on some of the twisting routes, and were deadlined for varying periods for repairs. Replacement vehicles were slow to arrive on all four routes because of the distance to vehicle pools, which were located in intermediate and base sections. In addition, there were the usual difficulties over communications, aggravated by frequent moves, and with bottlenecks at bridges, many of which were one-way ponton structures.[29]

Some observers believed that the organization was still defective in some respects. A few considered that the highway transport divisions were superfluous agencies, some that control of the POL tanker companies had not been clearly established, and others that the transport divisions should have been organized on the task force principle, with all ancillary units needed in the operation, such as signal, ordnance, and MP units, attached to the controlling headquarters.[30] But there was more general agreement that the command and administrative arrangements had been sound and effective. Colonel Warren, who headed the Motor Transport Service, considered the operations of the 2d HTD, which supported Third Army, particularly successful. The commander of the Advance Section had appointed the regulating officer at Third Army as his deputy for all ADSEC troops supporting General Patton's forces, and had arranged to have all HTD opera-

[26] Including relatively small tonnages to the First French Army in the last two weeks.

[27] Hist Rpt of TC ETO, VII, Ch. V, and App. 5.

[28] Drivers were provided another comfort on the Yellow Diamond Route in the form of hot food. The Special and Information Services of CONAD set up "GI Joe Dinners" at fifty-mile intervals, where truck drivers could exchange cold rations for hot at any time of day or night. CONAD History, pp. 151–52.

[29] Hist Rpt of TC ETO, VI, Ch. V, 14, and VII, Ch. V; CONAD History, pp. 188–94, 213–16.

[30] ADSEC Operations History, pp. 135, 139; Rpt of Truck Opns in the Battle of Germany, 6957th HTD (Provisional), in Hist Rpt of TC ETO, VII, Ch. V, 17–26.

tions in support of Third Army coordinated through that officer.[31]

Although the theater possessed much greater resources in the spring of 1945, and its logistic organizations greater know-how, the final offensive put a tremendous strain on transportation as had the drive across northern France in August 1944. The marshaling of fully three fourths of all motor transportation in the Communications Zone for the XYZ Operation was carried out only at the expense of other missions, particularly port clearance, with the results indicated in the preceding chapter. In mid-April officials in the combat as well as the communications zone were aware of the effect which the increasing withdrawals of truck units from the base areas would have on COMZ operations. But all were agreed that the end was near and that the sacrifice of essential rear area supply activities was justified.[32] Meanwhile the armies, each of which had about forty truck companies for its own use in the forward areas, imposed strict priorities on the use of trucks, and, as in the summer of 1944, they augmented their lift by forming provisional units with the organic transportation of field artillery and antiaircraft artillery units.[33]

The efficient operation of motor transport had of course depended in part on the availability of good highways. While railways had to be counted on to do most of the heavy-duty hauling, it was the existence of good road nets that gave the desired flexibility to tactical maneuver in the forward areas and, in the initial stages of a sustained offensive, made possible the logistic support of the combat elements.

On the whole the damage to roads, like bridge destruction, was not as great as expected in OVERLORD plans. Both road maintenance and bridge construction nevertheless placed a major call on engineer resources. Road repair in the combat zone was normally carried out exclusively by engineer units. The Communications Zone wherever possible used both civilians and prisoners of war under the supervision of engineers, and enlisted the aid of national and local highway organizations, usually providing them with POL, trucks, and some road equipment. At the peak of operations the Communications Zone maintained a road net for military purposes west of the Rhine of 7,700 miles and eventually built 337 bridges. Bridge building suddenly became a tremendous task at the German border, where almost all major bridges in the path of the advance had been destroyed. The largest tasks were in the north, where rivers were wide. The biggest problem was encountered at the Meuse, where fourteen highway bridges were constructed with a combined length of 6,751 feet. After mid-September the Advance Section alone built seventy-one semipermanent fixed bridges, many of them replacing floating or tactical fixed bridges emplaced by the armies.

Early in 1945 road maintenance temporarily became the largest single item of Army engineer work, first because of

[31] Ltr, Warren to OCMH, 23 Jun 54, OCMH.

[32] Diary of G–4 12 A Gp, 12 Apr 45; Ltr, CG COMZ to CG ADSEC, 24 Apr 45, sub: Adm Support of Forces Operating in Germany, ETO 390 RH Planning Directives.

[33] NUSA G–4 AAR, Apr 45; NUSA G–4 Periodic Rpts, Mar–May 45, NUSA 319.1 G–4 Periodic Rpts; FUSA Rpt of Opns, 23 Feb–8 May 45, Bk. II, pp. 63–65.

HIGHWAY BRIDGE OVER THE MEUSE RIVER, *built with Bailey bridge spans over river barges.*

snow and ice, and then, in February, because of an early thaw which extended over all of Belgium, Holland, Luxembourg, and northern France as far as the Seine, and which damaged about 1,250 miles of the military network. Unfortunately the thaw came at a time when heavy troop movements were being carried out in preparation for the closing to the Rhine. At that time the force employed in road maintenance in the Communications Zone rose to about 17,000, and remained at that level until the end of hostilities. Approximately 40 percent of this force consisted of military personnel, 25 percent of civilians, and 35 percent of prisoners of war. In the combat zone road maintenance became the major preoccupation of engineer units in these months. During the two-week thaw of February, First Army assigned all engineer battalions to road work, and in addition used 4,000 men from the 7th Armored Division and 1,300 men from the 3d Armored Division on drainage work, filling potholes, and repairing shoulders.

In March emphasis shifted back to bridge construction, the Rhine getting the major attention. Planning the Rhine highway spans, like planning rail bridges, had begun in the fall of 1944. In all, fifty-seven highway bridges were built over the Rhine, fifty-two of them by the armies. The major fixed bridges were at Wesel, Cologne, Neuwied, Mainz, Oppenheim, Gernsheim, Frankenthal, and Ludwigshafen. The 1,050-foot Oppenheim bridge, constructed by the 1301st General Service Regiment of Third

Army, was considered by the chief engineer to be the best designed and constructed bridge built by a field force unit. Most of the twenty-six treadway and heavy ponton bridges, having served their purpose, were no longer in use by V-E Day. Of the five structures built by the Communications Zone, two were not opened until 15 May.[34]

(3) Inland Waterways

The inland waterways experienced many operating difficulties throughout the winter, and handled disappointingly small tonnages until March and April. Most of the waterways were frozen tight during the severe cold of January, although a limited traffic was maintained on the Albert Canal by the use of sea mules as ice breakers. In February, early thaws created flood conditions, particularly on the Seine, where barge traffic remained at a standstill until late in the month. Supply movements by water improved considerably in March, and the Transportation Corps established inland ports at Paris, Reims, Lille, La Louviere, and Liège. Most of these were operated by regular Transportation Corps port companies, aided by civilian and prisoner of war labor. Port clearance by barge totaled 177,000 tons in March, which was equal to 8 percent of total clearances (2,090,000 tons). It rose to 410,000 tons the next month, equal to 23 percent of all clearances. The heaviest shipments were made out of Antwerp, Ghent, and Marseille. Military traffic on the inland waterways was not in as great volume as it might have been in the last months, for more and more of the capacity of the inland waterways was assigned to civilian movements. Such movements naturally aided military operations to the extent that they relieved the railways of this burden.[35]

(4) Air Transport

Air transport operations in 1945, like long-distance trucking operations, showed marked improvement over the previous summer, and made a notable contribution toward maintaining the momentum of the final offensive. At least two factors in addition to the obvious advantages of experience favored a smoother air support operation in the final drive into Germany. In contrast with the summer of 1944, continental airfields were plentiful in April, so that there was none of the competition over the use of fields for tactical or administrative purposes which had characterized the earlier airlift. Equally important, there was no withdrawing of aircraft for airborne operations such as had given the 1944 airlift its on-again-off-again character, although such a withdrawal of craft was considered at one point.

Supply by air had reached a low ebb after the emergency missions during the Ardennes counteroffensive. During February and March barely twenty-five sorties per day were flown to the 12th Army Group area, and deliveries, consisting almost exclusively of medical supplies, averaged only about fifty-five tons.

---

[34] *Final Report of the Chief Engineer, ETO,* I, 288–306; Road Maintenance and Highway Bridging, Hist Rpt 14, OCE ETO, pp. 67–79.

[35] Hist Rpt of TC ETO, VI, Ch. II, 65K, 161–64, Ch. VI, 16–18, 55–56, 65, and App., Chart 11; see also, Vol. VII, App., Chart 11; ADSEC Operations History, p. 141.

Early in March the 12th Army Group bid for a lift of 800 tons per week for itself and the Communications Zone, and asked in effect for a guarantee that planes allocated for this purpose not be withdrawn for airborne operations or training. Modest as this demand was, SHAEF turned it down on the ground that additional lift could not be allocated for needs that were not of a truly emergency nature.[36]

Before the end of the month, however, SHAEF had authorized the maximum possible use of air in support of forces beyond the Rhine, and on 27 March the Third Army made the first bid for air supply, asking for a 2,000-ton lift. Bad weather prevented the use of air for another two days, but on 30 March the IX Troop Carrier Command, using 329 planes, inaugurated the stepped-up lift with the delivery of 197,400 gallons of gasoline to General Patton's forces.[37]

On the following day General Crawford met with representatives of all interested agencies to review the administrative procedure for supply by air in order to ensure that all headquarters fully understood their responsibilities. Despite repeated attempts at clarification and simplification, the procedures involved in requisitioning supplies by air and in co-ordinating the actual shipments were defective even at this date. That they still left something to be desired was revealed on the very first day of the expanded airlift, when planes had arrived at forward fields and it was found that no arrangements had been made to unload them. On this particular occasion CATOR (Combined Air Transport Operations Room) had been at fault, for it had authorized the flights on the basis of reconnaissance reports on the condition of forward airfields without giving sufficient advance notice to either the Communications Zone or Third Army.[38]

The Communications Zone had only recently recommended certain changes in the handling of air supply. Although responsible for the support of the field forces, it lacked the direct control of air transportation which it considered essential to orderly and expeditious supply of the armies. In mid-March it asked SHAEF to place all aircraft allocated for air supply under the direct control of the Communications Zone for loading, unloading, and routing, proposing that such supervision be delegated to the chief of transportation, who exercised movement control over other forms of transportation. SHAEF gave little encouragement to this proposal, but the entire procedure was overhauled at the meeting of 31 March.[39] On General Crawford's suggestion it was agreed that the armies should henceforth submit all bids for

[36] Ltr, 12 A Gp to SHAEF, 6 Mar 45, sub: Airlift for Supply and Evacuation, and 1st Ind, 18 Mar 45, SHAEF AG 581.2-1 Supply by Air.

[37] All statistics on deliveries by air from CATOR Weekly Load Summaries, are either in SHAEF G-4 581.2 Transportation by Air of Supplies and Equipment, II, or in SHAEF Air Staff, A-3 505.46-7, Air University Library, Maxwell Air Force Base, except as otherwise noted. Memo for Record, Col Hopkins, Chief Supply Br QM 12 A Gp, 3 Apr 45, SHAEF 12 A Gp G-4 Memos for Record; Diary of G-4 12 A Gp, 27 Mar 45.

[38] Memo, Lt Col William A. Stephenson, COMZ Ln Off at 12 A Gp, for G-4 COMZ, 31 Mar 45, sub: Airlift Mtg at Crawford's Office, SHAEF, 31 Mar 45, 1400 Hours, 12 A Gp Supply by Air, No. 133.

[39] Ltr, Hq COMZ to SAC, 15 Mar 45, sub: Supply by Air, and Memo, SHAEF G-4 Mov and Tn Br for AG, 19 Mar 45, SHAEF AG 581.2-1 Supply by Air, No. 1.

supply by air directly to CATOR, designating the airfield at which they wanted delivery after clearing with the tactical air command. CATOR was to co-ordinate the entire operation, making arrangements with the Communications Zone and notifying the armies in advance of the estimated delivery time. Furthermore, because of the expected fluidity of operations in the forward areas and the inadvisability of dispatching ADSEC units to widely scattered airfields, it was agreed that the bidding agencies—normally the armies—should thereafter be responsible for unloading supplies from aircraft and clearing airheads of supplies, although they might call on the Advance Section for assistance. The Communication Zone itself expected to man certain airfields, particularly in the Giessen area, north of Frankfurt, where it planned to build up supply stocks by air. SHAEF issued a directive outlining the new procedure on 1 April.[40]

Supply by air expanded rapidly in the first days of April as the full resources of the IX Troop Carrier Command were committed to the airlift. Deliveries reached their peak in the second week, when more than 6,200 sorties were flown and more than 15,000 tons of supplies were set down on forward fields. As could be expected, the great bulk of this tonnage (about 80 percent) consisted of gasoline. By design, the lion's share of the lift went to the First and Third Armies, whose requests were given first priority.

Third Army had begun to plan early in March for the air support of its anticipated drive beyond the Rhine, negotiating directly with CATOR on the subject. One important feature of its plan was the arrangement to have the 2d Engineer Aviation Brigade follow closely behind advancing infantry and armor to rehabilitate landing fields. Aviation engineers were attached directly to each corps for this purpose. Once they reported a field ready to receive planes, a pilot of the Troop Carrier Command stationed with the IX Tactical Air Command then reconnoitered the field and checked its suitability. In this way Third Army alone used some thirty fields for supply and evacuation, in many cases abandoning them after using them only a few days. Motorized "flying supply points" moved from field to field as new ones were opened farther forward, often moving at night, and normally issued supplies to using units directly from the field.

General Patton's forces eventually pushed into Austria and Czechoslovakia, and therefore had the most extended supply lines of any of the armies. In accord with its greater requirements, Third Army was made the greater beneficiary of the airlift. Between 30 March and 8 May it received about 27,000 tons of supplies via air, more than half of all the tonnage moved by that means. Of its total receipts, 22,500 tons consisted of gasoline (6,000,000 gallons), and accounted for 22 percent of all the gasoline Third Army issued in that period. In addition, Third Army received an average of 50,000 rations per day by air, equal to 11 percent of its total issues, plus small tonnages of critical Class II and IV items like field wire, cable, dry cell batteries,

[40] Memo, Stephenson for G–4 COMZ, 31 Mar 45; Supply and Evacuation by Air, Gen Bd Rpt 26, pp. 22-23.

bogie wheels and tires.[41] Seventh Army called for supply by air only a few times when units were extended and supply routes were threatened. The main instances occurred on 9–10 April, when the 10th Armored Division was resupplied near Crailsheim, and on 26–27 April, when the VI Corps had to be supported by air.[42]

Although planes were never withdrawn from the airlift for other purposes during the final offensive, there was constant uncertainty as to the number of craft that would be available from day to day. The ETOUSA G–4 complained of this, but SHAEF refused to make a specific daily allocation for any particular force.[43] It followed the policy of making the maximum number of aircraft available for this purpose, however, and the IX Troop Carrier Command averaged more than 650 flights and 1,600 tons per day throughout the month of April. At the height of the airlift in September 1944 approximately 1,000 tons per day had been shipped by air, about half of it going to the First and Third Armies.[44] Aside from the large tonnages of gasoline which the airlift handled in April, its great value lay in the responsiveness to demand which it provided in the last sustained push, for it was able to meet urgent demands for specific items on much shorter notice than either rail or motor transport.

Airlift planes played an unprecedented secondary role in evacuation in the final month of operations. During April approximately 40,000 casualties were removed from the combat zone in the planes which had brought supplies forward. Even more spectacular was the evacuation of Allied prisoners, who were uncovered in increasing numbers as the armies overran enemy camps deep inside Germany. Seemingly endless "sky trains" of RAMPS (Recovered Allied Military Personnel) moved westward in the last days of hostilities. Third Army alone evacuated 135,000 men in the last month.[45]

(5) Forward Deliveries

Although the theater still faced serious problems in port clearance and depot storage in February 1945, its supply situation was more satisfactory than it had been for many months. Stocks of supplies in the Communications Zone had reached an all-time high of 4,027,250 tons, exclusive of backlogs in the ports and cargo awaiting discharge from ships. The distribution of these supplies was not yet ideal, since stocks in the U.K. depots still accounted for more than 40 percent of this tonnage. But some improvement had been made in the redistribution of stocks on the Continent, where too large a percentage had heretofore been held in the base areas, particularly in Normandy. The movements program of February brought about some

[41] Supply of Third Army by Air Supply and Evacuation During the Crossing of the Rhine and Beyond, 30 March to 9 May, by Gen Muller, TUSA G–4, and Memo, Kyser for Counts, 13 Apr 45, sub: Air Supply to Ninth Army, both in 12 A Gp Supply by Air, No. 133.

[42] SUSA Rpt of Opns, III, 783–84, 893.

[43] Cbl E–31233, ETO G–4 to SHAEF, 11 Apr 45, ETO Adm 402; Cbl FWD–19094, SHAEF to ETO G–4, 14 Apr 45, ETO Adm 414.

[44] See *Logistical Support, I*, pp. 580–82.

[45] 12 A Gp Rpt of Opns, XIII, 60; Gen Bd Rpt 26, pp. 37–38.

reduction in the Normandy stocks. Of the 2,323,000 tons of supplies on the Continent, approximately 313,000 tons were located in ADSEC and 100,000 in CONAD depots. In addition to COMZ stocks, 357,000 tons lay in army depots, ranging from 34,000 tons in the maintenance area of the First French Army to 95,000 in that of the Ninth U.S. Army.

In the forward areas the greatest concentrations of supplies were in the Liège–Namur–Charleroi–Lille area in the north, and in Verdun and the Dijon–Langres–Epinal area in the south. Stocks were not in proper balance in any of these installations, and a selective build-up of Class II and IV items was required. But supplies were considered adequate for the offensives planned for February. COMZ movement plans called for a heavy build-up in the Nancy–Toul area for the future support of the Seventh Army, the supplies for which were to come from the large classified stocks in Delta Base Section. This build-up got under way in February, although movements on the southern line of communications continued to be limited by inadequate rail transportation. The completion of a rail bridge at Avignon was expected to permit the opening of the line on the west bank of the Rhône, but this was again delayed early in February when a French barge crashed into the partially completed structure.

Deliveries nevertheless set a good record in February, totaling 25,000 tons per day to the five armies supported by the Communications Zone and averaging 5,600 tons for each of the four U.S. armies. At the end of the month army stocks had risen to 495,000 tons for a net gain of more than 130,000 tons. Stocks had risen to 414,000 tons in the Advance Section and 143,000 tons in CONAD.[46]

Forward movements continued in about the same volume to the combat zone in March, averaging more than 5,000 tons per day to each of the four U.S. armies, including the Third and Seventh, which carried out major offensives in closing to the Rhine. COMZ depot stocks meanwhile also continued to register net gains, rising to 4,790,000 tons by the end of the month, of which the two advance sections held an unprecedented 684,000 tons.[47]

Much of the heavy movement in March was in preparation for the offensive which was to carry Allied armies across the Rhine and deep into Germany. Plans provided that the Advance Section was to turn over its depots in the Namur–Liège–Maastricht area to Channel Base Section and that the Continental Advance Section should transfer its installations in the Verdun–Nancy–Toul–Metz area to the Oise Intermediate Section once the offensive got under way and then establish mobile dumps or depots in support of the armies. In accord with these plans, both advance sections turned over their installations in these areas immediately after the Rhine crossings and thereafter held only small tonnages in depots taken over from the armies just west of the Rhine, since

---

[46] Ltrs, Page to CG COMZ, 17 Feb and 7 Mar 45, sub: Estimate of COMZ Supply Situation (Rpts 1 and 2), EUCOM 381 Projected Operations of COMZ ETOUSA, 14 Mar 45; Weekly Rpts of Tonnages in Storage in U.S. COMZ Depots and Br GHQ Depots, G–4 SHAEF, SHAEF G–4 137/13/1 COMZ Tonnages in Storage.

[47] Ltr, Page to CG COMZ, 11 Apr 45, sub: Estimate of COMZ Supply Status (Rpt 3), SHAEF G–4 Estimate of Supply Situation Reports—COMZONE.

neither of them exercised territorial jurisdiction in Germany.

All the armies attempted to move supply stocks as far forward as possible before the Rhine crossings. The Ninth Army, for example, moved more than 100,000 tons of supplies and equipment from the Maastricht area to Muenchen-Gladbach in March, using more than forty assigned truck companies, plus several borrowed from the Advance Section and provisional companies which it formed from antiaircraft units. The First Army meanwhile had chosen Euskirchen as its next maintenance area in preparation for the Rhine crossings. The capture of the Remagen bridge on 8 March did not seriously disrupt these plans, and the army quickly shifted supply to exploit this success, putting in bridges and initially employing dukws to support the III Corps. Within two days it had established ration, POL, and ammunition supply points on the opposite shore.[48]

Both the Third and Seventh Armies carried out major offensives during March and then forced the Rhine without attempting a deliberate supply build-up in positions which would have afforded closer support. Support of the Third Army's drive had entailed supplying units on two axes—one through the Saar and one north of the Moselle, both of them over poor lines of communication. Twice after the first of the year the logistic support of Third Army was further complicated by lateral boundary changes. Late in January a shift in the boundary between the 6th and 12th Army Groups placed Nancy and Toul, then in the Third Army service area, in the area of the Seventh Army. Third Army turned over some of its installations in this area to the Continental Advance Section and evacuated others to its own maintenance area, which it then concentrated in the Metz–Thionville and Luxembourg City areas. In March a second boundary change placed Metz in the Seventh Army area, necessitating further adjustments in Third Army's administrative structure.[49]

In anticipation of the inevitable bottlenecks at the Rhine bridges all the armies attempted to ensure as high a degree of self-sufficiency as possible for forces operating beyond the river in the first few days. The Ninth Army, for example, specified that unit trains were to be loaded to capacity for the crossings, issued certain replacement items, such as tanks, in advance to meet initial losses, and restricted all traffic in the first forty-eight hours to tactical movements. As expected, the Wesel highway bridges, which were shared with the British, proved serious strictures to the desired build-up east of the Rhine, and carried the heaviest traffic in the north, partly because they were served by the most favorable road net to the east. At the height of the traffic on 9 April, the day the rail bridge was opened, a traffic count showed that the three highway bridges—a Bailey, a 25-ton ponton, and a treadway—carried more than 1,500 vehicles per hour.[50]

Although logistic support in the final month was more orderly and far more adequate than it had been in August and

[48] FUSA Rpt of Opns, 22 Feb–8 May 45, Bk. II, p. 59; FUSA AAR, March, pp. 59, 87–89; *Conquer: The Story of Ninth Army, 1944–45*, pp. 212, 221–25; NUSA AAR, G–4 Rpt for 1–15 Mar 45.

[49] TUSA AAR, II, G–4, 57–58.

[50] *Conquer*, pp. 255–56, 264, 320–21.

September 1944, it nevertheless had some of the characteristics of the earlier pursuit. Army supply installations were forced to deploy forward repeatedly and to adjust to the changing tactical situation and to altered missions, placing enormous demands on transportation. First Army initially chose Marburg, about seventy miles west of Bonn, for its maintenance center and began to move supply installations into that area. But the location soon proved unsatisfactory, and a new maintenance area was chosen in the vicinity of Warburg, fifty miles away, to support the northeastward drive. Later in April, with its axis of operations shifted southward, First Army developed a new maintenance area at Giessen, far to the south. Third Army's line of communications also changed twice during April, first extending northeastward from Frankfurt to Hersfeld, Eisenach, Erfurt, and Weimar, and then in mid-April, in response to the army's altered mission, shifting abruptly to the axis Frankfurt–Schweinfurt–Nuernberg–Regensburg.[51]

Under conditions of lengthening hauls and turnaround time, the staples of supply—rations, gasoline, and ammunition—naturally took priority over Class II and IV items. One result was a lack of spare parts, which in turn resulted in a high mortality in vehicles. Army reserves fell from 410,000 tons to 234,000 tons in the seven weeks between the Rhine crossing and V-E Day. While this caused some alarm in the armies, it never reached serious proportions. The Communications Zone maintained a steady flow of supplies to the armies throughout the final weeks, daily receipts averaging at least 5,000 tons in each army as compared with the 3,000–4,000 tons during the 1944 pursuit.[52] Perhaps the most significant commentary on logistic support in the final drive is the fact that neither the army groups nor SHAEF instituted tonnage allocations, as they had found it necessary to do in the summer and fall of 1944.

TABLE 11—COMBAT ZONE MAINTENANCE FACTORS, JUNE–OCTOBER 1944

[Long Tons per Divisional Slice]

Supply Class	Normal Combat	Regrouping or Negotiating Natural Obstacles	Rapid Advanec
Total	541	426	462
I	100	100	100
II and IV	117	117	117
III	144	144	180
V	180	65	65

Source: SHAEF G–4 Study, 27 Oct 44, sub: Maintenance, SHAEF G–4 400.22 Maintenance.

The consumption of supplies in the final offensive generally substantiated the factors which earlier experience had indicated would apply under similar conditions. SHAEF G–4 planners had studied the consumption experience of the period June–October 1944 and had arrived at the maintenance factors shown in Table 11.

On the basis of G–2 and G–3 estimates made in January, which assumed stiff enemy resistance and a slow advance, logistical planners at first calculated sup-

[51] FUSA AAR, April, pp. 1, 3; TUSA AAR, II, G–4, 60.

[52] The U.S. Fifteenth and First French Armies are not included. Rpt by G–4 Supply Br, 12 A Gp, 5 May 45, sub: Info on Supply Situation, 12 A Gp Supply—Reports of Status of, No. 131; Ltr, Page to CG COMZ, 13 May 45, sub: Estimate of COMZ Supply Situation (Rpt 4), ETO 381/400 Supplies (OVERLORD).

ply requirements for "normal combat" conditions, for which maintenance requirements were figured at 540 tons per division slice per day.[53]

By April it had become clear that operations would more nearly resemble those of a "rapid advance," and maintenance factors were accordingly revised. Based on a study of operations in March, a new factor of 450 tons rather than 462 was actually adopted, reflecting reductions in Classes I, II, and IV, and a greater allowance of ammunition than earlier mobile type operations had indicated. The new factors were:[54]

Class of supply	*Long tons*
Total	450
I	80
II and IV	90
III	180
V	100

Supply consumption in the final month varied from army to army, reflecting the differing operational conditions encountered. (*Table 12*) Consumption experience in the First and Third Armies closely approximated the revised factors in most classes of supply, although Third Army consumed somewhat higher quantities of gasoline (202 tons as compared with the planned 180). Ninth Army consumed both ammunition and Class II and IV supplies at a higher rate than was normal for rapid advance conditions (100 and 90 tons respectively), and correspondingly smaller quantities of gasoline (153 tons). For the three armies in the 12th Army Group, over-all maintenance rates averaged 455 tons per day as compared with the planned 450. Seventh Army's consumption record deviated most markedly from the planning figures, showing ammunition expenditures (171 tons) more nearly equal to normal combat scales (180 tons) and gasoline consumption at rates in excess of rapid advance scales (192 tons as against 180). Its over-all consumption rate was 513 tons per division slice as compared with 455 tons for the armies of the 12th Army Group. In the last month the average field strength of the two army groups, including the First French Army, was 1,525,700 men, and the average consumption per man was 30.38 pounds per day.[55]

The supply performance of the last three months of hostilities clearly demonstrated the greater maturity of the

[53] Of the supplies consumed in the combat zone under conditions of normal combat (541 tons) about 40 percent is consumed by the divisions themselves (223 tons as against 318 consumed by corps and army overheads). The Communications Zone accounts for another 201 tons, which raises the total theater slice (exclusive of air forces) to 742 tons per division. In other words, of the 742 tons per theater slice, the division accounts for 223 tons, or 31 percent, and overheads for 519 tons, or 69 percent. Under conditions of rapid advance the corresponding factors are 178 tons for the division itself, 284 tons for corps and army overheads, and 231 for COMZ overheads, and a total theater slice of 693 tons. Under conditions of regrouping or negotiating natural obstacles the factors are 166 tons for the division, 260 tons for corps and army overheads, and 201 for COMZ overheads, and a total theater slice of 627 tons.

[54] Operations in the period 24 February–23 March 1945 had revealed the following average consumption factors: Class I, 81.4 tons; Classes II and IV, 92.8; Class III, 123.9; Class V, 123.8, for a total of 422 tons per division slice per day. The factor of 450 tons represented an estimate which assumed slightly different conditions in the final drive. Consumption Rates of U.S. Forces in the Final Advance to the Rhine, 24 Feb to 23 Mar 45, prep by Statistical Sec G–4 SHAEF, 25 Apr 45, SHAEF G–4, G–4 Basic Statistical Reports 102/3/22.

[55] Study by Statistical Sec G–4 SHAEF, Consumption Rates U.S. Forces from the Rhine to the Elbe, 23 Mar–25 Apr 45, SHAEF G–4 Basic Statistical Reports 102/3/22.

TABLE 12—COMBAT ZONE CONSUMPTION EXPERIENCE
23 MARCH–25 APRIL 1945

[Long Tons per Divisional Slice]

Class	Average	NUSA	FUSA	TUSA	SUSA	FFA
Total	445.45	490.0	429.7	446.7	512.9	267.3
I	87.75	118.1	70.9	79.3	94.7	70.3
II and V	77.0	107.0	78.0	84.0	55.0	46.0
III	175.0	153.0	184.0	202.0	192.0	113.0
V	105.7	111.9	96.8	81.4	171.2	38.0

Source: Study by Statistical Sec G–4 SHAEF, Consumption Rates U.S. Forces from the Rhine to the Elbe, 23 Mar–25 Apr 45, SHAEF G–4 Basic Statistical Reports 102/3/22.

theater's supply organization and a tremendous improvement in the theater's over-all logistic potential. In the field of movements and distribution, in particular, the last three months contrasted markedly with the summer and fall of 1944. The period of relatively static operations had, of course, given the Communications Zone an opportunity to improve its operating procedures and to build the basic capacity in port discharge and better transportation. Eight months of operations had also given supply officers the experience which now was reflected in the greater self-confidence and expertness with which the final drive was planned and carried out. This was evident in all aspects of logistic support in the last few months, including marshaling resources, building the Rhine bridges, using motor and air transport, and extending rail service.

CHAPTER XVI

Supply in the Last Months

(1) Rations, POL, and Coal

In addition to being able to move huge tonnages in the last months, the logistic organization was in a much better position to meet the demand for specific items of supply. The U.S. Communications Zone was supporting 3,675,000 troops, plus 1,560,000 prisoners of war as hostilities came to an end early in May 1945.[1] On the whole, supply of these forces was better than it had been at any time since the beginning of the pursuit in the summer of 1944.

The supply of rations and POL presented no outstanding difficulties other than those normally accompanying highly mobile operations. The theater's ration position was good at the beginning of February. COMZ depots at that time held about 25 days of supply for all forces in the theater, and the armies held additional stocks ranging from 4 days in the case of the First Army to 9 days in the case of the Ninth. These levels represented approximately 400,000 tons in the Communications Zone (including some 30,000 in the United Kingdom) and about 23,000 tons in the combat zone. Nearly 140,000 tons, or 35 percent of the COMZ stocks, were well forward in the two advance sections.[2]

Between 80 and 90 percent of the issues in the combat zone continued to be of the bulk-type ration—either the A or B, depending on the availability of perishable items such as fresh meat, butter, vegetables, and fruit. A shortage of dehydrated yeast, caused by production difficulties in the United States, for a time threatened to limit the issue of fresh bread, but this deficiency was eventually resolved through local procurement. Meanwhile newly developed operational rations began to appear at the front. The Quartermaster Corps had made changes in all operational types—C's, K's, and 10-in-1's. The most welcome improvements came in the C ration, which, in addition to its original meat combinations, eventually included a variety of new ones such as meat and spaghetti; ham, eggs, and potatoes; meat and noodles; pork and rice; pork and beans; frankfurters and beans; ham and lima beans; and chicken and vegetables. Unfortunately the new combinations did

[1] The troop total included sixty-one U.S. and eleven French divisions. French forces supported by the United States totaled about 350,000, most of them in the 6th Army Group.

[2] Ltr, Page to CG COMZ, 17 Feb 45, sub: Estimate of COMZ Supply Situation (Rpt 1), EUCOM 381 Projected Operations of the COMZ ETOUSA, 14 Mar 45.

not appear in quantity before the end of hostilities.[3]

Theater ration levels continued to rise for several weeks, and by the time of the Rhine crossing were equivalent to forty-two days of supply. Army levels were somewhat erratic. Fluctuations in the army levels did not necessarily indicate a deterioration in the supply position. They reflected the changing tactical situation and, at times, the transfer of stocks to the Advance Section, as was the case in the period of rapid movement in March. Or, as was the case later, they resulted from the decision to defer receipts until direct rail shipments could be made to new depots. Sometimes they resulted from an increase in the ration strength without a corresponding build-up of reserves. At the time of the Rhine crossings reserves in the combat zone were excellent, totaling 31,365 tons and averaging five days of supply.[4]

Deliveries were good in the week after the crossings, and all the armies proceeded to build up a reserve east of the river. By the second week of April, however, army levels had begun to feel the effect of the lengthening lines of communications. Meanwhile the Advance Section turned over its large depot stocks to other COMZ sections and thereafter moved forward in close support of the armies, acting mainly as transporter and transfer agent and retaining no reserves under its own control. Reserves east of the Rhine consequently were never very great in terms of tonnages. Ration levels in the armies rose slightly in the third week, to 2.3 days in the First Army, 4.3 in the Third, 4.4 in the Ninth, and 4.1 in the Seventh, but the 12th Army Group considered Class I supply increasingly critical because of the tightness of transportation. In the final weeks the ration problem was aggravated by the necessity to feed large numbers of prisoners and displaced persons, although prisoners subsisted largely on captured stocks.

The rapid advance also brought the inevitable shift from bulk to operational rations, which were supplemented whenever possible with fresh bread, butter, and meat. The changeover was greatest in the First and Third Armies, in which operational rations accounted for 70 to 80 percent of all issues as compared with 10 to 15 percent in preceding months. In the theater as a whole operational rations comprised 26 percent of the total issues in April. Generally, while Class I levels in the combat zone threatened to become critical late in April, they never reached the precarious lows of August and September 1944, and U.S. forces did not suffer any want for subsistence. In the last week of operations no army had less than three days of supply on hand, and the First Army had built up a seven-day supply. Ration consumption in the final month had ranged from 71 tons per division slice in the First Army to 118 in the Ninth, and had averaged 87.75 tons as compared with the planned 80.[5]

POL supply was generally excellent

[3] Army and Army Group G–4 Periodic Reports for February 1945; QM Supply in the ETO, II (Subsistence), App. XXXI. For more detail on rations see Romanas *et al.*, Quartermaster Operations in the War Against Germany.

[4] Ltr, Page to CG COMZ, 11 Apr 45, sub: Estimate of COMZ Supply Situation (Rpt 3), SHAEF G–4 Estimate of Supply Situation Reports—COMZONE, No. 59.

[5] Army and Army Group G–4 Periodic Rpts for March and April; Report on Consumption Rates U.S. Forces From the Rhine to the Elbe, 23 Mar–25

throughout the last three months of operations. Reserve levels in the forward areas dropped somewhat in the final days of hostilities, as could be expected during a rapid advance, but shortages never actually caused the final drive to falter.

POL stocks had reached an unprecedented 560,000 tons in the Communications Zone by February, far exceeding the authorized sixty-day level. Combat zone depots held an additional 35,000 tons, with levels varying from 2.5 days in the First Army to 8.3 days in the Third. POL was being discharged at the rate of about 13,000 tons per day, mainly at Antwerp, Cherbourg, Le Havre, and Port du Bouc.[6]

Although the three main pipeline systems at this time operated only as far as Maastricht, Coubert, and Epinal, additional construction was then under way on both the Major and Southern Systems. (*Map 11*) On the Northern System, based on Antwerp, Maastricht remained the forward terminus until late in March. Extension of the system north-northeastward to Wesel was finally undertaken on 3 March and completed and in operation by the 28th. Construction difficulties were numerous on this section of the system. On the initial stretch from Maastricht to Sittard construction was slowed by deep mud, which forced engineers to hand-carry much of the pipe, by a shortage of 6-inch couplings, and by the necessity to weld and bury large sections of pipe in the towns. Beyond Sittard the right of way was found to be heavily mined. Finally, at Roermond, floods made it necessary to suspend the pipelines over the Roer by cables.

Extension of the Major System had been resumed at the end of January, and within a month all three 6-inch lines had reached Châlons-sur-Marne. In March engineers laid one 6-inch line beyond that city to Thionville.

Construction on the Southern System meanwhile had advanced the 4-inch line to Sarrebourg by mid-February. This brought a noticeable improvement in Seventh Army's gasoline supply. The 6-inch line had only reached St. Jean de Losne by that time and did not extend to Sarrebourg for another two months. Because of a shortage of 6-inch pipe the Southern System beyond that point took the form of three 4-inch lines. One of these was completed to Saaralbe by late March. Late in February all construction and operation of the Southern System had been taken over by the Military Pipeline Service. At the time of the Rhine crossings, therefore, the most advanced pipeheads of the three main pipeline systems were at Maastricht, Thionville, and Saaralbe.[7]

In preparing for the final offensive to be launched late in March logistic planners estimated POL requirements for sixty U.S. and French divisions at approximately 10,300 tons per day. To meet this requirement they planned eventually to extend all three pipeline

Apr 45, prep by Statistical Sec G–4 SHAEF, 11 May 45, 12 A Gp, Supplies—Backlog and Requisitions, No. 128.

[6] Ltr, Page to CG CZ, 17 Feb 45, sub: Estimate of COMZ Supply Situation (Rpt 1), SHAEF G–4 Estimate of Supply Situation Reports—COMZONE, No. 59; 12 A Gp G–4 Periodic Rpt for Week 4–10 Feb 45.

[7] Petroleum, Oil, and Lubricants, Hist Rpt 13, OCE ETO, pp. 94–105; Monthly Rpts, Construction Div OCE ETO, ADSEC 51–104; Seventh Army Rpt of Opns, III, 880.

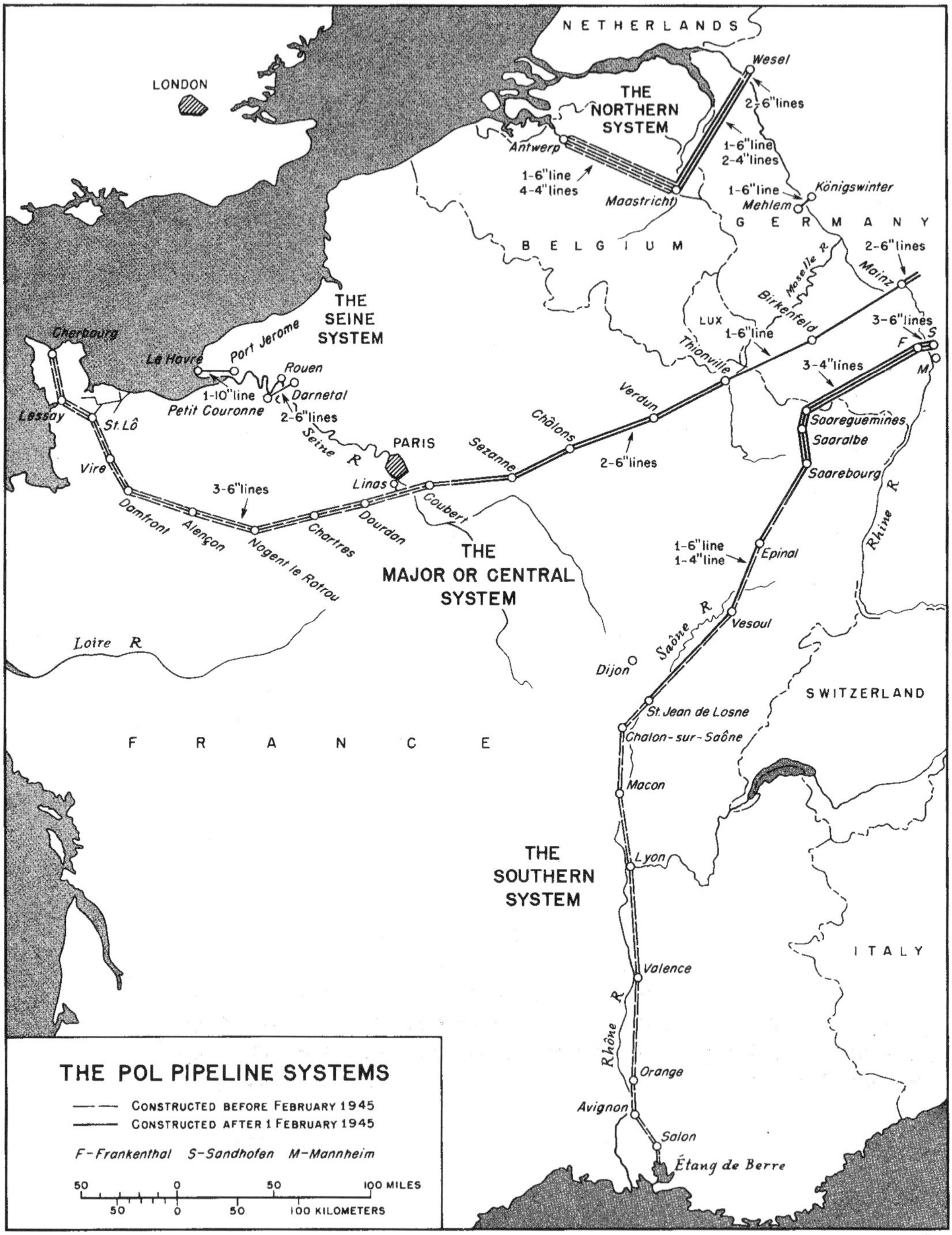
THE POL PIPELINE SYSTEMS
CONSTRUCTED BEFORE FEBRUARY 1945
CONSTRUCTED AFTER 1 FEBRUARY 1945
F-Frankenthal S-Sandhofen M-Mannheim
50 0 50 100 MILES
50 0 50 100 KILOMETERS
THE NORTHERN SYSTEM
THE SEINE SYSTEM
THE MAJOR OR CENTRAL SYSTEM
THE SOUTHERN SYSTEM
NETHERLANDS
BELGIUM
GERMANY
LUX
FRANCE
SWITZERLAND
ITALY
LONDON
PARIS
Wesel
Antwerp
Maastricht
Königswinter
Mehlem
Mainz
Birkenfeld
Thionville
Verdun
Châlons
Sezanne
Coubert
Linas
Dourdan
Chartres
Nogent le Rotrou
Alençon
Domfront
Vire
St. Lô
Lessay
Cherbourg
Le Havre
Port Jerome
Rouen
Darnetal
Petit Couronne
Saareguemines
Saaralbe
Saarebourg
Epinal
Vesoul
Dijon
St. Jean de Losne
Chalon-sur-Saône
Macon
Lyon
Valence
Orange
Avignon
Salon
Étang de Berre
Seine R
Loire R
Moselle R
Saône R
Rhine R
Rhône R
2-6"lines
1-6"line
2-4"lines
4-4"lines
3-6"lines
3-4"lines
1-10"line
1-4"line
D. Holmes, Jr.

MAP 11

systems across the Rhine—at Wesel in the north, at Mainz in the center, and at Worms in the south. Motor transport would of course have to deliver packaged POL to the Rhine bridgeheads in the first few days. To lessen the burden which this would impose on the Rhine highway bridges supply officers planned the construction of pipeline crossings at the above-named locations and also in the vicinity of Remagen before the links with the main pipeline systems were completed. Gasoline could then be shipped to all four crossings via tank car and pumped across to the east bank, thus alleviating congestion on the bridges.[8]

Such pipeline crossings were eventually constructed at all four locations. Engineers started the one at Mehlem, near Remagen, on 25 March and completed it three days later, using a footbridge to support the line. Gasoline began flowing across the Rhine before the end of the month, the western terminal being supplied from Maastricht first by tank truck and later by tank car. On the east bank the pipeline was eventually extended ten miles to the Autobahn, where storage tanks were erected and whence gasoline was shipped in tank trucks to Giessen and there decanted.

Construction of a similar crossing at Mainz could not begin until 30 March because of the tactical situation. The east bank pipehead there was completed on 8 April, two weeks before the line from Thionville reached the river. It began operating a few days later, carrying gasoline brought to the west bank terminal by rail from Thionville. The line was used only temporarily, for it had been laid on a wrecked rail bridge and proved to be unsafe. Two new lines were subsequently laid across the newly constructed rail bridge.

Farther south, in the Seventh Army area, engineers constructed three lines across the Rhine at Frankenthal, a few miles south of Worms, and established the east bank pipehead at Sandhofen, a few miles north of Mannheim. Construction of the first line, supported by a Bailey bridge, began on 7 April and the line started carrying gas across the Rhine on the 15th. Two additional lines were constructed later, one of which was laid on the river bed.

Construction of the Wesel crossing had begun on 21 March and was completed in the first week of April. But this line never operated independently of the main pipeline. By the time it was installed, the lines from Maastricht had also reached the Rhine, the first gasoline arriving at the tank farm on the west bank on 28 March. The two systems were immediately linked, and on 3 April gasoline began flowing across the Rhine to Wesel. Consequently gasoline did not have to be shipped to the west bank by rail, but could be pumped through the pipelines directly from Antwerp. The completed Northern System consisted of one 6-inch and four 4-inch lines from Antwerp to Maastricht, and one 6-inch and two 4-inch lines beyond that city to Wesel.[9]

---

[8] Memo, Petroleum Br G–4 SHAEF, 26 Mar 45, sub: Future Opns—POL Plan, SHAEF G–4 463.7 Gasoline and Motor Oil (Gen) 1945, II.

[9] Memo, Gilland for Crawford, 31 Mar 45, sub: Current POL Supply Program, SHAEF G–4 463.7 Gasoline and Motor Oil (Gen) 1945, II; Comd and Stf Conf Notes, COMZ, 5 May 45, EUCOM 337/3 Confs, Staff Weekly, I; FUSA Rpt of Opns, 22 Feb–8 Mar 45, Bk. III, p. 49; COMZ G–4 History, V, 28.

Work on both the Major and Southern Systems also went forward. On the Major System only one 6-inch line was laid to Mainz, arriving there on 22 April. A second line between Châlons and Thionville was laid by early May, completing the system. All the lines on the Major (or Central) System were 6-inch lines. One was extended to Mainz, two as far as Thionville, and three to Châlons. On the Southern System three 4-inch lines were eventually constructed forward from Saaralbe, the first arriving at Frankenthal on 20 April, the other two at the end of the month. By late April, therefore, gasoline was being pumped directly from the ports to pipeheads on the east bank of the Rhine—from Antwerp to Wesel in support of the Ninth Army, from Cherbourg to Mainz in support of the Third, and from Port du Bouc to Frankenthal in support of the Seventh. A fourth line, served by rail car, bridged the Rhine at Mehlem in support of the First Army. Delivery capacities at the pipeheads were 2,500 long tons per day at Wesel, 1,500 tons at Mainz, and 1,200 tons at Sandhofen, opposite Frankenthal. An additional 4,800 tons could be delivered to Maastricht, Thionville, Châlons, and Saarbourg. On its completion the continental network—including the Seine and Minor Systems, the latter of which was no longer in operation—had a route distance of 1,412 miles and had 3,577 miles of pipe. Storage facilities totaled 7,619,116 barrels.[10]

At the time of the Rhine crossings late in March both COMZ and army stocks of POL were the highest they had ever been. The Communications Zone at the time had nearly 600,000 tons of Class III supplies in its depots, of which the packaged portions alone—262,000 tons—constituted 22.8 days of supply. Army depots held another 65,000 tons, with levels ranging from 5 days of supply in the First Army to 9 days in the Ninth.[11] Supplying the armies with POL was therefore purely a transportation problem, and the attention which the planners had given to that problem paid good dividends. Deliveries were fully adequate in the first weeks. In the second week of April the armies had an average of six days of supply on hand, and each had succeeded in moving at least 1,000,000 gallons into forward areas east of the Rhine. The Ninth Army, which had a total of 4,500,000 gallons or 8.2 days of supply on hand, already enjoyed the benefits of the Wesel pipehead, and was building up a reserve east of the Rhine at a good rate. Daily issues of gasoline rose to between 600,000 and 700,000 gallons in each of the armies by mid-April, and on occasion exceeded 1,000,000 gallons. Deliveries came close to matching these requirements throughout most of the month. Much of the gasoline moved forward from the pipeheads in tank trucks and was decanted by the Advance Section in the forward areas. But a large portion also moved via tank car from the pipeheads west of the Rhine. Theater stocks had continued to rise, and totaled nearly 700,000 tons on 25 April. The armies at that time still had reserves totaling 47,000 tons.

Deliveries finally began to fall short of

[10] Monthly Rpts, Construction Div OCE ETO, ADSEC 51–104; Final Rpt of the Chief Engr, ETO, II, App. 33A–C.

[11] Ltr, Page to CG CZ, 11 Apr 45, sub: Estimate of COMZ Supply Situation (Rpt 3), SHAEF G–4 Estimate of Supply Situation Reports—COMZONE.

C–47 TRANSPORT PLANES BRINGING IN POL *for Third Army, 31 March 1945.*

daily consumption in the last days of April, and levels in the combat zone consequently dropped off. Air deliveries had been a major factor in meeting daily requirements until that time, but these also fell off at the end of the month. All the armies felt the effect of the increasing difficulties in transportation, and in the Seventh and Third Armies reserve levels for the first time fell below two days of supply. Third Army considered its POL situation critical enough to impose rationing for the last three days of operations. But POL shortages did not delay the final surrender.[12]

Despite the fairly consistent intensity of offensive activity in the last three months of operations, POL consumption experience brought a downward adjustment in the factor of 192 tons per division slice which had been adopted in January. Consumption in March had actually averaged barely 124 tons per division slice in the U.S.-supported armies, due mainly to low consumption in the Ninth U.S. and First French Armies. Logistic planners had ignored this statistic in estimating the requirements for the final drive, using the factor of 180 tons, which the experience of 1944 in-

[12] Army and Army Group G–4 Periodic Rpts for April; Memos for Record, Maj. James R. Howton, Supply Br 12 A Gp, reporting on visits to army quartermasters to survey Class I and III situation, 5 and 12 Apr 45, and Memos for Record, Hopkins, 12, 21, and 26 Apr 45, all in SHAEF 12 A Gp G–4 Memos for Record, File 77; Cbl C–3167, TUSA to SHAEF, 28 Apr 45, SHAEF G–4 463.7 Gasoline and Motor Oil (Gen) 1945, I; TUSA AAR, II QM, 26; FUSA Rpt of Opns, 22 Feb–8 May 45, Bk. III, p. 49.

dicated would apply in mobile operations. Consumption in the combat zone actually averaged 175 tons in the final month.[13]

The supply of coal saw some improvement in the final months, but remained unsatisfactory until the end of hostilities. Neither imports nor indigenous production had been satisfactory. The import program, which called for shipments of more than 200,000 tons per month, had fallen behind for a variety of reasons, including shortages of coal in the United Kingdom, bad channel weather, a lack of shipping, and shortages of rail cars on the Continent. Meanwhile, indigenous production faced endless difficulties. SHAEF had taken steps to prime local production, including the acquisition of mine supplies and equipment from the United States, and had established close liaison with national authorities through its Solid Fuels Section, which had a staff of mining experts.[14] But the industry of France, Belgium, and Holland was slow to recover, suffering from many shortages, particularly pit timber, from severe winter weather, which closed the canals, and from labor unrest resulting from bad living conditions. The most critical stage was reached in January, when a two-week strike by Belgian miners cut deeply into the meager reserves of coal and brought about the fall of the Belgian Government.[15]

The coal situation was aggravated from the start by the necessity to provide the minimum needs of the French economy, which was experiencing extreme hardship. Military requirements obviously had to take priority, and the French demands for greater concessions to the civil economy, including greater imports and priority for coal movements over purely military movements, had to be denied. The Communications Zone's requested allocations from French production, averaging 170,000 tons in February and March, were consistently pared down by SHAEF. Meanwhile, French efforts to restore coal production were disappointing; in February the French delivered only 58 percent of the tonnage allocated. Relations with the Allies over the matter of coal allocations became strained in April when the French threatened to withhold coal from U.S. and British forces. The COMZ reaction was a proposal to stop the flow of imports of raw materials for local industry and to end the allocation of POL to the French.[16]

[13] Included in this average is the low consumption experience of the First French Army, in which only 113.6 tons per division slice per day were consumed. Average consumption in the four U.S. armies was 183 tons. Consumption Rates of U.S. Forces in the Final Advance to the Rhine, 24 Feb–23 Mar 45, dated 25 Apr 45, and Consumption Rates U.S. Forces From the Rhine to the Elbe, 23 Mar–25 Apr 45, dated 11 May 45, both prep by Statistical Sec G–4 SHAEF, SHAEF G–4, G–4 Basic Statistical Rpts, 102/3/22.

[14] See also below Chap. XVIII, Sec. 3.

[15] Cbl G–4 SF–127, Belgian Solid Fuels Sub-Section to SHAEF, 7 Feb 45, ETO Adm 400; Min, CAO Mtgs, 26 Jan and 2 Feb 45, SHAEF AG 337–2 CAO Mtgs; Ltr, Crawford to CofS, 4 Mar 45, sub: Coal Production, SHAEF SGS 463.3 Coal Supply, II; Cbl FWD–17879, SHAEF to AGWAR, 15 Mar 45, App. to Current Opns Br, SHAEF G–4 War Diary/Jnl; Ltr, Plank to CG COMZ, 21 Jan 45, and 1st Ind, COMZ to ADSEC, 15 Feb 45, and Memos, Stearms for CG COMZ, 1 Feb 45, sub: Solid Fuel Situation, Belgium, Stearns for CG COMZ, 5 Feb 45, sub: Coal Situation, Belgium, and Stearns for Lord, 10 Feb 45, sub: Coal Situation, Belgium, all in EUCOM 463.3 Coal 1944–45, II.

[16] Ltr, Minister of Industrial Production to SAC, 2 Feb 45, sub: Coal Situation, and Memo, Gale for CofS, 8 Feb 45, sub: Coal Situation in France,

Some improvement in coal imports had occurred in February, when shipments to the Continent totaled 226,000 tons, equal to 90 percent of the target. Some of the difficulties which had hampered local production also eventually began to clear up. By mid-February, for example, a two and one-half months' stock of pit timber had been built up at the Nord and Pas-de-Calais mines. The supply of coal was far from ample, however, and military authorities tightened the conservation program urging greater use of wood, reducing the allowances for bathing and for hospitals, and ordering the suspension of space heating on 1 April.[17]

Allied authorities meanwhile had made plans to restore captured German mines to production. Early in December 1944 SHAEF had directed the 12th Army Group to occupy all mines uncovered in its area and to safeguard and maintain the mines and their equipment pending the resumption of operations, and offered technical assistance in this mission. Shortly thereafter it made the Communications Zone responsible for the actual operation of the mines, although it contemplated only a supervisory role for the military and specified that both management and labor would be German. The Communications Zone subsequently delegated this responsibility to the Advance Section, which immediately drew up plans for opening the mines and asked for the necessary technicians, supervisory units, and equipment. Not until late in March, however, did SHAEF actually designate the districts and mines to be operated, or begin to provide the necessary technical assistance. The Advance Section meanwhile organized a Coal Mining Division within its Engineer Section to serve as a staff supervisory agency, and a field organization known as the Rhine Coal Control to supervise and oversee actual operations at the mines. The necessary troop units were acquired very gradually, and by V-E Day consisted mainly of the Headquarters, 6th Engineer Special Brigade, augmented with various U.S. and Allied technicians, an engineer general service regiment, and a British Pioneer Group, plus various signal, transportation, and quartermaster units. The Continental Advance Section meanwhile similarly organized an Engineer Operating Group to reopen the Saar mines. In both the Saar and the Aachen–Ruhr areas these various efforts were just beginning to bear fruit as hostilities came to an end early in May.[18]

(2) Ammunition

The supply of field artillery ammunition improved only temporarily as a re-

SHAEF SGS 463.3 Coal Supply, I; Memo, COMZ G–4 for DCofS, 20 Mar 45, Ltr, Brig Gen Wayne R. Allen, GPA, to CofS COMZ, 1 May 45, sub: Coal, and Ltr, Larkin to SAC, 3 May 45, sub: Coal in France, all in EUCOM 463.3 Coal 1944–45, II.

[17] Min, CAO Mtg, 9 Mar 45; Ltr, Crawford to SHAEF Mission to France, 13 Feb 45, sub: Coal Situation in France, SHAEF SGS 463.3 Coal Supply, II; Memo, Broshous for G–4, 10 Mar 45, and Memo, G–4 COMZ for DCofS, 20 Mar 45, sub: Conservation of Coal, EUCOM 463.3 Coal 1944–45, II.

[18] Ltr, COMZ to ADSEC and Chief Engr COMZ, 29 Dec 44, sub: Opn of German Coal Mines, Ltrs, COMZ to SAC, 23 Jan and 16 Feb 45, same sub, and Cbl EX–25505, COMZ to ADSEC, 26 Mar 45, all in EUCOM 463.3 Coal 1944–45, II: ADSEC Operations History, pp. 166–73, 177–78; CONAD History, p. 244; COMZ G–4 History, V, 8.

sult of the Bull mission in December, and, for the field commanders at least, remained a cause for anxiety until after the Rhine crossings late in March. Both army groups complained of inadequate receipts early in February. The 6th Army Group had made heavy expenditures during January, caused in part by the Seventh Army's defensive operations and in part by the elimination of the Colmar Pocket, and asked for additional credits. But SHAEF answered that it could not authorize additional allocations in view of expected receipts from the U.S., and refused to release ammunition from its own reserve, which had not even reached the specified seven-day level and in any event was intended only for operational emergencies. The 6th Army Group consequently had no choice but to effect savings from its own resources, and for a time limited expenditures to half the authorized maintenance rates in order to rebuild its reserves.[19]

The 12th Army Group reported that its receipts had not even equaled the established SHAEF maintenance day of supply rates, and that consequently there were insufficient quantities of most mortar and artillery ammunition items available in forward depots to meet the command's accrued allowances. Deficiencies came to 31 percent in the case of 60-mm. mortar ammunition, 22 percent in the case of 8-inch howitzer ammunition, 39 percent in that of 240-mm. howitzer ammunition, and 84 percent in that of 4.5-inch gun ammunition. Only by strict rationing, it noted, had it been possible to establish a small reserve.[20]

These deficiencies could be attributed only partially to inadequate receipts from the United States. Shipments in some categories had actually exceeded the estimates provided General Bull, and theater stocks of most of the major types of artillery ammunition actually were at better levels at the end of January than at the end of December. The deficiencies could be traced in part to SHAEF's decision deliberately to withhold at least a portion of the receipts from the United States in order to build up working margins to levels recently authorized—that is, twenty-seven days in the army groups and seven days in SHAEF at current SHAEF maintenance rates of supply. At the end of January, however, the ASF admitted that shortfalls in December production of some of the heavier calibers—specifically 155-mm. gun, 8-inch howitzer, and 240-mm. howitzer ammunition—had necessitated revisions in its January allocation of those types. A few weeks later the Communications Zone pointed out that the quantities released for January loading had not been shipped as scheduled, and that releases for February and March shipment were not living up to expectations either.[21]

---

[19] TWX BX-23700, 6 A Gp to SHAEF, 30 Jan 45, and TWX S-77391, SHAEF to 6 A Gp, 1 Feb 45, Cbl Log, Smith Papers; Seventh Army Rpt of Opns, III, 882-83.

[20] TWX QX-26875, 12 A Gp to SHAEF, 10 Feb 45, and Ltr, 12 A Gp to CG COMZ, 11 Feb 45, sub: Ammo Status for Month of Jan, EUCOM 471 Allocation of Ammunition, III; Memo, Holtzworth for G-4, 19 Feb 45, sub: Ammo Forecast 19 Feb to 1 Mar 45, 12 A Gp 471/1 Ammunition Allocations.

[21] TWX Conf Between SHAEF and WD, 16 Mar 45, 12 A Gp 471 Ammunition General; Cbl WARX—29114, ASF to SHAEF, 30 Jan 45, SS&P Planning Div 201.02 Ammunition, A46-371; Memo, Styer for Lutes *et al.*, 13 Feb 45, sub: Production, ASF 200.02 Gen Somervell's Inspection Trip Eto, 333;

These developments were bound to play hob with theater maintenance supply rates—the rates at which the Communications Zone committed itself to deliver ammunition to the armies—and in turn with the tactical plans of the armies. The field commands had long desired guaranteed rates on which they could base tactical planning. Under current practice the Communications Zone computed its projected deliveries on the basis of stocks actually on hand plus War Department releases for future delivery. The War Department, in attempting to satisfy urgent theater demands, in turn based its shipping forecasts on estimated production, making no allowance for shortfalls in manufacture or unforeseen transportation difficulties. Since all critical items were shipped directly from assembly lines, any lag in production had an immediate effect on loading and shipping schedules. In this way shortfalls in December production had forced a cutback in January allocations, with inevitable repercussions on shipments to the European theater. The War Department's inability to maintain its December level of releases forced the Communications Zone to revise the maintenance day of supply in the theater, and had brought the expected protests.[22]

Once again experience had shown that there was many a slip twixt production forecasts and actual delivery at the front. No production or shipping schedule was completely reliable. Furthermore, the Communications Zone had found that, contrary to earlier assumptions, it could not predicate ammunition maintenance rates on the expectation that express shipments would make ammunition available in the forward depots in the month after its production. Unpredictable delays in loading, in convoy sailings, and in port discharge, or delays resulting from the necessity to segregate lots or arising from the manner of loading made it unwise to count on ammunition becoming available until the second month after its production. The lesson in this recent experience was but a variant of the one learned in October, when it was found that ammunition lying in ships off the continental ports could not safely be considered as part of COMZ depot stocks and therefore available for issue.

Early in March the theater submitted recommendations for new day of supply rates to the War Department, calling for reductions in some items which had been in relatively good supply, but for higher rates than ever for heavy artillery ammunition. In justification for the upward revisions it argued that the use of air observation posts had improved long-range observation and thus made the larger caliber weapons more effective in all types of combat, making it possible to strike deeper into the enemy's defensive zone against his command posts, supply installations, and so on. In addition, improved tactical mobility of the heavy caliber weapons and proficiency in firing had increased the employment of these weapons and increased the number of days they could be in action.[23] This re-

Cbl EX-98943, COMZ to WD, 15 Feb 45, SHAEF AG 471-1 Ammunition Status Reports; TWX E-13717, COMZ G-4 to SHAEF, 26 Feb 45, SHAEF G-3 O&E Ammunition.

[22] Cbl WX-40036, AGWAR to COMZ, 19 Feb 45, SHAEF AG 471-1 Ammunition Status Reports; TWX E-13717, COMZ G-4 to SHAEF, 26 Feb 45.

[23] Ltr, CG ETO to WD, 3 Mar 45, sub: Ammo Day of Supply, ADSEC 381B Day of Supply and Unit of Fire.

quest ran directly counter to current thinking in the War Department, which desired reductions rather than increases.

Shortly thereafter the War Department asked the theater for a careful review of its long-range ammunition needs. The War Department at this time faced a final decision as to whether it should drive ahead on the expansion program or hold productive capacity in line with what it believed would be reduced requirements in the European theater and, before long, with the requirements of a one-front war in the Pacific. The expansion of production facilities promised to cut deeply into the amounts of steel needed for other military purposes, particularly for tanks and trucks. Moreover, the War Department, which had been criticized in December for the ammunition shortage, was being subjected to increasing criticism from both Congress and industry for what was regarded as an unwarranted expansion of production facilities for ammunition.

In any event, much water had flowed over the dam since December, and General Lutes felt that the combat experience of recent months should have provided the theater with far more realistic data on which to base future ammunition requirements. The War Department's own study of ETOUSA expenditure reports for the period November through February appeared to substantiate the belief that the theater's requests were due for radical revision, for it indicated that actual expenditures in the critical types nowhere near approximated the figures presented by the Bull mission in December. In the case of 8-inch howitzer ammunition, for example, expenditures had come to only 22.2 percent of the estimated figures; for 240-mm. howitzer ammunition they had come to only 24 percent, for 155-mm. gun ammunition they were 51.7 percent, for 155-mm. howitzer ammunition 51.8 percent, and for 105-mm. howitzer 59.9 percent. According to War Department calculations the theater already had eight months of fire in sight in the 105-mm. category.

As usual, the theater challenged these figures. The War Department, as theater officials pointed out, had overlooked the fact that ammunition had been made available to the army groups at restricted rates in January and February so that reserves could be replenished. Expenditures consequently bore no relationship to the estimates made in December nor to the releases and shipments made by the War Department. Despite the many exchanges and visits, it appeared as late as March that the War Department and the theater had come little nearer speaking the same language on ammunition, for the theater continued to state its requirements in terms of *desired expenditures* and the War Department continued to argue from expenditure statistics representing *restricted firing*. Needless to say, ETOUSA officials were exasperated to find that the War Department was continuing to assess the theater's requisitions on the basis of rationed expenditures. In the opinion of one SHAEF staff officer, the basic difficulty in all the negotiations over the ammunition problem lay in the fact that the War Department had never really accepted the theater's increased expenditure rates.[24]

[24] TWX Conf Between SHAEF and WD, 16 Mar 45, 12 A Gp 471 Ammunition General; Ltr, Lutes to Larkin, 20 Mar 45, SHAEF G–3 O&E 471 Ammunition.

On the other hand the tactical outlook in the theater had changed substantially since the critical days of December and January, when the army groups had estimated their needs for breaching the Siegfried Line on the assumption that their attacks would continue to meet determined resistance. The request in March that ETOUSA take another look at its long-range ammunition needs proved fully justified by events. In accordance with that request, SHAEF on 18 March asked both army groups to reconsider their long-range needs, recognizing that supply in the past had not been all that was desired, but emphasizing the desirability of reducing requirements in view of the more optimistic tactical outlook.[25]

Both army groups were understandably reluctant to volunteer reductions in their estimates in light of the persistent deficits of past months. Both groups, moreover, were about to cross the Rhine, and were naturally hesistant to predict what the scale of resistance would be beyond that obstacle. Both commands again cited the handicap under which they had long operated. General Devers pointed out that his forces repeatedly had been forced to conserve ammunition for varying periods in order to permit proper support for major offensive operations. He noted that supply had rarely been adequate to permit the desired freedom of action to both armies simultaneously. Forced economies had persistently handicapped his forces, and for five weeks after the elimination of the Colmar Pocket 6th Army Group had restricted expenditures to about half the SHAEF maintenance rate so that it would have adequate stocks upon closing to the Rhine. General Devers asked that current rates not be reduced.[26]

General Bradley's response likewise emphasized the extent to which expenditures in the 12th Army Group had been curtailed by nonavailability and the necessity to create minimum reserves, and the extent to which his command had been forced to supplement its meager supply with captured ammunition and with ammunition and guns borrowed from the British.[27] The 12th Army Group commander volunteered certain reductions, but he repeated the point which the theater had emphasized again and again in its communications with the War Department, that past expenditures were no criterion for estimating future needs. Rationing, as one of his artillery officers noted, had constantly been in the background of all firing, and had made it impossible to say what ammunition expenditures might have been had they not been restricted.[28]

The limited availability of the major types of ammunition had in fact made it impossible to determine precisely what quantities the theater actually required. The army groups had previously submitted desired expenditure rates in an-

[25] TWX FWD-17983, SHAEF to A Gps, 18 Mar 45, G-3 O&E 471 Ammunition.

[26] TWX BA-25381, 6 A Gp to SHAEF, 23 Mar 45, SHAEF AG 471-1 Ammunition Status Reports, No. 2.

[27] The 12th Army Group had borrowed 100 25-pounders and 300,000 rounds of ammunition from the British and since 1 December had fired about 75,000 rounds of captured field artillery and mortar ammunition.

[28] TWX QX-31452, 12 A Gp to SHAEF, 24 Mar 45, SHAEF AG 471-1 Ammunition Status Reports, No. 2; Memo, Hedekin of Arty Sec 12 A Gp for G-3, 22 Mar 45, sub: Ammo Rqmts, Note 2, 12 A Gp 471 Ammunition General.

TABLE 13—AMMUNITION DAY OF SUPPLY BY TYPE: SELECTED PLANNED RATES, SEPTEMBER 1944–APRIL 1945, AND ACTUAL EXPENDITURES, JUNE 1944–FEBRUARY 1945

[Rounds per Weapon per Day]

Planned and Actual	Mortar		Howitzer				Gun		
	60-mm.	81-mm.	105-mm.	155-mm.	8-in.	240-mm.	4.5-in.	155-mm.	8-in.
Day of Supply									
WD Established Rate 1 September 1944[a]	10.0	15.0	40.0	28.0	25.0	25.0	20.0	15.0	7.0
ETOUSA Recommendation 3 March 1945[b]	8.0	20.0	45.0	28.0	35.0	25.0	25.0	15.0	15.0
SHAEF Maintenance Rate March 1945[c]	10.0	6.2	23.5	22.4	17.5	14.1	12.9	16.8	7.5
12th A Group Desired Rate March 1945[c]	6.1	15.0	40.5	26.6	31.9	22.8	25.0	15.0	15.0
6th A Group Desired Rate March 1945[c]	6.7	15.0	46.5	38.0	37.2	31.5	30.0	20.0	25.0
ETOUSA Recommendation 1 April 1945[c]	6.4	15.0	40.5	26.6	32.6	22.8	25.0	15.0	15.0
ETOUSA Proposed Rate 10 April 1945[d]	5.0	18.0	41.0	26.0	25.0	22.2	25.0	15.0	13.0
WD Established Rate 30 April 1945[e]	5.0	18.0	41.0	26.0	25.0	22.2	25.0	15.0	13.0
Actual Expenditures									
June 1944–February 1945[c]	4.7	7.8	19.3	13.2	12.9	11.8	6.2	4.8	4.3

Sources: [a] Ltr, WD to ETO, 1 Sep 44, Sub: Day of Supply of Ammo, ADSEC 381B Day of Supply and Unit of Fire.

[b] Ltr, CG ETO to WD, 3 Mar 45, Sub: Ammo Day of Supply, ADSEC 381B Day of Supply and Unit of Fire.

[c] Ltr, Bull to CofS, 2 Apr 45, Sub: Determination of Ammo Rqmts for U.S. Forces, SHAEF G–3 O&E 471 Ammunition.

[d] Ltr, ETO to WD, 10 Apr 45, Sub: Ammo Expenditures, Levels, and Day of Supply, EUCOM 471 Ammunition (General), IV.

[e] Ltr, WD to ETO, 30 Apr 45, Sub: Day of Supply for Ammo ETO, SHAEF AG 481-1 Ammunition Status Reports (Day of Supply), No. 2.

ticipation of continued heavy resistance, and the theater had recommended a single rate, examples of which are shown in Table 13. At the end of March the Communications Zone and Supreme Headquarters tentatively agreed on somewhat lower maintenance rates. In effect, the new rates, also shown in Table 13, did not actually represent very substantial reductions. General Bull, the SHAEF G–3, estimated that only about half of the available divisions would be engaged in later months against moderate resistance, and that the forces actively engaged could therefore be supplied at the desired rates. Consequently, current rates of production would easily meet ETOUSA's needs.

Allied forces were already across the Rhine at the time the new rates were discussed, and there were positive indications at last that enemy resistance was diminishing. These considerations undoubtedly encouraged General Bradley to concur in the proposed reductions. On 3 April SHAEF notified the War Department that it considered current production rates adequate to support future operations and that it would not require the completion of additional manufacturing facilities.[29] On 10 April the theater followed this up with new recommended day of supply rates and estimated future requirements. With previous difficulties in mind, it emphasized that its estimates represented minimum requirements and asked that they be approved and shipped without delay even if they might in some instances appear excessive. Justification could be provided later.[30]

The April requisition proved academic so far as the outcome of operations in Europe was concerned, but it illustrated the theater's continuing apprehension over the War Department's response to its requests. Actually, the entire ammunition picture had begun to brighten by late February, and supply proved adequate for all purposes in the final offensives. Theater stocks continued to rise even during the February–March offensives which carried all the armies to the Rhine. First and Third Army levels remained practically unchanged in this period despite firing in excess of current maintenance rates. Seventh Army gradually rebuilt its reserves, benefiting from periods of relative quiet and excellent receipts. Early in March, for example, issues for a time declined to a low of 585 tons per day for the entire army. By the middle of the month they temporarily rose to 2,335 tons per day coincident with the drive to the Rhine, but receipts about equaled expenditures even in that period.[31] On 22 March, with all the armies poised for the Rhine assault, ammunition stocks were excellent in all areas. Total theater stocks at the time came to approximately 843,000 tons, of which 675,000 were in COMZ depots and 168,000 in army installations. At the current SHAEF maintenance rates these tonnages represented an average of fifteen days of supply of all major types of ammunition in field force depots and an average of forty-five days in COMZ depots.[32]

[29] Ltr, Col Stewart to G–4 SHAEF, 26 Mar 45, sub: Determination of Future U.S. Ammo Rqmts for This Theater, SHAEF G–4 471 Ammunition 1945; Teletype Conf Between ETOUSA and WD, 30 Mar 45; Ltr, Bull to CofS, 2 Apr 45, sub: Determination of Ammo Rqmts for U.S. Forces, and Cbl FWD–18510, SHAEF to Somervell, 3 Apr 45, SHAEF G–3 O&E 471 Ammunition.

[30] Ltr, ETO to WD, 10 Apr 45, sub: Ammo Expenditures, Levels and Day of Supply, EUCOM 471 Ammunition General, IV.

[31] 12 A Gp Rpt of Opns, XII, 151, VI, 61, SUSA G–4 Periodic Rpts for Feb and Mar 45.

[32] Army group personnel reports actually show army reserves averaging about twenty-five days of supply for most types. One possible explanation for the discrepancy: army group figures may include obligated stocks still in COMZ depots. Ltr, Page to COMZ-G–4, 11 Apr 45, Estimate of COMZ Supply Situation (Rpt 3), EUCOM 381 Projected Operation of COMZ ETOUSA, 14 Mar 45; Memo, OCOO for CG COMZ, 22 Mar 45 sub: Days of Supply in COMZ Depots and Armies as of 22 March, EUCOM 471 Ammunition (General); 12 A Gp G–4 Periodic Rpt, 18–24 Mar 45.

The last heavy expenditures of ammunition occurred at the Rhine crossings themselves, and in the case of the Ninth Army proved to be the most tremendous outpouring of any action during the war in Europe. Ninth Army had carefully conserved its ammunition for several weeks before, as it had for the Roer crossing operation in February, and then poured forth its savings with crushing weight in a short period in order to achieve the maximum support of the two divisions in the assault. In a period of twenty-nine hours on 24–25 March Ninth Army artillery units fired 6,470 tons of ammunition at five to twelve times the current maintenance rates of supply. Its 105-mm. howitzers, for example, fired at the rate of 132 rounds per gun per day, its 155-mm. guns at the rate of 166 rounds, 8-inch guns at 112 rounds, 240-mm. howitzers at 89 rounds, and 25-pounders (borrowed from the British) at the rate of 563 rounds.[33]

Any anxieties which field commanders still had over ammunition supply in March disappeared with the successful crossing of the last major obstacle. Ammunition was no cause for worry in the final month of operations except for the means of transporting it. All armies reported complete satisfaction with deliveries in the days immediately after the Rhine assault, and in the first week of April the 12th Army Group noted that ammunition supply was adequate for the first time since August 1944.[34] As in August and September 1944, Class V requirements were small during April. As could be expected, the need for the heavier caliber ammunition declined rapidly, and as early as 31 March Third Army asked that no further allocations of 8-inch gun and 240-mm. howitzer ammunition be made.[35] First Army's total expenditures averaged only 900 tons per day in the first half of the month, and Seventh Army's fluctuated between 500 and 1,200, depending on the resistance encountered. Expenditures throughout the four U.S. armies had actually averaged 115.3 tons per division slice in April, slightly exceeding the planning factor of 100 tons adopted at the beginning of the drive beyond the Rhine. Seventh Army's expenditures had far exceeded those of the other armies, totaling 171.2 tons per division per day. (*See Table 12.*) Deliveries in most cases matched issues, and at the end of April stocks in army depots were only slightly lower than at the start of the offensive, totaling 130,000 tons.[36]

COMZ stocks had continued to increase, meanwhile, rising to nearly 800,000 tons by mid-April.[37] Imports had averaged more than 9,200 tons per day since the first of February as against average expenditures of only 5,090 tons. Despite this excellent position the theater at first opposed any cancellations of ship-

[33] Gen Bd Rpt 58, pp. 38–39.

[34] 12 A Gp Ord Sec Jnl, 27 Mar 45; Monthly Rpt of Activity, Arty Sec 12 A Gp, 8 Apr 45, 12 A Gp 319.1 Reports.

[35] 12 A Gp Ord Sec Jnl, 31 Mar 45.

[36] Ltr, Page to CG COMZ, 13 May 45, sub: Estimate of COMZ Supply Situation (Rpt 4), EUCOM 381 Projected Operation of COMZ ETOUSA, 14 Mar 45; 12 A Gp G–4 Periodic Rpt, 22–28 Apr 45; Consumption Rates U.S. Forces From the Rhine to the Elbe, 23 Mar–24 Apr 45, prep by Statistics Sec, G–4 SHAEF, 11 May 45, SHAEF G–4 Basic Statistical Reports, No. 31.

[37] On the basis of the current weapons population the authorized seventy-five-day reserve permitted the theater to have about 900,000 tons of ammunition.

ments from the United States, particularly of the traditionally "critical" items.[38] After having been urged repeatedly by the War Department to review its stock position, however, the Communications Zone on 20 April recommended that loadings in the United States be reduced from 6,300 to 4,500 tons per day beginning on 15 May, the balance to be held to the credit of the theater. Representatives of SHAEF, the army groups, and Communications Zone accepted this proposal at a conference at Supreme Headquarters on the 24th. Army Group stocks were higher at this time than they had ever been, and it was proposed that these could be safely reduced. General Devers' spokesman refused to concur, despite the fact that the 6th Army Group's level at that time was equal to forty-five days of supply. Both army groups indicated they would be satisfied if expenditures could be replaced. The Communications Zone estimated that it could do this with the reduced imports. The restriction on loadings entailed a cancellation of approximately 60,000 tons for the period 15 May–15 June.[39]

Meanwhile the War Department, pointing out that theater stocks would soon reach maximum levels if expenditures followed their current trend, asked the theater for more outright cancellations of accrued credits. Within the next week the Communications Zone in turn asked SHAEF's approval to cancel all ammunition which had been deferred plus all future shipments scheduled to arrive after 1 June, and asked for authority to declare as excess to theater requirements all ammunition exceeding the seventy-five day level, which at the end of April totaled 98,000 tons. SHAEF immediately approved the latter; it also authorized the Communications Zone to release to the War Department half of all future production scheduled for release to ETOUSA, the remainder to be held to the theater's credit pending later instructions.[40]

Distribution problems had plagued ammunition supply until the last. Stock accounting and allocation were complex administrative problems at best, and they did not always guarantee the delivery or distribution of ammunition as intended. In February, for example, activity factors of 45, 30, and 25 percent assigned to the Ninth, First, and Third Armies had given the smallest share of ammunition to General Patton's forces. In actual fact, however, Ninth Army had been inactive early in February and Third Army had done the heaviest firing of any of the armies. As a result, Third Army's reserves had been quickly drained while

[38] Memo, Gilland for G–4 SHAEF, 14 Apr 45, sub: Reduction of Shipments to ETOUSA on Advent of V-E Day, SHAEF G–4 400.22 Shipments, Exchanges and Transfers, I.

[39] TWX EX–35274, COMZ to SHAEF, 20 Apr 45, TWX FWD–19863, SHAEF to COMZ, 24 Apr 45, Conf at SHAEF Fwd, 24 Apr 45, Cbl EX–36910, COMZ to AGWAR, 25 Apr 45, all in SHAEF G–3 O&E 471 Ammunition; Memo, Stewart for Reference Reduction of Ammo Shipments, SHAEF G–4 471 Ammunition.

[40] Cbl WARX–72683, ASF to COMZ and SHAEF, 24 Apr 45, and Cbl EX–37413, COMZ to WD, 26 Apr 45, both in SS&P Planning Div 201.2 Ammunition, A46–371; TWX EX–38866, COMZ to SHAEF, 29 Apr 45, TWX E–39064, COMZ to SHAEF, 30 Apr 45, and Cbl EX–39114, COMZ to AGWAR, 30 Apr 45, all in SHAEF G–3 O&E Ammunition 471; Ltr, Hq SHAEF to CG COMZ, 3 May 45, sub: Future Ammo Rqmts, SHAEF AG 471–1 Ammunition Status Reports (Day of Supply) No. 2.

those of the other armies, particularly the Ninth, had risen. Third Army's stocks of 155-mm. gun ammunition, for example, had fallen to 5.4 days by mid-February while Ninth and First Armies possessed between 23.3 and 25.9 days; its stocks of 81-mm. mortar ammunition stood at 2.4 days while those of the Ninth and First Armies respectively stood at 12.7 and 16. Army Group recognized this inequity by the middle of the month and acted to resolve it by substantial revisions in the activity factors for the remainder of the month.[41]

The constantly changing weapons strength, on which ammunition allocations were based, also complicated accounting procedures. Ground rules were not always clear on how the weapons population was to be counted. In January, for example, 12th Army Group discovered that the Communications Zone had not been including in its weapons basis the 81-mm. mortars on tank recovery vehicles and protested the discrepancy. The Communications Zone did not consider these as combat weapons. In any case, it pointed out that their inclusion in the weapons basis would not make any more ammunition available, since the maintenance day of supply rate was already computed on the basis of every round on hand or due to arrive. Their inclusion would simply result in the publication of a new day of supply at a lower rate, and would only upset the formula then being used and to no one's advantage. SHAEF accepted the logic of this argument and ruled against the inclusion of the disputed mortars.[42]

The Communications Zone's attempt to calculate the SHAEF and army group reserves alone involved much complicated bookkeeping in view of the periodic revisions in the maintenance day of supply. In mid-March the Communications Zone proposed to ease its task somewhat by stabilizing the current twenty-seven day army group reserve and ten-day SHAEF reserve at the existing level based on the March weapons strength and maintenance day of supply rate.[43] Using these figures, it proposed to specify a fixed number of rounds due each of the army groups, which would not be altered except upon sizable transfers of troops from one to another.[44] But the distribution of reserves between the armies, army group, and SHAEF, and the physical disposition of ammunition between army depots and the forward and rear depots of the Communications Zone, plus the

[41] Monthly Rpt of Activity (Feb), Arty Sec 12 A Gp, by Col Hedekin, 13 Mar 45, 12 A Gp 319, Rpts; Memo, Arty Sec 12 A Gp for G-3, 12 Feb 45, sub: Ammo Allocs, Memo, Col Holtzworth, Supply Br G-4 12 A Gp, for G-4, 14 Feb 45, sub: Revised Ammo Forecast, 9 to 19 Feb 45, Memo, Arty Sec for G-3 12 A Gp, 18 Feb 45, sub: Ammo Estimate, and Memo, Arty Sec 12 A Gp for G-3, 22 Feb 45, all in 12 A Gp 471/1 Ammunition Allocations.

[42] TWX QX-26507, 12 A Gp to COMZ, 30 Jan 45, TWX EX-93877, COMZ to 12 A Gp, 2 Feb 45, TWX Ex-10532, COMZ to SHAEF, 19 Feb 45, TWX FWD-17224, SHAEF to COMZ and A Gps, 21 Feb 45, all in SHAEF AG 471-1 Ammunition Status Reports (Day of Supply), 1945.

[43] Early in March SHAEF raised the reserve under its own control from seven to ten days in order to provide an additional means of influencing a particular operation, particularly in the event of a sudden strengthening of one of the army groups. Memo, Holtzworth for G-4 12 A Gp, 10 Mar 45, 12 A Gp 471/1 Ammunition Allocations.

[44] Ltr, COMZ to SAC, 17 Mar 45, sub: Stabilization of A Gp and SHAEF Reserve, SHAEF AG 471-1 Ammunition Status Reports (Day of Supply) No. 2.

fact that large amounts were always in the pipeline, made stock accounting difficult at all times. The vagaries of the system were further illustrated as late as the end of April when SHAEF found it necessary to instruct the army groups to have the various armies report not only the status of ammunition physically on hand in their own depots but the quantities on credit to them in the Communications Zone, for COMZ reports included only unobligated stocks on hand in COMZ depots. Unless obligated stocks were reported by the armies they were not reported at all.[45]

The ammunition allocations procedure underwent one final change in mid-March designed in part to provide a uniform system for the two army groups. Up to this time the armies of the 6th Army Group had followed the practice of submitting requisitions twenty days in advance, the requests being filled by Continental Advance Section as ammunition became available. The 12th Army Group wrote credits for only the ammunition which it expected would be on the ground at the time the armies called ammunition forward from the Advance Section. On 14 March, after discussing proposed changes with representatives of the army groups and the two advance sections, the Communications Zone outlined a new procedure. Beginning on 21 March the Communications Zone was to allocate specific quantities of ammunition to the two army groups for the ten-day periods 1–10, 11–20, and 21 to the end of the month, the allocations to be published so as to reach them three or four days before the start of the period—that is, on the 27th, 7th, and 17th respectively—and to present estimates or forecasts of the amounts to be allocated for succeeding periods up to ninety days. Allocations were to be based, as before, on weapons strengths as computed by the Communications Zone. The army groups were to specify the amounts to be credited to each army, and the Communications Zone was then to set up credits in specific advance depots and notify the armies of the location of the ammunition.

In effect, the new system represented the theaterwide application of the credit and forecasting procedure which the 12th Army Group had adopted for a while in November. Opinions were divided as to the efficacy of the new system. The 6th Army Group found the new procedure something less than satisfactory, claiming that allocations were habitually late in arriving and the channels for acquiring credits too devious. The 12th Army Group was satisfied that ammunition supply had finally been placed on a firm basis. Supply was so satisfactory in April that the army group discontinued the practice of furnishing the armies with estimates of future supply at the end of the month. A few weeks earlier the improved supply prospects had led the Communications Zone to propose a change in the allocation period from ten to thirty days and also the elimination of the time limitation by which credits expired at the end of thirty days if not used, in both cases to ease its bookkeeping burden. But a decision was postponed, and the recom-

[45] Ltr, Col Stewart, Chief of Current Opns Br G–4 SHAEF to G–4, 1 Feb 45, sub: Ammo Status Rpts, SHAEF G–4 471 Ammunition; TWX FWD–20178, SHAEF to 6 A Gp, 29 Apr 45, SHAEF AG 471–1 Ammunition Status Reports (Day of Supply) No. 2.

mendations were not adopted before V-E Day.[46]

Operations in April, when ammunition was plentiful, were hardly representative of the entire eleven months' experience. Since ammunition was rationed for most of that period there was no adequate test of the validity of War Department day of supply rates, and the theater never really determined precisely what quantities would have been satisfactory. Possibly such figures were beyond determination anyway, for it is axiomatic that field commanders rarely consider that they are adequately supplied with ammunition. In any case, the theater did not cease until V-E Day to emphasize to the War Department that releases of ammunition must not be predicted on past expenditures where such expenditures had been limited by nonavailability.[47]

(3) *Equipment*

Class II and IV supply, like the supply of rations and POL, was largely a transportation and distribution problem in the final months. Shortages which had reached genuinely critical proportions in December, attributable to production difficulties in the United States, had either been resolved by February or were being overcome as the result of the various measures taken in December and January. Shortages of one kind or another plagued U.S. forces until the last day of fighting, but relatively few of the hundreds of thousands of Class II and IV items of supply were classed as "critical" after the beginning of March.

As before, the main supply problems concerned the engineer, signal, and ordnance services. The demand for engineer equipment was unusually high in the last few months because of the numerous water obstacles and because of additional special requirements for the Roer and Rhine crossings. For the Rhine crossing, as an example, the 12th Army Group's bill of materials included the following: 120 Bailey bridge sets of various types, 20 treadway bridges, 40 25-ton ponton bridges, 1,200 assault boats, 200 utility power boats, 1,800 outboard motors, 46 ponton barges, 47 LCM's, 72 LCVP's, and 490 anchors. Much of this equipment had to be obtained through special procurement projects (PROCO), which the theater submitted to Washington in the fall of 1944. Early in January the deputy chief engineer returned to the United States to expedite the delivery of several items, including assault boats, utility power boats, and ponton boats. Late in January the engineer service contracted with five French firms for 700 plywood storm boats, which were built in record time.

Assembling the needed equipment in forward parks proved an unprecedented

[46] Ord Stf Offs Mtg, OCofOrd ETOUSA, 24–25 Feb 45, SUSA File 34, Supply Sec G–4, Supply Rpt From ETOUSA; 12 A Gp Ord Sec Jnl, 27 Feb and 18 Apr 45; Memo, 12 A Gp G–4 for Ord, Arty, and G–3, 13 Mar 45, sub: Forecast of Ammo Supply to the Armies, 12 A Gp 471 Ammunition General; Ltr, ETO to Major Comds, 14 Mar 45, sub: Alloc of Ground Force Ammo to Major Comds, SHAEF AG 471–1 Ammunition Status Report No. 2; 6 A Gp G–3 Sec Final Rpt, pp. 104–05; Memo, 12 A Gp Ord for G–4, 28 Apr 45, 12 A Gp 471/1 Ammunition Allocations, 12 A Gp Rpt of Opns, XII, 153; Gen Bd Rpt 58, pp. 63–66.

[47] Ltr, CG ETO to Gen Heileman, Div of Supply ASF, 14 Apr 45, 1st Ind to Ltr of 21 Mar 45, sub: Projected Supply Status of Ammo, EUCOM 471 Ammunition (General), III: ETO G–4 Quarterly Periodic Report for quarter ending 31 Mar 45, 6 May 45, EUCOM 319.1 G–4 Periodic Rpts ETO Quarterly.

undertaking because of the out-of-gauge size of much of it. With few exceptions, however, the supplies were on hand in time for the assaults. The main deficiency was in maintenance supplies for truck-mounted cranes, which were in great demand in all the armies. Third Army stocks had to be brought all the way from Luxembourg. Movements east of the Rhine later entailed a heavy commitment of truck transport. First Army, for example, used seven 2½-ton companies, two 10-ton semitrailer companies, three engineer dump truck companies, the tractor trucks of an engineer heavy ponton battalion, and 100 5- to 7-ton tractor trailers to haul bridging materials. The 5- to 7-ton trailers were loaded and spotted in forward dumps and thus constituted a moblie dump and were unloaded only as needed. Third Army was one of the heaviest users of river bridging equipment in the last three months, building 225 bridges in February, 341 in March, and a peak number of 394 in the final month, which saw the crossing of the Danube and the Inn. Most of the timber for fixed bridges was procured locally, and steel I-beams came from mills operated under the supervision of the Advance Section.[48]

In signal supply the shortage of field wire remained chronic throughout February, and the armies continued to enforce strict rationing to ensure active units an adequate supply. This deficiency gradually eased in March, when the ASF began releasing about 100,000 miles per month. Forward depot stocks remained low in March, but the armies received large quantities and made heavy issues as the result of improved deliveries, which averaged 1,100 miles per day in the First Army and about 800 in the Third. The supply of wire, and of radios and spare parts, was often unpredictable in the final month, causing nervousness in the field commands. But this was fairly characteristic of all supply in a highly mobile situation, since it was rarely possible to build up substantial forward stocks. Shortages sometimes became momentarily "critical" overnight, as on 20 April, when Third Army's reserve of wire dropped to 238 miles. Three days of supply (1,900 miles) arrived on the following day, however, and another seventeen day's supply was en route. At the end of April all the armies had from twelve to eighteen days of wire either on hand or en route. Some of the most urgent needs —batteries, for example—were met via air. On the whole, signal supply was better in April than it had been for several months.[49]

Ordnance Class II and IV supply problems in the last few months were concerned largely with maintenance. Shortages of major items, such as tanks and general purpose vehicles, which had become so critical in the fall, were largely eliminated by February. The losses of the Ardennes had also been replaced by

[48] *Final Report of the Chief Engineer ETO*, I, 223; Supply, Hist Rpt 3, OCE ETO, p. 43; TUSA AAR, II, Engr, 25–26, 35, 40; NUSA G–4 AAR, 15–31 Mar 45; FUSA Rpt of Opns, 22 Feb–8 May 45, Bk. II, pp. 175, 179; 12 A Gp Rpt of Opns, VI, 61, 66, 76; 12 A Gp G–4 Periodic Rpts for Feb and Apr; SUSA G–4 Periodic Rpts, Feb–Apr.

[49] FUSA Rpt of Opns, 22 Feb–8 May 45, Bk. II, p. 217; TUSA AAR, II, Sig, 16, 18, 20; Comd and Stf Conf Notes, COMZ, 9 Mar 45, EUCOM 337/3 Conf, Staff Weekly, I; 12 A Gp G–4 Periodic Rpts for Feb–Apr 45; SUSA G–4 Periodic Rpts for Feb–Apr 45.

that time. In March the armies were reporting for the first time in months that items such as small arms, artillery weapons, and fire control and sighting equipment were in sufficient supply to cover losses. The main shortages that were to persist to the end were in spare parts, accessory items, major replacement assemblies such as truck engines, axles and transmissions, hot patches, brake fluid and lining, batteries, and, to some extent, tires and tubes. Shortages of tank tracks and bogie wheels became especially acute in the final week. Lacking spare parts, forward maintenance units made it a common practice in the final months to strip salvage vehicles of all serviceable parts. Ninth Army, for example, obtained about 30 percent of all the automotive spare parts requirements for Operation GRENADE by systematically stripping damaged vehicles.[50]

The supply of combat vehicles was unusually good in the final months. The medium tank shortage had been solved once and for all by the allocations which General Lutes had arranged in December and by the increase in the replacement factor. Some improvement was already noticeable by early February, when theater resources totaled 6,374 medium tanks against total theater requirements for 7,095. Against a T/E requirement for 5,255 there were actually 5,434 on hand in the armies. The armies thus possessed only a small reserve of 179 tanks, but an additional 940 were in theater pipelines. The theater as a whole had a reserve of 1,119 tanks and lacked 721 to meet its full authorized reserve of 1,840 tanks.[51]

In January the War Department had refused the theater's request for an increase in its on-hand reserve of medium tanks to 70 percent. It agreed instead to a 35-percent reserve on the basis of the new 14-percent replacement factor, and agreed to create an additional 35-percent reserve in the zone of interior on which the theater could draw if necessary. The theater in turn had authorized the armies to have the bulk of the reserve—28 percent, or a sixty-day level—in their own areas.[52] Under these arrangements the War Department had scheduled shipments of about 1,200 medium tanks per month beginning in February. Receipts under this program greatly improved the tank situation in the theater. On the eve of the Rhine crossings theater stocks totaled 7,620, only 159 tanks short of the total T/E and reserve requirement of 7,779. Against a T/E requirement for 5,477 the armies actually had 6,606 on hand, giving them a reserve of 1,129 against an authorized reserve of 1,535. While the armies' stocks included nearly 600 unserviceable tanks, the armies had another 600 in reserve, and additional tanks intended for the army reserves were in theater pipelines.[53]

---

[50] TUSA AAR, II, Ord, 24; *Conquer; The Story of the Ninth Army*, pp. 162–63; NUSA Ord AAR, 1–15 Mar 45; FUSA Rpt of Opns, 22 Feb–8 May 45, Bk. III, pp. 7, 9–10.

[51] Ltr, Craword to Smith, 7 Feb 45, sub: Weekly Rpt on Lutes' Memo, SHAEF G–4 319.1 Rpts Gen Lutes, 1945, I.

[52] Cbl E–84778, Somervell to Styer, 10 Jan 45, and Cbl WARX–20645, AGWAR to Somervell, 14 Jan 45, both in EUCOM 470.8 Combat Armored Cars and Tanks.

[53] Rpt on Status of Medium Tanks, prep by AFV&W Sec ETO, 20 Mar 45, SHAEF G–3 O&E 470.8 Tanks, III.

PERSHING TANKS M26 *mounting the new 90-mm. gun move through Wesel, Germany.*

In light of this increasingly favorable situation the War Department notified the theater in mid-April that it contemplated eliminating the 35-percent reserve being created for ETOUSA in the United States. The theater first opposed such action, pointing out the difficulty of maintaining the full 28-percent reserve in the army areas in view of the large percentage of total theater stocks that was constantly in the pipeline or under repair. It also predicted heavier losses under current operating conditions. In the end, however, ETOUSA agreed to accept the War Department's proposed elimination of the zone of interior reserve on condition that the War Department ensure continued deliveries at the rate of the past three months.[54]

Contrary to expectations, tank losses in the 12th Army Group actually came to only 9.7 percent in March and were to drop to 5.2 percent in the final month. Losses in the 6th Army Group came to about 11 percent for the last six weeks. In mid-April the COMZ G-4 reported to SHAEF that sufficient medium tanks were then available to meet the full authorized reserve of 28 percent in the armies and 14 percent in the Communications Zone. The downward trend in losses had in fact led the theater to re-

[54] Cbl E-33545, G-4 ETO to AGWAR, 16 Apr 45, ETO Adm 402.

assess its needs with a view toward canceling certain releases not already shipped.[55]

Field commanders had long looked forward to the arrival of the new 90-mm. gun T26, or Pershing, in hopes that they would at last have a tank to match German armor. The first shipment to the theater contained only twenty tanks, all of which went to the 3d and 9th Armored Divisions in the First Army and first saw action on 23 February. While no additional receipts were expected before April, General Eisenhower specified in mid-March that all T26's arriving in April should be assigned to the 12th Army Group, and that a portion of May receipts would go to the 6th Army Group. General Bradley in turn allocated the expected 126 tanks in approximately equal numbers to his three armies. By mid-April 185 of the new tanks had arrived, of which 110 were then with armored divisions. Whether the remaining 75, or additional tanks, saw action before V-E Day is unknown.[56]

Even before the first token number had been tried in battle the theater had urged the War Department to provide the maximum number of the new heavy tanks, as the T26's were then classified, and the War Department had assured it that both production and shipment of the new tank was being given the highest priority. Two weeks after the first commitment of the new weapon Maj. Gen. Gladeon M. Barnes, one of the ASF ordnance officers who had gone to the theater to introduce the T26 and observe its performance, reported an enthusiastic reception for the tank among combat commanders and asserted that its 90-mm. gun was far superior to any tank gun the Germans then had in use. There was little doubt that the new weapon represented a great improvement over the Sherman. But Brig. Gen. Joseph A. Holly, head of the theater's armored section, felt constrained to correct any impression that the T26 was superior to the German Tiger and Tiger Royal. The Pershing's 90-mm. gun, he pointed out, was superior to the German 88 only when firing the new high velocity armor-piercing (HVAP) ammunition, and was still definitely inferior to the enemy's gun when using ordinary armor-piercing ammunition. In extension of this sobering fact, he noted that the better ammunition was being manufactured in very limited quantities because of the shortage of tungsten carbide, a critical metal used in the core of the projectile. As if to deflate

[55] A report by the Armored Fighting Vehicle and Weapon Section, ETOUSA, gives conflicting although incomplete figures: 5,626 tanks on hand in the armies as against a total requirement for 5,129 which would give the armies a reserve of 497 as against an authorized one of 1,436. Another source indicates there were 1,100 additional tanks in COMZ depots, but these, plus the 5,626 on hand in the armies (6,726) would not have matched the over-all theater requirement of 7,283. Rpt, Status of Medium Tanks, AFV&W Sec ETO, 20 Apr 45, SHAEF G–3 O&E 470.8 Tanks, III, and Cbl EX–39080, COMZ to 6 A Gp, 30 Apr 45, SHAEF AG 480.8–1 Allocation of Tanks, 1945; Memo, Gilland for Crawford, 20 Apr 45, sub: Lutes' Memo to the Supreme Commander, dated 25 Dec 44, SHAEF G–4 400.192 Supply Reports, I; 12 A Gp AAR, XI, 67; Losses of Medium Tanks for Period 13 Mar–28 Apr 45, Tab C to missing Ltr, 6 A Gp G–3 Armd Sec 252, Medium Tank Correspondence, RG 914.

[56] Ltr, Bull to CofS, 15 Mar 45, sub: Alloc of Heavy Tanks, TWX QX–28443, 12 A Gp to SHAEF, 28 Mar 45, and Ltr, Bull to CofS, 17 Apr 45, sub: Alloc of Heavy Tanks T26, SHAEF SGS 470.8 Tanks, II; Ltr, Bull to Twitchell, 26 Mar 45, SHAEF G–3 O&E 470.8 Tank III; Teletype Conf Between Maj Gen G. M. Barnes and Maj Gen Levin H. Campbell, 5 Mar 45, EUCOM 470.8 Combat Armored Cars and Tanks, II.

M24 Light Tanks *with the 75-mm. gun advance along a forest road in Germany.*

any unwarranted optimism as to the availability of the new tanks, he observed, moreover, that they were not scheduled for production in large numbers until the third and fourth quarters of 1945.[57]

While these observations were pertinent to any eventual evaluation of the new tank, they proved of little consequence to the outcome of the fighting in Europe, for hostilities soon came to an end. First Army, which had been the first to receive the new tanks, later concluded that the delayed arrival and commitment of the T26's plus the fact that German armor had already been crippled, precluded any adequate test of their worth. On the other hand, the M24, the new light tank with the long-barreled 75-mm. gun, had been furnished in larger numbers and tried over a longer period, and had proved its excellence more conclusively.[58]

Meanwhile the theater continued trying to replace the obsolescent 75-mm. gun medium tank with 76's, and to get more HVAP ammunition for the latter. Production difficulties limited the shipments of the improved ammunition, however, to less than two rounds per gun

[57] Ltr, Barnes to Campbell, CofOrd, 6 Mar 45, sub: Rpt of Heavy Tanks Mission, and Ltr, Holly to Deputy Theater Comdr, 2 Apr 45, sub: Comments on Rpt of Heavy Tank Mission, EUCOM 470.8 Combat Armored Cars and Tanks, II.

[58] FUSA Rpt of Opns, 22 Feb–8 May 45, Bk II, p. 147. The new M6 gun on the M24, although a 75-mm. gun, was commonly referred to as a 76-mm., to distinguish it from the short-barreled 75-mm. on the medium tank.

per month. Late in February the theater made plans to convert 75-mm. gun tanks into 76's. But the program was slow to get under way, and the theater subsequently canceled the project in view of the high rate of receipts from the United States and the declining loss rate. At the time of the Rhine crossings about 40 percent of all medium tanks in the theater were of the 76-mm. gun type.[59]

A project for installing 17-pounder guns in some of the Shermans, characterized by several false starts, was never completed. U.S. tank reserves had finally improved sufficiently by mid-February to permit the release of M4's for this purpose, and arrangements were made with the British to convert 160 tanks, the maximum number which the supply of 17-pounder ammunition would support. The first of the modified tanks were delivered in March and were assigned to the Ninth Army. But in mid-April the program was cut from 160 to 80, again because of ammunition shortages, and few of the tanks ever saw action.[60]

Aside from the complications which the development and introduction of improved weapons presented, the supply of tanks typified fairly well the general improvement in Class II and IV supply by February 1945. On the whole, supply in all classes was better in the last three months than it had been at any time since July 1944. Testifying to this improvement was the statement which appeared with increasing regularity in the G–4 periodic reports of the field commands: "There are no . . . items in critical short supply which will adversely affect operations."

The marked improvement had in fact led theater officials themselves to consider the possibility that ETOUSA was requisitioning and stocking supplies in excess of its needs. Early in March the G–4 instructed the supply services to review their position with a view toward reducing imports. The War Department had long suspected the theater of overstocking. As late as April it pointed out that ETOUSA was asking for quantities of signal, engineer, medical, and ordnance supplies far in excess of consumption rates.[61] The memory of past difficulties made supply officials cautious to the last, however, and there were few cancellations until the end of hostilities was a certainty.

[59] Ltr, Holly to Deputy Theater Comdr, 2 Apr 45; Ltr, ETO to A Gps, 25 Apr 45, sub: Cancellation of 76-mm. Gun Conversion Program for 300 75-mm. Gun Tanks, M4 Series, EUCOM 470.8 Combat Armored Cars and Tanks, II; 12 A Gp AAR, XI, 54.

[60] Ltr, Crawford to Maj Gen J. A. M. Bond, Dir of Ln and Munitions, War Office, 16 Apr 45, SHAEF G–4 471.1 Cannon, Mortars, and Ammunition; Final Hist Rpt, AFV&W Sec ETO, p. 22, ETO Adm 540; 12 A Gp AAR, XI, 54.

[61] Cbl WX–71085, AGWAR to COMZ, 21 Apr 45, SHAEF S S 400.3/1 Supply Problems of Allied Advance.

CHAPTER XVII

End of the Replacement Problem

(1) The Turning Point

Early in February the replacement problem, like the shortages of major supply items such as tanks and field artillery ammunition, was still a major cause of anxiety. Estimates of future deficits varied from week to week, depending in part on casualty forecasts, but they indicated throughout January and February that the theater would still be short between 20,000 and 27,000 infantry riflemen and anywhere from 2,400 to 5,400 armored replacements on 1 March. The shortage of riflemen, it was predicted, would not be overcome until some time in April, and the shortages in armored force replacements would continue for several months.[1]

This pessimistic outlook resulted mainly from the theater's forecasts of expected casualties. These forecasts had risen steadily in the light of the actual loss experience of November and December.[2] To make matters worse, the War Department early in March informed ETOUSA that its shipments that month would be considerably smaller than previously estimated, owing in part to the use of infantrymen to provide additional armored force replacements, and in part to the shipment of additional units to the theater which the War Department had counted on using as a source of replacements.[3]

In actual fact, signs of improvement were already evident. The shortages within the army groups had begun to recede in February. By the middle of March, Colonel Shannon, chief of the theater Manpower Section, reported that there were more than sufficient infantry replacements available in army depots to meet all reported shortages. On the strength of this development, and on Shannon's estimate that the situation would continue to improve in March, SHAEF finally suspended the allocations of infantry replacements between the two

[1] Memos, Col D. J. Renfroe, Actg G–1 ETO for G–1 SHAEF, 13 and 18 Jan 45, sub: Reinforcements, Memo, Franey for G–1 SHAEF, 24 Jan 45, sub: Reinforcements, Memo, Twitchell for G–3, 26 Jan 45, sub: Reinforcement Situation, Memo, Franey for Lear, 5 Feb 45, sub: Reinforcement Data, and Memo, Shannon for Lear, 21 Feb 45, sub: Reinforcement Data, all in SHAEF G–3 370.092 Reinforcements, 1944.

[2] The casualty forecast for March 1945, which in September 1944 was estimated at 45,000, had risen to 91,000 by February. See EUCOM 320.2 Strength Reports 1944–45 for the theater's periodic forecasts of losses and requirements.

[3] Cbl, WD to ETO, 6 Mar 45, ETO 200.3 Manpower.

army groups effective at the end of the month.[4]

In its communications with the War Department regarding requirements from the United States, meanwhile, ETOUSA pursued a more cautious course. Early in the month the War Department asked the theater to release 19,000 men per month from scheduled shipments beginning in May in order to bring units in the Pacific area up to strength and also build up stockages which would permit the release of men there when redeployment under the point system went into effect.[5] ETOUSA objected strongly to the proposed cut, insisting that it could not get along without the scheduled shipments. The theater admitted that casualties had fallen off substantially, but it argued that an urgent new need had arisen for units to control the masses of displaced persons and prisoners of war which the armies were uncovering in their advance. At this time approximately 30,000 prisoners were being captured every day, and between 800,000 and 1,000,000 displaced persons had also been uncovered. As an emergency solution the theater had already begun withdrawing from the Replacement System infantrymen whose training had been only partially completed to provide overstrengths for MP, antiaircraft artillery, and heavy field artillery units being used for control duties.[6]

The War Department considered the needs of the Pacific area more urgent. It pointed out, moreover, that ETOUSA would still get 26,000 men in May even after allowing for the diversion of 19,000, and that 16,000 of this number would consist of infantrymen to be used in building up its MP units for military government duties. It noted that the theater was already employing combat divisions for such duties. In view of these developments it informed the theater that it planned to go ahead with the diversions unless there were other reasons for not doing so.[7]

ETOUSA agreed to cut the 19,000 for May. But it asked the War Department to reconsider carefully any additional diversions. On the basis of current casualty rates it estimated that its own resources —that is, the present stockage and the output of retraining—would be exhausted by late August, for it could not continue the withdrawal of specialists from service units for retraining without impairing the efficiency of those units. It therefore wanted the flow of men from the zone of interior maintained.[8]

At the very moment that ETOUSA was painting this relatively gloomy picture it was accumulating replacements far in excess of its immediate needs, and actually faced a problem of how to use the surpluses. Two developments had

[4] Ltr, Barker to G–3, 19 Mar 45, sub: Summary of Present Reinforcement Situation, Ltr, Shannon to Lear, 20 Mar 45, sub: Alloc of Reinforcements for Apr 45, and Ltr, Bull to Smith, 23 Mar 45, sub: Reinforcement Situation, all in SHAEF G–3, 370.092 Reinforcements, 1944.

[5] Cbl WAR–65819, Marshall to Eisenhower, 10 Apr 45, ETO 200.3 Manpower.

[6] Cbl FWD–19314, SHAEF to AGWAR, 16 Apr 45, ETO Adm 414 Cable Files; Memo, Lear for Smith, 16 Apr 45, sub: Reinforcements, and Memo, Shannon for Howland, 9 Apr 45, both in ETO 200.3 Manpower.

[7] Cbl WAR–72080, AGWAR to Smith, 23 Apr 45, ETO Adm 414 Cable Files.

[8] Cbl FWD–19883, SHAEF to AGWAR, 25 Apr 45, SHAEF AG 200.3–1 Reinforcements (Requisitions, Availabilities, Training).

operated simultaneously beginning in February to bring about the improvement. The most important was the very substantial drop in the number of casualties. Battle casualties came to 39,400 in February, as compared with losses of 69,100 in January and 77,700 in December. Nonbattle losses continued fairly high, totaling 52,100 in February as compared with 67,600 in January. In fact, they exceeded battle casualties in February, and were to do so again in April. But total casualties came to only 91,500 in February, as against 136,700 in January and 134,400 in December. Battle casualties rose to 53,200 in March, and total casualties that month came to 101,000. In April battle casualties dropped to 41,000 and total casualties that month came to 87,100. Losses were consistently lower than anticipated. (*See Table 10.*)

The theater's stepped-up retraining program had begun to produce results. Conversion training for enlisted men under the program announced in January did not get under way until early the next month, and the results of the expanded program therefore did not appear until March. Under the impetus of events in December, however, the theater had already begun to retrain greater numbers of men. The output in January totaled 7,685 infantry riflemen, representing some gain over November and December conversions, which had come to only 1,325 and 5,751 respectively. The February output, totaling 8,193, was only slightly higher and, likewise, did not yet show the results of the expanded program. In the course of that month, however, the Replacement System accepted more than 32,000 men for retraining—about 17,700 from the Communications Zone and 14,500 from the Air Forces. The fruits of these large withdrawals finally began to be realized in March, when the output of the conversion training centers totaled 17,152.[9]

(2) *Withdrawals Are Stopped*

The expanded conversion program had been in operation barely two months when theater officials began to consider the curtailment or complete cessation of the entire program. At the beginning of April the theater's program called for withdrawals totaling 88,450 men between 1 February and 14 May—55,180 from the Communications Zone, 32,920 from the Air Forces, and 350 from SHAEF. In the case of both the Communications Zone and the Air Forces the planned withdrawals comprised approximately 9 percent of their total strength. Plans tentatively called for the withdrawal of an additional 10,500 from the Air Forces and 22,400 from the Communications Zone in the period 15 May–15 June.[10]

Lt. Gen. Carl Spaatz's request early in April for a cut in the Air Forces quota led to a re-examination of the entire program. General Eisenhower was reported to favor the cut in view of the combat situation and the smaller number of casualties. General Lear's first reaction was to disapprove any reduction in withdrawals on the ground that neither the Communications Zone nor the Air Forces had yet suffered any net loss of strength because the withdrawals had been re-

[9] [Henderson] Procurement and Use of Manpower in the ETO, p. 117.

[10] Memo, Osborne for Lear, 29 Mar 45, sub: Withdrawal of Pers From COMZ, ETO AG 319.1/1 Daily Progress Reports, II; Memo, Twitchell for Bull, 30 Mar 45, sub: Inf Reinforcements, SHAEF G–3 370.092 Reinforcements, 1944.

placed with limited assignment men, and because the prospects of getting replacements from the zone of interior were deteriorating. As a safety measure, he preferred to ensure the maximum flow of replacements for at least a few months. Nevertheless, Lear asked for a restudy of the entire replacement picture, specifying that it take into consideration the above factors, plus such factors as the smaller rate of hospital returnees in recent months, the need to continue retraining in order to permit the absorption of limited assignment men becoming available in numbers in excess of normal replacement needs, and the possibility of heavier casualties.[11]

COMZ officials generally favored a suspension of the retraining program. There was no denying, as both General Eyster and General Larkin pointed out in their appraisal of the situation, that casualties were now running considerably below the estimates on which replacement requirements were being calculated. They thought there was no longer much danger of another strong counteroffensive by the enemy, or even the probability of a very determined defense. Replacement requirements were therefore definitely on the decline.

The main consideration in deciding whether retraining should continue was an estimate as to the effect which withdrawals were having on the service forces. There was no question as to the number of qualified troops available for retraining. The main question was whether a sufficient number of replacements could be supplied to offset withdrawals and also ensure that efficiency of the service forces would not suffer. COMZ officials claimed that the efficiency of many service units had already been impaired by the exchange of untrained limited assignment personnel for trained general assignment men, and insisted that the units could not stand additional losses. Furthermore, they argued that service units would have to operate at maximum efficiency even after V-E Day, and that the theater would have to furnish well-trained units for early deployment to the Far East.[12]

Section commanders and the technical services provided ample documentation for these arguments. In fact, to judge from the tenor of their claims it would appear that the entire Communications Zone was approaching collapse. General Plank claimed that over-all efficiency in the Advance Section had already been reduced by an estimated 18.8 percent in the supply services, and by as much as 30 percent in the signal and ordnance services. The Office of the Chief Quartermaster stated that quartermaster units had suffered a 40-percent decrease in efficiency through the exchange of limited assignment for general assignment men. It cited as a specific example bakery companies, in which heavy lifting required able-bodied men.

For the most part the withdrawal of

[11] Memo, Col Paul Porch for Lear, 4 Apr 45, sub: Tel Conv Between Porch and Barker on Inf Retraining Program, and Ltr, Lear to Spaatz, 4 Apr 45, both in ETO 200.3 Manpower; Memo, Porch for Lord, 6 Apr 45, ETO AG 319.1/1 Daily Progress Reports, II.

[12] Ltr, Eyster to Larkin, 16 Apr 45, sub: Curtailment or Cessation of the Present Retraining Program for Inf Riflemen, ETO 353/1 Retraining; Ltr, Larkin to Eisenhower, 14 Apr 45, sub: Withdrawal of Pers From COMZ Units for Retraining as Inf, SHAEF AG 200.3-4 Reinforcements (Reconversion Program—Policy) 1945.

critical specialists had been avoided. There were instances where a single section of an installation took the heaviest loss of men because the other sections were staffed with specialists. In a single depot company, for example, the loss fell almost completely upon the storage and issue sections because all the men in the repair section were classed as key personnel.[13]

Meanwhile the claim was made repeatedly that the limited assignment men who were supposed to replace the men withdrawn did not arrive until long after vacancies had occurred. The ADSEC commander claimed at the beginning of March that white units in the Advance Section were understrength by an average of 16 percent, and that some units were as much as 34 percent understrength.[14] Plans had provided that limited assignment replacements should be assigned to service units as much as thirty days before the actual withdrawal of the men they were to replace, and that they be given on-the-job training. But these intentions were not always fulfilled. Withdrawals of general assignment men actually exceeded allocations—to say nothing of actual deliveries—of limited assignment replacements until late in March, and there were over-all shortages in both the Communications Zone and the Air Forces until at least that time. Consequently many replacements did not get on-the-job training from the men whose jobs they inherited. An attempt to provide specialist training in certain job categories, such as truck drivers, cooks, and bakers, at schools in Shrivenham, England, and at the 19th Replacement Depot on the Continent, also fell short. These training centers simply lacked the capacity to handle the thousands of men thrust upon them, beginning in February, and the pressure to fill the vacancies in units led to the assignment of many men before they were prepared to occupy positions.[15]

The loss of efficiency occasioned by the exchange of untrained limited assignment replacements for trained general assignment men was, at best, difficult to measure. The Communications Zone itself tended to reject the claims of the sections as unfounded, insisting in the case of the Advance Section that it had actually gained rather than lost in the exchange of personnel,[16] that its quotas thus far had necessitated no surrender of specialists, and that the number of key specialists in at least two technical services—ordnance and signal—had actually increased.[17]

COMZ officials nevertheless supported the general argument that further withdrawals would be harmful, and recommended that they be discontinued, and that all limited assignment men in excess of normal requirements be assigned to

[13] Ltr, Plank to CG COMZ, 31 Mar 45, sub: Release of Gen Assignment Pers, EUCOM 322 Replacement Units, II (a); [Henderson] Procurement and Use of Manpower in the ETO, pp. 105–08.

[14] TWX R–58561, ADSEC to Theater Manpower Section, 1 Mar 45, ETO Adm 410.

[15] [Henderson] Procurement and Use of Manpower in the ETO, pp. 115–18; Memo, Renfroe, G–1 ExO, for Lord, 29 Mar 45, EUCOM 220.3 Assignment and Transfer 1943–45.

[16] It claimed that the Advance Section had received 6,437 limited assignment men in exchange for 5,480 withdrawn up to 31 March.

[17] Ltr (1st Ind to Ltr of 31 Mar 45), COMZ to ADSEC, 12 Apr 45, EUCOM 322 Replacement Units, II (a).

service force units as overstrength.[18] Colonel Shannon urged a more conservative course, arguing that the curtailment or cessation of the retraining program would involve too great a risk in view of the higher casualties which the army groups had most recently forecast. He also again called attention to the problem of absorbing limited assignment men if withdrawals were discontinued. He recommended that the withdrawals continue as scheduled until 15 May.[19]

But the surpluses which had begun to accumulate in the Replacement System provided a compelling argument for suspending the retraining program. In mid-April the theater Manpower Section reported a stockage of 50,000 infantrymen in the Replacement System in excess of shortages reported by the field commands.[20] Vacancies in branches other than infantry totaled about 20,000. Even assuming that these would have to be met with infantry replacements, there were still 30,000 men available in excess of needs.

Colonel Shannon thought that about 20,000 of this surplus could safely be used to meet the rising requirements for prisoner of war guards and for military government. About 5,000 had already been provided 12th Army Group for this purpose. The theater provost marshal had indicated a need for an additional 10,000 men to meet prisoner of war guard requirements. One corps reported that it had already diverted about 6,000 men to military government duties, and both army groups reported that their combat effectiveness was being dissipated by the necessity to divert combat units for occupational duties. The use of infantry replacements for this purpose obviously involved some risk, and General Bull at first opposed it. But in the end he approved the idea and arranged for the release of 10,000 replacements to the two army groups in proportion to their divisional strength. An additional 10,000 were earmarked for use as prisoner of war guards as required. But Bull informed the army groups that they would have to use the overstrengths furnished in this way to meet their combat losses in the event they could not be met by the Replacement System. On 21 April General Lear ordered the suspension of withdrawals effective at the end of the month.[21]

With the manpower problem apparently solved, the theater Manpower Section became more or less superfluous. Early in May the theater announced that the general assignment men already withdrawn from the Communications Zone and Air Forces would complete their

[18] Ltr, Eyster to Larkin, 16 Apr 45, sub: Curtailment or Cessation of Present Retraining Program for Inf Riflemen, ETO 353/1 Retraining; Ltr, Larkin to Eisenhower, 14 Apr 45, sub: Withdrawal of Pers From COMZ Units for Retraining as Inf, SHAEF AG 200.3-4 Reinforcements (Reconversion Program—Policy) 1945.

[19] Stf Study, Shannon, 20 Apr 45, ETO 353/1 Retraining.

[20] Memo, Shannon for Barker, 13 Apr 45, sub: Reinforcements, SHAEF G-3 370.092 Reinforcements.

[21] Memo, Shannon for Lear 17 Apr 45, summarizing SHAEF Conf on Pers of 14 Apr 45, ETO 200.3 Manpower; Ltr, Bull to Smith, 19 Apr 45, sub: Alloc of Reinforcements to Meet Manpower Rqmt of Armies, TWX FWD-19470, Bull to Lear, 19 Apr 45, SHAEF G-3 370.092 Reinforcements; Cbl BAX-25788, 6 A Gp to SHAEF, 13 Apr 45, SHAEF G-4 381 War Plans General, 1945, II; Memo, Lear for Porch, 21 Apr 45, ETO 200.3 Manpower; Memo, Porch for Shannon, 23 Apr 45, ETO 353/1 Retraining; TWX EX-36982, Lee to Secs, 25 Apr 45, GFRC 353 Training, May 44 to Jan 45.

conversion training. At the same time it ordered the theater Manpower Section to revert to its original status as a section of the G-1 Division of the theater general staff effective 15 May. The G-1 was thereafter to assume responsibility for the staff supervision of all matters concerning the morale, welfare, and proper use of manpower, responsibilities which had been exercised by the deputy theater commander, General Lear. Meanwhile the trend of events largely nullified the work of the ETO Manpower Board, which was still conducting detailed job analyses in the theater designed to save personnel. On 4 May General Lear approved a proposal to dissolve the Manpower Board upon the completion of the surveys it was then conducting, estimated at about 1 June.[22]

The relatively low casualties in April, plus a record retraining output of nearly 25,000 that month, had resulted in the continued accumulation of replacements in the depots of the Replacement System. Toward the end of April the stockage reached a total of 86,000 men, and was beginning to exceed the capacity of the Replacement System's accommodations. Partly to relieve the burgeoning Replacement System, partly to meet the needs of the armies in their occupational role, the theater late in April made two additional allocations to the 12th Army Group as overstrengths. On 27 April it ordered the release of 10,000 infantry replacements for attachment to the 106th Division to serve as prisoner of war guards, and two days later it released another 15,000 to the 12th Army Group for similar duty.[23]

While the expanded retraining program thus more than met the need for infantry riflemen, the shortage in at least two other rather critical categories—armored force enlisted men and infantry officer replacements—was never completely eliminated. The shortage of armored replacements, particularly tank crewmen, became especially acute in March and April 1945, when deficits ranged from 3,000 to 5,000. The War Department insisted that it could no longer furnish armored replacements in the numbers desired except by diversions from infantry training as well as diversions of replacement tanks, conditions which the theater was reluctant to accept. The Replacement System itself was ill-equipped to offer the required training. A proposal by General Eyster, the COMZ G-3, to use recently arrived armored divisions like the 16th and 20th as training units, or to withdraw other armored units for use as training cadres, found little favor with the field commands. Rather than release units for that purpose, both army groups preferred to conduct their own training, converting either their own armored infantrymen or other infantry replacements into tank crewmen. Divisions in the First Army attempted to meet their requirements by transferring men to their attached tank battalions for training. But this proved impracticable, and the army

[22] Ltr, Hq ETO to Major Comds, 4 May 45, sub: Duties and Responsibilities of the Deputy Theater Comdr, Ltr, Rodgers to Lear, 3 May 45, sub: ETO Manpower Board, ETO 322.01/3 Manpower Board and Manpower Section; History of the Theater Manpower Section, p. 10, ETO Adm 560.

[23] Memo, Theater Manpower Section for Lear, 27 Apr 45, SHAEF G-3 370.092 Reinforcements; Teleprinter Conf, ETO/SHAEF, 28 Apr 45, ETO Adm 402; TWX FWD-20096, SHAEF to Lear and 12 A Gp, 27 Apr 45, and TWX FWD-20181, SHAEF to 12 A Gp, 29 Apr 45, ETO 200.2 Manpower.

finally established its own training center for the conversion of new infantry replacements into tank crewmen.[24]

The training of infantry officer replacements had got under way late in February, the officer candidate school at Fontainebleau beginning on the 23d and the course for the conversion of officers of other branches on the 24th. The theater planned to meet by far the greatest portion of its needs through the officer candidate school, which was scheduled to start three classes of 240 men each per week, each class graduating about 200 men. The officer conversion course was planned to start classes of approximately 200 officers each at two-week intervals.

Early in March the outlook for eliminating officer shortages was still gloomy. Officials predicted a shortage of nearly 5,000 as late as the end of May, even taking into consideration the eventual output of the training program. Personnel officers saw only two possibilities of reducing the deficit earlier—by shortening the officer candidate course, and by increasing the number of direct commissions. Any results from the first would obviously be deferred several weeks. Nevertheless, General Lear in March ordered the officer candidate school course shortened from twelve to eight weeks, effective with the first class. At the same time the theater commander again urged on both army group commanders the necessity to increase the number of direct appointments as the only means of quickly alleviating the acute shortage.[25] The 6th Army Group had possessed an especially good source for appointees to officer candidate school in the several thousand ASTP students and preflight cadets who had been distributed among several of its divisions before they left the United States.[26]

The outlook on infantry officers soon took a turn for the better, mainly because of lower casualties in the last months. The number of battlefield appointments never lived up to expectations, despite the change in policy which permitted units to retain the officers they appointed. Meanwhile the entire officer training program proved too belated to have any effect in relieving the theater's shortage. Neither the officer candidate school nor the conversion program, both of eight weeks' duration, produced any infantry officer replacements until the last week of April, less than two weeks before the end of the fighting in Europe.[27]

---

[24] Memo, Eyster for Lear, 11 Apr 45, Memo, Shannon for Lear, 14 Apr 45, sub: Conf with WD G-1 re Reinforcements, ETO 200.3 Manpower; Stf Study, Bull for Smith, 15 Apr 45, sub: Armd Replacements, Cbl WX-23158, AGWAR to COMZ, 19 Jan 45, and Cbl WX-27725, AGWAR to COMZ, 27 Jan 45, all in SHAEF G-3 370.092 Reinforcements; [Henderson] Procurement and Use of Manpower in the ETO, pp. 74-76; FUSA Rpt of Opns, 1 Aug 44-22 Feb 45, Bk. II, p. 142.

[25] Memo, Col F. J. de Rohan, Chief Training and Experiments Sec G-3 SHAEF, for G-3, 22 Feb 45, sub: Battlefield Appointment, Ltr, Barker to Smith, 7 Mar 45, sub: Off Replacements, Ltr, Eisenhower to A Gps, 11 Mar 45, Memo, Shannon for G-1, 22 Mar 45, sub: Estimated Off Situation Through Aug 45, and Memo, Twitchell for Bull, 20 Mar 45, sub: Off Shortages, all in SHAEF G-3 370.092 Reinforcements; Memo, Lear for G-1 SHAEF, 26 Mar 45, sub: Combat Off Reinforcements, ETO AG 319.1/1 Daily Progress Reports, II.

[26] The developing manpower shortage had led the War Department to discontinue the Army Specialized Training Program and transfer 10,000 students, plus 30,000 preflight cadets to combat formations about ten months earlier. Ltr, Devers to Eisenhower, 17 Mar 45, SHAEF G-3 370.092 Reinforcements.

[27] [Henderson] Procurement and Use of Manpower in the ETP, p. 133.

(3) Results of the Retraining Program

Statistically, the results of the emergency conversion program were impressive enough. Between 1 February and 30 April, when ETOUSA suspended the withdrawal of general assignment men from the Communications Zone and the Air Forces, 78,708 men were transferred to the Replacement System for retraining as infantry riflemen. About 30,000 of these had come from the Air Forces, and 49,000 from the Communications Zone. Approximately 70,000 out of this total eventually completed retraining, at least half of them after V-E Day. Adding the conversions carried out between August 1944 and February 1945 raises the cumulative withdrawals to 147,600 men and the total completing retraining to 113,700.[28]

The training of infantry officers, unlike the enlisted conversion program, was almost wholly a product of the emergency program begun in February 1945. The officer candidate school at Fontainebleau, the twenty-ninth and last class of which finally completed training on 14 July, commissioned a total of 4,171 second lieutenants out of a total of 6,614 candidates. In addition, the Replacement System gave conversion training to 560 officers of other branches, organized into five classes, and indoctrination courses to about 500 men commissioned in the field, organized into four three-week courses.[29]

Organized and expanded in great haste and under pressure, the conversion program suffered several shortcomings which it would not have suffered had planning and preparation been more deliberate. Not the least of these was the haste with which men were withdrawn from service units before their replacements could be properly trained. In most cases there was no time for formal training in the schools established for that purpose, and in any case the facilities for such training were limited. When the conversion program got under way in earnest the demand for replacements in service units became so great that men had to be assigned immediately and given on-the-job training.

The noncommissioned officers withdrawn from service units posed a special problem. In many cases the noncommissioned officer held his rating because of technical qualifications. He often had no experience in command except in an administrative setting. As a trainee under the conversion program he had to learn to be an infantryman and also how to lead a squad or platoon. In the early stages of the program in 1944 noncommissioned officers were given every opportunity to retain their ratings. If they did not meet the minimum standards aften ten weeks of training they were turned over to classification and assignment officers for reassignment. Under the accelerated and expanded program in 1945 noncommissioned officers were examined at the end of six weeks. Those who did not meet the minimum standard were immediately enrolled for another period. Those who did not qualify after the second period were subject to reduction in grade. Many of the retrained noncommissioned officers did not meet the test of battle.[30]

Opinions vary as to the quality of the men trained under the emergency pro-

[28] *Ibid.*, p. 140.

[29] *Ibid.*, pp. 132–37.

[30] *Ibid.*, pp. 157–59

gram of 1945. Both the officer candidate program and the enlisted conversion program were organized in great haste and under definite handicaps. The fact that none of the officers trained after January 1945 were tested in combat makes it hazardous to guess at the state of their training and preparation to lead other men in combat. Since the candidates for the officers' course were selected men, many of whom had had experience in combat; since instruction was modeled on that given at the Infantry School at Fort Benning; and since instruction was actually given by an experienced cadre dispatched from that center, it is reasonable to assume that the products of the school at Fontainebleau would have met the test of combat creditably.

Opinions vary as to the caliber of the infantrymen trained under the conversion program. Divisional commanders generally agreed that it did not match that of replacements trained by the replacement centers in the United States.[31] The two factors mainly responsible for this were: (1) the lower standard of training which the Replacement System was obliged to offer, and (2) the poorer material which the Communications Zone and Air Forces released for such training. Training was not one of the major duties originally assigned the Replacement System, for it had been established policy that the theater should avoid duplicating the training function performed in the United States. The Replacement System had been called upon to offer some conversion training in the summer of 1944, and thus had had some experience. But it was not prepared, from the point of view of either equipment, personnel, or sites, for the sudden expansion early in 1945. The Replacement System at that time had a training capacity of approximately 18,000 men. The new program required that it increase its capacity for infantry retrainees to 40,000, and, in addition, provide facilities to train nearly 6,000 officer candidates and officers simultaneously. General Matchett estimated a need for 3,200 training personnel for the expanded program. Two provisional depots, the 6900th and 6960th, were in fact established early in 1945, but the Replacement System was consistently handicapped by lack of both training personnel and equipment.[32]

There is no doubt that the quality of manpower released for retraining to infantry was distinctly inferior to the average of personnel sent to the replacement training centers in the United States. Commanders of both the Communications Zone and the Air Forces saw in the emergency retraining program an opportunity to rid their units of misfits and undesirables, and the absence of any qualifications other than physical fitness and a maximum age limitation led many a commander to make the most of the opportunity to "clean house." The results which such practices could have were strikingly illustrated by a survey of replacements in one of the First Army's

[31] Reinforcement System and Reinforcement Procedures in ETO, Gen Bd Rpt 3, pp. 24–25.

[32] Ltr, Matchett to CG U.K. Base, 10 Jan 45, and Ltr, Matchett to Lee, 10 Jan 45, both in History of GFRS, Ch. V, pp. 72–73, 77; Ltr, Matchett to Lee, 15 Jan 45, sub: Retraining Program, and Ltr, Matchett to Lee, 15 Jan 45, sub: Pers Rqmts for Activation of Units, GFRC 353 Training, 1945; [Henderson] Procurement and Use of Manpower in the ETO, pp. 126, 147–54.

forward battalions early in April, which revealed that 514 men released by the Air Forces had a total of 231 court-martial convictions. Later that month General Lear complained to General Spaatz that about 22 percent of all Air Forces personnel released for retraining had records of court-martial convictions. USSTAF admitted that as a result of "undue haste, overzealous and faulty administration" it might have transferred a disproportionately large number of men with inferior records, and promised to prevent a repetition of the error. But by that time all withdrawals had ceased.[33]

[33] Memo, Lear for Spaatz, 30 Apr 45, sub: Pers Allocated by Air Force for Retraining as Infantrymen, and Ltr, Brig Gen E. P. Curtis, CofS USSTAF, 5 May 45, ETO 353/1 Retraining; [Henderson] Procurement and Use of Manpower in the ETO, pp. 99–100, 143.

CHAPTER XVIII

Local Procurement on the Continent, June 1944–August 1945

(1) Purpose and Policy

U.S. forces foresaw the possibilities and advantages to be gained from the procurement of labor, supplies, and services within the theaters of operations early in the war, and made plans to exploit continental resources long before the invasion took place. There was ample precedent and justification for such a policy. U.S. forces had already made extensive use of local resources in the United Kingdom.[1] The use of indigenous labor could mean substantial savings in military manpower. Procurement of supplies and services on the ground could relieve the strain on production facilities in the United States. In some cases it could meet emergency requirements the requisition of which on the zone of interior would involve unacceptable delays. Perhaps most important, it could conserve shipping, of which there was a chronic shortage throughout the war.[2]

While U.S. forces could and did pay directly for some supplies and services procured locally, the amounts acquired in this manner were negligible. The great bulk of supplies and services so acquired were arranged for under the terms of reciprocal aid agreements concluded with the various governments involved, according to which no payments were made in cash. Reciprocal aid, or reverse lend-lease, was a part of the lend-lease device for pooling Allied resources, and provided a means by which countries receiving lend-lease benefits could repay the United States "in kind or property, or any other direct or indirect benefit which the President deems satisfactory."

The first reciprocal aid agreements negotiated with the governments-in-exile of the continental nations in the summer of

[1] See *Logistical Support*, I, 253–58.

[2] Except as otherwise noted, this chapter is based on two monographs: [Henry G. Elliott] Local Procurement, U.K. and Continental, Part X of the Administrative and Logistical History of the ETO, Hist Div USFET, March 1946, MS; and [Royce L. Thompson] Local Procurement in the ETO, D Day to V-E Day, December 1954, MS, both in OCMH. The former was written from preliminary studies and histories prepared in the Office of the Chief of Engineer, ETO, by the General Purchasing Agent, ETO, and from COMZ section reports and records of ETO headquarters. The latter is based on the History of the General Purchasing Agent, ETO, histories of the COMZ sections and technical services, the Army after action reports, reports to the President on Lend-Lease operations, and certain SHAEF and EUCOM files on procurement.

1942 generally applied only to procurement in areas then under the control of those governments. Agreements were later signed covering U.S. procurement on the Continent. The agreement covering Metropolitan France was concluded on 28 February 1945, retroactive to 6 June 1944; those with Belgium and the Netherlands were signed on 18 and 30 April 1945 respectively, and were also retroactive and superseded earlier agreements. The United States never entered into a formal reciprocal aid agreement with Luxembourg. U.S. forces received aid from it under the sponsorship of the Belgian Government, which paid Luxembourg for supplies and services furnished by that country.[3]

All the agreements covering the Continent recognized the limitations of the economies of the occupied countries, and foresaw that local procurement might defeat its own purpose if it necessitated compensating imports to support the civilian populations. They specifically prohibited local requisition of certain items like medical supplies and soap, most foodstuffs, and POL, and expressly forbade Allied personnel to purchase food in restaurants. A preliminary "Memorandum Relating to Lend-Lease and Reciprocal Aid" entered into by French and U.S. authorities in Washington included a provision that supplies and materials in short supply for the civilian population requiring replacement by import through dollar purchase would not be furnished as reciprocal aid except in minor quantities or as component parts. This clause subsequently caused difficulties for U.S. authorities in the procurement of items such as coal, which was in critically short supply. The French later insisted that coal used by U.S. forces either be purchased with cash or replaced through imports from the United States, even though such shipments were precluded by the shortage of shipping.[4]

Experience in the United Kingdom had indicated some of the administrative problems involved in the procurement of both labor and supplies, and ETOUSA attempted to establish policy on such matters several months before the invasion and to estimate its needs. At the Allied level SHAEF established a Combined Military Procurement Control Section within its G–4 Division early in April 1944 to formulate policy on local procurement and to supervise its execution. Later that month it laid down policy on local procurement in neutral, liberated, and occupied areas, directing the maximum resort to local procurement of "supplies, facilities, billets, and services."

The actual implementation of local procurement policy lay with the respective national forces, each operating its own program on the Continent. On the U.S. side over-all supervision of local procurement was the responsibility of the theater General Purchasing Agent (GPA), first designated in 1942.[5] The GPA held special staff status in the Services of Supply until the consolidation of January 1944, when the position became subordinate to the G–4 Section. Brig.

[3] History of Reciprocal Aid, 9 May 1941–31 December 1945, prep by International Div ASF, pp. 1–6, 50, 52–53, 99, MS OCMH.

[4] *Ibid.*, pp. 45–46.

[5] See *Logistical Support*, I, 80, 254.

Gen. Wayne R. Allen, a purchasing official for railroads in civilian life, and the deputy director of purchases for the ASF before going to Europe, had succeeded Col. Douglas MacKeachie in the position in January 1943 and was to continue in the post until deactivation of the agency late in 1945.

In general the functions of the GPA were to formulate procurement policies and procedures, negotiate standard arrangements with designated representatives of the European governments for the purchase of needed supplies and procurement of labor, and issue regulations to ensure co-operation between purchasing agents of the United States in the theater and thus prevent competition among them for locally procured supplies. Within the office of the GPA a Board of Contracts and Adjustments was established to prepare agreements between the supply services of the Communications Zone and the governmental agencies concerned regarding the rental or use of materials, shipping, docks, railways, and other facilities, to assist contracting officers in negotiating contracts, and to aid in the settlement of obligations arising out of agreements with government or private agencies.[6]

Acting on experience gained in the United Kingdom and the North African theater, the office of the GPA began drafting a set of operating procedures on the use of indigenous labor and the local procurement of supplies and services early in 1944. On the GPA's recommendation General Lee charged the Engineer Service with responsibility for the actual procurement of labor, partially on the ground that it would be the first agency requiring civilian labor on the Continent. The Office of the Chief Engineer, ETOUSA, in collaboration with the G–1 and G–4, then drafted an outline policy on all procurement, utilization, and administration of civilian labor in both liberated and occupied areas, which was published as ETOUSA SOP 29 on 26 May 1944. In essence, the SOP declared that local civilian labor was to be utilized to the greatest possible extent "consistent with security," that labor was not to be procured by contract with a civil agent or labor contractor, and that wages were to be paid directly to the individual laborer.

Labor officials in the Engineer Service soon realized the difficulties in attempting to anticipate the problems involved in the use of civilian labor. Estimating requirements alone was hardly better than guesswork, and there were many unknowns regarding the availability of labor in the occupied countries, the rates of pay, and the difficulties in the administration of labor whose customs and laws on social benefits, insurance, and the like were strange.

Late in May the theater engineer published a detailed plan for organizing and administering the whole civilian labor program. Essentially it called for the administration of the program through a civilian labor procurement service, which was to operate through regional and local labor offices in the various COMZ sections. In line with the rest of the OVERLORD administrative plans, the engineer labor procurement service set up regional organizations for each of the COMZ sections then planned. Each was to have a town major office located

[6] History of Reciprocal Aid, pp. 8–9.

in the vicinity of the heaviest concentration of troops or supply installations, and suboffices in other locations where U.S. installations were likely to have need for civilian labor. Town majors, who were to handle the real estate and labor functions for the Engineer Service, were initially assigned to several French towns in the original lodgment area in the vicinity of the beaches and Cherbourg, and were given specific instructions in their duties at schools in the United Kingdom. Many of the town majors were engineers or lawyers.

The engineer labor service anticipated some of the problems which were to arise in connection with the employment of civilian labor, but it could not possibly foresee all of them. For procurement purposes, two types of labor were defined—static and mobile. Static labor comprised those workers who were employed in the area of their residence and for whom U.S. forces expected to assume no responsibility for food, clothing, or housing. Mobile labor comprised those workers who were organized into units and could be moved from place to place as needed, and for whom U.S. forces took the responsibility for providing food, clothing, and shelter. The basic unit was to be the mobile labor company, consisting of a military cadre of 5 officers, 23 enlisted men, and 300 laborers.

On the method of procurement the engineer manual specified that until a definitive administrative procedure was established, representatives of the using services should procure labor from any available source. Eventually, however, labor was to be procured through local employment offices designated by the regional labor office of the civilian labor procurement service. Local and regional offices, it specified, were to investigate all static and mobile labor, prepare employment contracts before turning the workers over to the using service, and prepare individual records containing information on illnesses, accidents, forfeitures, deductions, and so on. Mobile workers were, in addition, to be given physical examinations to guard against communicable diseases and physical handicaps which might result in claims against the U.S. Government.

Labor was to be allotted to the various using services, which in turn would requisition against such allotments through the nearest regional labor offices and determine where the men would actually be used. Emergency labor could be employed on an oral basis; all other workers were to be employed on individual written contracts setting forth conditions of employment and signed by the using services on behalf of the U.S. Government. Practically all civilian labor initially was to be regarded as unskilled. Using services were to classify workers after their skill had been determined on the job.

Setting wage scales promised to be a troublesome task even before the first experience on the Continent. In general, theater and SHAEF at first refrained from declaring any definite wage scales, announcing simply that every effort would be made to conform to local rates of pay and conditions of work. Job assignments were to determine the rate of pay irrespective of qualifications, and workers were to be classed as soon as possible according to ability. Just before D Day, however, ADSEC published maximum and minimum wage rates for four classes of workers according to zone, communities of less than 20,000 persons having one scale of wages and communities of more than 20,000 another. Rates for women were set at 75 percent of those for men, and monthly rates were set for clerical and supervisory workers and for hotel, mess, and hospital employees. The policy was generally adopted that food and clothing would not be provided workers. Finally, the General Purchasing Agent compiled an estimate of U.S. requirements for civilian labor, which totaled nearly 50,000 at D plus 41, and rose to 273,000 at D plus 240.

The GPA's draft of operating procedures on the procurement of supplies and services appeared as ETOUSA SOP 10 on 1 April 1944. It made clear that the GPA was to act primarily as a staff agency, supervising and co-ordinating the purchases by the operating services and major headquarters, and disseminating information on the possibilities of local procurement. The actual procurement of supplies was to be undertaken by the technical services and COMZ sections, and by the field commands, including the army groups, armies, corps, and divisions, all of which were authorized to engage in local procurement. Each of the supply services of the SOS organized procurement offices early in their history, and each of the COMZ sections likewise named officers to serve as purchasing agents. In general, the purchasing offices of the supply services and the COMZ sections dealt primarily with long-range requirements, known normally as headquarters procurement. Meeting short-range needs was known as field procurement, and was handled by purchasing and contracting officers who were specially designated by the field commands and sometimes by the COMZ sections. The First Army made local procurement a responsibility of its G–5, or Civil Affairs Section. The Third Army assigned similar responsibility to the Fiscal Branch of its G–4 Section.

By D Day, then, U.S. forces had taken important steps in preparation for local procurement on the far shore. SHAEF had established general policy at the Allied level, declaring its intention of exploiting local resources to the maximum within the restraints designed to protect the war-torn economies; it had negotiated agreements with several of the refugee governments regarding the requisition of supplies, facilities, and services within their borders; ETOUSA had issued regulations regarding local procurement and had a supervisory staff agency, the GPA, in being; and many purchasing officers had received at least basic instruction in the methods and regulations on local procurement via the schools operated by the GPA at army and various logistical headquarters.

It remained to be seen, of course, to what extent the Allied forces might be able to draw on the local economies for support. The area to be liberated—France, Belgium, the Netherlands, and Luxembourg—was one of the most highly industrialized regions of the world. But all four countries had already been through nearly five years of war and occupation; much of their industry and transport had been bombed, first by the Germans, and then in some cases more intensively by the Allies. Labor could be expected to be relatively plentiful, although the French were eventually to make their own claims on manpower for the rearmament which they planned. In general, food, coal, transportation, and raw materials were the most serious deficiencies in all the areas the German Army had occupied. The shortage of food at first overshadowed all others. France had been nearly 90 percent self-sufficient in food products before the war. With the German occupation, however, imports of food, fertilizer, and animal feeds stopped, heavy requisitions were imposed, and shortages in agricultural labor developed. In some areas inadequate transportation aggravated the situation further.

The movement of grain, meat, and potatoes from surplus areas to the cities became a problem of the first order. The transport system naturally became a priority target for Allied bombers, particularly in the months just before the invasion. Combined with the German requisitions, the Allied attacks reduced the locomotive and freight car population to less than half of its prewar total. Greatly reduced auto manufacturing and the shortage of spare parts and gasoline depleted the highway transport capacity even more drastically. Part of the transport problem could be laid to the shortage of coal. Although a leading coal producer, France imported more than one third of her needs in peacetime. Despite prodigious efforts to restore the industry following the liberation, French mines were to produce less than two thirds of their peacetime rate by late 1944. The shortage of this vital commodity was to have direct bearing on potentialities for producing military supplies. Finally, raw materials, particularly those which had to be imported, were at low levels if not entirely depleted.

With minor variations, the deficiencies noted above applied also to Belgium, the Netherlands, and Luxembourg. Of the three, the Netherlands, most of which was occupied by the enemy until the end of the war, had the least to offer. The Dutch people had resisted the Nazi invader bitterly, and the Germans retaliated savagely, destroying ports, removing industries and transportation, and systematically starving the people. Belgium suffered an even greater food shortage than France, although in some other respects the enemy dealt less harshly with Belgium. Ports, for example, were left largely intact, and industrial plants, coal mines, and public utilities were ready to resume operations, given raw materials. The rail system had suffered badly, however, losing more than half of its locomotives and one third of its freight cars.

U.S. forces had of course planned to provide aid to the liberated areas, at least to meet their most urgent needs. But these plans were based on purely military requirements, which were shaped largely in terms of the "disease and unrest" formula. For the most part they did not provide items needed to rehabilitate the economies of the liberated areas. This policy was to have serious results for local procurement.

(2) The Use of Nonmilitary Labor

The first use of indigenous manpower occurred four days after the landings in Normandy, when the Provisional Engineer Special Brigade at OMAHA Beach hired thirty-seven civilians to pick up salvageable scrap. Their only pay was food. The first major need for labor arose with the capture of Cherbourg, where civilian labor was essential to the huge task of clearing away the debris, enlarging the port's capacity, and handling incoming supplies. Within two days of the capture of the city civil affairs officials moved through the streets of Cherbourg in a sound truck announcing the opening of a branch of the French labor office and offering jobs. In the first week the labor office placed about 200 men in jobs with military units.

The lack of governmental authority in the lodgment area made it difficult to follow policy regarding procurement through government agencies. Military officers consequently hired French civilians directly and without resort to the labor exchange, or neglected to notify the exchange that laborers sent them by the exchange had been hired. The 4th Port, which operated Cherbourg, preferred to deal with French contractors for stevedore labor, in direct violation of procurement policy laid down before D Day. Civil affairs officials regarded this as the simplest and most satisfactory method. Under it, workers were paid the rates prevailing in the area, and the contractor received a flat fee per ton of all cargo handled as payment for taking care of the recruiting and preparing the payrolls. Meanwhile, an engineer real estate and labor office opened in Cherbourg on 12 July and eventually established a uniform procedure for requesting labor.

The engineer office also began to allot the available labor to using services, for the supply was soon inadequate. By the beginning of August a shortage of about 2,500 men had developed against current requests by the various using agencies. The use of prisoners of war in August gradually relieved the shortage, and by the end of September the deficit was completely eliminated. From that time on there was no attempt to replace French laborers who left their jobs.

Cherbourg was virtually the only city in northern France in which the labor shortage became serious. After the breakout from Normandy early in August the rapid liberation of the rest of France and Belgium uncovered a more than adequate supply of labor. In general, requirements did not even approach those estimated in plans, partly because of the large numbers of prisoners of war who became available, and partly because of the course of operations. In Brittany, for example, the decision not to restore the port of Brest and develop a major logistic base as planned almost completely voided plans for the use of indigenous labor. Instead of the 12,000 men which the using services originally estimated they would need in Brittany by D plus 90, therefore, less than 2,000 civilian workers were actually employed on that date. At the peak, in December 1944, Brittany Base Section employed less than 4,000 men.

In other areas, meanwhile, the number of civilian employees steadily rose, even though it did not reach the expected volume. The Advance Section, whose area of operations was forever changing, employed an average of less than 3,000 throughout the summer and fall. But in the spring of 1945, when its activities embraced parts of Belgium, Holland, Luxembourg, and Germany, it employed as high as 17,000 men. Seine Section, which included COMZ headquarters, was by far the biggest user of civilian labor. In the fall of 1944, when the total

TABLE 14—CIVILIANS EMPLOYED IN THE COMMUNICATIONS ZONE IN SELECTED WEEKS, 1944–1945

Area	22 July	9 Sep	11 Nov	16 Dec	10 Feb	24 Mar	28 Apr
ADSEC	4,386	2,925	4,101	6,379	9,260	16,946	14,406
Normandy Base	(a)	5,163	7,378	7,679	19,553	20,552	22,672
Brittany Base	(a)	1,717	2,134	3,427	2,866	(a)	(a)
Seine Section	(a)	(b)	29,564	28,653	42,716	52,218	75,654
Channel Base	(a)	(a)	7,733	4,095	29,081	39,639	49,305
Loire Section	(a)	(a)	1,094	(a)	(a)	(a)	(a)
Oise Section	(a)	(a)	(b)	4,274	8,975	21,033	34,153
CONAD	(a)	(a)	(b)	(b)	(b)	6,335	9,515
Delta Base	(a)	(a)	(b)	(b)	(b)	(b)	23,706

[a] Base not in existence.
[b] Figures lacking.

Source: [Henry G. Elliott], the Local Procurement of Labor and Supplies, UK and Continental, Part X of The Administrative and Logistical History of the ETO, Hist Div USFET 1946, MS, OCMH' pp. 99–100.

number of civilian workers in the Communications Zone was averaging about 45,000, Seine Section accounted for fully half, and at the end of the war was employing between 75,000 and 80,000 persons.

Southern France was the major exception to the general rule that labor was adequate. Continental Advance Section, in particular, suffered shortages, and employed an average of only about 5,000. On the whole, however, France, with its economy at low ebb for lack of raw materials and power, and with no military establishment to speak of (the French divisions raised in North Africa were largely equipped and supplied by the United States) provided a relative abundance of manpower. Table 14 indicates the number of civilians employed in the COMZ sections at selected dates.

Administrative problems involved in the use of French labor eventually proved to be much more worrisome than any shortages. The two most onerous were related to wage scales and methods of payment. In these fields, as in the matter of procurement, U.S. forces found it necessary to accommodate the policies laid down before D Day to the realities of local conditions. Preinvasion policy directives had specified, for example, that only minimum wages would be paid. A wage committee began to reconsider this policy within the first two weeks after D Day. It also proposed that the Allied forces pay family allowances, an important feature of the French pay structure, so that civilians working for the Allied forces would enjoy the same advantages as other civilian workers.

The system of family allowances made the wage structure a complex one, and U.S. officials were reluctant to undertake the additional administrative burden involved in paying allowances. While French labor was hired through the engineer service, payment was the responsibility of the employing unit, which lacked the personnel and facilities to handle the involved computations. SHAEF nevertheless ordered the payment of family allowances, specifying that payments begin on 1 August. French officials agreed to administer the system, making payments from funds equal to 10 percent of the total payroll, which U.S. forces were to provide.

New wage rates, established in agreement with the Provisional French Government, also went into effect at the same time, and on the basis of six zones rather than four. Zone 1 comprised metropolitan Paris and paid the highest wages in recognition of the higher living costs in that area. Six grades of workers were established, ranging from "unskilled," which required no special training nor physical aptitude, to "very skilled," which required professional training. Regulations on overtime were also changed at this time to provide for higher rates only where employees worked more than forty-eight hours in one week. An agreement reached late in August specified that overtime rates would also be paid for work on the approved French holidays, ten in all.

At the time the new wage rates went into effect Cherbourg was still the only city where civilian workers were employed in important numbers, and the new rates actually involved a reduction in wages for most workers. But these reductions were offset by more liberal rates for night work and by the payment of family allowances. Failure to inform workers of the latter change led to a two-hour strike on 6 August in an ordnance project near the port.

The change by which the French took over the administration and payment of indigenous labor was a more gradual one. SHAEF and the French Ministry of Labor agreed late in September that the administration and payment of all French civilians employed by U.S. forces should be assumed by the French as soon as possible, and ETOUSA made plans for the transition beginning in mid-October. Whenever possible, the engineers were thereafter to procure laborers through the representatives of the French Services à la Main d'Oeuvre. French authorities were to begin the changeover by paying social insurance, workmen's compensation, and family allowances.

Putting the agreement into practice proved a long and involved process. French administrative machinery, for one thing, simply did not exist for the purpose in many areas, with the result that U.S. officials had no choice but to continue performing the main administrative tasks connected with employing French civilians. Secondly, wage rates proved to be a major stumbling block to a smooth transition. In many cases it was found that the official SHAEF rates were still badly out of line with local rates. In the case of hotel employees, U.S. policy, which specified "no tipping," was completely at variance with French custom whereby employees received a low wage and depended on tips for a substantial portion of their income. Hotel managers resisted proposals to raise wages to compensate for the loss of tips for fear that employees would not be willing to return to the lower rate after U.S. forces turned hotels back. This particular dilemma was finally resolved by transferring all hotel workers to the employ of the French Ministry of Labor.

In the first stages of the transition U.S. forces continued to meet payrolls, and the French initially took over the payment of only social insurances and taxes.

French officials gradually took over the job of paying, but it was December 1944 before they assumed full administrative responsibility in the Paris area, and February 1945 before they assumed responsibility in eastern France.

U.S. forces no longer determined wage rates once the French assumed responsibility for the administration and payment of workers. Theoretically, the French Government enforced maximum pay rates for all skills, so that U.S. forces would not be at a disadvantage in competing with French industry for the available labor. In actual practice, French officials were often unable, and sometimes unwilling, to keep wage scales within the established ceilings, and substantial discrepancies developed between the rates paid by the U.S. forces and by French industry, with a resulting high rate of turnover. Some private employers circumvented regulations by offering other inducements, such as bonuses and meals.

ETOUSA policy directives had originally specified that neither food, clothing, nor shelter would be provided static workers. But U.S. forces eventually were forced to make exceptions, in part because of the above-mentioned practice of French industry. Late in November 1944 the theater directed that static employees might be furnished one meal per day when they were unable to obtain meals through normal civilian facilities, when malnutrition resulted in loss of effective work output, or when the civilian food ration was such that it could not be broken down to permit employees to carry meals to the place of work. Demands for clothing were more successfully resisted.

Experience with civilian labor in Belgium, Holland, and Luxembourg generally paralleled that in France. In Belgium and Holland, in particular, the main problems were those connected with the establishment of wage rates, and with the transfer of administrative responsibility to national authorities. SHAEF published wage rates for Belgium in July 1944. But these were intended to serve only as a temporary guide, and ETOUSA made efforts soon after Allied forces entered Belgium in September to have the Belgian Government take over responsibility for the payment of civilian workers employed by U.S. forces. Belgian authorities agreed to begin paying employees of U.S. forces on 15 October. But the government issued no over-all guide or directives on pay scales, and many burgomasters were hesitant or unwilling to establish rates without guidance from above.

In practice, therefore, differences over wages and working conditions had to be ironed out locally, and this was often done in meetings attended by local government officials, trade union representatives, U.S. and British military representatives, and an official representative from the interested ministries of the central government in Brussels. Not until early in March 1945 did ETOUSA issue a comprehensive directive covering the procurement and handling of Belgian labor. It announced that procurement was to be effected wherever possible through the local burgomaster, who would also be responsible for the administration and payment of wages, including social insurance, family allowances, and taxes. It also established policy on the provision of meals and overtime. At the same time the Belgian govern-

FRENCH CIVILIANS EMPLOYED BY U.S. FORCES *repairing a railroad in southern France.*

ment announced its intention of assuming responsibility for administering payments to the civilians employed by Allied forces. By the end of April it had assumed responsibility for fixing the rates of pay, had set up regular committees composed of government and trade union representatives to handle labor problems, and was gradually taking over responsibility for the administration and payment of all civilian labor.

One feature of civilian employment in Belgium not common to other countries was the payment of bonuses for workers in areas which were the target of large air raids or the new V-weapons. The Belgian government first offered "danger pay" to civil servants in an attempt to induce them to work in Liège, which became the target of German V-weapons. Municipal authorities at Antwerp at about the same time offered a 25-percent increase in wages for hazardous work in that area because of attacks on the docks. Danger pay was difficult to administer because it was based on a zoning system and because it was suspended whenever aerial bombardment ceased for a specific number of consecutive days. Opinions were mixed as to its worth or wisdom. Civil affairs officials reported that danger pay encouraged inflationary tendencies, undermined the wage structure, made it more difficult to get men to work overtime, and encouraged every locality under aerial attack to demand the additional compensation. In at least one instance failure to pay the additional compensation led workers in Antwerp to strike.

Workers in Belgium struck for other reasons as well—at Brussels, because of the refusal of a midshift meal, at Ghent because of the practice of employment through contractors, and at Antwerp for a long list of grievances, including insufficient transportation to and from the dock area, failure to provide a noonday meal, and failure to pay insurance benefits to workers injured as the result of enemy actions. The strike at Antwerp, in mid-January 1945, was of vital concern to the Allies. Although of short duration, it was ended only after the Allied authorities and the burgomaster gave firm assurances of action on the grievances, particularly as to the provision of coal and food.

In the Netherlands and Luxembourg, as in France and Belgium, SHAEF published a temporary guide to wage rates for labor, with the intention that adjustments would be made later and the actual administration and payment taken over by national or local authorities. The liberation of Holland proved much slower than that of France, and Allied forces administered and paid for civilians employed by them for fully six months after they entered the country. Dutch authorities unertook to procure labor for U.S. forces through the burgomasters in January 1945, and to administer the payment of social insurance, family allowances, and workmen's compensation, but they were unable to administer and pay wages. The usual regulations on the provision of meals applied, but Allied authorities found it necessary in February 1945 to provide overcoats from military stocks to keep Dutch workers on their jobs. U.S. forces never employed more than about 2,500 civilians in Holland as compared with 13,000 employed by British forces. U.S. forces employed about 2,600 at the peak in Luxembourg.

SHAEF laid down the basis for the employment of indigenous labor in Germany in mid-September 1944, only a few days after U.S. forces crossed the German border, and in mid-October ETOUSA issued a detailed directive on the procurement and payment of labor recruited within Germany. Lacking specific data on German wage scales, ETOUSA issued temporary wage schedules based on a zoning system and the division of various classes of labor into grades, as SHAEF had done in the case of liberated countries.

German labor was not used in important quantities until January 1945, when the Advance Section began operating within the enemy's boundaries. The Geneva Convention and Rules of Land Warfare restricted the use of enemy nationals to activities not directly connected with combat operations. In any case, U.S. forces had an immediate source of labor in the many displaced persons uncovered. By January a study of local wage rates had revealed that the temporary rates established in October were higher than those in effect before the Allied entry. Moreover, it was discovered that there was no national system of wage zones such as existed in France. Instead, labor trustees in twelve regional labor offices had considerable freedom in determining working conditions and rates of pay, thus producing substantial wage differentials among various localities.

These discoveries eventually led Allied authorities to publish revised schedules based on the most recent German wage tariffs covering personnel in public employment. As progress was made in repatriating displaced persons, U.S. forces gradually employed more and more Germans. But the numbers so employed were not very great until after the end of hostilities. As of 1 May 1945, the Advance Section was employing approximately 14,000 persons in Germany, displaced persons included.

Plans to organize mobile labor units, as contrasted with the static labor described above, were not very successful. The entire program got off to a slow start. The Advance Section organized about 600 men as a mobile labor force in

July, but this force consisted entirely of former Todt Organisation workers obtained through prisoner of war channels. The first efforts to organize French civilian workers actually met with failure. French authorities estimated in August 1944 that they could furnish between 20,000 and 35,000 men for mobile units. But the offer carried the condition that U.S. forces provide clothing and various supplies, which the War Department refused to allocate for that purpose. Recruiting for mobile labor units went ahead in the Brittany area. Only kitchen equipment and rations were made available. Sufficient laborers to form one company were medically examined and processed by early September. By that time, however, the demand for labor was already being fairly adequately met either through static labor at the places where it was needed, or through the increasing number of prisoners. There were distinct advantages in using mobile labor units, the most obvious being that they could be moved about as needed. They also possessed an advantage over prisoners in that they required no guards and could be used in handling ammunition. But there were continuing difficulties in organizing such units, particularly in getting sufficient equipment, and the number of men thus employed did not exceed 4,000 until February 1945. The theater found greater use for such labor in the late stages of the war and, via a Military Labor Service which it had activated in December 1944, it had organized about 20,000 men into 108 mobile companies by the time hostilities ceased. Ample quantities of civilian labor became available in the closing stages of the war through the liberation of thousands of displaced persons. Their use was discouraged, however, except where prisoners could not be used, because of the greater expense involved and because of the likelihood of losses through repatriation.

Establishing wage scales was a problem with mobile as well as with static labor. SHAEF decreed a method of establishing wage scales for mobile labor in August, based on the zone system which had been adopted. But French authorities challenged these rates immediately, and both U.S. and British forces agreed to revisions. Early in 1945, in response to complaints over the practice of basing pay rates on the zone in which workers were hired, U.S. forces agreed to pay workers transferred from one zone to another at the scale of the higher zone.

Difficulties also arose from the desire of U.S. forces to move mobile labor units across national boundaries. The problem first arose when U.S. forces transferred interpreters hired in France to Belgium. The Belgian Government would assume no responsibility for paying these workers under reciprocal aid. Pending settlement of the issue, therefore, U.S. forces undertook to pay such workers. More serious difficulties arose when U.S. forces attempted to move civilian workers into Germany, the problem first arising in the case of Belgians employed by the First Army and the Advance Section. Negotiations with the governments concerned through the various SHAEF missions eventually resulted in authorization to move the nationals of Belgium, the Netherlands, and Luxembourg into Germany. But French labor officials declined such permission except for the workers who had already accompanied U.S. forces across the border, totaling about 2,300. The ruling was particularly hard on civilian censorship detachments, which needed nearly 5,000 workers with a knowledge of German to censor civilian mail in the U.S.-occupied zone after V-E Day and had already begun to train French civilians for those jobs. Paying such workers also posed a problem, for their respective governments declined to exchange the reichsmarks in which they were paid. The General Purchasing Agent eventually made arrangements with the various Allied governments to pay workers a small part of their wages in reichsmarks to meet immediate personal expenses and for the government concerned to pay the balance to dependents.

Italian service units, formed from prisoners taken in North Africa, provided a unique and substantial supplement to the labor resources of the European theater. The North African theater had first undertaken the formation of such units in the fall of 1943, after Italy was accorded the status of a cobelligerent. Italian soldiers meeting the required physical and mental qualifications were organized into service units with specified T/O&E's similar to those of U.S. units. ETOUSA first had occasion to use such units in the summer of 1944, when the War Department ordered the shipment of 7,000 laborers from the North African theater to the United Kingdom in response to a request for prisoners to meet the Quartermaster's labor requirements there. The number of Italians employed by such U.S. forces in the United Kingdom built up to nearly 8,000 by November, when disagreements between U.S. and British authorities over rates of pay and treatment led to their gradual withdrawal. All such units had been transferred to the Continent by the end of April 1945.

ITALIAN SERVICE UNIT MEN LOADING CASES OF RATIONS *at a Quartermaster Depot near Marseille.*

Meanwhile Allied authorities had obtained the approval of the de Gaulle government to transfer 30,000 (later 38,000) enlisted Italians to France in anticipation of the need for labor at Marseille. The movement to southern France began in September, and by March 1945 the strength of such units there had nearly reached the total number authorized. Most of the units were initially employed in southern France, although Oise Intermediate Section also became a heavy user, and along with Delta Base Section eventually accounted for 28,000 of the 38,000 employed in the last month of hostilities.

Italian co-operators, as they were called, enjoyed a special status. Legally they remained prisoners of war; but they were released from stockades (although placed in custody of American officers attached to their units), and certain provisions of the Geneva Convention were waived, notably the restrictions against the use of prisoners on work having a direct connection with combat operations. In general, they could be used in any type of work except in actual combat, in handling classified materials, in close proximity to other prisoners, or in areas where they would be in danger of capture by the enemy. Among the types of units formed and transferred to the ETO were engineer general service regiments, ordnance heavy automotive maintenance units, and quartermaster service companies. U.S. supervisory personnel were attached to each unit, the senior U.S. officer acting through Italian officers and noncommissioned officers. Special regulations were issued on such matters as pay, remittances to Italy, compensation for injury, the dispatch of mail, off-duty privileges, and so on.

In addition to the Italian service units, twenty-seven Slav service units were also eventually shipped to France, most of them in the last month of hostilities. These units were made up of Yugoslavs who had served in the Italian army and had not been taken prisoner. Among them was a large number of artisans, particularly skilled mechanics, whose productivity and efficiency were generally rated as superior to those of the Italians. All of the Slav units were used in Delta Base Section.

Prisoners of war constituted an even larger source of labor than civilians, and plans for their use proved far too conservative. Before D Day the services estimated requirements for only 14,000 prisoners up to D plus 90. On that date they were already employing nearly 40,000. At the end of April 1945 U.S. forces were using approximately 260,000, which comprised nearly one quarter of the total work force in the Communications Zone.

As in the case of civilian labor, Cherbourg was the first place where U.S. forces employed prisoners in large numbers. Approximately 1,000 prisoners arrived at the port early in August, and were immediately put to work with engineer units in reconstruction of the port. In accordance with the Geneva Convention and Rules of Land Warfare, ETOUSA had specified that prisoners would not be employed on work involving the handling of arms or munitions, nor in fact any work the primary purpose of which was the support of combat operations. ETOUSA did not lay down detailed instructions on the organization and use of prisoner labor until October 1944. It then specified that prisoners should be formed into labor companies of approximately 250 men each, and it outlined the responsibility of section commanders and the using services with regard to the administration of prisoner of war units, including such matters as enclosures, medical care, purchase of post exchange items, and the handling of escapees and deaths. The original tendency had been for using units to form 212-men companies modeled roughly on the T/O of the quartermaster service company. Such units provided 160 common laborers in addition to supervisory and overhead personnel and training cadres.

In the first months after the landing most of the prisoners came from combat units. Few of these men possessed technical skills, and they were used mainly on tasks calling only for common labor. The wholesale captures of later months extended to enemy service units and uncovered more men with the desired technical training. The need for such trained men became especially great with the inauguration of the conversion program in January 1945, when large numbers of men were withdrawn from service units for conversion to infantry. ETOUSA then encouraged the formation of technical service units from prisoners and sought War Department approval for equipping such units under T/E's similar to those under which Italian service units were being formed. Plans had originally anticipated that the bulk of the prisoners captured would be shipped to the United Kingdom and the United States. The discovery that prisoners were an excellent source of labor and that they could even fill the need for technical service units caused the theater to alter quite radically its earlier estimates regarding the number to be retained on the Continent. The decision to use prisoners more extensively naturally created a much greater demand for accommodations and placed an unexpected burden on commandants of prisoner enclosures. Oise Intermediate Section and Normandy Base Section were by far the biggest users of prisoner labor as hostilities ended, employing 92,000 and 87,000 respectively and accounting for 84 percent of the total employed at the time.

U.S. plans did not at first contemplate the employment of British civilians on the Continent. But with the approach of D Day one agency after another pleaded the need for the continued use of British civilians on the Continent on the ground that highly skilled and experienced workers could not be replaced. SHAEF agreed to authorize the transfer of civilians to France, and in agreement with British officials established 1,000 persons as the maximum number that might accompany U.S. forces to the Continent. Theater headquarters early in August 1944 specified the manner of selection, in general limiting it to key specialists for whom no replacements from either military personnel or continental civilians might be found, the age limits of employees, salary scales, leave, medical and post exchange privileges, and so on. British civilians were required to pass physical examinations similar to those for enlistment in the U.S. Army, were required to wear a prescribed uniform, and were subject to the Articles of War. Several problems arose in carrying out these provisions. A suitable uniform was not available in sufficiently large quantities, for example, and arrangements had to be made for a supply of British ATS uniforms and for paying for them. British insurance and compensation laws, it developed, did not cover personnel in foreign countries, and British civilians had to be put under the protection of the U.S. Employees Compensation Act of 1916.

ETOUSA began processing British

employees for movement in the first week of August, and about 500, mostly women, were transported to France by air within the next two months. U.S. forces in the United Kingdom continued to employ about 40,000 British civilians there, although this figure began to fall off during the fall as port and storage operations shifted to the Continent.

Utilizing prisoner of war and civilian labor posed exasperating administrative problems. Some of them would have been obviated had decisions been made earlier as to the planning responsibilities, and had better information been assembled as to continental wage scales and employment regulations and customs. Some of the problems attending the use of civilian labor, such as the shortage of food and clothing, were an inescapable product of wartime conditions. Nevertheless, both prisoner of war and civilian labor served an important need which probably could not have been filled any other way in view of the increasing U.S. manpower shortage and made an important contribution to the Allied victory. By V-E Day, civilian and prisoner of war labor and Italian service units working for U.S. forces totaled 540,000 and comprised 48 percent of the total COMZ force of 1,121,650 men. Civilian laborers alone totaled 240,000 at the end of the war, and added 4,000 men to the normal division slice of 40,000. Civilians, prisoners of war, and Italian service units combined raised the size of the ETOUSA division slice by nearly 9,000 men.

In addition to meeting a vital manpower requirement, civilian labor provided an important means by which Allied nations were able to pay the United States for wartime aid. At the end of the war approximately 172,000 of the 237,000 employed by U.S. forces were being paid by their respective governments under reciprocal aid or reverse lend-lease.

(3) Local Procurement of Supplies

Compared with the eventual scale of local supply procurement, acquisitions in the first three months of operations on the Continent were relatively insignificant. U.S. forces, leaving little to chance in the early months, brought with them nearly everything they needed that was transportable. Moreover, the area initially uncovered was primarily an agricultural region, and had relatively little to offer. Even Normandy, however, could provide certain very necessary and useful items. These at first took the form mainly of real estate, certain fixed installations, particularly signal communications facilities, and construction materials.

Procurement of supplies and facilities began within the first few days after the landings, when engineers requisitioned a sawmill at Isigny and Signal Corps officials took over telephone facilities and the underground cable, plus various signal supplies such as copper wire and cable. Real estate was by far the most important single type of acquisition in the first months, a natural development arising from the requirements for bivouac areas for troops pouring into the bridgehead and for land and buildings for dumps, depots, and headquarters. By the end of June property had been acquired for one purpose or another in 110 towns in the restricted beachhead, and within three months of the landings U.S. forces had requisitioned upwards of 15,000 separate pieces of property. In addition,

GERMAN PRISONERS OF WAR FILLING 50-GALLON OIL DRUMS *from railroad tank cars.*

engineer procurement embraced sizable quantities of construction materials in the form of crushed stone, gravel, sand, and timber. Subsistence items bulked large in quartermaster procurement, especially vegetables, fresh fruits, eggs, and fresh meat, since these products were surplus in the Normandy area and could not be marketed in any other way because of the lack of transportation. In all, U.S. forces contracted for purchases or rentals with a total value of less than $300,000 in the first three months. By far the greatest part of this was accounted for by the engineer service, which had responsibility for real estate requisition. Practically all acquisitions thus far had been in the category of field procurement.

The events of August 1944 and early September brought a radical alteration in both the requirements and potentialties for local procurement. For one thing, the drive to the German border suddenly uncovered a vast industrial and transportation complex, infinitely richer in productive capacity. Meanwhile, the pursuit itself had generated an urgent requirement for just the kind of products and fabrication or processing services which the industry in that area was equipped to offer. Furthermore, the stabilization of the battle line in the fall and the gradual emergence of a French national authority made it possible to plan and arrange for headquarters procurement, in which requirements were consolidated by the supply services and submitted to the French government through the GPA.

The biggest beneficiary of this development was the Ordnance Service, whose requirements had been most affected by the rapid advance to the German border. Mid-September found all U.S. units which had taken part in the drive across northern France badly in need of maintenance and replacements, far beyond the capabilities of the ordnance heavy maintenance units available to the theater. This applied particularly to combat vehicles and trucks. Hundreds of engines had to be rebuilt as quickly as possible. Thousands of trucks were threatened with deadlining for lack of replacement tires which the United States could not supply in the numbers now needed. In addition, a heavy requirement had developed for spare parts and other ordnance items, and for weapon overhaul and modification. One of the most prominent modification projects, as it turned out, was the fabrication for tanks of a device known as an extended end connector, which was designed to widen tank tracks and thus give tanks better flotation in soft terrain. This project, the engine-rebuild program, and the attempt to manufacture tires constituted the "Big Three" of the ordnance local procurement program on the Continent.

Engine rebuild for combat vehicles, the first of the major industrial projects on the Continent, was undertaken by a tactical command—the First Army—rather than at theater or COMZ level. First Army found its tank situation critical as it approached the German border early in September. More than 200 engines had already been evacuated to the United Kingdom for rebuild and an additional 170 awaited evacuation. COMZ heavy maintenance units did not yet possess the necessary fifth echelon rebuild capability. Investigating facilities in the Paris area, First Army found the Gnome-Rhône motor works well equipped for the task, and negotiated a contract with the plant for the overhaul of about 200 continental-type radial engines for the medium tank, which was to include disassembly, inspection of gaskets and rings, reassembly, and run-in. Tank transporters delivered the first engines, together with gas and oil for testing, to the Paris plant within forty-eight hours after the contract had been signed, and the plant overhauled a total of 252 engines.

With the completion of First Army's contract early in October the Ordnance Section of the Communications Zone assumed responsibility for further work, and overhaul continued at the Avenue Kellerman Branch of the Gnome-Rhône plant. Under the early contracts only "top overhaul" was performed, partly because of the urgency of the program and partly because of the lack of spare parts. When only a small fraction of the engines so rebuilt met a test, civilian officials decided that a more thorough overhaul would have to be given, including the inspection of bearings, bushings, and seals. Meanwhile, Ordnance Service had surveyed rebuild and maintenance requirements for other combat vehicles and for various types of trucks, and decided to expand the program. With the help of information provided by the Automotive Society of France, Ordnance soon enlisted the services of other plants in the Paris area, including such well-known names as Citroen, Renault, Simca, Peugeot, Hotchkiss, Salmson, and General Motors, and gave top priority to the overhaul of such work horses as the 2½-ton 6x6 truck, the jeep, the ¾-ton weapons carrier, and later to armored cars, scout cars, half-tracks, and tractors.

The program got under way slowly, mainly because of spare parts shortages. Many an engine arrived at the rebuild plant ingeniously cannibalized. Eventually Ordnance contracted with a large number of small firms in both France and Belgium to produce such items as pistons, rings, and gaskets, and rebuild firms were also encouraged to subcontract for parts. The rebuild rate was low in the first few months, but it gained momentum early in 1945. The production rate, which barely exceeded 800 engines per week early in January, nearly doubled by the end of the month, and in the weeks just preceding V-E Day began to exceed 3,000. A grand total of slightly more than 45,000 engines of all types eventually was overhauled through local procurement.

The three months of uninterrupted offensive operations since D Day had resulted in a high mortality in tires as well as engines. Theater ordnance officials, taking into account the poor prospects of relief from the zone of interior, where tire production was already proving inadequate, and the shortage of repair and retreading units in the theater, predicted a deficit of some 250,000 tires by the end of January 1945 and feared that as many as 10 percent of the theater's vehicles might be deadlined by that time. Early in October the ASF, fully aware of the developing shortage, and concerned over its own inability to meet the theater's needs, sent Brig. Gen. Hugh C. Minton, director of its Production Division, to Europe to explore local procurement as a possible solution. General Minton, aided by another officer who was experienced in the rubber field and had long lived on the Continent, surveyed eight major rubber plants, six in France and two in Belgium, checking mold inventories and condition of machinery, and consulting with other American and British tire experts and with French management. General Minton concluded that local production was practicable and, on approval by the ASF in Washington, took steps to get the program under way as promptly as possible. On his recommendation the theater immediately formed a Rubber Committee, consisting of the theater ordnance officer as chairman, the General Purchasing Agent, and representatives of the G–4 and G–5 divisions of both the Communications Zone and SHAEF and of the SHAEF Mission to France, and made an agreement with the French on the allocation of the finished products between U.S. Army and French civil requirements. The actual supervision of the program was placed in the hands of a Rubber Branch organized within the Industrial Division of the Ordnance Section of the Communications Zone.

The key factor in the entire project was the supply of raw materials. At the time of the survey the eight plants were operating with materials left by the Germans, but these stocks were nearly exhausted. It was clear from the start that production would continue only through the importation of raw materials from the United States and Britain. The Minton mission recognized this and proposed both the quantity and schedules of shipments as part of its recommendations.

RENAULT PLANT *(background), in the Paris area, used for rebuilding combat vehicle engines.*

About one third of the rubber was to be synthetic.

Late in December a small quantity of raw materials was flown in from England, and a trial run was made at the Renault plant in Paris. By the first week in January sufficient material was on hand to begin a production program on a small scale in two plants, and within the next month all eight plants were in operation. Priority was given first to the manufacture of tires for the 2½-ton 6x6 (size 750x20) and the ¼-ton 4x4 jeep (size 600x16).

The tire production program had only a limited success insofar as U.S. forces were concerned. All raw materials had to be imported, and some components, like carbon black, were in short supply even in the United States. Furthermore, a large share of the production eventually went to meet critical French needs rather than U.S. military requirements. The original agreement with the French Ministry of Production had allocated 50 percent of the tire production to the French. But the dire straits of the French transport led to a revision of this allocation by which only a third of the entire production was to go to U.S. forces, the remaining two thirds to be shared equally by French armed forces and French civil transport. Reliable statistics on the production and distribution of tires to the end of hostilities are not available, but it appears that something less than 200,000 tires were produced under the program.

More successful than the tire program and probably more publicized than

BELGIAN WORKERS IN A RUBBER PLANT, *Liège. Members of U.S. rubber industry watch as a tire is removed from a mold.*

either of the other major ordnance projects was the local procurement of extended end connectors for tanks, a project which had an importance hardly suggested by the size and simple design of the item. The extended end connector, or "duck bill," was merely a piece of steel about four inches square, designed to be welded to the end connectors of each track, of which there were 164 on each track of a medium tank. Its importance lay in the fact that in widening the track it added greatly to the flotation of a tank by reducing ground pressure, and therefore increased traction in mud and soft terrain.

The problem of reducing ground pressure in the medium tank had been recognized for some time, and the expedient of the extended end connector had been tested in the United States. The zone of interior tests were not entirely successful, but the field commands in ETOUSA were faced with an immediate requirement for some expedient which would improve the flotation and maneuverability of their tanks and asked that the item be shipped in sufficient quantity to modify their tanks. The theater soon learned that the zone of interior could not provide the device in sufficient quantity. As was frequently the case in local procurement programs, a field command—in this case the Third Army—took the first steps to acquire the duck bills, contracting with local firms for their fabrication. Shortly thereafter the Advance Section undertook a greatly expanded program, contracting with twenty plants in the Liège–Charleroi area and with a similar number in the Paris area, for a total of nearly 1,000,000 of the plates. The Ordnance Section of the Communications Zone eventually assumed responsibility for the entire program, and in the end contracted for more than a million and a half.

Two frabricating processes were possible in manufacturing the extended end connector. By a casting, it could be made from a single piece of steel. This method had obvious advantages, but was relatively slow. The fastest and least complicated process was simply to weld an extension to the existing end connector. The disadvantage in this method lay in the fact that the manufacturers had to have end connectors to which they could affix the extensions. Theoretically, field units were supposed to disengage tracks, remove the 164 end connectors from each of them, and then attach extended end connectors provided by the makers, so that there would be no loss of time. Under combat conditions the field units found this installation process difficult enough in itself, and often failed to forward the removed end connectors. Consequently factories frequently exhausted their stock of end connectors, while building up large stocks of the extensions. As a result, modifications were not accomplished as rapidly as planned. In the case of the First and Ninth Armies some of the trouble was eventually eliminated by the inauguration of a shuttle service between the armies and the Fabrique Nationale wherein an exchange took place on a one-for-one basis. The delays in getting end connectors to the factories generally constituted a bottleneck. The program was highly successful, nevertheless, and was largely completed by the end of February. All together, the French and Belgian factories turned out more than a million and a half extended end connectors for the medium tank, about 70 percent of which were of the welded type, the remainder of the cast type.

Similar attempts to modify the track of the light tank were unsuccessful. After tests by the First Army in November, the Communications Zone placed orders with a Belgian firm in Charleroi, first for 250,000 of the duck bills, and then for an additional 150,000. When actually employed in combat late in December the extensions were found to be impracticable, either causing the track to be thrown or chewing up the rubber on the bogie wheel and thus creating a bad maintenance problem. Attempts to further modify the duck bill for the light tank failed, and the armies decided to scrap the bulk of them.

While engine rebuild and the manufacture of tires and duck bills constituted the most ambitious ordnance local procurement projects, the ordnance sections at both field command and COMZ level engaged in a multitude of smaller procurement undertakings designed to meet urgent requirements fulfillment of which would have entailed unacceptable delays if requisitioned from the zone of interior. Many of these projects were undertaken by the armies. The First Army was particularly favored in the procurement of ordnance items because of its proximity to the highly developed weapons industry in Belgium. Faced with a shortage of 60-mm. mortars early in the fall, it turned to the weapons firm of J. Honres in Charleroi and contracted for 220 complete weapons, which were delivered in

December. It also contracted with the same firm to rebuild all 81-mm. mortars in the army in line with modifications recommended by ordnance officials who had studied defects and deficiencies of the weapons in the field. First Army claimed much greater accuracy of fire and a much smaller maintenance problem as a result. Similarly, First Army eliminated a critical shortage of a vital component of the firing mechanism of the 155-mm. gun—the gas check pad, which usually failed at about one third of its normal life in the hands of inexperienced units—when it found a superior pad in captured Germans guns and succeeded in getting a tire manufacturer in Liège to duplicate it. First Army found relief for another problem—a serious shortage of tires for tank transporters—in a rather unexpected manner. In capturing Malmedy, Belgium, it fell heir to about 50 tons of German Buna and two tons of Japanese gum rubber and promptly made use of the windfall by putting the well-known firm of Englebert in Liège to work retreading tires. Without this help First Army estimated that it probably would not have been able to muster more than half of its tank transporters during the winter months.

The Third, Seventh, and Ninth Armies, like the First, all resorted to local procurement in varying degree to meet similar requirements. The Third Army, for example, procured a multitude of spare parts for tanks, machine guns, rifles, grenade launchers, and artillery pieces. After the German offensive in December it arranged with local manufacturers to reinforce the armor of the M4 tank by welding plates salvaged from destroyed tanks in the Ardennes area.

SOLDIERS EQUIPPING MEDIUM TANK TRACKS *with extended end connectors.*

Similarly, the Seventh and Ninth Armies contracted for a variety of items, such as spark plugs, bushings, and so on. Meanwhile, the Ordnance Section of the Communications Zone had taken over theater-wide procurement of such items as grenade launchers (30,000), modification kits for the conversion of the carbine to automatic fire (10,000), trigger adapters for the M1 rifle (75,000), spark plugs (100,000), defrosters (75,000), hydrometers (4,500), and storage batteries (1,000), and had contracted with the Ford Motor Company to box nearly 25,-

000 general purpose vehicles and 7,500 trailers at the port of Antwerp.

The Chemical Warfare Service turned to plants in France, Belgium, and Luxembourg to produce a variety of parts for the popular 4.2-inch mortar, including base plates, shock-absorber slides, base cups, and so on, and contracted with eight French firms to convert some 350,000 gas masks, replacing defective synthetic rubber face pieces which lost their pliability in cold weather, with natural rubber face pieces. One of its most important finds was the discovery of a plant that could manufacture premixed flame-thrower fuel, which obviated the need for mixing in the field.

Of the various quartermaster procurement efforts, the most outstanding were those involving subsistence and winter camouflage garments. Local purchase of foodstuffs continued in sizable amounts after the front had moved far from Normandy. For the most part, these purchases involved fresh fruits and vegetables, such as potatoes, apples, cabbages, onions, carrots, and turnips, and did not violate the spirit of SHAEF injunctions regarding the purchase and consumption of food, particularly where transportation was not available to move perishables to city markets. The quartermaster eventually also contracted with both French and Belgian firms for the delivery of salt and yeast, and with food processing plants for the roasting and grinding of coffee. In the main these represented surplus items and facilities, and in some cases U.S. forces provided items like sugar, lard, and coal in exchange. Allied personnel were permitted to shop for certain categories of luxury items in French shops—perfumes, cosmetics, handicrafts, jewelry, and books—and both individuals and clubs bought substantial amounts of liqueurs. But U.S. personnel were expressly forbidden to eat in French restaurants. The restriction did not apply to night clubs.

There were violations of the restrictions on food purchases, to be sure, particularly among forward units, which sometimes could not resist the opportunities to obtain fresh beef or veal. It was such acts, undoubtedly, that inspired some of the grumbling in the editorial columns of certain French newspapers that Allied forces were aggravating France's serious food shortages. In actual fact the opposite was true, for American deliveries to the French through civil affairs channels far exceeded the amounts procured locally. In April 1945 French authorities finally interceded to scotch the rumors.

The ETOUSA quartermaster attempted to institute the manufacture of winter clothing on the Continent, but like most of the local procurement projects which depended on the importation of raw material—in this case wool—it was largely unsuccessful. Only about 50,000 wool trousers, 40,000 wool headgear, 2,000 wool jackets, and 25,000 blankets were turned out locally. The same was true of the items requiring cotton, such as towels, handkerchiefs, and duck yardage, requirements for which ran into the millions. Except for projects to manufacture tents and sleeping bag liners, under which 20,000 and 200,000 were delivered respectively, most remained unfulfilled.

Far more successful was the project to produce winter camouflage garments. Plans of the Engineer Service, like those

FIRST DIVISION TROOPS WEARING WINTER CAMOUFLAGE GARMENTS *move along a snow-covered road in Belgium, January 1945.*

of the Quartermaster Service, had not anticipated an advance into the "cold-wet" areas in the winter of 1944–45 or that snow camouflage would be a problem. But U.S. forces had already entered such an area in September, and by mid-November it became evident that they would continue to operate in an area of fairly heavy snowfall for some time. The effort to procure snow camouflage clothing was basically a "crash program," which suddenly acquired great urgency. Both the Engineer and Quartermaster Services participated in the program, although the Quartermaster early assumed responsibility for the program at the theater level, acquiring white cloth from meager French stocks and arranging for the manufacture of garments by many civilian firms. Under the Quartermaster's program, about 130,000 garments were eventually produced, either of the short snow cape type with hood, the snowsuit type, which consisted of a jacket and trousers, or the long cape or "nightgown" type. Of the three, the last was least favored because it hampered movement.

In the meantime the field armies organized their own programs. The First Army assigned the job to the 602d Engineer Camouflage Battalion, which was already supervising civilian factories in Verviers and Liège in the manufacture of various camouflage materials like nets. Third Army employed both military units and civilian factories to fabricate snowsuits. Military units consisted primarily of a chemical maintenance company and several quartermaster units, including a salvage repair company which turned out 700 capes in less than twenty-four hours on one occasion. The Ninth Army, like the First, relied mainly on civilian firms to produce camouflage garments, using factories in Holland, Belgium, and Germany. It acquired the cloth partly by purchase in Belgium and partly by having military government personnel drive through German towns in its area and call on civilians via a public address system to turn in white sheeting, for which receipts were given for later redemption. Through these various means the three armies in the 12th Army Group produced nearly 170,000 suits.

Other Quartermaster procurement projects included the manufacture of tent stoves and mess gear, field ranges, lanterns, immersion heaters, and jerricans. But these projects, most of them involving the importation of raw materials, were only partially successful and in some cases were almost a complete failure.

For the Engineer Service the major item of local procurement continued to be real estate. This was to be expected, for U.S. forces, exceeding 3,000,000 at the height of the build-up, required either land or buildings for a variety of purposes, including headquarters, depots and dumps, bivouacs, repair shops, hospitals, rest and leave centers, replacement depots, and training sites. Undoubtedly the most outstanding case of real estate acquisition took place in the Paris area after it was freed at the end of August 1944. Paris itself, with its excellent housing and accommodations, facilities, and supply of labor, was a logical site for a major headquarters, and the COMZ headquarters and its subordinate headquarters area command, the Seine Section, lost no time in establishing themselves there. Initially, the Communications Zone simply requisitioned most of the hotels and other facilities which the Germans had occupied. Eventually it took over considerable additional properties, and by October the COMZ and Seine headquarters occupied nearly 1,100 pieces of property, including 300 hotels, hospitals, schools, theaters, warehouses, vehicle parks, and so on.

SHAEF headquarters took over a large part of Versailles, establishing offices for the general staff in the Trianon Palace, for the special staff in the Grandes Ecuries, for the air staff in the Petites Ecuries, and for miscellaneous agencies in other hotels like the Reservoir, the Royale, and the Vittel. Satory Camp and several schools accommodated troops, and officers were billeted in homes in neighboring villages. In all, Allied forces took over about 1,800 pieces of property in Versailles and nearby towns, in which some 24,000 officers and men were housed. American town majors meanwhile found living quarters for thousands of troops in other French cities, the port cities alone providing accommodations for at least 110,000 men.

The need for leave centers and rest camps resulted in the acquisition of additional buildings, particularly hotels, in many cities. First Army established the first of such centers at Barneville, on the west coast of the Cotentin in Normandy. But after the breakout at the end of July this facility was quickly left far to the rear, and no attempt was made to set up other centers until the front again became relatively static in September. Most of the major field units, including divisions, thereafter situated rest camps near the front, and major leave centers were established at places like Spa, Dinant, Liège, Namur, and Brussels in Belgium. The favored spots were of course Paris and, later, the French Riviera. By February 1945, 8,400 U.S. and 700 British troops were arriving daily in Paris on seventy-two-hour passes. Excellent entertainment was provided in some of the finest Parisian theaters. The American Red Cross maintained at least ten clubs for enlisted men and four for officers, the best known rendezvous for the former being the Hotel de Paris, better known as "Rainbow Corner."

U.S. forces also needed sizable tracts of land. Two of the more highly developed tracts were the Red Horse Staging Area and the Assembly Area Command. Channel Base Section organized the Red Horse Staging Area in November. It consisted of Camps Lucky Strike, Twenty Grand, and Old Gold, in the Le Havre–Rouen area. The three camps, with a capacity of 138,000 men, were intended primarily as staging areas for units arriving from the zone of interior.

The Communications Zone also began to plan for the redeployment of troops at the end of hostilities. For this purpose it activated an Assembly Area Command in April 1945 in the vicinity of Reims, embracing an area 50 by 100 miles. Plans called for laying out seventeen camps, each with a capacity of about 15,000 men and named for an American city. Establishment of these enormous temporary installations involved the construction of some 5,000 huts and the erection of more than 30,000 tents, in addition to the construction of roads and hardstandings. But it was soon apparent that the Assembly Area Command would not be ready when hostilities came to an end, and the Communications Zone therefore planned alternative facilities. In mid-April it began negotiating for additional land in the neighborhood of the Red Horse Staging Area. Resistance from the French over release of the land was finally overcome and two camps were added to those already existing in the Le Havre–Rouen area. The two camps, along with those which until then had served in staging units arriving from the United States, were also named after well-known brands of cigarettes—Camps Philip Morris, Pall Mall, and so on—and soon began staging American troops in the opposite direction—that is, either home or to the Far East.

Most of the Engineer Service's other local procurement efforts related to its various construction activities. One of the more outstanding projects was the fabrication of steel I-beams for use in highway and bridge construction. After investigating several possibilities the engineers chose the Hadir Steel Works at Differdange, Luxembourg, partly because of its location near the source of iron ore and the areas where steel was needed, and partly because of its capacity to roll beams of the desired gauge and length. The Hadir plant began operations in October 1944 and operated steadily thereafter except for temporary interruptions caused by labor troubles and by enemy artillery during the December fighting.

Early in December the deputy GPA established an office in Brussels to co-ordinate the various procurement programs in Belgium, the Netherlands, and Luxembourg and took over responsibility for co-ordinating the steel requirements of the supply services, officially representing U.S. forces in the Belgian Steel Production Control Committee. Production of steel was restricted to only a few mills because of supply and transportation difficulties. The Hadir Works alone eventually produced about 50,000 tons of beams, enough for all army, ADSEC, and British projects, including the bridging of the Rhine, Elbe, and Danube Rivers, plus some civilian projects. In addition, U.S. forces obtained rails, light structural shapes, wire, sheets for oil drums, stoves, jerricans, and hutting from the mills operating in the area.[7]

Lumber and crushed rock comprised two other major items of engineer procurement. Quarries were first opened near Cherbourg to meet port reconstruction needs and for the construction of railway marshaling yards south of the port. Later in the fall the Advance Section alone placed about twenty quarries in operation, mainly to meet the need for crushed rock in road repair. Lumber was needed for many purposes, particularly in the fall when the engineers began to replace floating bridges with fixed bridges over the Marne and Meuse Rivers. ADSEC engineers alone operated some thirty sawmills, employing forestry companies and general service regiments, and the armies also used engineer units to run sawmills in the combat zone, the Ninth Army operating as many as thirty-five mills at one time in January. All together, it is estimated that about 284,000,000 board feet were produced by these various means. In addition, the Engineer Service procured a variety of other supplies and equipment in varying

---

[7] History of the Office of the GPA, May 1942–October 1945, prep by Samuel I. Katz, Hist Off, OGPA, pp. 125–28, ETO Adm 556.

amounts, including such items as pickets for airfield landing mats, plywood storm boats, compressors, asphalt kettles, pile drivers, drafting instruments, transformers, fire fighting equipment, hospital hutting, cranes, wrenches, jacks, forestry tools, paint, and various hardware and mill supplies..

The Signal Service's main local acquisition was of course the use of the French communications network, including the underground cable net. In addition, the Signal Corps procured a limited quantity of supplies, the major purchases of which were signal ground panels (10,000), vacuum tubes (17,000), telephone repeaters (300), 35-mm. film (5 million feet), and telegraph repeaters (220). Local procurement of medical supplies, restricted by policy, was limited to certain biologicals, vaccines, and antitoxins, and totaled less than $250,000 in value in France, Belgium, and Luxembourg.

While the GPA acted mainly as a staff agency to supervise and co-ordinate procurement by the supply services and the purchasing and contracting officers, it operated directly as a procurement agency for certain organizations which had no machinery for that purpose, such as the American Red Cross, the Office of Strategic Services, the U.S. Information Office, Special Services, Information and Education, and the U.S. Navy. In this capacity it obtained supplies and equipment for Red Cross clubs, "doughnut dugouts," motion picture units, public address equipment, broadcasting and printing facilities, musical instruments, phonograph records, and the like.

The GPA also engaged in direct procurement to obtain certain supplies for export to the United States, the most notable case being sodium metal. Before it could get production under way, however, it first had to provide several components, through imports from the United States, and to transfer clothing, steel drums, and chemicals from Army stocks in the theater.

Finally, SHAEF in October also gave the GPA responsibility for the procurement of coal mine supplies which were so vital to getting the mines back in production. Items required included rubber belting, miners' lamps, bearings, power transmission belts, medical supplies, tires, cables, tools, chemicals, and clothing. GPA obtained some of these supplies from theater stocks, some from Britain, and some, notably mine timber, from neutral countries in southern Europe. GPA procured some 82,000 tons of supplies in this way, about half of it for the Nord and Pas-de-Calais mines.[8]

U.S. procurement in countries outside the area of ground operations, such as Switzerland, Spain, Portugal, and Sweden, did not achieve much volume until the spring of 1945, and contributed little to the support of U.S. forces in the period before V-E Day. U.S. purchasing missions were established in Berne, Switzerland, and Lisbon, Portugal, late in January 1945, and later in Madrid, Spain. Each consisted of a representative of the GPA and a representatives of the U.S. Commercial Company, a civilian governmental agency already operating in other areas of the world. Requests for purchases in the neutral countries were transmitted to the GPA representative on the appropriate purchasing mission, which investigated bids, negotiated con-

[8] *Ibid.*, pp. 163–67.

tracts, and supervised payments. By the end of August 1945 contracts had been let for goods valued at nearly $60 million —$24 million in Spain, $18 million in Switzerland, $7 million in Portugal, $5 million in Sweden, and $5 million in Denmark. The purchases involved a variety of items, including watches, blankets, prefabricated barracks, lumber, map paper, vaccine, butter, eggs, fiber board, fruit, tenting, handkerchiefs, towels, bedding, tank cars, and spirits.[9] Spanish brandy, better known as "Franco's Revenge," began to appear in officers' liquor rations late in 1945, and received no acclaim as a substitute for Scotch whiskey.

Since no specific long-range goals were ever established for local procurement in terms of monetary value, tonnage, or numbers of items, it is difficult to measure the success or failure of the program. Local procurement did not live up to expectations if one measures its accomplishments against the numerous programs initiated. Many of these reflected an unwarranted optimism, and proved far too ambitious. Local procurement was beset with many difficulties, resulting directly from the dislocations caused by the enemy occupation and the subsequent Allied invasion.

Not the least of these was the lack of raw materials and coal. In some manufacturing it was possible to employ substitutes. This was the case in the manufacture of certain spare parts in which steel, say, could serve in place of brass or bronze. But there were outstanding failures in instances where a program depended on the importation of sizable quantities of raw materials. A conspicuous case was the attempt to fabricate tires (rubber), jerricans (sheet steel), clothing (wool), and tentage (cotton). Only one jerrican was turned out, for example, for every fifty contracted for. The scarcity of raw materials affected procurement in the neutral countries as well as in the liberated areas. Switzerland, Portugal, and Spain were all reluctant to commit themselves to manufacturing items without assurance of replacement of raw materials. As a result, procurement in Switzerland was limited to items which did not involve substantial imports. Procurement of mine timber in Portugal depended on the shipment of coal from the United States. Since this was impossible, the project was abandoned. In the final analysis, many of the difficulties over procurement could be attributed to the shortage of coal. The requirement for coal was in fact the criterion by which numerous projects were screened and as a result eliminated. A substantial cement-making program would have been undertaken, for instance, had coal been available for the purpose. In the end, the shortages of coal and raw materials voided much of the local procurement program, since the importation of either in large amounts would have canceled the gains to be derived from local procurement.

Lack of transportation and power were also major deficiencies which could have either direct or indirect influence on local capabilities. U.S. Army trucks, hauling coal from northern France, helped get the engine-rebuild program under way in the Paris area. In other cases Ordnance trucks hauled food to the Paris

[9] *Ibid.*, pp. 155–59.

area so that employers could provide workers with a noon meal. It was also necessary for the Ordnance Service to haul steel to the Paris area for the expanded end connector program. An engineer project to procure wooden hutting virtually failed for lack of means to move raw material, and a shortage of coal delayed the opening of two Belgian tire factories for several weeks. In some cases French manufacturers agreed to commit themselves to production for the U.S. Army only on the promise of gasoline, oil, and tires, of packing and crating materials, or of other supplies.

Added to the economic paralysis which characterized much of western Europe was the factor of political instability. On the whole, the continental governments were eager to contribute to the Allied effort, but they did not immediately possess effective control over their respective economies, and there were many frustrating changes in personnel and policies within the various ministries. One of the most troublesome administrative difficulties centered on the payment of suppliers. The problem became particularly acute in Belgium, where long delays in payment led some suppliers to withhold deliveries. It appears that the main difficulty arose from the attempt to make payments through local burgomasters, who were unprepared to assume the administrative burdens involved. Arrangements were eventually made whereby burgomasters were to be responsible for the payment of labor only; the payment for supplies was transferred to the Office of Mutual Aid in the central government. Similar difficulties in France led to an arrangement under which payment could be advanced pending certification of invoices. In this way manufacturers could repay short-term loans which had enabled them to undertake production. In a few instances the GPA authorized the advance of funds from U.S. sources to keep production going, as on two occasions at the Hadir Steel Works. But this practice was generally avoided.[10]

The shortages of coal and raw materials undoubtedly placed the severest limitations on potentialities for local procurement. In addition, they were partly responsible for preventing the French from producing the equipment to which they had committed themselves for the Metropolitan Rearmament Program agreed to late in 1944. The performance in local procurement raises the question as to whether Allied plans for aiding the liberated countries should have gone beyond the provision of pure relief and included rehabilitation supplies needed to revive industry, agriculture, transport, and communications. Whether such a program could have been carried out in view of the shipping situation is of course questionable. The chronic shipping shortage and the lack of adequate port discharge facilities on the Continent certainly precluded any expansion of imports before the end of 1944. Deficiencies in these two fields prevented even the purely military needs from being met, and resulted in the backlogging of civil relief needs. Late in 1944 the French, anxious to make an early start in rehabilitating their economy and to assume a larger role in the war, presented an import program and

[10] *Ibid.*, pp. 141–44.

asked for an allocation of shipping to carry it out. SHAEF considered some concessions to the proposal after the opening of Antwerp. But it had not reckoned with the shipping shortage, which remained as critical as ever. Early in 1945 SHAEF agreed to allocate some tonnage to both the French and Belgian national import programs. Until V-E Day, however, the major emphasis remained on the import of relief items as opposed to rehabilitation supplies, with the result that the economies of the liberated areas were too weak to make any major contribution in meeting Allied supply requirements or in implementing the rearmament program.

Whatever its failings and frustrations, local procurement undoubtedly made a substantial contribution to the Allied victory. Lend-lease reports have estimated that procurement on the Continent had a total value of about $945 million through August 1945—$760 million in France, $174 million in Belgium, $8.3 million in Luxembourg, and $111 million in the Netherlands.[11] The dollar value is, of course, a poor measure of the true value of local procurement to military operations. The projects inaugurated on short notice to meet critical needs, such as those for camouflage garments, extended end connectors, and engine repair, obviously had a value and importance disproportionate to the monetary expenditure. Moreover, local procurement represented hundreds of items and thousands of tons which did not have to be loaded into scarce shipping in the zone of interior, run the gantlet of submarines in the Atlantic, or be squeezed through the ports of northwest Europe. Finally, it provided a means for both Britain and the continental Allies to repay at least a part of the tremendous contribution which the United States was making by way of lend-lease.

---

[11] As of 2 September 1945 Belgium had furnished more aid than it had received from the United States. The lend-lease reciprocal aid balance on that date showed a credit of $90 million in Belgium's favor. History of Reciprocal Aid, pp. 50–51.

CHAPTER XIX

Retrospect

A renowned British soldier has observed that for every ten military students who can tell how the Battle of Blenheim was fought there is only one who has any knowledge of the administrative preparations that made the march to Blenheim possible. In most military books, he noted, strategy and tactics are emphasized at the expense of logistics. He left no doubt of his conviction that the emphasis should be reversed. The principles of tactics and strategy, he said, can be apprehended in a short time by any reasonable intelligence. It is in the field of movement and supply that "most critics and many generals go wrong." [1]

Developments of the last hundred years have caused an inexorable encroachment by logistics on tactics and strategy. Resistance to this trend has been futile, and commanders had reason to regret attempts during operations in Europe in 1944–45 to ignore the iron laws of logistics, notably in such cases as the effort to acquire combat strength without giving sufficient consideration to the logistic requirements for its support, as in the U.K. build-up in 1942, and in the premature acceleration of the divisional build-up, as on the Continent in the summer and fall of 1944.

Operations in Europe demonstrated with probably greater force than ever before the extent to which logistic factors have entered into all strategic and tactical planning. To cite only the most obvious examples, logistic considerations dominated the selection of the place for the Normandy invasion, determined the time when it could be launched, and clearly influenced the tactical decisions of September and October 1944.

Recognition of logistic requirements was obviously sufficient to achieve victory in Europe. Some outstanding feats were accomplished in the field of supply. The most noteworthy probably were the logistic aspects of the assault itself and the subsequent support over the Normandy beaches. It can hardly be claimed that U.S. forces operated on a shoestring. In fact, few armies in history have been as bountifully provided for.

The European theater nevertheless experienced many difficulties in supplying its forces, and at times the shortages reached truly critical proportions. It might be said that two fundamental problems explain the theater's logistic difficulties. One was the problem of the theater's command and organizational

[1] Field Marshal Earl Wavell, in a lecture on "Generals and Generalship," delivered at Trinity College, Cambridge, 1939, published as part of his *Soldiers and Soldiering* (London: Cape, 1953), pp. 21–22.

structure, which defied solution to the very end. The other, and more important, is to be found in "limiting factors," which also defied elimination to the end.

In retrospect, it is safe to say that the thorniest command and organizational problem which the theater had to face was that of the proper place of the logistical function vis-à-vis other functions of the theater. This would have been a vexing problem in any case, but it was complicated by the presence of U.S. forces in what was predominantly an Allied theater of operations, and by General Eisenhower's dual role as American theater commander and Supreme Allied Commander.

In 1942–43 the basic problem was the relationship between theater headquarters and the SOS in controlling the logistical function during a time when the theater was mainly occupied in preparing for combat operations. A viable solution to this problem had been reached early in 1944 in the consolidation of the theater and SOS headquarters with General Lee designated the deputy theater commander for supply and administration.

The introduction of SHAEF as an over-all Allied headquarters under an American who was also the U.S. theater commander altered the position of the new ETOUSA-SOS headquarters and in effect relegated it to a more subordinate position, even within the strictly U.S. command setup. Since the top U.S. commander resided at SHAEF, and since he had with him a staff of influential and high-ranking U.S. officers, many decisions on U.S. supply and administrative matters were made at Allied headquarters. SHAEF consequently assumed more and more the aspect of a U.S. theater headquarters. The ultimate result was really to divide the theater headquarters between ETOUSA-SOS (or ETOUSA-COMZ) and SHAEF, leaving the former in a difficult position in which it might be said that it ceased to be a genuine theater headquarters.

General Eisenhower apparently recognized some of the difficulties inherent in this arrangement and at the time when the transfer to the Continent began took pains to emphasize that the staff members resident with him at SHAEF were not to be regarded as members of the theater staff. He obviously intended to preserve as far as possible the integrated control of supply and administrative matters in the theater which had been achieved through the consolidation of the ETOUSA and COMZ headquarters. But the intent was at least partially voided by the simultaneous revocation of General Lee's status as deputy theater commander for supply and administration, which relegated the Communications Zone to a position coequal with that of the top air force and ground force commands.

The position of ETOUSA-COMZ headquarters thereafter was anomalous. In the sense that it contained the official theater staff, including the chiefs of technical services, who exercised theater-wide control in their respective fields, it was a true theater headquarters. General Lee continued to issue theaterwide instructions on administrative matters in the form of ETOUSA orders and circulars. But he actually held no position in the strictly theater setup, for his authority as deputy theater commander had been terminated. His staff, meanwhile, held a dual position, as both the COMZ staff and the theater staff. In the latter capacity Lee's staff members were in the strange situation of occupying positions in a headquarters in which their chief in the COMZ command had none.

The field commands preferred to regard ETOUSA-COMZ as a COMZ headquarters only—that is, as a co-ordinate or coequal command—and not as a true theater headquarters speaking with the authority of the theater commander. The tendency of the U.S. officers at SHAEF to render many decisions normally made by a theater staff, combined with the lack of confidence in the Communications Zone, often openly expressed, encouraged the field commands to look to SHAEF as the ultimate authority even in purely American matters.

This point of view is clearly evident in the prolonged argument between the armies and the Communications Zone over the control of scarce items and the apportionment of service troops, which culminated in the Lutes-Somervell criticisms of the requisitioning of Class II and IV supplies. The armies generally took the view that the Communications Zone exercised too much control. They preferred that control in such matters be exercised by the theater commander through the technical service chiefs, and that the Communications Zone act only as a freight-handling agency. The armies, in other words, desired that SHAEF assume the role of theater headquarters even more completely. The argument was interesting, for it pointed up the fact that no genuine theater headquarters existed when the chiefs of services were physically separated from the theater commander.

The U.S. component at SHAEF, on the other hand, was in no position to provide detailed guidance in logistic matters. The result was that no theater headquarters existed which could properly enforce uniform policy on administrative matters. Unfortunately theater directives, well intentioned as they were, consistently fell short of clear-cut delineations of authority and responsibility, always leaving room for contention, so that constant efforts had to be made to keep the organizational mechanism running with some degree of harmony.

The shortcomings of the system can be clearly seen in the matter of manpower management. Enforcement of the War Department's policies on manpower conservation was obviously a job for the highest theater command level, since it involved all three major subcommands. While the Communications Zone undoubtedly was remiss in effecting the necessary manpower economies and conversions in its own house, failure to forestall the predicted manpower crisis was definitely a theater responsibility, and the problem was not resolved until unequivocal action was taken at the highest theater level of authority—that is, by General Eisenhower.

It is ironic that the goal toward which General Lee had worked so long—the theaterwide integration of all supply and administration under one headquarters—and which appeared to have been achieved with the consolidation early in 1944, was substantially voided by the subsequent revocation of the authority by which it might have been more effectively attained. It is understandable that the field commands should object to having the Communications Zone, a co-ordinate command and thus an interested party in all logistic matters, control the supply and administrative matters in which they had a vital stake. Whether a true theater headquarters, completely independent of the Communications Zone, should have been established is debatable. Personalities affect the way in which any command and organizational system actually operates, and it is conceivable that the system which evolved in ETOUSA might have worked better in other hands. But denying ETOUSA-COMZ the authority of a theater headquarters was surely inconsistent with the accepted goal of complete integration of all supply and administrative activities under one authority.

If the history of command and organizational difficulties in the European theater contains any lesson, it is to reaffirm the long-accepted principle that responsibility and authority must be clearly defined and understood and that the grant of authority must be consistent with the missions and responsibility assigned.

Command and organizational difficulties of another kind within the Communications Zone had their repercussions on theater logistics. The conflict between territorial and functional organization and control is an old administrative problem. The principle of centralized control and decentralized operation to which General Lee was firmly committed was often difficult to follow, particularly where operations were intersectional, as in the case of transportation. Failure to define adequately the COMZ-wide "technical control" which technical service chiefs were authorized to exercise inescapably led to the friction with base section commanders which characterized the first year or two in the United Kingdom. Constant efforts at co-ordinating the activities of the technical service chiefs with the section commanders eliminated much of this friction. But technical service chiefs generally felt that the cellular structure of the Commuications Zone hampered the centralized control of their respective supply depots and made them too dependent on command action via the circuitous route through G–4 or on personal diplomacy.

Important as was the command and organizational problem, its effect on logistical support was never as far-reaching or real as the various "limiting factors" which constantly threatened to restrict the tactical operations of the field armies by creating supply and manpower shortages in the combat zone. In general, these limiting factors were of three types: (1) supply and manpower shortages resulting from the War Department's inability or refusal to meet the theater's request, (2) movement difficulties within the theater, arising from port discharge, port clearance, or long-distance transportation deficiencies, and (3) faulty "administration," which prevented supplies available in theater depots from reaching units needing them even when transportation capacity was adequate for movement.

Shortages in the zone of interior were almost inevitable in view of the global commitments of the United States and the necessity to plan production many months in advance of actual need. War Department replacement factors and reserve allowances, first based on empirical data from World War I and later on experiential data from other theaters of war in World War II, did not always fit experience in the European theater. On the whole, replacement and reserve factors were designed to meet average expenditures and losses over a long period. The War Department, its demands always subject to close Congressional scrutiny, was understandably reluctant to increase allowances in response to every demand from the theater, particularly when it suspected that the demands did not represent average expenditures of losses over an extended period. Moreover, supplying a force of several million men over sea lines of communication several thousand miles long involved tying up huge quantities of supplies in transit. The order and shipping time for most supplies averaged about 120 days. The time involved between placing a requisition for a tank and delivering it to a using unit required that 135 days of supply be in the pipeline at all times. With a replacement rate of 11 percent and a T/O&E allowance of approximately 4,000, this meant that nearly 2,000 tanks were constantly tied up in the pipeline.

Unfortunately, relations between the War Department and the theater often suffered from misunderstandings or arguments arising from conflicting interpretations of ground rules. In more than one case the War Department had ample reason to believe that the theater's reading and application of War Department policy and regulations were contrary to the original intent. This was most clearly evident in such matters as the policy on ammunition reserves, the application of the term *critical* as used in requisitioning supplies under the express shipping procedure inaugurated early in 1945, and the interpretation of War Department policy with regard to the replacement pool. On some issues, notably ammunition and replacements, the argument dragged on for months, and it appeared that there would never be a meeting of minds. In some cases the theater had just ground for exasperation, as, for example, the War Department's persistence in editing theater requirements for ammunition on the basis of past expenditures that had been limited by nonavailability, and its habit of ignoring certain requests for increases in replacement and consumption factors.

Despite the uniformity of Army training and indoctrination, relations between the theater and the War Department at times betrayed a surprising need for a common language.

Normally, while insisting that the theater follow established doctrine and policy, the War Department leaned over backward in trying to meet the theater's demands and, in accord with long-standing command policy, granted the theater a large degree of independence of decision, particularly in matters where the theater presumably was in a better position to judge its needs. In at least two instances it had reason to regret its indulgence. In both the winter clothing controversy and the replacement problem, the War Department unquestionably possessed superior knowledge based on experience in other theaters. ETOUSA ignored its advice in both cases, and important lessons learned in other areas consequently went unheeded.

The theater's logistic difficulties were often aggravated by the unexpected course of tactical operations. The unpredictability of the course of operations is one of war's inherent characteristics and has always posed its greatest dilemma. One of the central features of war in the twentieth century, with its tremendous material requirements, is the long lead time required in all logistic planning. The requirements for a major operation must be anticipated and requisitioned from eighteen to twenty-four months in advance, a factor which the ASF never ceased to impress upon the various theaters of operations. But the experience in Europe taught logistical planners that they could not expect to have anything but the vaguest operational data by the time orders had to be placed for the bulk of supplies. Lack of firm planning factors plagued logistical planners constantly in preparing for the OVERLORD operation, as was evidenced in such matters as estimating ammunition requirements. Technical data on which to base estimates of requirements under a given set of assumptions are usually available—for example, the material requirements to rebuild a given length of quay—and in the European theater the schedules of materials and labor required for particular units of work served admirably well. But determining what the assumptions should be is ever a matter of judgment based on a cold analysis of all known factors. There is obviously no way of predicting with any certainty the requirements for a distant battle. The best plans went sour as operations took an unexpected turn, and the sudden needs generated by unexpectedly heavy fighting, an unexpectedly rapid advance, or an unexpected reverse, often threw consumption and expenditure factors, and therefore "production" plans, askew, whether for material items, like ammunition and tanks, for manpower in the form of trained replacements of the proper SSN, or for services such as transportation.

The difficulty in anticipating future needs, and particularly in attempting to schedule supply deliveries, is clearly illustrated in the effort to preschedule supply shipments from the United Kingdom to the Continent for the first ninety days of the OVERLORD operation. Fortunately, sufficient resiliency was built into logistic plans to permit adaptation to the needs of tactical operations in the first few months. But the scheduling of supplies so far in advance proved unwise and resulted in a wasteful immobilization of shipping because of the resort to selective unloading necessitated by the limited port capacity.

The most disrupting event of the entire eleven months of operations was the breakout from Normandy and the subsequent chase across northern France. It is doubtful whether many persons even in the Communications Zone immediately foresaw how far-reaching and lasting the effect of this explosive lengthening of the lines of communications would be. The immediate effects were obvious in the restrictions imposed on the combat forces. But these restrictions were only the most apparent. The pursuit had cumulative effects on capabilities for future logistic support from which the Communications Zone did not recover for several months and which explain to a large degree the logistic difficulties that crippled all tactical planning in the fall of 1944.

There is no questioning the wisdom of the strategic decisions of August 1944 which led to a major tactical victory and at the same time upset logistic plans so drastically. The pursuit which followed entailed a sacrifice of future capabilities, however, that eventually had to be rectified. The damage to logistical capabilities was eventually recognized in such things as: (1) the costly attrition of equipment, particularly motor transport which had to be relied on heavily for long-distance hauling at the sacrifice of proper maintenance; (2) the failure to establish a proper depot system; (3) the suspension of other activities, such as port clearance, with its inevitable repercussions on related activities such as ship unloading; and (4) the inability to establish stocks in forward areas. For several months thereafter the logistical structure had practically no resiliency to meet unexpected demands, and was in fact hard put to meet minimum operational needs. The events of August and September precluded an orderly development of the Communications Zone, and at the end of 1944 it still suffered from major weaknesses as the result of its premature exertions, an important one being the lack of a proper depot system.

It was this last deficiency, in fact, which largely accounted for the Communications Zone's inability to ensure the kind of selectivity and certainty of resupply so important to the field armies. The test of a good supply system is not in amassing huge quantities, but in the certainty of replenishment and in selectivity. Nothing concerned the field commands quite so much as the lack of assurance that they could count on getting the items they wanted within the tonnages allocated. Uncertainty over future deliveries inevitably led the armies to requisition supplies beyond their needs as insurance against possible future supply failures.

The varying demands of the combat forces and the shifting course of tactical operations naturally aggravated the eternal problem of achieveing proper balance in the logistic structure. The efficiency with which the supply mechanism operated depended to a large degree on the efficiency with which its various components or parts were employed in relation to each other. The capacity of the system could never be measured in terms of any one of its parts, and the capacity of any one component, like port discharge, was always the sum of many variables. Port discharge might be affected by a combination of many factors, including congestion in the British ports, lack of berths on the far shore, improper stowage, or shortages of amphibious craft, service troops, or shipping. The shortages, in turn, could be laid to a chain of factors, such as low serviceability rate (that is, repair time on ships), long turnaround, selective unloading, or the use of ships as floating storage. Port clearance, in turn, depended on a multitude of related capacities, such as the availability and condition of rail transportation from the quays, the maintenance of trucks and the availability of drivers, the efficiency of traffic control, the prompt unloading and release of rail cars or trucks and trailers at the destination, and so on.

The movement of supplies from ship to army depot, in other words, entailed a series of highly synchronized functions, the failure of any one of which could have a resonant effect, reverberating along the entire line of communications.

At no time were all the components of the structure in perfect balance. Indeed, the elimination of one limiting factor sometimes created another at a different point. The elimination of the deficiency in rail transport in November, for example, made the forward depots one of the main strictures, for they were unable to receive the large tonnages which the railways had become capable of forwarding. The history of logistic operations in the European theater, in fact, seems characterized by a succession of alarums over one critical deficiency or another, and the theater was occupied at all times with efforts to eliminate some bottleneck and to bring the system into balance.

The problem of balance applied with equal force to the troop basis. The objective at all times, of course, was to maintain the highest possible ratio of combat to service forces in order to achieve the greatest possible combat potential. The War Department, always fearful that the theater might become top-heavy in service troops, never stopped urging the theater to "comb its tail" and "sharpen its teeth." But achieving the perfect balance was an elusive goal. The ratio naturally will vary with circumstances. Combat commanders, although recognizing that developments in warfare of the past century have reduced the proportion of a total force that can be put into the front line, never ceased to demand a larger slice of the total manpower allocation, as was evidenced in the premature acceleration of the divisional build-up on the Continent in the late summer and fall of 1944. Within the Communications Zone, meanwhile, each technical service, concerned primarily with its own mission and desirous of providing perfect service, naturally tended to exaggerate its own needs and asked for the largest slice of the manpower pie which it could justify. The sum total of "minimum" requirements invariably exceeded the authorized troop ceiling. Resolving such conflicting demands usually calls for an arbitrary decision. Unfortunately the wisdom of the allocation must always await the test of operations.

The difficulty of achieving balance in the theater's logistic structure and troop basis and the time element involved in preparing for large-scale operations underscore two fundamental characteristics of the modern system of logistic support for large ground forces—its size and complexity. The logistic base required for the support of U.S. forces in Europe was a huge complex of ports, depots, rail lines, repair and maintenance shops, and hospitals, much of it of a semipermanent nature and representing a prodigious effort in planning and organization and an enormous material investment. It was also an intricate system, requiring the synchronization of many activities and the satisfaction of unpredictable demands.

Organizing and operating such a system has all the attributes of "big business" and requires managerial ability of the highest order. Success in battle now depends as much on the "generalship" of those who plan and organize the logistic system and control its various functions as on the command exercised in the field. That this may not have been fully recognized in the European theater is evidenced by the complaint often heard in the Communications Zone that it was used as a dumping ground for officers who did not measure up in combat. Those officers were often no better suited for assignments in supply and administration, in which training and competence are also exacting, than in combat. The premium which officers have traditionally placed on "troop duty," usually considered more rewarding in terms of standing and promotion, has hardly been conducive to providing the corps of personnel with training in the field of logistics so badly needed under the conditions of modern warfare.

Field Marshal Erwin Rommel, commander of the Germans' famed *Afrika Korps,* may have exaggerated when he said that battles today are fought and decided by the quartermasters before the actual fighting begins. But adequate logistic support undoubtedly constitutes a far more critical ingredient of victory on the battlefield than ever before. By the same token the failure to provide such support is an almost certain invitation to defeat.

The increasing dependence of the combat elements on their supply base has naturally conflicted with the eternal desire for greater mobility. Its huge appetite has made the present-day combat force dependent on continuous replenishment, practically on a daily basis. Self-sufficiency, or self-containment, has become impossible except for a few days at a time. In effect, therefore, the modern ground army has become shackled to its base, unable to venture far afield because it cannot risk severance of its line of communications. Despite all its vehicles, the modern field army's mobility is actually extremely limited, for its knapsack is relatively small in terms of the days of supply it can carry. The supply base on which it depends, ponderous and immobile, cannot be shifted with ease. Consequently it is the position of the base that largely determines the line of advance and the eventual course of operations. In effect, under this system of logistic support the battle must be fought much as it is planned.

Glossary

AAR	After Action Report
ACofS	Assistant Chief of Staff
ACofT	Assistant Chief of Transportation
Actg	Acting
ADCofS	Assistant Deputy Chief of Staff
Adm	Administrative, administration
ADSEC	Advance Section, Communications Zone
AFHQ	Allied Force Headquarters
AFV&W Sec	Armored Fighting Vehicle and Weapon Section
AG	Adjutant general
A Gp	Army group
Alloc	Allocation
Ammo	Ammunition
ANCXF	Allied Naval Commander Expeditionary Force
ASF	Army Service Forces
Arty	Artillery
ASTP	Army Specialized Training Program
ATS	Auxiliary Territorial Service
Bd	Board
CAO	Chief administrative officer
CATOR	Combined Air Transport Operations Room
Cbl	Cable
CCS	Combined Chiefs of Staff
CE	Corps of Engineers
CGRS	Commanding General, Replacement System
Cir	Circular
COS	Chiefs of Staff (British)
CofT	Chief of Transportation
COMZ	Communications Zone
CONAD	Continental Advance Section, COMZ
CONBASE	Continental Base Section
Conf	Conference
Corresp	Correspondence
COSBASE	Coastal Base Section
Dev	Development
DQMG	Deputy Quartermaster General

Dukw	2½ ton 6 x 6 amphibian truck
ETO	European Theater of Operations
ETOUSA	European Theater of Operations, United States Army
EUCOM	European Command, successor to USFET
Exec	Executive
ExO	Executive Officer
FFRS	Field Force Replacement System
FUSA	First U.S. Army
FUSAG	1st U.S. Army Group
Fwd	Forward, forwarded
G–1	ACofS for personnel
G–2	ACofS for intelligence
G–3	ACofS for operations
G–4	ACofS for supply
GFRC	Ground Force Replacement System
GPA	General Purchasing Agent
Hist	Historical
Hq	Headquarters
HVAP	High-velocity, armor piercing
I&E	Information and Education
IG	Inspector General
Info	Information
JAG	Judge Advocate General
Jnl	Journal
LCT	Landing craft, tank
LMAB	London Munitions Assignment Board
Ln	Liaison
Log	Logistical
LST	Landing ship, tank
Ltr	Letter
Mat	Matériel
MGA	Major General of Administration (British)
Mil	Military
Min	Minutes
Mov and Tn Br	Movements and Transportation Branch
MRS	Military Railway Service
MT ship	Liberty ship converted for maximum vehicle-carrying purposes
MT80	Motor transport gasoline, 80-octane
MTB	Motor Transport Brigade
Mtg	Meeting
MTOUSA	Mediterranean Theater of Operations, United States Army
NATO	North African Theater of Operations

NATOUSA	North African Theater of Operations, United States Army
NOIC	Naval Officer in Charge
NUSA	Ninth U.S. Army
NYPOE	New York Port of Embarkation
OCOE	Office, Chief of Engineers
OCofOrd	Office, Chief of Ordnance
OCOO	Office, Chief Ordnance Officer
OCQM	Office, Chief Quartermaster
OCofT	Office, Chief of Transportation
Off	Officer
OGPA	Office, General Purchasing Agent
Opns	Operations
Ord	Ordnance
Orgn	Organization
P&O	Plans and Operations Div, WD
PC&R	Port construction and repair
Plng	Planning
PLUTO	From "pipeline under the ocean"—a cross-Channel underwater pipeline planned for bulk POL deliveries to the far shore
PMG	Provost Marshall General
POL	Petrol, oil, and lubricants
prep	prepared
Pur Div	Purchasing Division
R&D	Research and Development
R&SC Div	Requirements and Stock Control Division
RAMPS	Recovered Allied Military Personnel
Repls	Replacements
Rqmt	Requirement
SAC	Supreme Allied Commander
SCAEF	Supreme Commander, Allied Expeditionary Force
Sec	Section
SGO	Surgeon General's Office
SGS	Secretary, General Staff
SHAEF	Supreme Headquarters, Allied Expeditionary Force
SOLOC	Southern Line of Communications
SOP	Standing Operating Procedure
SPOBS	Special Observer Group
SSN	Specification Serial Number
Stf	Staff
SUSA	Seventh U.S. Army
Sv (s)	Service, Services
SvC	Service Command
TC	Transportation Corps

T/E	Tables of Equipment
Tn	Transportation
T/O&E	Tables of Organization and Equipment
TSFET	Theater Service Forces, European Theater
TUSA	Third U.S. Army
USAFE	United States Air Forces in Europe
USFET	United States Forces in the European Theater, successor command to ETOUSA
USSTAF	United States Strategic Air Forces
WD	War Department

Code Names

ALPHA	Beach in DRAGOON Operation; landings in the vicinity of Toulon and Marseille
ANVIL	Plan for the Allied invasion of southern France, finally executed as Operation DRAGOON in August 1944
BOLERO	The build-up of U.S. troops and supplies in the United Kingdom in preparation for the cross-Channel invasion
CAMEL	Beach in DRAGOON Operation; landings in the vicinity of Toulon and Marseille
DELTA	Beach in DRAGOON Operation; landings in the vicinity of Toulon and Marseille
DRAGOON	*See* ANVIL
GRENADE	Ninth Army supporting attack for Operation VERITABLE
MARKET-GARDEN	Airborne operation intended to establish a bridgehead across the Rhine in the Netherlands, September 1944. Operation MARKET involved seizure of bridges in the Nijmegen–Arnhem area, and Operation GARDEN was to open a corridor from Eindhoven northward toward Germany.
MULBERRIES	The artificial harbors constructed off the Normandy beaches
NEPTUNE	Code word for the cross-Channel operation, naming the specific assault area and target date, and for which a special security procedure known as BIGOT was developed
OVERLORD	Code word which came to be applied to the general concept of a cross-Channel invasion in 1944
PHOENIXES	Concrete caissons towed across the English Channel and sunk to form the main breakwaters for the artificial harbors
RHUMBA	Plan for reversing BOLERO and transferring U.S. forces, supplies, and logistic structure from the United Kingdom to the Continent
TOMBOLA	A flexible 6-inch underwater pipeline designed to discharge POL tankers anchored offshore at Ste. Honorine-des-Pertes
UNDERTONE	Seventh Army operation to breach West Wall and establish bridgehead over Rhine in Worms area, March–April 1945
VERITABLE	21 Army Group plan for a Canadian attack between the Maas and the Rhine, January–February 1945
WHALE	Flexible steel roadway, made of bridge spans and resting on pontons, forming the piers for the artificial harbors

Bibliographical Note

The task of the administrative historian is probably made somewhat easier than that of the historian reconstructing the story of tactical operations by the fact that the records of the relatively more settled headquarters were physically better preserved, and by the fact that fewer important decisions were lost through having been transmitted orally. But these advantages are at least partially offset by the manner in which administrative records were scattered after hostilities ended, and by the almost complete absence of the type of interview material which was collected from combat units in the field during the war and which helped fill important gaps in the record.

The official records of the various administrative headquarters in the European theater were never collected under one roof and, to make matters worse, were retired to U.S. repositories in piecemeal fashion over a period of several years. Research in ETOUSA records consequently was also piecemeal, proceeding neither by subject nor by chronology. Army regulations to the contrary, moreover, the records of the various technical services were not handled consistently. Some were sent to the main Army repository at St. Louis, Mo.; some were retained by the technical service chiefs and transferred directly to the respective technical service schools or camps in the United States. A number of officers retained official records for personal use.

The deficiency in interview material was remedied in part during the preparation of this volume by seeking the testimony of the principal commanders and staff officers who by reason of their participation possessed first-hand knowledge of events. Their testimony was secured through personal interviews conducted by the author, through correspondence on specific questions, and through comments made at the author's request on the manuscript in its first draft.

Primary Sources

Primary sources consist mainly of the official records of the various headquarters involved. They take the form of correspondence, interoffice memorandums, staff studies, cables, plans, minutes of conferences, journals, diaries, message files, and various periodic reports filed in accordance with the AGO decimal classification system. For the theater the main collections are those of SHAEF (principally those of the Adjutant General, Secretary of the General Staff, and the G–3 and G–4 Sections, including the War Diary of the last with key documents attached), and the papers of its predecessor, COSSAC; ETOUSA, SOS, and their successors ETOUSA-SOS and ETOUSA-COMZ; 12th Army Group; and the Advance Section. After the end of hostilities the U.S. Army command in Europe was successively renamed USFET and EUCOM, and some of the wartime records are filed under those designations.

The records of SHAEF, 12th Army Group, and the operational records of the armies are in the custody of the Operations Reports Section, Departmental Records Branch, AGO, in Alexandria, Va. The records of the more strictly administrative headquarters—ETOUSA, SOS, and their subordinate commands such as the Ground Force Reinforcement Command, the Advance Section, and certain of the technical services—were consulted at the Records Administrative Center in St. Louis, Mo., but have since been transferred to Kansas City, Mo. Certain planning files of First and Third Armies were also consulted in St. Louis. The COSSAC papers are in the SHAEF SGS files.

Two "unofficial" collections which proved valuable in reconstructing the history of the War in Europe were the files referred to in footnotes as *ETO Adm* and *ETO Preinvasion*. These consist of miscellaneous planning papers, cable files, and correspondence, which for the most part were rescued from destruction by personnel of the Historical Section, ETO, and were transferred intact to the Departmental Records Branch, AGO, in the War Department in 1946.

Two bodies of primary source material originating in the War Department and proving highly useful were the correspondence files of the Army Service Forces, which threw particular light on the role of Generals Somervell and Lutes in the support of the U.S. forces in Europe, and files in the War Department Operations Divisions, including logs of incoming and outgoing cables and decimal files on the subject of the troop basis and troop flow. The author had access also to the papers collected by Lt. Gen. Walter Bedell Smith, consisting mainly of "Eyes Only" cables, which have been deposited with the Department of the Army Library in the Pentagon. Limited use was made of the diary kept for General Eisenhower by his naval aide, Capt. Harry C. Butcher, and cited in this volume as Diary Office CinC. Excerpts from this diary were published by Captain Butcher in the volume *My Three Years with Eisenhower* (New York, 1946). The author had complete access to all official records relevant to this history regardless of classification.

Secondary Sources

An extensive body of unpublished secondary material exists covering the activities of the U.S. Army in Europe, the most important of which are the following:

(1) Histories of the technical services, the staff sections of Headquarters, ETOUSA-COMZ, the base sections, and the Ground Force Reinforcement Command, all required by ETO regulation. These vary in quality, the most useful being those of the Office, Chief of Transportation, Office of the Chief Engineer, Office of the Chief Surgeon, the COMZ G-4, the Ground Force Reinforcement Command, the Advance Section, and Normandy Base Section. They are filed in the ETO Administrative File, Operations Reports Section, Departmental Records Branch, AGO, in Alexandria, Va.

(2) The Administrative and Logistical History of the European Theater of Operations, eleven studies on logistics and administration, prepared in the Historical Section, ETO, under the author's supervision, and based for the most part on primary source materials available in the theater in 1945-46. The most useful of these preliminary histories consulted in the preparation of the present volume

are The Predecessor Commands: The Special Observers (SPOBS) and United States Army Forces in the British Isles (USAFBI), by WOJG Henry G. Elliott; Organization and Command in the European Theater of Operations, by Robert W. Coakley; and NEPTUNE: Training for and Mounting the Operation, and the Artificial Ports, by 1st Lt. Clifford L. Jones. All are on file in the Office, Chief of Military History.

(3) General Board Reports, 131 studies covering all aspects of the war in the European theater by a special board of officers appointed after V-E Day. These are uneven in quality, but some are extremely helpful, particularly in their critical analyses of plans, preparations, and methods and techniques of operations. They are on file in the Office, Chief of Military History.

(4) After Action Reports of the First and Third Armies, the 12th Army Group, and in some cases of the divisions. They were consulted for the supply story from the point of view of the field commands.

(5) Miscellaneous monographs on a wide range of subjects prepared by personnel of the Historical Section, ETO, by historians of the ASF, the Transportation Corps, and the Quartermaster Corps. Included are such studies as Overseas Supply Policies and Procedures, by Richard M. Leighton, and those on Quartermaster supply in the ETO prepared at the Quartermaster School, Camp Lee, Va.

Published histories, including memoir literature, have had occasional usefulness. The principal works cited are: Butcher's *My Three Years with Eisenhower;* General Omar N. Bradley's *A Soldier's Story* (New York, 1951); Robert E. Sherwood's *Roosevelt and Hopkins* (New York, 1948); Lt. Gen. Frederick Morgan's *Overture to Overlord* (New York, 1950); and Wesley Frank Craven and James Lea Cate (editors), *The Army Air Forces in World War II* (Chicago, 1948–55).

The service journals of the United States and Great Britain contributed first-hand accounts of experience as well as research articles. Most important were *The Journal of the Royal United Service Institution, Royal Engineers Journal, Military Review, The Quartermaster Review, Army Ordnance,* and *Army Transportation Journal.*

UNITED STATES ARMY IN WORLD WAR II

The following volumes have been published or are in press:

The War Department
- *Chief of Staff: Prewar Plans and Preparations*
- *Washington Command Post: The Operations Division*
- *Strategic Planning for Coalition Warfare: 1941–1942*
- *Strategic Planning for Coalition Warfare: 1943–1944*
- *Global Logistics and Strategy: 1940–1943*
- *Global Logistics and Strategy: 1943–1945*
- *The Army and Economic Mobilization*
- *The Army and Industrial Manpower*

The Army Ground Forces
- *The Organization of Ground Combat Troops*
- *The Procurement and Training of Ground Combat Troops*

The Army Service Forces
- *The Organization and Role of the Army Service Forces*

The Western Hemisphere
- *The Framework of Hemisphere Defense*
- *Guarding the United States and Its Outposts*

The War in the Pacific
- *The Fall of the Philippines*
- *Guadalcanal: The First Offensive*
- *Victory in Papua*
- *CARTWHEEL: The Reduction of Rabaul*
- *Seizure of the Gilberts and Marshalls*
- *Campaign in the Marianas*
- *The Approach to the Philippines*
- *Leyte: The Return to the Philippines*
- *Triumph in the Philippines*
- *Okinawa: The Last Battle*
- *Strategy and Command: The First Two Years*

The Mediterranean Theater of Operations
- *Northwest Africa: Seizing the Initiative in the West*
- *Sicily and the Surrender of Italy*
- *Salerno to Cassino*
- *Cassino to the Alps*

The European Theater of Operations
- *Cross-Channel Attack*
- *Breakout and Pursuit*
- *The Lorraine Campaign*
- *The Siegfried Line Campaign*
- *The Ardennes: Battle of the Bulge*
- *The Last Offensive*

The Supreme Command
Logistical Support of the Armies, Volume I
Logistical Support of the Armies, Volume II

The Middle East Theater
The Persian Corridor and Aid to Russia

The China-Burma-India Theater
Stilwell's Mission to China
Stilwell's Command Problems
Time Runs Out in CBI

The Technical Services
The Chemical Warfare Service: Organizing for War
The Chemical Warfare Service: From Laboratory to Field
The Chemical Warfare Service: Chemicals in Combat
The Corps of Engineers: Troops and Equipment
The Corps of Engineers: The War Against Japan
The Corps of Engineers: Military Construction in the United States
The Medical Department: Hospitalization and Evacuation, Zone of Interior
The Medical Department: Medical Service in the Mediterranean and Minor Theaters
The Ordnance Department: Planning Munitions for War
The Ordnance Department: Procurement and Supply
The Ordnance Department: On Beachhead and Battlefront
The Quartermaster Corps: Organization, Supply, and Services, Volume I
The Quartermaster Corps: Organization, Supply, and Services, Volume II
The Quartermaster Corps: Operations in the War Against Japan
The Quartermaster Corps: Operations in the War Against Germany
The Signal Corps: The Emergency
The Signal Corps: The Test
The Signal Corps: The Outcome
The Transportation Corps: Responsibilities, Organization, and Operations
The Transportation Corps: Movements, Training, and Supply
The Transportation Corps: Operations Overseas

Special Studies
Chronology: 1941–1945
Military Relations Between the United States and Canada: 1939–1945
Rearming the French
Three Battles: Arnaville, Altuzzo, and Schmidt
The Women's Army Corps
Civil Affairs: Soldiers Become Governors
Buying Aircraft: Materiel Procurement for the Army Air Forces
The Employment of Negro Troops

Pictorial Record
The War Against Germany and Italy: Mediterranean and Adjacent Areas
The War Against Germany: Europe and Adjacent Areas
The War Against Japan

Index

Aachen, 6, 19, 22, 255
Abbott, Brig. Gen. Oscar B., 336
ABC Haul, 141–43, 146, 147, 396, 397
Adcock, Brig. Gen. Clarence L., 297, 299
ADSEC. *See* Advance Section, Communications Zone.
Advance Section, Communications Zone (ADSEC), 12, 27, 33, 35, 37, 90, 113, 183
  ADSEC North, 380
  ADSEC South, 380
  ammunition build-up, 254
  ammunition handling, 275
  and Ardennes battle, 181–86
  bridge construction, 149, 422
  and civilian labor, 473–74, 476, 481
  and coal mining, 441
  depots, 113, 181, 254, 356, 398, 428, 434
  effect of manpower withdrawals, 462, 463
  engineers, 71, 406, 407
  local procurement, 498
  motor transport, 138, 421
  operations east of the Rhine, 375, 379, 380, 434
  POL stocks, 206, 208, 208n
  POL storage, 207
  rail restoration, 406–07
  ration levels, 188–89
  and regulating stations, 414, 421–22
  relations with field commands, 349
  requisitioning procedures, 358–59
  supply reserves, 19, 178, 180, 428, 434
  territorial changes, 35, 380
  tonnage allocations, 173, 176, 177
AFHQ. *See* Allied Force Headquarters.
Air attacks, enemy
  on ADSEC installations, 154, 186, 209
  at Antwerp, 113–15, 396
  on POL facilities, 183, 208, 209
Air Force commands and units
  Air Force, Ninth, 28, 234
    supply reserves build-up, 19
    tonnage allocations, 175
  Group, 46 (Br.), 162
  Tactical Air Command, IX, 426
  Tactical Air Command, XII, 29
  Troop Carrier Command, IX, 162, 163, 425, 427
Air Forces, U.S., men withdrawn for retraining, 467, 468–69. *See also* Air Force commands and units.
Air Priorities Board (SHAEF), 162
Air transport, 6, 10, 18, 161–65, 424–27
  administrative procedures for, 163, 425, 426
  bombers used for, 161, 162
  during Ardennes battle, 163
Air transport—Continued
  emergency uses, 163n
  in evacuation role, 427
  in last offensive, 426, 453
  parachute drops, 163, 165
  of POL, 161, 426
Airborne divisions
  13th, 286
  82d, 28
  101st, 163
Airborne operations, 28, 139–40. *See also* First Allied Airborne Army; MARKET-GARDEN.
  command organization for, 28
Albert Canal, 104, 113, 166–68, 424
Alençon, 195, 201
Allen, Brig. Gen. Wayne R., 472
Allied Container Advisory Committee, 203
Allied Force Headquarters (AFHQ), 24, 29, 40, 41
  Advance Detachment, 29
Allied Naval Commander, 114
Allocations System. *See* Tonnage allocations.
ALPHA Beach, 120, 122
Ameri-can, 201n
American Red Cross, 497
American School Center, 311
Ammunition, 22, 25, 174, 178, 180, 184, 247–75, 441–52. *See also* Ammunition day of supply rates; Ammunition maintenance rates.
  40-mm., 256
  60-mm. mortar, 257, 442
  81-mm. mortar, 257, 258, 259, 260, 263
  3-inch, 256
  75-mm., 256
  76-mm., 256
  90-mm., 256
  4.5-inch gun, 442
  105-mm. howitzer, 255, 257, 259, 263, 266, 266n, 268, 270, 271, 274, 444
  155-mm. gun, 257, 259, 266n, 268, 271, 442, 444
  155-mm. howitzer, 255, 258, 259, 266, 268, 270, 271
  8-inch gun, 257, 258, 258n, 259, 268, 271, 448
  8-inch howitzer, 248, 257, 258, 258n, 259, 271n, 442, 444
  240-mm. howitzer 248, 257, 258n, 259, 442, 444, 448
  accounting practices, 261–62, 449–50
  activity factors for allocations, 263–64, 449
  allocations, 247, 248, 249, 252, 253, 263–64, 270, 450, 451
  and Bull Mission, 268, 442
  captured, use of, 256, 270–71, 445
  expenditures, 247, 255, 256, 258, 259, 264, 267, 270, 270n, 446, 447, 448

Ammunition—Continued
handling at Antwerp, 114–15
losses during Ardennes battle, 183
lot segregation, 274–75
Ninth Army expenditures at Rhine, 448
October crisis, 247–57
order and shipping time, 261, 261n, 262, 443
port discharge of, 249, 250, 254
production in United States, 265, 269, 271
rationing and control system, 251–54, 272, 450–52
requirements estimated, 257, 265, 275, 400, 445, 449
reserves, 247, 442, 447
reserves policy, 272–73, 442, 448n
and September decisions, 12
SHAEF maintenance rate, 272, 273, 442
SHAEF reserve, 272
theater level authorized, 261–63
War Department contention with the theater, 257–63, 443–47, 450n
Ammunition Allocations Committee, 251
Ammunition day of supply rates, 260, 260n, 267, 272, 442, 443, 446, 447, 452
Ammunition maintenance rates, 267–68, 272, 273, 442, 443, 446, 447
Amphibious Truck Company, 821st, 72
Amsterdam, 51
Antiaircraft artillery cancellations, 291, 291n, 292
Antwerp, 5, 10, 13, 16, 18, 21, 24, 38, 48, 49, 50, 52, 104–15, 424
and ABC Haul, 141, 142, 396
administration of, 108, 109
air attacks on, 113–15, 186, 396
allocation of facilities, 105, 109
and Ardennes battle, 181–86
backlogs, 396–97
civilian labor, 480
clearance, 110–11, 141, 396
discharge performance, 110, 113, 115–16, 393, 396
estimated opening date, 18
facilities described, 104–05
and Memorandum of Agreement on, 108, 141
planned discharge, 110, 111
as POL port, 199–201
role in logistic plans, 9
as solution to logistic dilemma, 49
ANVIL, 117. *See also* DRAGOON.
Anzio, 219
Ardennes counteroffensive, 25–26, 113, 115, 181–87
and ADSEC depots, 181
air supply in, 163
ammunition expenditures, 256
and depot requirements, 182
effect on port operations, 132, 184–85
and embargo on forward movements, 182, 186
and evacuation of supplies, 182–83
and First Army installations, 181, 182
marshaling of truck transport in, 148
Ardennes counteroffensive—Continued
and northern depots, 181, 398
and POL stocks, 208–09
and Third Army's redeployment, 184
Argentan, 5
Armed Forces Network, 245, 386
Armies, U.S. *See* First Army; Third Army; Fifth Army; Seventh Army; Eighth Army; Ninth Army; Fifteenth Army.
Armored Division (Fr.), 2d, 298
Armored Divisions
2d, 237
3d, 165, 237, 423, 456
4th, 234
7th, 17, 423
9th, 281, 369, 456
10th, 281, 427
11th, 280
12th, 286, 293, 295, 298n, 465
13th, 285, 286
14th, 280, 299
16th, 286, 465
20th, 287, 298n
Armored Group, 10th, 238
Army Ground Forces, 343
Army Group, 6th, 16, 24, 26, 41, 229, 371. *See also* Devers, Lt. Gen. Jacob L.; Southern Line of Communications.
ammunition shortage, 250–51, 266, 270, 272, 442, 445, 451
control of transportation, 158
division slice, 301, 302, 303
and divisional build-up, 284
establishment, 29
operations in October, 24
relationship to SOLOC, 41, 46
replacements, 333, 334
service troop basis, 293, 297–303
strength in divisions, 25, 176, 302, 303
Army Group, 12th, 9, 17, 18, 26, 27, 170, 173, 247 251, 293, 349, 387. *See also* Bradley, Lt. Gen. Omar N.
and air supply, 161, 162, 163, 425
ammunition allocations, 247, 248, 251–54, 263–64, 451
ammunition shortage, 251, 260, 266, 267, 442, 448
becomes operational on Continent, 26
casuals, policy on, 344
in command plan, 27, 29
and diversions through Marseille, 297
division slice, 301, 303
forward reserves, 180
maintenance requirements, 173, 431
plans for Rhine crossing, 373
priority over 6th Army Group, 299, 300
relationship to COMZ, 27, 163, 349, 362, 387
release of divisions to 6th Army Group, 17
replacements, 333, 465

Army Group—Continued
and service troop basis, 293, 295, 299, 300, 303, 371
strength in divisions, 18, 19, 20, 25, 173, 174, 175, 176, 177, 371
supply consumption rates, 431–32
tactical plans, 9, 371
tank strength, 237–38
tonnage allocations, 170–71, 173, 174
and winter clothing, 234
Army Group, 21. *See also* Montgomery, Field Marshal Sir Bernard L.
air transport of POL to, 162
and Antwerp, 107, 110
and Ardennes battle, 182
in command plan, 27
and MARKET-GARDEN, 15, 139
and loan of artillery and ammunition, 271
and loan of tanks to U.S. forces, 240–41
October operations, 24
operations east of the Rhine, 372
November offensive, 25
and rail allocations over the Rhine, 412
and Red Lion Route, 139
and Rhine crossing, 370
strength in divisions, 176n
tactical plans, 8, 10, 14
on use of ports, 59, 98, 106, 114
Army Group B (Ger.), 371
Army Service Forces (ASF), 151, 268, 352, 354
Army Specialized Training Program. *See* ASTP students.
Army Talks, 230
Arromanches-les-Bains, 58, 60. *See also* MULBERRY B.
Assembly Area Command, 383, 385, 497
ASTP students, 466
Attrition rates, 235–36. *See also* Replacement factors.
Aurand, Maj. Gen. Henry S., 38, 380
Aviation gasoline. *See* Gasoline, aviation.

Back ordering, 359
Bad Kreuznach, 406
Barfleur, 51, 60, 61
Barker, Maj. Gen. Ray W., 306, 323, 324, 327, 388
Barnes, Col. Elmer E., 195
Barnes, Maj. Gen. Gladeon M., 456
Barr, Maj. Gen. David G., 298
Base 901 (Fr.). *See* Base Sections, Base 901.
Base Sections
Base 901 (Fr.), 379
Brittany, 28, 32, 34, 37, 378, 476. *See also* Base Sections, No. 1.
Central, 35
Channel, 45, 101, 106, 190
activation and mission, 36, 38
and Antwerp, 38, 108, 109
Base Sections—Continued
Channel—Continued
and express routes, 140, 142
and Le Havre, 101
and Red Horse Staging Area, 497
and Rhine crossings, 375, 379
Cherbourg, 34
Coastal (COSBASE), 38. *See also* Base Sections, Continental.
command problems, 380–82
Continental (CONBASE), 40
Delta, 40, 45, 123, 301, 428, 484
Eastern, 35
intermediate, 33, 35, 36. *See also* Base Sections, Loire; Base Sections, Oise.
Loire, 34–35, 37, 378
No. 1, 32, 34
No. 2, 32, 33, 34
No. 3, 32, 33, 34, 36
No. 4, 36
No. 5, 35
Normandy, 37, 45, 58, 85, 90
and Cherbourg's clearance, 87–88
command changes, 38, 88, 380
establishment, 34
and express routes, 140
territorial changes, 378
use of prisoners of war, 485
Northern Ireland, 35
Oise, 36, 37, 45, 375, 379, 428, 484, 485
planned organization, 32, 34–36
Seine, 32, 35, 45, 138, 148, 379, 476–77, 496–97
Southern, 35, 36
United Kingdom (U.K.), 33, 35, 36, 45, 379, 380
Western, 35, 36
Bastogne, 165
Battle of the Bulge. *See* Ardennes counteroffensive.
Bayeux, 139
Beaches, 17, 46, 51, 52, 53–60. *See also* ALPHA Beach; CAMEL Beach; DELTA Beach; OMAHA Beach; St. Raphael; St. Tropez; Southern France beaches; UTAH Beach.
Belfort, 24, 157
Bell, Col. Frank F., 99
Berlin, 10, 11, 15
Black market, 3, 31
Blankenberge, 106
"Boleros," 152
Bombers for air transport, 161, 162
Bond, Brigadier D. H., 198
Bonesteel, Maj. Gen. Charles H., 342, 384
Boulogne, 51, 106
Bradley, Lt. Gen. Omar N., 27, 104, 170, 371, 372. *See also* Army Group, 12th.
and ammunition shortage, 250, 254, 255, 266, 267, 273, 445, 447
attitude on logistics, 349

Bradley, Lt. Gen. Omar N.—Continued
  and build-up acceleration, 283–84
  and September decisions, 11–12, 16
  and service troop basis, 293
  and tank allocations, 456
  and tonnage allocations, 170
  and winter clothing, 217–18, 220, 221
Bremen Port Command, 385
Brereton, Lt. Gen. Lewis H., 28. *See also* Air Force commands and units, Air Force, Ninth.
Brest, 11, 15, 16, 28, 34, 36n, 47, 50, 51, 52
British Chiefs of Staff, on MULBERRY B, 59
Brittany. *See also* Base Sections, Brittany.
  and DRAGOON, 118
  and OVERLORD plan, 4, 37
  and Third Army missions, 6
Brittany Base Section. *See* Base Sections, Brittany.
Brittany ports, 5, 7, 13, 15, 16, 46, 47, 89, 94–95, 118. *See also* Brest; Cancale; Granville; Lorient; Morlaix; Nantes; Quiberon Bay; Roscoff; St. Brieuc; St. Malo; St. Michele-en-Grève; St. Nazaire.
Brussels, 139
Brussels Conference, 106, 108
Build-up, Allied forces, 9
Build-up, U.S. forces, 5, 276–86
  acceleration of divisions, 276–86
  advance shipment of regiments, 283–84, 285–86
  arrival dates, divisions, 282–83
  days of combat, 282–83
  divisions on Continent, 5, 12, 276, 281, 285, 286
  divisions in United Kingdom, 276, 285, 287
  effects of acceleration, 278, 284, 290, 299
  plans for (W, X, Y, and Z), 6, 276, 278, 281, 285
  service troops. *See* Service troops.
  theater strength, 276, 281, 285, 286, 288, 296, 433
  training status of divisions, 279–80
Bull, Maj. Gen. Harold R., 266, 291, 292, 293–94, 295, 297, 306, 388, 446, 464
Burgundy District. *See* Continental Advance Section (CONAD).
Burpee, Brig. Gen. Clarence L., 82, 392
Byrd, Adm. Richard E., 235n

C Ration. *See* Rations, operational.
Caen, 210
Caissons, 100, 101, 102. *See also* MULBERRY; PHOENIXES.
Calais, 51, 106
Cambrai, 155
CAMEL Beach, 120, 122
Camouflage clothing, 496
Camps
  Lucky Strike, 497
  Old Gold, 497
  Pall Mall, 498
  Philip Morris, 498
  Twenty Grand, 497
Cancale, 47, 51, 89
Cannibalization, 454, 489
Captured ammunition. *See* Ammunition, captured, use of.
Carentan, 60, 61, 62
Caserta, 38, 40, 42
Casuals, 340, 343–46
Casualties
  battle, 307, 315, 316, 317, 321, 331, 333, 343n, 459n, 461, 462, 465
  evacuation by air, 427
  hospitalization experience, 343n
  nonbattle, 316, 317, 321, 333, 461
  from trenchfoot, 230, 316
CATOR. *See* Combined Air Transport Operations Room.
Central Base Section. *See* Base Sections, Central.
Châlons-sur-Marne, 195, 435, 438
Channel Base Section. *See* Base Sections, Channel.
Channel ports, 15, 106. *See also* Amsterdam; Antwerp; Blankenberge; Boulogne; Calais; Dieppe; Dunkerque; Ghent; Gravelines; Nieuwpoort; Ostend; Rotterdam; Tréport, Le; Zeebruggé.
Charleroi, 142, 149, 181, 184, 491
Chartres, 83, 146, 195, 356
Cherbourg, 4, 33, 46, 48, 51, 52, 53, 62–89, 476
  civilian labor, use of, 476
  clearance, 79, 81–84, 88
  condition on capture, 63–68
  discharge performance, 52, 78, 84, 88, 89, 393, 394
  mining of harbor, 69
  organizational problems, 79–81, 87, 88
  organized as Area No. I, 33
  planned capacity, 63, 71, 77, 78, 85, 87, 88, 394
  as POL port, 197–98, 198n
  prewar capacity and use, 63, 66
  prisoner of war labor, use of, 484
  reconstruction, 71–77
  salvage operations, 68–71
Cherbourg Command, 33, 34
Chief Engineer, 196, 472. *See also* Moore, Maj. Gen. Cecil R.
Chief Quartermaster, 294, 462. *See also* Littlejohn, Maj. Gen. Robert M.
Chief of Staff, Communications Zone. *See* Lord, Maj. Gen. Royal B.
Chief of Staff, Supreme Headquarters, Allied Expeditionary Force. *See* Smith, Lt. Gen. Walter B.
Chief of Staff, U.S. Army. *See* Marshall, Gen. George C.
Chief of Transportation, 138, 160, 382, 393. *See also* Ross, Maj. Gen. Frank S.
Chorpening, Col. Claude H., 78
Cigarette camps. *See* Camps.
Civil imports (Fr.), 166–67
Civil relief, 130
Civilian labor, 102, 297, 423, 424, 472, 475–86

Civilian labor—Continued
Belgian, 105, 115, 168, 440, 479–81
British, 485–86
Dutch, 481
French, 89, 102, 103, 124, 137, 475, 476, 482
German, 481
Luxembourg, 481
mobile units, 481–82
number employed in COMZ, 477, 486
policy on procurement, 472–73, 476, 482
Slav service units, 484
strikes, 115, 440, 480–81
wage scales, 473–74, 477–79, 481, 482
Clark, Lt. Gen. Mark W., 29n, 314
Clarkson, Col. R. Hunter, 123
Clay, Maj. Gen. Lucius B., 38, 88, 88n, 221, 266, 268, 385, 388
Clearance of U.K. depots. *See* Reverse BOLERO; RHUMBA.
Clothing, 218–28, 507
committee on winter clothing, 234
development of the 1943 uniform, 219
the "Eisenhower" vs. the M1943 jacket, 220–22, 226–27
emergency requisitions for, 225
field tests of, 234
footwear. *See* Footwear.
investigation of inadequacy of, 232–33
maintenance and replacement factors, 224–25, 226
tactical forecasts and clothing plans, 223–24
Coal
continental receipts, 210, 212, 213–14, 440–41
discharge at Granville, 90, 92, 210
discharge at St. Brieuc, 93–94, 210
effect of shortage on local procurement plans, 475, 500
local procurement
German, 441
labor unrest, 213
in liberated countries, 210–12, 213, 440–41
pitwood shortage, 212, 441
and reciprocal aid agreements, 471
mine supplies provided, 499
requirements and allocation of, 209–10
and U.K. coaster shortage, 125
Coal mining. *See* Coal, local procurement of.
Coastal Base Section (COSBASE). *See* Base Sections, Coastal.
Coasters, 100, 102–03
OVERLORD requirements, 124
return of to British, 152–53
and Reverse BOLERO, 402
shortages, 125, 212, 217
U.K. requirements for, 125
and use of small ports, 61, 91, 94, 96–97
Cobb, Col. Charles, 363
Cole, Col. Walter C., 317
Colglazier, Col. R. W., 363
Collins, Brig. Gen. Leroy P., 34, 35. *See also* Base Sections, Loire.
Colmar, 25, 26, 157, 445
Cologne, 423
Combined Air Transport Operations Room (CATOR), 425, 426
Combined Chiefs of Staff, 10, 14, 124, 126, 283, 374
Combined Military Procurement Control Section, 471
Command and organization, 350, 504–06. *See also under major commands.*
COMZ relations with field commands, 349, 505
COMZ-ETOUSA staff, 350, 504–05
Lee's position in theater command, 350, 504–06
SHAEF's position in theater command, 350–51, 504
Commodity loaders, 129, 132, 217
Commodity loading, 110
Communications Zone (COMZ), European Theater of Operations (ETOUSA)
and air transport, 162, 163, 425
ammunition shortage, 247–75, 442–52
ammunition stocks, 447, 448
and Ardennes counteroffensive, 113, 155, 182–86
boundary changes, 389
civilian labor, 478, 479, 481, 484, 485
command and organization, 26, 31–45, 347, 382–89, 415, 504–06
delivery capability, 18, 169, 174, 175, 176, 177, 178, 179, 180
depot plans, 398
and diversions through Marseille, 297–98
division slice, 299, 302
and divisional build-up, 277–89, 290
effect of pursuit on, 8, 86
effect of withdrawals on, 463, 467
establishment on Continent, 26–27, 31
general staff. *See* European Theater of Operations, G–1, G–3, G–4.
incorporation of southern France, 40–45
local procurement, 471–75, 489
Lutes-Somervell mission, 348–63
move to Paris, 31
movement control, 390–93, 413–14
pipelines, 198
planes for Rhine crossings, 373, 375
POL stocks, 194, 206, 208, 208n
and port plans, 7, 47, 49–50, 85–86, 98, 101, 102, 106, 114
posthostilities planning, 385
ration levels, 188–93
relations with field forces, 27, 179, 349–51, 362, 386–88, 505
replacements, 305–47, 459–69
requisitioning procedures, 358–59
responsibility for 6th Army Group support, 42–43, 263n
and Reverse BOLERO, 402–04

Communications Zone—Continued
role as theater headquarters, 328, 347, 350, 386–88, 504–06
separation of COMZ from ETOUSA, 388
and service troop basis, 291, 295, 296–97, 299, 300, 302, 303
and shipping problem, 127
standing operating procedures. *See* Standing Operating Procedures, COMZ-ETOUSA; SOP's 7, 10, 29.
and supply expedients, 359–63
and supply shortages, 3, 4, 244. *See also* Supply shortages.
territorial organization, 32–40, 378–80
tonnage allocations, 170, 172, 173, 175, 176, 177
withdrawals for retraining, 467, 468–69
Communications Zone, Mediterranean Theater of Operations (COMZ-MTOUSA), 42
COMZ-ETOUSA. *See* Communications Zone (COMZ), European Theater of Operations (ETOUSA).
COMZ-MTOUSA. *See* Communications Zone, Mediterranean Theater of Operations.
CONAD. *See* Continental Advance Section.
Connery, George, 218n
Construction
bridge, 149, 157, 216, 407, 408–10, 422–24, 453
pipeline, 196, 199, 200
port, 71–77, 90, 99–100, 102, 122
railway 81–83, 149, 155–57, 406–10, 412
road, 422–24
staging areas, 497
Consumption factors. *See* Maintenance rates.
Continental Advance Section (CONAD), 40, 44, 46, 159, 301, 378
and Burgundy District, 378, 379–80
civilian labor, use of, 477
coal mining, 441
control of transport, 159
depots, 428
operations east of the Rhine, 375, 379, 380
rail congestion, 186
and regulating stations, 414n
supply reserves, 180, 428
Continental Base Section (CONBASE). *See* Base Sections, Continental.
Conversion training. *See* Replacements, conversion training of general assignment men.
Corps, V, 30, 367
Corps, VI, 257, 427
Corps, VII, 264
Corps, VIII, 28, 34, 92, 170n, 183, 184
Corps, XII, 216, 264, 370
Corps, XV, 17, 186, 251, 298
Corps, XVIII Airborne, 28, 234
Corps, XX, 256, 264
Corps XXX ((British), 264
Corps of Engineers, 191, 290
Coubert, 195, 205
Couville, 82–83
Crain, Maj. Gen. James K., 241
Crawford, Maj. Gen. Robert W., 60, 98, 174, 177, 221, 254, 260, 294, 297, 306, 348, 388, 425
Crerar, Lt. Gen. H. D. G., 24. *See also* First Canadian Army.
Cress, Col. James B., 72
Crothers, Col. James A., 85

"Danger pay," 480
Davis, Brig. Gen. Benjamin O., 384, 385, 389
De Gaulle, Gen. Charles. *See* Gaulle, Gen. Charles de.
De Lattre de Tassigny, Gen. Jean. *See* Lattre de Tassigny, Gen. Jean de.
Decanting. *See* POL, decanting of.
Dehydrated rations. *See* Rations, dehydrated.
Delta Base Section. *See* Base Sections, Delta.
DELTA Beach, 120, 122
Depots, 6, 84, 128, 181, 187, 355, 375
advance, 6, 154, 155
and Ardennes battle, 181–86, 398
base, 8, 356, 397–98
control of, 382
effect of pursuit on, 8, 356
inadequacy of system, 397–98, 509
intermediate, 6, 137, 187, 355, 357
and movement plans, 390–93
Normandy, 84, 86, 97
OVERLORD plans for, 356
plans for echeloned structure, 398
plans for final offensive, 375, 428
replacement. *See* Replacement depots.
Devers, Lt. Gen. Jacob L., 16, 17, 29, 272, 280, 302, 306, 314, 333, 345, 372, 384, 445. *See also* Army Group, 6th.
Dieppe, 98, 106
Differdange, 149, 498
Dijon, 40, 42
Displaced persons, 460, 482
Divette, 61
Division slice, 289, 296, 486, 510
6th and 12th Army Groups compared, 301–02
effect of accelerated build-up on, 278
northern and southern LofC compared, 299, 299n, 301–02
plans for, 289, 299, 299n
size in July and August, 290
supply requirements of, 18, 169, 173, 373, 374, 430, 431. *See also* Maintenance rates.
Divisions. *See also* Airborne divisions; Armored divisions; Infantry divisions.
arrival dates, 282–83
build-up of. *See* Build-up, U.S. forces.
build-up east of the Rhine, 373, 374, 375, 376, 377
days of combat, 282–83

Divisions—Continued
  diversion of through Marseille, 17, 280–81, 284, 297
  number supported, 5, 6, 12, 173, 286
  number deployed in World War I, 286n
  planned employment in Berlin thrust, 11
  training status of, 279–80
  undeployed reserve, 279–80, 280n
Domfront, 195
Doriot, Col. Georges F., 221, 222
Dourdan, 205
DRAGOON, 28, 29, 117, 118, 118n, 294, 299
Driver shortages, 147n, 291, 292, 293
"Duck bill." *See* Extended end connectors.
Duisburg, 410, 413
Dukws, 56, 57, 80, 102
Dunkerque, 106

Eastern Base Section. *See* Base Sections, Eastern.
Eighth Army, 27n
Eisenhower, Gen. Dwight D., 3, 25, 27, 107, 108, 211, 220, 262, 263, 276, 284, 371, 372, 374, 402, 456, 461
  and Antwerp's role, 107–08
  and August plan for Third Army thrust, 8–9
  and Berlin thrust proposal, 10–11
  and build-up schedule, 276–77, 284, 285
  on COMZ move to Paris, 31
  on DRAGOON, 118
  dual command role, 504
  on manpower problem, 312, 323, 326, 331, 336, 505
  and Reverse BOLERO, 402
  and Rhine crossing plans, 372–75
  and Schelde Estuary decision, 107–08
  and September decisions, 3, 9, 11, 12, 13–14, 15
"Eisenhower jacket," 220–28, 227n
Embargo on movements, 182, 186
Engineer Aviation Brigade, 25, 426
Engineer Combat Group, 1186th, 58
Engineer Camouflage Battalion, 602d, 496
Engineer General Service Regiments, 290, 484
  332d, 168
  342d, 73
  355th, 168
  373d, 99
  1301st, 423
Engineer Groups
  A, 407, 410
  B, 408, 410
  C, 407
Engineer Port Construction and Repair (PC&R) Groups
  1051st, 122
  1055th, 90, 166
  1056th, 72, 168, 407, 409
  1057th, 94, 166
  1058th, 90
Engineer Service, 494–96, 498
Engineer Shore Regiment, 531st, 58
Engineer Special Brigades
  1st, 58
  5th, 58, 110
  6th, 58, 441
  Provisional, 475
Englebert of Liège, 493
Epinal, 156
ETO Jacket. *See* "Eisenhower jacket."
ETOUSA. *See* European Theater of Operations, U.S. Army; Communications Zone, European Theater of Operations.
European Theater of Operations, U.S. Army (ETOUSA), 26, 27, 251, 265. *See also* Communications Zone, European Theater of Operations.
  boundary changes, 29, 389
  command and organization, 26–45, 378–80. *See also* Command and organization.
  G–1. *See* Franey, Col. James M.
  G–3. *See* Eyster, Brig. Gen. George S.
  G–4. *See* Stratton, Brig. Gen. James H.; Gilland, Brig. Gen. Morris W.
  incorporation of southern France, 40–45
Evans, Col. George R., 305, 335
Evans, Col. Ira K., 383
Express shipping service. *See* REX shipping procedure.
Extended end connectors, 491–92
Eyster, Brig. Gen. George S., 278, 295, 385, 462, 465

Falaise, 5
Field Artillery, proposed inactivations, 293, 294, 295, 296, 297
Field Artillery Brigade, 32d, 270
Field Force Replacement System. *See* Replacement System.
Field jacket, M1943, 219–28, 234. *See also* Clothing.
Fifth Army, 218, 235, 314, 315, 372
Fifteenth Army, 30, 371, 378
First Allied Airborne Army, 28, 161
First Army, 6, 11, 24, 176, 369, 370, 371, 372, 378, 410
  and air transport, 426
  and ammunition supply, 248, 249, 251, 255, 260, 264, 270, 274, 447, 449–50
  and Ardennes battle, 26, 181–83, 208–09
  bridge construction, 216
  captured rations, use of, 188
  local procurement, 474, 488, 492–93, 496
  logistics in final offensive, 430
  POL stocks, 194, 205, 206, 208, 208n, 209, 438
  preparations for final drive, 429
  rail service for, 406, 412
  rail transfer point, 137

First Army—Continued
ration levels, 189, 189n, 434
and Red Ball routes, 134, 135
relations with COMZ, 349
replacements, 337, 465, 468–69
September tactical plans, 14
strength in divisions, 19
supply consumption rates, 431, 432
supply priorities, 11, 170, 174, 175
supply reserves, 19, 172, 176, 179, 429
truck transport support of, 418, 419, 422
and winter uniform, 234
First Canadian Army, 24, 99, 107, 367
First French Army, 24, 29, 157, 264, 301, 371, 421n. *See also* Lattre de Tassigny, Gen. Jean de.
in November offensive, 25
service troop shortages, 301, 302
supply consumption rates, 432
Folliard, Edward T., 218n
Fontainebleau, 36, 331, 467
Footwear, 228–32
and trenchfoot incidence, 229–31
Forward deliveries. *See* Supply deliveries to armies.
Forward Echelon, Communications Zone (FECOMZ), 27, 31
Forward reserves. *See* Supply reserves.
Fourth Army, 27, 27n
"Franco's Revenge," 500
Franey, Col. James M., 318, 323
Frankenthal, 423, 437, 438
Frankfurt, 9
Franklin, Brig. Gen. John M., 130
French forces. *See* First French Army; Base Sections, Base 901; Service troops, for French forces.
Fuel. *See* Coal; POL; Wood fuel.

Gale, Lt. Gen. Sir Humphrey M., 59, 175
Garside, Col. Charles, 232
Gasoline, 8, 12, 174, 180, 183, 193–201. *See also* Pipeline Systems; POL.
aviation (AVGAS), 205n, 208
motor vehicle (MT80), 195, 200, 204, 205n, 207
Gasser, Maj. Gen. Lorenzo D., 326, 326n, 384
Gaulle, Gen. Charles de, 379
General assignment personnel. *See* Replacements, conversion training of general assignment men.
General Inspectorate Section, ETOUSA, 384, 389
General Purchasing Agent, 382, 471–72, 473, 474, 483, 499, 501
Geneva Convention, 481, 484
Germany, plans for advance into, 8–9, 10–11, 11–15
Gerow, Maj. Gen. Leonard T., 30
Ghent, 114, 184, 394, 395, 424
"GI Joe Diners," 421n
Gilland, Brig. Gen. Morris W., 40, 361, 363, 383, 399, 400
Gnome-Rhône motor works, 488
Goodfriend, Maj. Arthur, 342n
Goodrich tire plant, 245
Gorby, Col. Alvin L., 230
Graham, Maj. Gen. Miles H., 108
Grand divisions, 150n
Grandcamp-les-Bains, 51, 60, 61
Granier, Brig. Gen. Georges, 379
Granlieu, 67, 70–71
Granville, 51, 53n, 60, 89–92, 210, 212, 394
Granville raid, 394
Gravelines, 106
Gray, Brig. Gen. Carl R., 156, 161, 415
Green Diamond Route, 140–41, 159
Gregory, Maj. Gen. Edmund B., 221, 223, 233, 235. *See also* Quartermaster General, The.
GRENADE, 367
Grenoble, 156
Gross, Maj. Gen. Charles P., 243
Ground Force Reinforcement Command. *See* Replacement System.
Ground Force Replacement System. *See* Replacement System.
Group Control Council, U.S., 385, 389
Grower, Brig. Gen. Roy W., 35, 378
Gullatt, Col. Doswell, 110
Guns, captured, 270–71

Hadir Steel Works, 149, 498
Halifaxes in air supply role, 162
Hanau, 412
Handy, Lt. Gen. Thomas T., 314, 315, 320, 323, 324
Havre, Le, 19, 36, 38, 48, 49, 50, 51, 60 97, 98, 99, 100, 103, 184, 356, 395, 497, 498
allocation of discharge capacity, 17
as POL port, 199
Hennecke, Konteradmiral, Walther, 63
Henry, Maj. Gen. Stephen G., 319, 323
High velocity armor piercing ammunition. *See* HVAP ammunition.
Highway Transport Divisions
1st HTD, 418, 419
2d HTD, 418, 419, 421
Highways, 422
rights at Antwerp, 109
Hinds, Brig. Gen. John H., 253
Hirson, 135
Hitler, Adolf, 9
Hodges, Lt. Gen. Courtney H., 221, 349. *See also* First Army.
Hog Islanders, 403
Hoge, Brig. Gen. William M., 100, 100n
Holly, Brig. Gen. Joseph A., 456
Houffalize, 26
Hughes, Maj. Gen. Everett S., 347, 384, 389, 404
HVAP ammunition, 456, 457–58

Indochinese labor, 124

Infantry divisions
1st, 342n
3d, 219
10th (Light Mountain), 280n
26th, 170, 234, 281
35th, 257
42d, 285, 302, 322, 334
44th, 281
63d, 285, 302, 322, 334
65th, 285
66th, 285
69th, 285, 322
70th, 285, 302, 322, 334
71st, 286
75th, 283, 284
76th, 285, 289
79th, 17, 298
84th, 264, 287
86th, 286
87th, 283, 284, 302
89th, 285
95th, 163n, 170
97th, 286
99th, 280
100th, 280, 299
102d, 281
103d, 280, 299
104th, 170, 281
106th, 283, 284, 465
Infantry School, 331, 466, 467, 468
Infantrymen
casualties, 307, 321
conversion (retraining) to, 310, 311, 315, 316, 318, 321, 324, 326, 328–34, 461, 463, 465
planning factors, 308, 309, 316, 319
stockage, 316, 317, 319, 331
Information and Education, 386
Inland waterways, 104–05, 165–68, 424
Inland Waterways Committee, 165
Intermediate Section. *See* Base Sections, intermediate; Base Sections, Loire; Base Sections, Oise.
Isigny, 5, 60, 61
Is-sur-Tille, 158, 159
Italian service units, 301, 302, 483–84

Jacobs, Brig. Gen. Fenton S., 36, 108, 142
James B. Weaver, 110
JB–2 missile, 271
Jerrican, 201, 201n
procurement on continent, 204
procurement in United Kingdom, 202, 203, 204
requirements, 202–03, 203n
recovery program, 204
Jodl, Generaloberst Alfred, 372
Joint Chiefs of Staff, 130, 131, 286
Joint Military Transportation Committee, 130, 131

K Ration. *See* Rations, operational.

Karlsruhe, 410
Katy mine, 69–70
Koenig, Brig. Gen. Egmont F., 380
Kruisschans Lock, 110

Labor. *See* Civilian labor; Prisoner of war labor.
Labor strikes
at Antwerp, 115, 480–81
in Belgian mines, 440
Labor, U.S., for munitions, 246
Ladd, Brig. Gen. Jesse A., 380
Laon, 149
Larkin, Maj. Gen. Thomas B., 38, 42, 43, 301, 302, 379, 383, 404, 462
Lattre de Tassigny, Gen. Jean de, 28. *See also* First French Army.
Layman, Brig. Gen. Walter G., 310, 334, 336
LCM's, 416
LCT's, 48, 100
Lead time (procurement), 352–53, 508. *See also* Requisitioning procedures.
Lear, Lt. Gen. Ben, 327, 328, 383–84, 387, 388, 461, 462, 464, 465, 466, 469
Leave centers, 497
Lee, Lt. Gen. John C. H., 27, 31, 43, 50, 98, 105, 244, 362, 379, 472, 504. *See also* Communications Zone, European Theater of Operations.
and ammunition shortage, 250
and Antwerp's planned use, 105, 108
and Ardennes battle, 184
command problems, 350, 382–89, 504–06
and COMZ move to Paris, 31–32
in dual role, 504
and Lutes-Somervell visit, 348–63
relations with field commands, 362, 505
and replacement problem, 306, 308, 329
and Reverse BOLERO, 403
review of port plans, 49–50, 98
on shipping problem, 128
Le Manoir, 167
Le Mans, 35, 83, 311, 356
Lend-lease, 130, 502
Liberty ships, 94, 102, 122, 131, 403
berths for discharge of, 61, 74, 75, 77, 78, 101, 105, 123
and shipping tie-up, 129
Lichfield, 334
Liège, 111, 142, 148, 154, 181, 207, 209, 252, 406, 407, 491, 493
Lighterage operations, 80, 127
Lille, 113, 155, 182
Limited assignment personnel. *See* Replacements, limited assignment men.
Limiting factors, 506–10
Linas, 205
Lines of communication, 4, 511
capacity, 49
length of, 4–5, 7, 13, 15, 16, 49, 115, 419, 421

Lines of communication—Continued
strictures or bottlenecks, 133
U.S. and British cross, 36, 37, 101, 143
Little Red Ball, 143
Littlejohn, Maj. Gen. Robert M., 192, 213, 217
and clothing controversy, 218–35
Local procurement, 213, 244–45, 440, 470–502. *See also* Civilian labor; General Purchasing Agent.
agreements with countries, 470–71
by Chemical Warfare Service, 494
by Engineers, 452, 486, 487, 496
extent and value, 502
of food, 193
limitations on, 474–75, 490, 500–502
from neutral countries, 500
of ordnance items, 488–94
payments for, 501
policy established, 471
potentialities for, 474–75
of quartermaster items, 487, 494–96
of real estate, 486, 496, 497
requirements for, 488
of signal items, 486, 499
Locomotives. *See* Rolling stock, locomotives.
Logistic estimates for thrusts into Germany, 8–15
Logistic planning, 508
for Rhine crossing, 374–75
Logistical Plans Branch, SHAEF, 173, 174
London Munitions Assignments Board, 241
Longwy, 184
Lord, Maj. Gen. Royal B., 31, 33, 171, 362, 383, 400, 403
Lorient, 16
Lot segregation of ammunition, 273–74
LST's, 48, 51, 74, 77, 85, 92, 93, 106, 125, 217, 403
Ludwigshafen, 423
Luetterode, 207
Luneville, 157
Lutes, Lt. Gen. LeRoy, 233, 348–63
and ammunition shortage, 444
criticism of depot structure, 355–57, 397
and rapid shipping procedure, 352, 401
on supply of clothing, 233
on tank replacement factor, 238–39
visit to theater, 243, 348–63
Lutes-Somervell mission, 348–63, 382
Luxembourg City, 184

M24 tank, 457, 457n
Maastricht, 200, 201, 435, 437
MacKeachie, Col. Douglas, 472. *See also* General Purchasing Agent.
McNair, Lt. Gen. Lesley J., 327
McNamara, Col. Andrew T., 207
McNarney, Lt. Gen. Joseph T., 29n, 240, 305, 306, 307, 308, 312, 313, 330, 389
McSherry, Brig. Gen. Frank J., 388
Magnatex Limited, 202n
Maintenance of vehicles, 144–46, 421
Maintenance rates, 169, 171, 176n, 373, 374, 377, 420, 431. *See also* Attrition rates; Replacement factors.
east of the Rhine, 373, 374, 377, 430, 431, 431n, 432
prospects for in October, 19
requirements in October, 18, 20, 20n, 173
Mainz, 410, 412, 423, 437, 438
Maizieres les Metz, 256
Major System. *See* Pipeline Systems, Major.
Malta Conference, 374
Mancieulles, 208
Mannheim, 410
Manpower, 304–07, 308, 323–25. *See also* Replacements; Troop basis.
theater overstrength, 311–12
MARKET-GARDEN, 15, 22, 139, 140
Marseille, 116, 119–24
condition on capture, 121
described, 119
discharge performance, 123, 393, 394
plans for use, 121–22
port clearance, 123, 133, 424
role in support of northern forces, 15, 108
ship diversions to, 132, 133
Marseille-Rhône Canal, 119
Marshaling yards
at Antwerp, 104, 109, 141
at Cherbourg, 82–83
Lison Junction (OMAHA Beach), 83n
Marshall, Gen. George C.
on ammunition requests, 262, 263
on closing out U.K. depots, 402
on manpower conservation, 304–05, 336–37
offer to advance ship regiments, 283, 284
and reliance on Antwerp, 107–08
on replacement system, 384
on Rhine crossing plans, 374
Matchett, Brig. Gen. Henry J., 341, 342n, 468
"Meat Ball Express," 406n
Mediterranean Theater of Operations, U.S. Army (MTOUSA), 42, 42n, 43, 330
aids European theater with replacements, 330
aids European theater with tanks, 240
Memorandum of Agreement (Antwerp), 141
Mersey River, 78
Metropolitan Rearmament Program, 501
Metz, 19, 22, 25, 149, 255–56
Meuse River, 25, 181
Middleswart, Brig. Gen. William H., 383
Middleton, Lt. Gen. Troy H., 93
Military Affairs (Mead) Committee, 246
Military government, 464
Military Labor Service, 482
Military Pipeline Service, 196, 199, 205, 435
Military police, 293
Military Railway Service, 150, 161, 415

Military Railway Service—Continued
1st MRS, 156, 158, 160, 410, 412, 415
2d MRS, 82, 148, 160, 412, 415
Ministry of Labor (Fr.), 478
Minor System. *See* Pipeline Systems, Minor.
Minton, Brig. Gen. Hugh C., 244, 489
Mons, 142, 148, 184
Montgomery, Field Marshal Sir Bernard L. *See also* Army Group, 21.
and Antwerp, 107
and Ardennes battle, 26
and October operations, 24
and September decisions, 15
Moore, Maj. Gen. Cecil R., 408
Morlaix, 47, 51, 94
Moselle River, 12, 22
Moses, Brig. Gen. Raymond G., 13, 18, 53, 162, 171, 172, 267, 268, 272, 273
Motor transport, 5, 6, 10, 49, 134, 144, 145, 146, 147, 415–24. *See also* ABC Haul; Motor Transport Service; Red Ball Express.
operations east of Rhine, 374, 377, 415–24
in port clearance, 12, 55, 56, 84
requirements for advance to Rhine, 10
role in the pursuit, 5
Motor Transport Brigade, 138
Motor Transport Service, 138–39, 146, 416, 417, 418
Movement control, 154–55, 159, 160, 357, 382, 390–93, 412, 413–14
Movements and Transportation Committee (Antwerp), 109
MT80. *See* Gasoline, motor vehicle.
MT ships, 48, 60, 125
MTOUSA. *See* Mediterranean Theater of Operations, U.S. Army.
Muenchen Gladbach, 412
MULBERRY A, 53, 124. *See also* OMAHA Beach.
MULBERRY B, 58–60, 100, 124, 129
Murrill, Col. Hugh A., 391

Namur, 111, 149, 155, 181, 192, 405
Nancy 19, 22, 49
Nantes, 16, 47
Napier, Maj. Gen. Charles S., 98
NATOUSA. *See* North African Theater of Operations, U.S. Army.
Naval Harbor Master (Cherbourg), 79
Naval Officer in Charge (NOIC), 109
Negro troops and units, 147, 322–23, 384–85
New York Port of Embarkation (NYPOE), 44, 131, 190, 353, 399, 400, 401
Nieuwpoort, 106
Ninth Air Force. *See* Air Force commands and units, Air Force, Ninth.
Ninth Army, 107, 134, 367–68, 370, 372, 378, 410. *See also* Simpson, Lt. Gen. William S.
and ammunition allocations, 264
ammunition expenditures, 448
Ninth Army—Continued
ammunition levels, 447
and Ardennes battle, 26, 184
formation, 27
local procurement, 493, 496, 498
move to Continent, 28
November offensive, 25, 176
Operation GRENADE, 367–68
POL stocks, 206, 208, 208n, 438
preparations for final offensive, 429
rail service to, 406, 412
rail transfer point, 137
ration levels, 189, 189n, 434
relations with COMZ, 349
strength in divisions, October, 19
supply consumption rates, 431, 432
supply reserves, 19, 429
tonnage allocations, 170, 176
truck transport support, 418, 419
and winter uniform, 234
Nord mines, 441, 499
Normand, 67, 70
Normandy, 8. *See also* Base Sections, Normandy.
Normandy Base Section. *See* Base Sections, Normandy.
Normandy ports, 47, 51, 53, 60–89. *See also* Barfleur; Beaches; Carentan; Cherbourg; Grandcamp; Isigny; MULBERRY A; MULBERRY B; Port-en-Bessin; St. Vaast.
North African Theater of Operations, U.S. Army (NATOUSA), 28, 29, 263, 280. *See also* Services of Supply, NATOUSA.
boundary changes, 29
experience with manpower, 306, 336
replacement problem, 306, 313–14
support of 6th Army Group, 41–42
Nijmegen, 367
Northern Ireland Base Section. *See* Base Sections, Northern Ireland.

Office of The Quartermaster General, 219, 221, 222, 223, 225, 228, 232, 234, 235. *See also* Quartermaster General, The; Gregory, Maj. Gen. Edmund B.
Officer Candidate School. *See* Officer replacements.
Officer replacements, 325, 329, 330–31, 466, 467–68
O'Hair, Pvt. William G., 73
O'Hare, Brig. Gen. Joseph J., 323
Oise River, 166
OMAHA Beach, 51, 53–60. *See also* MULBERRY A.
OMAHA Beach Command, 58
"Operation OVERBOARD," 46
Oppenheim, 370, 423
OQMG. *See* Gregory, Maj. Gen. Edmund B.; Office of The Quartermaster General.
Order and shipping time, 238, 239, 261, 261n, 262, 351, 353, 354, 507
Ordnance Service, COMZ, 488, 492, 493

ORELOB. *See* Reverse BOLERO.
Organization and command. *See* Command and organization.
Osborn, Maj. Gen. Frederick H., 386
Ostend, 51, 106, 114, 184, 199
"OVERBOARD." *See* "Operation OVERBOARD."
OVERLORD, 4, 7. *See also* Post-OVERLORD plans.
  command and organization, 27, 34
  and DRAGOON, 117, 118
  logistic plans, 4
  port plans, 46, 53, 61
  predicted development, 4, 5
Overshoes. *See* Footwear.

Page, Col. Carter, 363
Parachute supply, 163. *See also* Air transport.
Paris, 5, 7, 31–32, 210, 356, 496–97
Parsons jacket, 219, 221, 222, 227
Pas de Calais beaches, 17
Pas de Calais mines, 441, 499
Patch, Lt. Gen. Alexander M., 28, 370. *See also* Seventh Army.
Patton, Lt. Gen. George S., 12, 183, 349, 370. *See also* Third Army.
Persian Gulf Command, 417
Peterson, Maj. Gen. Virgil L., 312, 313
Petit Couronne, 199
Petroleum. *See* POL.
PHOENIXES, 59, 60
Pilferage, 415
Pilotless aircraft. *See* V-weapons.
Pipeline Systems
  air attacks on, 208
  cross-Channel. *See* PLUTO.
  delivery capability, 438
  Major System, 77, 195–96, 197, 208, 435, 438
  Minor System, 196, 197, 198
  Northern System, 199–201, 435, 437
  PLUTO. *See* PLUTO.
  and Rhine crossings, 436–37
  Seine River System, 199
  Southern System, 435, 438
  Western System, 197n
Pitprops. *See* Pitwood.
Pitwood, 211, 212, 441
Plank, Maj. Gen. Ewart G., 33, 113, 349, 462. *See also* Advance Section, Communications Zone (ADSEC).
PLUTO, 198n
POL, 8, 12, 174, 180, 183, 193–209. *See also* Gasoline; Pipeline Systems.
  air attacks on distribution facilities, 208, 209
  air transport of, 161–62, 163, 426–27
  in Ardennes battle, 148, 183, 208–09
  bulk shipments of, 201, 438
  consumption rates, 194, 194n, 438, 439–40
  decanting of, 201, 205, 207, 208, 437
POL—Continued
  distribution system, 204–05, 207, 373, 374, 376, 438
  intake facilities, 77, 97, 98, 106, 122, 123, 198, 205, 207, 435
  levels of supply, 194, 194n, 435, 438
  pipelines for Rhine crossing, 373, 374, 375, 376, 436–37
  planning factors, 194
  rationing, 194
  and September decisions, 12
  shortage in pursuit, 8
  storage, 104, 198, 199, 207, 438
Port clearance, 7, 12, 79, 81, 96, 101, 123, 132, 148, 424n. *See also individual port names.*
Port Construction and Repair Groups. *See* Engineer Port Construction and Repair (PC&R) Groups.
Port discharge. *See also individual port names.*
  capacity, 4, 7, 13, 18, 47, 48, 49, 50, 51, 124, 128, 130, 132, 248
  effect of accelerated build-up on, 278–79
  performance, 12, 46, 50, 96, 103, 116, 124, 132, 133, 254, 393
  requirements, 4, 13, 14, 47, 48, 96–97
  September decisions on port development, 50–51
  shift to Antwerp, 49–50, 52–53
Port du Bouc, 119, 122, 123, 438
Port-en-Bessin, 51, 97, 197
Port Executive Committee, Antwerp, 109, 114
Port Jerome, 199
Porter, Maj. Gen. Ray E., 30, 324
Ports, 4, 5, 10, 15, 46–95, 96–97, 108, 117–18, 132. *See also individual port names;* Brittany ports; Channel ports; Port clearance; Port discharge.
  4th, 33, 78, 476
  5th, 93, 94, 110
  6th, 122, 123
  11th, 90, 102
  12th, 78
  13th, 110
  16th, 92, 93, 94, 100, 102
  17th, 395
  52d, 102
  backlogs, 396
  Brussels Conference on, 106
  changes in planned use, 49–53
  role in DRAGOON plan, 117–18
Posthostilities planning, 385–86
Post-OVERLORD plans, 6
Portugal, local procurement, 500
Potter, Col. Harold E., 331
Potter, Col. W. E., 86, 111
Pounder, Capt. William F., 219, 222
Priorities for supply. *See* Tonnage allocations.
Prisoner of war labor, 89, 102, 103, 124, 169, 212, 213, 297, 301, 423, 424, 476, 484–85, 486. *See also* Italian service units.

Prisoners of war
  number supported, 433, 460
  supplies for, 225
PROCO. *See* Projects for continental operations.
Production of munitions, U.S., 246, 265, 269, 271
Projects for continental operations, 452

Quartermaster Corps, 190, 191, 232
Quartermaster General, The, 219, 221, 226, 235. *See also* Gregory, Maj. Gen. Edmund B.; Office of The Quartermaster General.
Quartermaster Group, 469th, 419
Quartermaster Service, 496
Quartermaster Truck Company, 3884th, 73
Querqueville, Digue de, 77, 198
Querqueville Bay, 69
Quiberon Bay, 5, 16, 34, 46, 47

Rail transfer points, 135, 137, 139
Railways, 6, 10, 14, 21, 49, 81, 83, 84, 86, 87–88, 148–61, 213, 374, 405–15. *See also* Construction, railway; Marshaling yards; Military Railway Service; Rolling stock.
  at Antwerp, 109, 113
  in Ardennes battle, 186
  Belgian, 104
  bridging the Rhine, 408–10
  command and control problems, 87–88, 160–61, 415
  east of the Rhine, 374, 375, 376, 377, 410, 412–13
  extent of lines, 148–49, 157–58, 406, 410
  in northern France, 148–56
  operating units, 150n
  plans for restoration and use, 149–50, 407
  in southern France, 156–61
  tie-up east of Paris, 153–56, 348
  tie-up in southern France, 158–59, 160
  transfer points, 135, 137, 139
"Rainbow Corner," 497
RAMPS. *See* Recovered Allied Military Personnel.
Ramsay, Adm. Sir Bertram H., 60
Ratay, Brig. Gen. John P., 40
Rationing. *See also* Tonnage allocations.
  of ammunition, 247–75, 444–52
  of transport, 169–80
Rations, 178, 180, 433, 434
  10-in-1. *See* Rations, operational.
  A, 189, 191, 192, 433
  B, 189, 190, 191, 192, 433
  balancing problem, 190, 191, 191n, 193
  C. *See* Rations, operational.
  consumption rate, 189, 430, 431, 432, 434
  dehydrated, 193
  K. *See* Rations, operational.
  operational, 189, 191, 192, 433, 434
  perishable, 191–93
  stocks in combat zone, 188, 433, 434
  theater stocks, 433, 434
Real estate, 486, 496, 497
Reciprocal aid, 470–502. *See also* Civilian labor; Local procurement.
Recovered Allied Military Personnel (RAMPS), 427
Red Ball Express, 134–39, 144
  performance record, 135, 137
  operational problems, 138–39
  route changes, 134–35, 135n
  vehicle attrition rate, 236
Red Horse Staging Area, 497, 498
Red Lion Route, 139
Redeployment, 385
Reefers
  rail cars, 192
  ships, 191, 192
  vans, 193
Refrigerated transport. *See* Reefers.
Regulating stations, 413–14, 414n
  25th, 182
  attitude of various commands toward, 414n
Reims, 19, 356
Rennes, 28, 34, 184, 356
Replacement battalions
  51st, 337
  54th, 338
Replacement depots
  2d, 337, 339
  3d, 337
  9th, 338
  10th, 338
  11th, 338
  12th, 338
  14th, 338, 339
  15th, 338
  16th, 338
  17th, 337, 339
  18th, 337
  19th, 338, 463
  6900th, 338, 468
  6960th, 338, 468
Replacement factors, 354, 507
  clothing, 224–25, 226
  tanks, 236, 237, 237n, 238, 239
  tires, 244
  trucks, 242
  vehicles, 145–46
  weapons, 236, 354
Replacement System, 310n, 334–47, 384, 389
  battalions. *See* Replacement battalions.
  casuals, policy on, 343–46
  conversion program. *See* Replacements, conversion training of general assignment men.
  depots. *See* Replacement depots.
  operation of, 338–47
  organization and command, 334–38, 347
  planning for, 334–37
  retraining courses, 321
  Stilwell proposal, 343

Replacement System—Continued
stockage, 311, 312, 315–16, 317, 319, 464, 465
training centers, No. 1, 338
Replacements, 304–47, 459–69, 507. *See also* Replacement System.
allocations to field forces, 333, 334, 465
armored, 315, 333, 334, 459, 465
Bull mission, 319–20
casuals, policy on, 343–46
command problems, 505
conversion training of general assignment men, 305–07, 311, 315, 316, 317, 320, 324, 326, 328–30, 332, 461, 463, 464, 465, 467
December crisis, 321–23
infantry. *See* Infantrymen.
limited assignment men, 305, 307, 308, 309, 311, 317, 321, 330, 462, 463
North African theater experience, 306, 336
officers, 325, 329, 330–31, 466, 467
planning factors, 308, 309, 316, 319
requisitioning policy, 345–47
results of retraining program, 467–69
shipments from zone of interior, 307–08, 313, 315, 317, 318, 320, 324, 325, 333, 459, 460
stockage, 316, 317, 319, 464, 465
War Department conference, 323–25
Requisitioning objective, 353, 353n, 354
Requisitioning procedures
back ordering, 359
intratheater, 358–59, 360
on War Department and New York Port, 353. *See also* Lead time (procurement); Order and shipping time.
Retraining. *See* Replacements, conversion training of general assignment men.
Reverse BOLERO, 33, 402–04. *See also* RHUMBA.
Reverse lend-lease. *See* Reciprocal aid; Local procurement.
REX shipping procedure, 352, 401–02
Rex Theater (Antwerp), 115
Rhine bridges
Highway, 423–24
Railway
allocation of running rights, 412
bottlenecks, 412
at Duisburg, 410
at Karlsruhe, 410
at Mainz, 410, 412
at Mannheim, 410
plans for, 408–09
at Wesel, 409–10, 412–13
Rhine Coal Control, 441
Rhine River, 8, 9, 14, 15, 25, 369. *See also* Rhine bridges.
plans foi crossing, 8, 9, 14–15, 367, 372–73
plans for logistic support across, 372–78
COMZ outline plans, 375–76
crossing sites, 372–73, 375
Rhine River—Continued
plans for logistic support across—Continued
requirements laid down, 373–74
SHAEF directive, 373–75
Rhône-Saône, 168
RHUMBA, 286, 287, 402–04. *See also* Reverse BOLERO.
Ridgway, Maj. Gen. Matthew B., 28. *See also* Corps, XVIII Airborne.
Robinson, Maj. Gen. Clinton F., 232, 362, 392
Rodgers, Brig. Gen. Robert C., 326n
Roer River, 25, 264, 367, 369
Rogers, Brig. Gen. Pleas B., 35
Rolling stock, 150–56
freight cars, 150, 153
Allied imports to Continent, 414
at Antwerp, 105, 111
discharge facilities on Continent, 74, 75, 153
effect of Ardennes battle, 154–55
effect of shortage on port clearance, 84–85
effect of shortage on port discharge, 210
at Marseille, 123, 212
requirements, 87, 123
tie-up of loaded cars, 153–55, 158–59, 413
for "Toot Sweet Express," 405
locomotives, 150–53, 414
Rommel, Field Marshal Erwin, 511
Roosevelt, Franklin D., 118, 131
Roscoff, 94
Ross, Maj. Gen. Frank S., 82, 85n, 86, 126, 151, 155, 404
Rotterdam, 51
Rouen, 19, 38, 48, 51, 97, 100, 101, 102, 103, 356, 395, 497, 498
Royal Marine Engineers, 98, 99
Rubber Committee, ETOUSA, 489
Ruhr
in final offensive, 367, 370, 371
in OVERLORD plans, 10, 372
in September plans, 11, 14, 15, 24
in source of coal, 441
Rules of Land Warfare, 481, 484

Saar, 11, 441
Saaralbe, 435
St. Brieuc, 47, 51, 93–94, 212
St. Lô, 5
St. Malo, 6, 46, 47, 51, 89
St. Maxime, 120
St. Michele-en-Grève, 53, 92
St. Nazaire, 16, 47
St. Raphael, 119, 120, 156
St. Tropez, 119, 120, 156
St. Vaast-la-Hougue, 51, 60, 61, 62, 78
St. Vith, 155
Sandhofen, 437, 438
Saverne, 24, 157
Sawbridge, Brig. Gen. Ben M., 315

Schelde estuary, 24, 107, 110
Schroeder, Col. August H., 78
Seine ports, 4, 5, 13, 16, 49, 51, 97, 98. *See also* Havre, Le.; Rouen.
Seine River
  and tactical developments, 5
  and tactical plans, 4, 6, 48
  as waterway, 166, 424
Seine River System. *See* Pipeline Systems, Seine River System.
Seine Section. *See* Base Sections, Seine.
Selective discharge, 125, 128, 131, 402
Service troops, 289–303
  and accelerated build-up of divisions, 278, 290
  allocation between north and south, 297–303
  demand for additional, 7, 290, 292, 294, 296
  for French forces, 301–03
  planned troop basis, 7, 294, 296
  on southern line of communications, 186, 297–303
Services à la Main d'Oeuvre, 478
Services of Supply (SOS), 334
Services of Supply (SOS), NATOUSA, 38, 43
Seventh Army, 5, 6, 26, 119, 156, 157, 158, 219, 370, 371. *See also* Patch, Lt. Gen. Alexander M.; Southern Line of Communications (SOLOC).
  and air transport, 427
  ammunition shortages, 250–51, 264, 447
  and DRAGOON, 28
  and local procurement, 493
  and November offensive, 25
  POL stocks, 439
  preparations for final offensive, 429
  replacements, 314, 323, 333
  supply consumption rate, 431, 432
  supply reserves, 180
  truck transport support, 418
  winter clothing, 231, 235
Seventh Army (Ger.), 5
SHAEF. *See* Supreme Headquarters, Allied Expeditionary Force.
SHAEF Maintenance Rate (ammunition), 272
SHAEF Mission to France, 211
Shannon, Col. Lyle T., 323, 327, 384, 459, 464
Sheppard, Lawrence B., 234n
Shipping, 108, 124–33, 277–78, 351–52, 398–402. *See also* Coasters; Commodity loaders; LCT's; LST's; Liberty ships; MT ships.
  cancellations of, 399–401
  effect of shortage on local procurement, 501–02
  express procedure (REX), 352
  selective discharge, 125, 128, 131, 402
  theater requirements, 126, 128, 133, 399, 400
  tie-up of ocean shipping, 124–33, 398–401
  turnaround of, 125, 127, 130, 402–03
Shipping Control Board, ETOUSA, 131
Shoepac. *See* Footwear.
Shrivenham, 311, 463
Sibley, Col. Cleland C., 33, 78, 84, 85, 85n
Siegfried Line, 11, 15, 25, 369. *See also* West Wall.
Simpson, Lt. Gen. William S., 25, 349. *See also* Ninth Army.
Siple, Maj. Paul A., 235, 235n
Sisteron, 156
Smith, Lt. Gen. Walter B., 31, 32, 221, 315, 348, 374
Soissons, 134, 149, 184, 252, 356
Solbert, Brig. Gen. Oscar N., 386
Solglint, 67, 70–71
Solid Fuels Section (SHAEF), 440
Sombernon, 29
Somervell, Lt. Gen. Brehon B.
  and ammunition shortage, 265, 268, 271
  criticism of depot structure, 355–57, 397
  on ETOUSA command practices, 390
  and release of servicemen to foundries, 246
  and shipping tie-up, 129, 130
  and tank shortage, 241
  and vehicle shipments, 243
  visit to theater, 348–63, 390, 397, 403
Sommesous, 134, 149, 190
SOP 7, 358
SOP 10, 474
SOP 29, 472
SOS, NATOUSA. *See* Services of Supply, NATOUSA.
Sottevast, 82, 83
Southern Base Section. *See* Base Sections, Southern.
Southern France. *See* DRAGOON; Army Group, 6th; Seventh Army; First French Army; Southern Line of Communications.
Southern France beaches, 118, 119. *See also* ALPHA Beach; CAMEL Beach; DELTA Beach; St. Raphael; St. Tropez.
Southern France ports. *See* ALPHA Beach; CAMEL Beach; DELTA Beach; Marseille; Port du Bouc; Toulon.
Southern Line of Communications (SOLOC), 42, 301, 383
  control of transportation, 159
  dissolution, 46, 378, 383
  division slice, 301, 302
  forward reserves, 178, 180, 186
  incorporation with COMZ, 40–45
  relations with field commands, 349
  relationship with other commands, 42–45, 263, 303
  role in support of 12th Army Group, 17, 40, 123, 186, 263n, 297–98, 299
  service troop basis, 297–98, 299, 302
  tonnage allocations, 173, 186
Spa, 182
Spaatz, Lt. Gen. Carl, 461, 469
Spain, local procurement, 500
Spare parts supply, 8, 236, 430, 453, 454
  local procurement, 489
Special Services, 386

Standing Operating Procedures, COMZ-ETOUSA, 139, 163. *See also* SOP's, 7, 10, 29.
Stars and Stripes, 230, 245, 342n, 386
Stavelot, 209
Stilwell, Gen. Joseph W., 343
Stock control, 358
Storage, 111–13, 182. *See also* Depots.
Strasbourg, 26
Stratton, Brig. Gen. James H., 18, 87, 162, 217, 278, 352, 383
Strikes, 213
Sullivan, Commodore William A., 69
Supply, 3–4, 8, 12, 22, 188, 215, 351, 433, 458. *See also* Rationing; Requisitioning procedures; Supply deliveries; Supply reserves; Supply shortages; Tonnage allocations.
  Class I. *See* Rations.
  Class II and IV, 108, 177, 215–46, 452–58
  Class III, 178, 180, 434–41. *See also* Coal; Gasoline; POL.
  Class V. *See* Ammunition.
  expedients, 359–63
  priorities for operations, 11, 173, 175, 176
  requisitioning procedures, 353, 358–59, 360
Supply "bricks," 165
Supply deliveries to armies, 169–80, 182, 186, 428, 430
  by air, 424–25, 426, 427
  allocations system, 20–21
  estimated for October, 18, 19
  by motor transport, 416, 419, 421
  by rail, 154, 158, 410–12
  requirements for October, 20
  in September, 12
Supply requirements. *See* Maintenance rates.
Supply reserves
  in COMZ, 427, 428
  in forward areas, 8, 108, 172, 177, 178, 179, 180, 206, 390–93. *See also* Advance Section, supply reserves; Continental Advance Section, supply reserves.
Supply shortages, 3–4, 169, 257, 351. *See also* Spart parts; Tanks; Tires; Trucks; Winter clothing.
  engineer, 216, 452–53
  ordnance, 235–46, 453–58
  quartermaster, 216–35
  signal, 215, 453
Supreme Headquarters, Allied Expeditionary Force. *See also* Eisenhower, Gen. Dwight D.
  and accelerated division build-up, 280, 287
  and air transport, 162, 163, 425, 426
  and ammunition shortage, 254, 259, 272, 273, 442–47
  and Antwerp, 114, 181
  and Ardennes battle, 182
  and attrition rates, 235–36
  on capabilities of SOLOC, 17
Supreme Headquarters, Allied Expeditionary Force—Continued
  Chief of Staff. *See* Smith, Lt. Gen. Walter B.
  and civilian labor, 478, 479, 481, 482, 485
  and coaster problem, 126
  command relationships, 28, 386–89, 504–06
  and diversions through Marseille, 297–98
  establishment on Continent, 27
  estimate of Berlin thrust plan, 10, 11
  estimate of logistic capabilities, 13–14, 20
  estimate of supply requirements, 19
  G–1. *See* Barker, Maj. Gen. Ray W.
  G–3. *See* Bull, Maj. Gen. Harold R.
  G–4. *See* Crawford, Maj. Gen. Robert W.
  Logistic Plans Branch, 98
  OVERLORD logistic estimates, 4
  port plans, 50, 51, 59, 85–87, 98, 104, 105, 106, 108
  and reciprocal aid, 471, 474, 489
  relations with COMZ, 350, 386–88
  and replacements, 322, 326, 327, 333, 334, 459
  and Rhine crossing plans, 8–9, 10–11, 372–73
  and service troop basis, 292–97
  Solid Fuels Section, 210, 211, 212
  and southern France forces, 29, 41–45
  supply allocations system, 20–21
  tactical plans, 8, 9, 367, 372–73
  tonnage allocations, 172, 173, 174, 175, 176, 205
  as a U.S. theater headquarters, 505
Surge pool, 141, 142
Switzerland, local procurement, 500

T26 tank (Pershing), 456–57
Tactical Air Command, IX. *See* Air Force commands and units, Tactical Air Command, IX.
Tactical Air Command, XII. *See* Air Force commands and units, Tactical Air Comamnd, XII.
Tactical plans
  for advance into Germany, 10, 14, 17, 173, 175, 176, 367, 372–73, 374–75
  as affected by ammunition shortage, 255–57, 443, 444, 445
  as affected by pursuit, 508
  logistic implications, 9, 17
  for Rhine crossing, 8, 9, 367
Talley, Col. Benjamin B., 85n
Tancarville Canal, 166, 175
Tanks, 22, 236
  M24, 457
  T26 (Pershing), 242, 456–57
  conversion of 75-mm. gun, 457–58
  engine rebuild, 488
  German. *See* Tanks, Tiger (Ger.) vs. T26.
  loan of by Mediterranean theater, 240
  local procurement of parts, 488, 491
  losses, First Army, 239–40
  order and shipping time, 507
  reduced T/O&E's, 236

Tanks—Continued
replacement factor, 236–39, 454
replacement of 75-mm. gun, 241, 457–58
theater losses, 455
theater reserves, 236, 237, 238, 454, 456n
Tiger (Ger.) vs. T26, 456
Tarascon–Beaucaire bridge, 212
Temple, Commander John B. G., 70
Terneuzen Canal, 395
Thackrey, Capt. L. A., 85, 87
Thaon, 323
Theater Manpower Board, 306, 307, 465
Theater Manpower Section, 326, 327, 329, 384, 389, 464, 465
Theater Service Forces, European Theater (TSFET), 389
Theater troop strength
by component, 288
in divisions, 288, 433n
Thionville, 435, 438
Third Army, 3, 6, 16, 24, 107, 299, 369–70, 371, 372, 410. *See also* Patton, Lt. Gen. George S.
air transport, 163, 425, 426, 427
ammunition, 248, 255, 264, 270, 447, 448, 449–50
and Ardennes battle, 26, 183–84
bridge construction, 216, 453
and Brittany, 7
captured rations, 188
clothing, 227, 234
local procurement, 474, 491, 493, 496
logistics of final drive, 430
mission in September, 14, 16
and motor transport support, 416, 418, 419, 422
November offensive, 25
and OVERLORD plans, 6
POL stocks, 193–94, 205, 206, 208, 208n, 438, 439
preparations for final offensive, 429
rail service for, 406
rail transfer point, 137
ration levels, 189, 189n, 434
and Red Ball route, 134
relations with COMZ, 349
replacements, 322, 323
supply consumption rates, 321, 432
strength in divisions, 19
supply priorities, 11, 170, 174, 176
supply reserves, 19, 172
tactical plans, 8
Thompson, Brig. Gen. Paul W., 386
Thrasher, Brig. Gen. Charles O., 36
Tiger tanks (Ger.), 456
Tires, 243–44, 488
local procurement, 244–45, 489–90
preventive maintenance program, 245
Tirlemont, 142
TOMBOLAS, 197, 198
Tonnage allocations, 169–80, 206, 215, 217, 430
"Toot Sweet Express," 352, 405–06
Torgau, 372
Toulon, 118–19, 120, 121, 122, 123
Track router, 149
Transportation, 6, 10, 84, 134–68, 169, 251, 351–52, 373, 405–27. *See also* Air transport; Inland waterways; Motor transport; Movement control; Rail transport; Tonnage allocations.
embargo on, 113
and OVERLORD estimates, 7
and Rhine crossing, 373, 374
rights at Antwerp, 109
tonnage allocations system, 169–80
Transportation Corps, 78, 82, 83, 146, 147, 191, 192, 390, 405, 415, 416, 417, 424. *See also* Ports.
Trenchfoot, 229–32
Tréport, Le, 106
Troop accommodations, U.K., 287–89
Troop basis, 279, 286, 289, 290, 291, 292, 294, 295, 300. *See also* Build-up, U.S. forces; Service troops.
Troop build-up. *See* Build-up, U.S. forces.
Troop Carrier Command, IX. *See* Air Force commands and units, Troop Carrier Command, IX.
Troop strength, ETOUSA, 288
Truck driver augmentation, 292, 293, 294, 295
Trucks
assembly at Antwerp, 243
maintenance, 144–46, 421
replacement factors, 242
shortages, 242–43
Somervell arranges releases, 243
TSFET. *See* Theater Service Forces, European Theater.
Turnaround of ships, 125, 127, 130, 402–03
Twickenham Ferry, 75, 77

United Kingdom
effect of accelerated build-up on, 286–89
and Reverse BOLERO, 33, 402–04
and RHUMBA plan, 287–89
United Kingdom Base Section. *See* Base Sections, United Kingdom (U.K.).
United States Air Forces, Europe (USAFE), 389
United States Commercial Company, 499
United States Forces, European Theater (USFET), 389
United States Group Control Council. *See* Group Control Council, U.S.
United States Strategic Air Forces (USSTAF), 329, 469
USAFE. *See* United States Air Forces, Europe.
USFET. *See* United States Forces, European Theater.
USSTAF. *See* United States Strategic Air Forces.
UTAH Beach, 51, 53–60
UTAH Beach Command, 58

V–1's. *See* V-weapons.

V–2's. *See* V-weapons.
V-weapons
at ADSEC installations, 186, 480
at Antwerp, 100, 113–14, 115, 480
and "danger pay," 480
at Liège, 154, 209, 407, 480
Valenciennes, 148, 184
Valognes, 31
Vaughan, Brig. Gen. Harry B., Jr., 35, 380, 385, 403
Vehicle discharge, 55, 57
Vehicles
engine rebuild, 488
maintenance and attrition, 144–46
re-equipping program, 146, 417
replacement factor, 145–46
theater truck strength, 146, 417
Verdun, 149, 154, 184, 207, 252, 405
VERITABLE, 367
Versailles, 497

War Department
and ammunition shipment cancellations, 449
and ammunition shortages, 257–63, 267–71, 443–49
and build-up of U.S. forces, 276–86
and changes in troop basis, 289, 292, 293, 295–96
disagreements with theater, 507
jerrican procurement, 203
and munitions production, 246, 265
and replacement factors, 226, 236–37, 239, 507
on replacements, 305–13, 317, 323, 324–25, 330–31, 459, 460
and requisitioning procedures, 353
and responsibility for 6th Army Group support, 263n
and Southern Line of Communications, 44
and supply procurement planning, 352–53
War Department—Continued
on use of shipping, 126, 127, 128, 130, 131, 132, 192, 398–401
and winter clothing, 220, 221, 223, 225, 226, 229, 231, 232, 233, 234
War Department Manpower Board, 304, 326, 384
War Manpower Commission, 246
War Production Board, 246, 269
War Shipping Administration, 130, 277, 403
Warren, Col. Ross B., 418, 421
Washington Post, 218, 218n
Weed, Col. Thomas J., 100n
Wesel, 200, 201, 409–10, 412–13, 423, 435, 437, 438
West Wall, 26
Western Base Section. *See* Base Sections, Western.
Whipple, Col. William, 7n, 49, 107, 174, 278, 279, 298, 299, 300
White Ball Route, 140, 148
Willis, Pvt. Charles I., 72
Wilson, Maj. Gen. Arthur R., 38, 44
Wilson, General Sir Henry Maitland, 28, 118
Winter clothing. *See* Clothing.
Winterization program, 134, 217–18
of MULBERRY B, 59–60
Winterization supplies, 171, 173, 217
Wood fuel, 213
Wylie, Brig. Gen. Robert H., 126
Wyman, Col. Theodore, 33, 38, 78, 85n, 88

XYZ Operation, 415–22

Yank Magazine, 386
Yellow Diamond Route, 419, 421, 421n
Younger, Col. James W., 234
Yserburg Bridge, 113, 168

Zeebrugge, 106

☆ U.S. GOVERNMENT PRINTING OFFICE: 1995 382–043

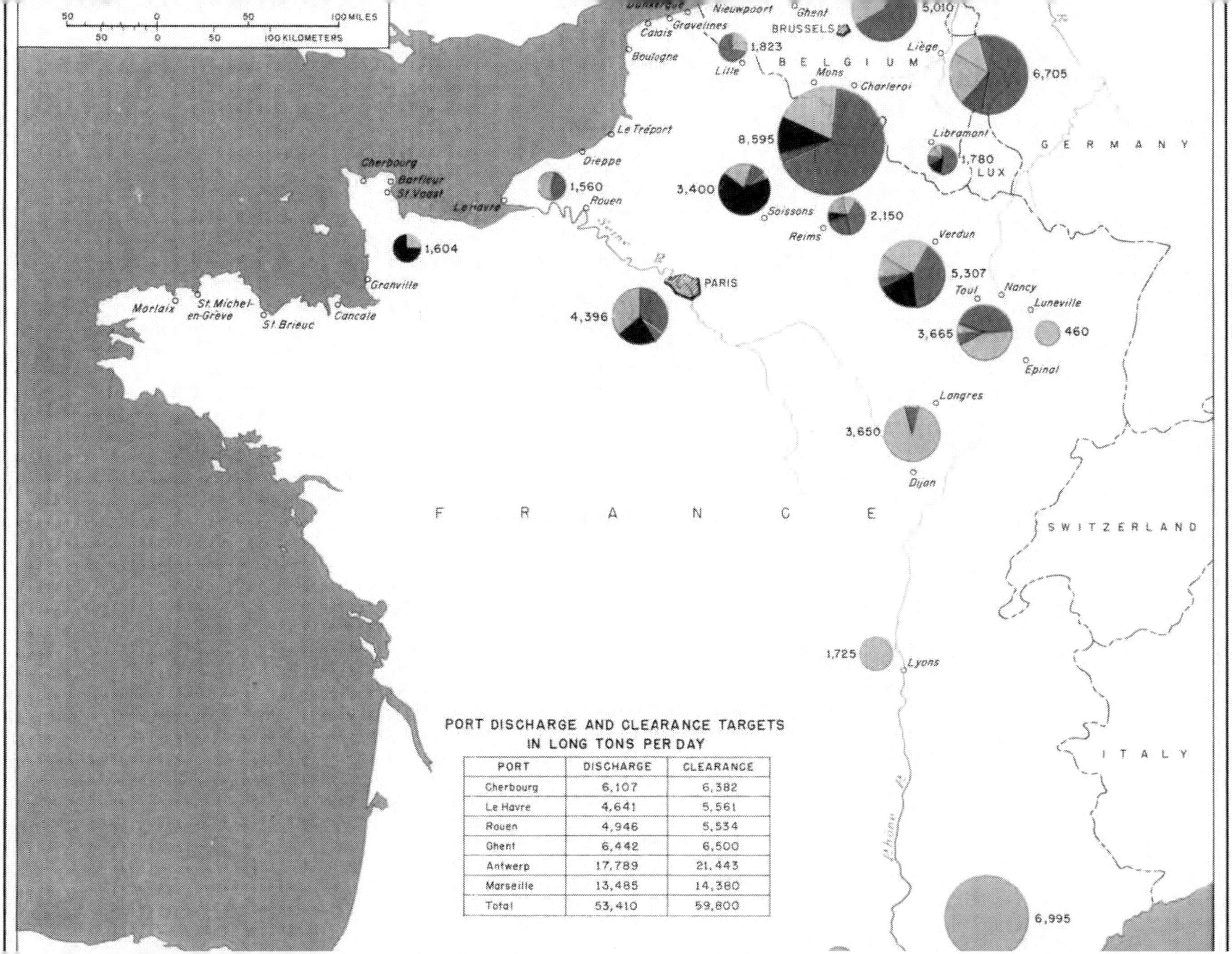

PORT DISCHARGE AND CLEARANCE TARGETS IN LONG TONS PER DAY

PORT	DISCHARGE	CLEARANCE
Cherbourg	6,107	6,382
Le Havre	4,641	5,561
Rouen	4,946	5,534
Ghent	6,442	6,500
Antwerp	17,789	21,443
Marseille	13,485	14,380
Total	53,410	59,800

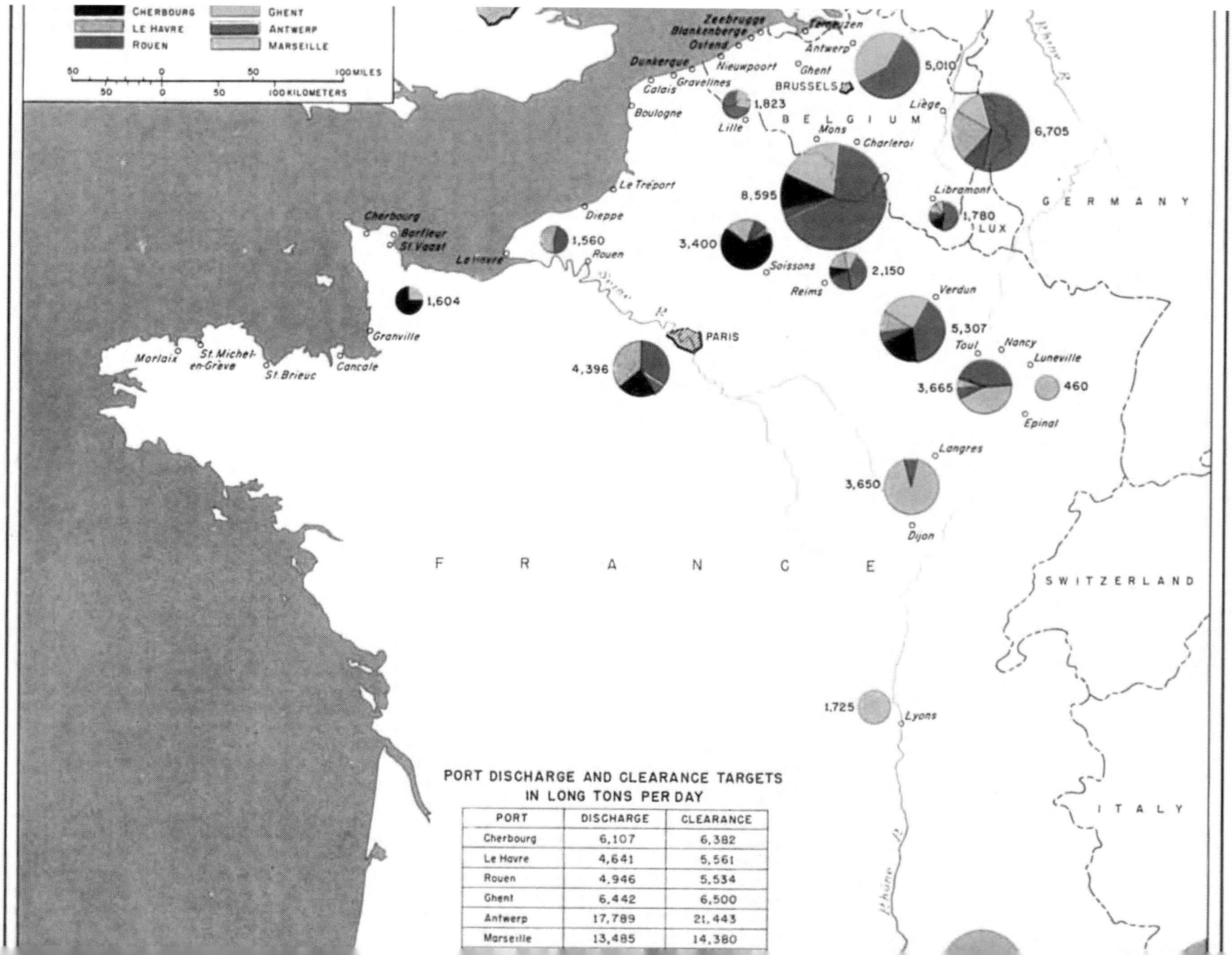

PORT DISCHARGE AND CLEARANCE TARGETS
IN LONG TONS PER DAY

PORT	DISCHARGE	CLEARANCE
Cherbourg	6,107	6,382
Le Havre	4,641	5,561
Rouen	4,946	5,534
Ghent	6,442	6,500
Antwerp	17,789	21,443
Marseille	13,485	14,380

Made in United States
Orlando, FL
10 November 2026

90498890R00308